# NEXT GENERATION NUCLEON DECAY AND NEUTRINO DETECTOR

# Related Titles from AIP Conference Proceedings

**531**  Particles and Fields: Seventh Mexican Workshop
Edited by Alejandro Ayala, Guillermo Contreras, and Gerardo Herrera, July 2000,
1-56396-954-8

**528**  Acceleration and Transport of Energetic Particles Observed in the
Heliosphere: ACE 2000 Symposium
Edited by Richard A. Mewaldt, J. R. Jokipii, Martin A. Lee, Eberhard Möbius, and
Thomas H. Zurbuchen, July 2000, 1-56396-951-3

**526**  Gamma-Ray Bursts: 5$^{th}$ Huntsville Symposium
Edited by R. Marc Kippen, Robert S. Mallozzi, and Gerald J. Fishman, June 2000,
1-56396-947-5

**516**  26$^{th}$ International Cosmic Ray Conference: ICRC XXVI, Invited,
Rapporteur, and Highlight Papers
Edited by Brenda L. Dingus, David B. Kieda, and Michael H. Salamon
May 2000, 1-56396-939-4

**515**  GeV-TeV Gamma Ray Astrophysics Workshop: Towards a Major
Atmospheric Cherenkov Detector VI
Edited by Brenda L. Dingus, Michael H. Salamon, and David B. Kieda,
May 2000, 1-56396-938-6

**490**  Particles and Fields: Eighth Mexican School
Edited by Juan Carlos D'Olivo, Gabriel López Castro, and Myriam Mondragón,
November 1999, 1-56396-895-9

**444**  Particle Physics and Cosmology: First Tropical Workshop/High Energy
Physics: Second Latin American Symposium
Edited by José F. Nieves, September 1998, 1-56396-775-8

To learn more about these titles, or the AIP Conference Proceedings Series, please
visit the webpage **http://www.aip.org/catalog/aboutconf.html**

# NEXT GENERATION NUCLEON DECAY AND NEUTRINO DETECTOR

NNN99

Stony Brook, New York    23–25 September 1999

EDITORS
**Milind V. Diwan**
Brookhaven National Laboratory

**Chang Kee Jung**
State University of New York at Stony Brook

Melville, New York, 2000
AIP CONFERENCE PROCEEDINGS ■ VOLUME 533

**Editors:**

Milind V. Diwan
Brookhaven National Laboratory
Building 510A
Upton, NY 11973
USA

E-mail: diwan@bnl.gov

Chang Kee Jung
HEP Group
Department of Physics and Astronomy
State University of New York
Stony Brook, NY 11794
USA

E-mail: alpinist@ale.physics.sunysb.edu

Authorization to photocopy items for internal or personal use, beyond the free copying permitted under the 1978 U.S. Copyright Law (see statement below), is granted by the American Institute of Physics for users registered with the Copyright Clearance Center (CCC) Transactional Reporting Service, provided that the base fee of $17.00 per copy is paid directly to CCC, 222 Rosewood Drive, Danvers, MA 01923. For those organizations that have been granted a photocopy license by CCC, a separate system of payment has been arranged. The fee code for users of the Transactional Reporting Service is: 1-56396-956-4/00/$17.00.

© 2000 American Institute of Physics

Individual readers of this volume and nonprofit libraries, acting for them, are permitted to make fair use of the material in it, such as copying an article for use in teaching or research. Permission is granted to quote from this volume in scientific work with the customary acknowledgment of the source. To reprint a figure, table, or other excerpt requires the consent of one of the original authors and notification to AIP. Republication or systematic or multiple reproduction of any material in this volume is permitted only under license from AIP. Address inquiries to Office of Rights and Permissions, Suite 1NO1, 2 Huntington Quadrangle, Melville, NY 11747-4502; phone: 516-576-2268; fax: 516-576-2450; e-mail: rights@aip.org.

L.C. Catalog Card No. 00-105241
ISSN 0094-243X
ISBN 1-56396-956-4

Printed in the United States of America

# CONTENTS

Preface .................................................................................... vii

## NUCLEON DECAY

Opening Remarks at the Next Generation Nucleon Decay and Neutrino Detector
Workshop (NNN99) ....................................................................... 3
    M. Goldhaber

A Proton Decay Detector Using Pb/scintillator Calorimeter ................................ 6
    A. Konaka

Nucleon Decay Studies in a Large Liquid Argon Detector ................................. 12
    A. Bueno, M. Campanelli, A. Ferrari, and A. Rubbia

A New Underground Laboratory in Finland ............................................... 18
    J. T. Peltoniemi for the CUPP Collaboration

Study of 1 Megaton Water Cherenkov Detectors for the Future Proton Decay Search ....... 21
    M. Shiozawa

Comments on Future Proton Decay Experiments ........................................... 25
    Y. Suzuki

Feasibility of a Next Generation Underground Water Cherenkov Detector: UNO ............ 29
    C. K. Jung

## THEORY

Discovery of Proton Decay: A Must for Theory, a Challenge for Experiment .............. 37
    J. C. Pati

From Neutrino Masses to Proton Decay .................................................. 54
    P. Ramond

Radical Conservatism and Nucleon Decay ................................................ 62
    F. Wilczek

Matter Effects on Long Baseline Neutrino Oscillation Experiments ...................... 74
    I. Mocioiu and R. Shrock

$SO(10)$ and Large $\nu_\mu \to \nu_\tau$ Mixing ...................................... 80
    S. M. Barr and C. H. Albright

Nucleon Instability and (B-L) Non-Conservation ........................................ 84
    Y. Kamyshkov

## ASTROPHYSICAL NEUTRINOS

Solar Neutrinos ....................................................................... 91
    J. N. Bahcall

HELLAZ—The New Generation Solar Neutrino Experiment
to Measure the Spectrum of ($\nu_{pp}$ and $\nu_{Be}$) ............................... 103
    T. Patzak

Status of the BOREXINO Solar Neutrino Experiment ..................................... 106
    L. Oberauer

Progress on HERON: A Real-time Detector for P-P Solar Neutrinos ...................... 112
    J. S. Adams, A. Fleischmann, Y. H. Huang, Y. H. Kim, R. E. Lanou, H. J. Maris,
    and G. M. Seidel

The Sudbury Neutrino Observatory .................................................... 118
    J. Heise on Behalf of the SNO Collaboration

A Novel Supernova Detector .......................................................... 124
    D. B. Cline

SNEWS and Future Supernova Detectors ................................................ 128
    K. Scholberg

**OMNIS, The Observatory for Multiflavor Neutrinos from Supernovae** .................... 132
    R. N. Boyd and A. S. Murphy

**Atmospheric Neutrino Fluxes** ................................................................ 135
    T. K. Gaisser

**Results from Atmospheric Neutrinos** ......................................................... 139
    J. G. Learned

**Neutral Current ($\pi^0$) Production and the ($\nu_\mu \to \nu_\tau / \nu_\mu \to \nu_s$) Debate** .................... 155
    S. B. Boyd

**The ANTARES Project** ....................................................................... 159
    S. Navas-Concha

**Baseline Concept for a Precise Measurement of Atmospheric Neutrino Oscillation** ....... 165
    M. Aglietta, M. Ambrosio, E. Aprile, G. Bologna, M. Bonesini, G. Bencivenni,
    M. Calvi, A. Castellina, A. Curioni, W. Fulgione, P. L. Ghia, C. Gustavino,
    R. P. Kokoulin, G. Mannocchi, F. Murtas, G. P. Murtas, P. Negri, M. Paganoni,
    L. Periale, A. A. Petrukhin, P. Picchi, A. Pullia, S. Ragazzi, N. Redaelli, L. Satta,
    T. Tabarelli de Fatis, F. Terranova, A. Tonazzo, G. Trinchero, P. Vallania, and B. Villone

**Radio Detection of Ultra High Energy Neutrinos ($E_\nu > 10^{18}$ eV)** ..................... 171
    D. Seckel

**Studies and Site Characterisation for a $km^3$ Scale Underwater Neutrino Telescope in the Mediterranean Sea** ........................................................... 175
    G. Riccobene

## REACTOR AND ACCELERATOR NEUTRINOS

**Muon Storage Rings — Neutrino Factories** ................................................... 181
    Z. Parsa

**Matter Effects Study in Very Long Baseline Neutrino Oscillation Experiments** ........... 196
    A. Bueno, M. Campanelli, and A. Rubbia

**The Booster Neutrino Experiment: BooNE** .................................................... 205
    R. Tayloe

**Neutrino Oscillation Experiments at Nuclear Reactors** ..................................... 211
    M. Grassi

**ICANOE Imaging and Calorimetric Neutrino Oscillation Experiment** ........................ 220
    A. Rubbia

**OPERA: A Long Baseline ($\nu_\tau$) Appearance Experiment in the CNGS Beam from CERN to Gran Sasso** .................................................................. 233
    P. Migliozzi

**Summary of Detector and Beam Parameters Working Group from Lyon** ...................... 240
    D. A. Harris

**Summary of Accelerator Neutrino Experiments Working Group** ............................. 244
    R. Bernstein

**Contributions** ............................................................................... 247
**Author Index** ................................................................................ 249

# PREFACE

The atmospheric neutrino results from Super-Kamiokande have had an electrifying effect on high energy physics. There are also many other intriguing reports on neutrino physics from other experiments. It is clear that we have many exciting years ahead with more data expected from existing and new facilities, notably, K2K, SNO, KAMLAND, Boone, MINOS, Borexino, and CERN-Gran Sasso.

Nevertheless, the complexity of physics issues and the scale of experiments are such that it is important to start planning beyond the current round of experiments. The goal of observing proton decay remains elusive. Even after the spectacular progress in the neutrino sector many questions remain unanswered, and other questions have emerged. Neutrino astronomy has emerged as a new exciting field that needs new facilities.

Therefore, as a first step, this workshop was organized to stimulate discussion and create consensus in the high energy physics community about future large nucleon decay experiments.

During the three day workshop at Stony Brook, the current status of neutrino oscillations and nucleon decay searches was discussed. We also discussed theoretical interpretations of the current experimental results and motivation for a future nucleon decay detector. Many technical ideas were exchanged on massive underground cavities, underwater volumes, neutrino detection, and photodetection techniques.

The design for any new detector for nucleon decay has to address two issues: Is it sensitive enough for proton decay ? And is the physics agenda for such a detector sufficiently broad ? These questions as well as the questions of cost and technical feasibility are complex. Water Cerenkov detectors with the resolution of Super-Kamiokande (SK) will provide the necessary broad sensitivity; however, a new design is needed to scale SK up by more than an order of magnitude. It could also be argued that the new detector should focus on proton decays preferred by Super-symmetry inspired models: $p \to K^+ \bar{\nu}$, $p \to K^o \mu^+$. In this workshop studies of both a larger Water Cerenkov detector and a lead-scintillator detector were presented.

Any facility of the size we are contemplating must have a broad physics agenda. This is clear from the experience of IMB, Kamiokande and Super-Kamiokande. All sources of astrophysical neutrinos – solar, supernova, atmospheric, and high energy – must be explored in finer detail. These considerations as well as the idea of using a large underground detector with accelerator (muon storage ring) generated neutrinos were discussed.

Regardless of the exact answer about the type and size of the next detector, this project will be much larger than any of the previous projects of this type. Therefore we need to create consensus in our community. This workshop was the first step in this exercise; before the publication of these proceedings we will have held two additional smaller meetings.

As Jogesh Pati has observed: "The discovery of proton decay would undoubtedly constitute a landmark in the history of physics. It would provide the last, missing piece of gauge unification and would shed light on how such a unification may be extended to include gravity." We hope that we converge on the correct experimental program to make this discovery.

Milind V. Diwan
Brookhaven National Laboratory

Chang Kee Jung
State University of New York at Stony Brook

**Table 1.** NNN99 working groups and organizers

| | |
|---|---|
| Nucleon Decay | Chang Kee Jung |
| | alpinist@superk.physics.sunysb.edu |
| Atmospheric and HE Astrophysics | Dave Casper |
| | casper@master.ps.uci.edu |
| Solar and Supernova Neutrinos | Bob Svoboda |
| | svoboda@phlash.phys.lsu.edu |
| Accelerator Neutrinos | Bob Bernstein |
| | rhbob@fnal.gov |

# Acknowledgments

Tireless efforts of the workshop organizers and the working group leaders made this a truly enjoyable workshop. We would also like to thank all the students from Stony Brook who helped efficiently to copy the transparencies, distribute various materials, and perform many other tasks that are often overlooked.

Special thanks are due to Brett Viren who found time to help us edit and organize the computer files for these proceedings while writing his Ph.D. thesis on proton decay.

The workshop was sponsored by the State University of New York at Stony Brook, Brookhaven National Laboratory, and Hamamatsu corporation.

The work was also supported by DOE grant DE-AC02-98CH10886 and DE-FG02-92ER40697.

**Table 2.** International Advisory Committee (IAC)

| | |
|---|---|
| J. Bahcall | IAS |
| R. Cowsik | IIAP |
| L. DiLella | CERN |
| G. Feldman | Harvard |
| T. Gaisser | Bartol |
| M. Goldhaber | BNL |
| F. Halzen | Wisconsin |
| W. Haxton | Washington |
| P. Langacker | Penn |
| W. Marciano | BNL |
| L. Moscoso | CEA/Saclay |
| K. Nakamura | KEK |
| J. Peoples | Fermilab |
| F. Sciulli | Columbia |
| H. Sobel | UCI (Chair, IAC) |
| C. Spiering | DESY/Zeuthen |
| P. Strolin | Napoli/CERN |
| Y. Totsuka | ICRR |
| F. Wilczek | IAS |
| S. Wojcicki | Stanford |
| C.N. Yang | StonyBrook |

**Table 3.** NNN99 Organizing Committee (OC)

| | |
|---|---|
| D. Casper | UCI |
| M. Diwan | BNL(Co-chair, OC) |
| C.K. Jung | Stony Brook (Co-chair, OC) |
| R. Hahn | BNL |
| T. Kajita | ICRR |
| R. McCarthy | Stony Brook |
| C. Mcgrew | Stony Brook |
| J. Napolitano | Stony Brook (Secretary) |
| K.K. Ng | Stony Brook (Computing) |
| A. Rubbia | ETH/Zurich |
| D. Schamberger | Stony Brook(computing) |
| R. Shrock | Stony Brook |
| B. Svoboda | LSU (Chair, Scientific Program) |
| C. Yanagisawa | Stony Brook |

# NUCLEON

# DECAY

# Opening Remarks at the Next Generation Nucleon Decay and Neutrino Detector Workshop (NNN99) State University of New York at Stonybrook, Sep. 23-25, 1999

Maurice Goldhaber

*Department of Physics, Broohaven National Laboratory, Upton, NY, 11973*

**Abstract.** In this workshop we are trying to look ahead at what new detectors can do to improve answers to questions concerning proton decay and the masses and mixing of neutrinos.

I have been asked to make a few remarks about the problems confronting the ever-growing community of researchers interested in two crucial questions which have often been investigated symbiotically in large-scale experiments, and which, as we shall hear at this workshop, are presumably also theoretically related:

- Does the proton decay and, if so, what are the branching ratios for the various potential decay modes?
- What are the exact masses and mixing angles of the three neutrinos, $\nu_e$, $\nu_\mu$, $\nu_\tau$?

At present we know only upper limits for many potential two-body decay modes of the proton, as well as for the values of the neutrino masses. We have some knowledge of the parameters governing neutrino mixing ($\Delta m^2$ and $sin^2 2\theta$), though for solar, i.e. electron neutrinos, our knowledge is still incomplete.

At the beginning of the trend to test conservation laws was the questioning of proton stability. The search for proton decay was begun explicitly 45 years ago, following indications from indirect arguments for a fairly long minimum proton lifetime, well exceeding the time our universe has existed. By using a large scintillation counter which Reines and Cowan had built at Los Alamos to look for atmospheric neutrinos, we could give improved proton lifetime limits. Thus, the search started out parasitic to a search for atmospheric neutrinos. With the development of Grand Unified Theories (GUTS), which predicted lifetimes for some proton decay branches, large water Cherenkow detectors were built by the Irvine-Michigan-Brookhaven (IMB) and the Kamiokande collaborations. Different, smaller, counters were built by other collaborations. The SU(5) GUT was found to predict too short a lifetime. We shall have a discussion of predictions of a modified theory (SUSY GUT) at this workshop. The large proton decay detectors also accumulated data on atmospheric neutrinos which indicated deficiencies for $\mu$ neutrinos, compared with expectations. With improved statistics, SuperKamiokande announced last year evidence for $\mu$ neutrino oscillations.

The solar neutrino search was started about a third of a century ago. Two of the pioneers of that search, Ray Davis and John Bahcall, are present today. The experiment was designed to demonstrate the predicted existence of solar neutrinos. Pontecorvo warned early that the possibility of neutrino oscillations might lead to a lower than expected value for the recorded neutrino flux. To remove doubt to what extent neutrino intensity deficiencies can be ascribed to the Standard Solar Model or to neutrino oscillations, one is awaiting signals typical of oscillations, e.g. the day-night effect. Oscillations will also be further searched for with reactor neutrinos.

One can hope that the remaining questions concerning oscillation parameters for electron and muon neutrinos will be settled within the near future by the many experiments which are either planned, or in preparation, or recently started, as will be discussed at this workshop. Since neutrino oscillations experiments yield only $\Delta m^2$ values, a special effort is still needed to obtain absolute neutrino masses. Are they near zero or near a larger value? See figure 1. An answer accurate enough to distinguish between the two possibilities is likely to prove an important guide to a theory of neutrino masses and mixing. Since we already know that the masses of the three neutrinos are close to each other, we must await a measurement of the absolute mass of at least one of the neutrinos for which the electron neutrino is the most promising candidate. An upper limit of a few eV is known for its mass. It is worth noting that the average masses of known elementary particles increase as new interactions are switched on, starting with only weak interactions for neutrinos, adding electromagnetic interactions for the charged leptons and finally strong interactions for the quarks. See figure 2.

For the two rare phenomena, proton decay and supernova neutrinos, we may just as well let our imagination go wild and suggest large scale, long-term efforts which are clearly needed. With the new millennium upon us, can we hope that true peace will break out in the not too distant future, and that research will become a candidate for the long hoped for "moral equivalent of war", allowing us to think big and to build detectors appropriate to the problems to be solved? If we think very big, perhaps the goal should be approached in stages, or divided up for parallel efforts. Some parallel efforts might be strategically placed to study the effects of different neutrino paths through the Earth.

To reduce the influence of neutrino background it might be desirable to have detectors with more free protons relative to bound ones to avoid the "smearing" of the decay products due to Fermi motion and to secondary interactions of decay products inside nuclei. Some candidate materials are given in table 1, although some of them may raise questions of safety.

Looking far ahead, can we perhaps ultimately design detectors on Mars where frozen water has been reported to exist, and perhaps find out, in case there are still ambiguities left among solar neutrino solutions for $\Delta m^2$ and $sin^2 2\theta$, which of them are "just-not-so" and incidentally also reduce the importance of atmospheric neutrinos? A detector on the moon would reduce the atmospheric neutrino background, except for the few which reach it from the Earth, and then might be of value for long baseline investigations.

It would often be useful to have detectors which can distinguish positive from negative particles. Are magnetic fields practical for very large detectors? Icarus might detect annihilation $\gamma$'s from stopping positrons.

The neutrino field is haunted by two further questions with which this workshop will try to grapple:

- Do we deal with Majorana or Dirac neutrinos, or some combination?

- Is there a role for sterile neutrinos?

If one of the three neutrinos were found to oscillate into a sterile neutrino which does not participate in weak interactions but can oscillate back into its parent neutrino, then universality suggests the possibility that there are three distinct sterile neutrinos. Some of these could, however, be mixed in which case they would be able to oscillate into more than one active neutrino. Neutrino oscillations would then be more complex than so far considered.

My thanks are due to Milind Diwan for producing the table and figures and to W.J. Marciano for discussions. This work was supported by DOE contract no. DE-AC02-98CH10886.

Note added on April 26, 2000; recent work at Super-Kamiokande has established with 99% confidence that sterile neutrinos do not arise from $\nu_\mu$ oscillations (see Ph.D. thesis by Kenji Ishihara, ICRR-Report-457-2000-1).

**Table 1.** Some possible materials for a large proton decay detector. A large ratio of free protons to bound protons will aid in background suppression. The computation of the relative light yield does not account for losses due to absorption different from water.

| Material | Density gm/cc | Index of refraction | Cher. angle | Light yield (rel.) | Free Protons |
|---|---|---|---|---|---|
| $H_2O$ | 1.00 | 1.33 | 41.2 | 1.00 | 0.20 |
| Liquid $H_2$ | 0.071 | 1.112 | 25.9 | 0.44 | 1.00 |
| Liquid $CH_4$ | 0.424 | $\sim 1.2$ | $\sim 33.6$ | 0.71 | 0.40 |
| Octane $C_8H_{18}$ | 0.703 | 1.397 | 44.2 | 1.12 | 0.27 |
| Ethanol $C_2H_5OH$ | 0.79 | 1.36 | 42.7 | 1.05 | 0.23 |

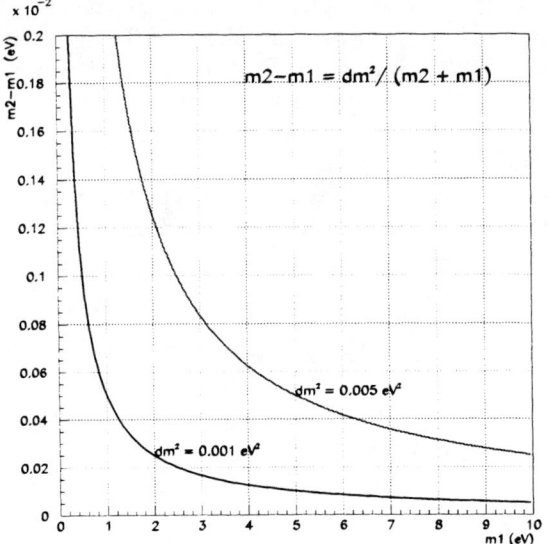

**FIGURE 1.** The difference between neutrino masses, $m_1$ and $m_2$, as a function of the mass $m_1$ given two candidate values of $\Delta m^2$ from the observation of neutrino oscillations. It is obvious that the two masses become highly degenerate as $m_1$ increases.

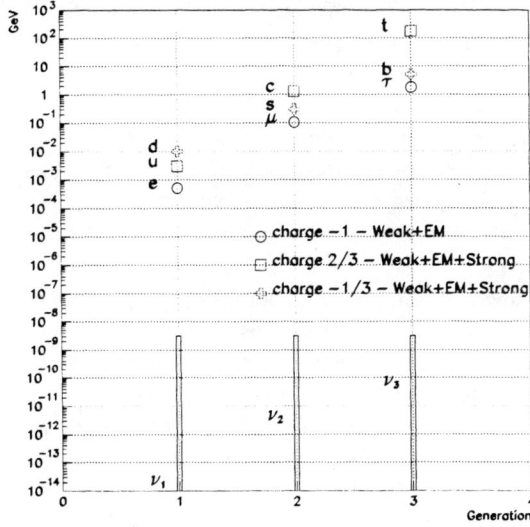

**FIGURE 2.** Masses of leptons and quarks in 3 generations. The possible range of neutrino masses is indicated. The current direct limit on the electron neutrino mass as well as the various oscillation hints can be accommodated.

# A proton decay detector using Pb/scintillator calorimeter

Akira Konaka

*TRIUMF, 4004 Wesbrook Mall, Vancouver, British Columbia, Canada V6T2A3*

**Abstract.** An idea of a general purpose proton decay/neutrino detector based on finely segmented Pb/scintillator calorimeter is described. Extruded scintillator with wave length shifter fiber readout enables us to build a large scale detector cost effectively. Proton decay lifetime, particularly the SUSY-GUT mode $p \to K^+ \bar{\nu}$, could be studied up to $\tau_p = 10^{35}$ years. Sufficient background suppression and reliable measurement of the background can be performed by detailed kinematic measurements (range, energy, decay topology) combined with delayed coincidence between $K^+$ and its decay products. Fine segmentation of the detector with good position and angular resolution makes it ideal for the atmospheric/accelerator neutrino experiments to study neutrino oscillation parameters, including $\nu_e - \nu_\tau$ mixing, matter (MSW) effect, and CP violation. A large enhancement in the $\nu - Pb$ cross section at lower energy also makes this detector effective for supernova neutrino detection. Prototyping and calibration of the detector would provide rich physics outputs along the way.

## INTRODUCTION

The recent discovery of neutrino oscillation(1) has opened up a new exciting era in neutrino physics. The existence of neutrino mass indicates a high energy scale (see-saw mechanism) which can cause a nucleon decay. Particularly, the SUSY-GUT proton decay mode $p \to K^+ \bar{\nu}$ may be experimentally accessible (2). In this paper, we suggest a general purpose detector based on Pb/scintillator fine grained calorimeter as a next generation nucleon decay and neutrino experiment.

A fine grained sampling calorimeter based on plastic scintillator and Pb has a great advantage of good spacial/energy/timing resolutions and compactness. The main obstacle has been the cost of the detector, which was much higher compared to wire chamber based fine grained calorimeters (e.g. SOUDAN2) or water Čerenkov detectors (e.g. Super-Kamiokande). In the last several years, a technique of extruded scintillator with wave length shifter (WLS) readout has become a mature technology (3). The cost of the scintillator, which was dominated by the casting/polishing cost, has dropped dramatically by the extrusion technique. The short attenuation length of $\sim 10 cm$ is overcome by converting the light in the WLS fibers whose attenuation length is as long as 4m. Developments of quality fibers and optimum matching between WLS fibers and scintillator enabled flexible and cheap, scintillation detectors with good energy resolution. Per-channel cost of pixel photodetectors for readout has also dropped dramatically thanks to the multianode photomultiplier (PMT) and hybrid photo diode (HPD) technologies (4). A large scale fine grained calorimeter based on plastic scintillator and Pb is no longer out of question.

In the following section, we will introduce the basic idea of the general purpose nucleon decay and neutrino detector. Next, we focus on the SUSY-GUT proton decay mode $p \to K^+ \bar{\nu}$, and see how reliably we could discover the signal if the lifetime is within theoretical prediction ($\tau < 10^{34}$ years). Then, possible neutrino experiments are discussed, which addresses various neutrino oscillation parameters as well as astrophysical neutrino detection. Finally, a scenario for prototyping and calibration with rich physics outputs is presented.

## THE DETECTOR

Figure 1 shows the structure of the detector. The basic fine grained calorimeter module consists of a series of 7m by 7m wide and 3mm thick extruded scintillator plates sandwiched with 1mm thick lead plates. Photons from individual WLS fibers running along the grooves on the scintillator are read out by multianode PMTs or HPD. The spacing of the grooves is 2cm and its direction is alternated layer by layer to provide 3 dimensional position information. One super-layer consists of a $7m \times 7m$ wide and 3mm thick Pb/scintillator sandwich supported by $7m \times 7m$ wide and 1m thick magnetized iron calorimeter which is a sandwich of 1 inch-thick magnetized iron and 5mm extruded scintillator plates. The iron calorimeter also allows momentum and charge measurements. The total mass is 213kton out of which 127kton is the Pb/scintillator calorimeter. The fiducial mass of

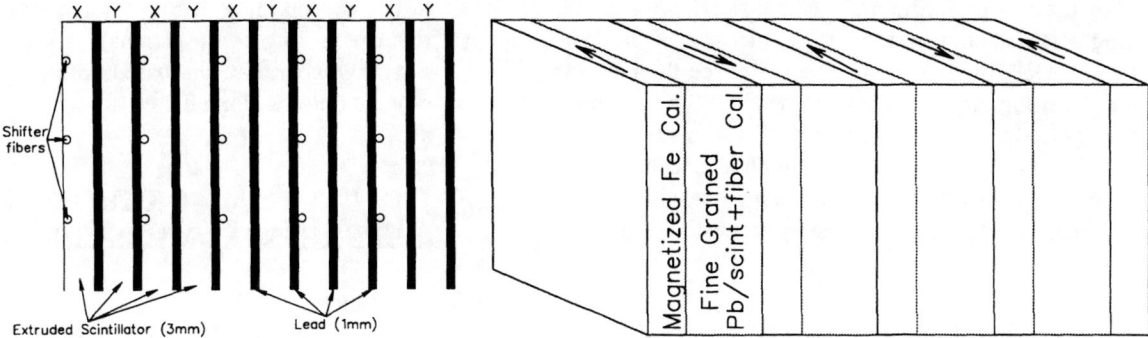

**FIGURE 1.** End views of the detector. The Pb scintillator sandwich calorimeter consists of layers of 1mm Pb and 3mm extruded scintillator (left). The Pb/scintillator calorimeters are supported by magnetized iron calorimeters (right).

Pb/scintillator section is 100kton, which corresponds to 3 p-decays/year for a proton lifetime of $\tau_p = 10^{34}$ years The total volume is the same as the super-Kamiokande detector.

A similar detector has been tested as a prototype veto detector of KOPIO(BNL-E926) experiment(5). A timing resolution of $\sigma_t \leq$ 1nsec and a light yield of 12 photoelectrons/MeV have been obtained (8). With a groove spacing of 2cm, a position resolution is expected to be $\sigma \sim$ a few mm. Figure 2 shows a track pattern expected from $p \to K^+ \bar{\nu}$ decay. A heavily ionizing stopping $K^+$ track is followed by a nearly minimum ionizing track of $\mu^+$ or $\pi^+$ from $K^+ \to \mu^+ \nu, \pi^+ \pi^0$ about $\tau_{K^+} \sim$ 12nsec later. The range of $\mu^+$ is monochromatic and is 26cm, which can be measured with a resolution of $\sigma_{R_\mu}/R_\mu \sim 5\%$. In the case of $\pi^+$, the track is also associated with 2 photons from $\pi^0$ whose conversion points are detected accurately to make good $\pi^0$ reconstruction. Good $\pi^0$ reconstruction also helps non-SUSY GUT mode $p \to e^+ \pi^0$ and neutrino experiments such as neutral-current ($\pi^0$)/charged-current ($e$) separation.

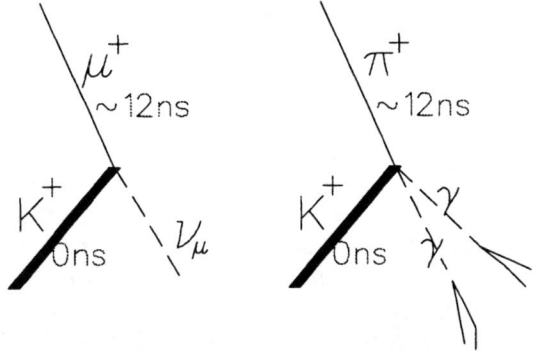

**FIGURE 2.** Decay topologies of $p \to K^+ \bar{\nu}$ followed by $K^+$ decays $K^+ \to \mu^+ \nu, \pi^+ \pi^0$ at rest.

In the case of water Čerenkov detectors (eg. super-Kamiokande), only the decay products of $K^+$ are observed because the velocity of $K^+$ from proton decay is below the Čerenkov threshold. The main background comes from atmospheric $\nu_\mu$ producing $\mu$ through charged-current reaction. In the case of a wire chamber (TPC) based calorimeter (SOUDAN2), both $\mu$ and $K^+$ tracks are identified. Heavily ionizing proton and nearly minimum ionizing $\mu^-$ from the atmospheric $\nu_\mu$ reaction $\nu_\mu n \to p \mu^-$ mimic $K^+$ and $\mu^+$ tracks. SOUDAN2 observed one event which is consistent with a background at a single event sensitivity of $\tau_{p \to K^+ \bar{\nu}} = 10^{32}$ years. The $\mu^+$ track for the signal is delayed by the kaon lifetime of around $\tau_K \sim$ 12nsec whereas there is no delay in the background and thus a delayed coincidence cut would have a large rejection. Unfortunately, the timing resolution of the TPC calorimeter is not good enough to identify the $K^+$ decay time. Like the SOUDAN2 detector, the Pb/scintillator calorimeter has an additional advantage of larger fiducial fraction and higher density ($\rho \sim 3.5 g/cm^3$), which keeps the detector volume small. The energy resolution is however worse than for water Čerenkov, which makes it harder to detect very low energy solar neutrinos.

The cost of Pb/scintillator calorimeter has been reduced because of the extruded scintillator, pixel photodetector (multianode PMD and HPD), and CMOS electronics technologies. The long attenuation length of the WLS fibers ($L_{att} \sim 4m$) makes a large detector feasible. The flexibility of the fibers allows multiplexing of distant fibers (more than 45cm away) to reduce the number of readout channels. An estimate of the material cost of a detector with a 100kton fiducial mass would be ~$500M:

450M m fiber × $0.3/m ~ $135M
27kton Scint. × $3.0/kg ~ $81M
100kton Pb × $0.5/kg ~ $50M
1.3M channels × $100 (PMT/electronics) ~ $130M
Magnetized iron calorimeter ~ $60M
Excavation of the site ~ $25M

The cost may be further reduced considering the continuing cost reduction in the PMT's, electronics and WLS fibers. One might also be able to reduce the cost further by using larger spacing between WLS fibers and/or thicker lead plates. One could also consider starting with 1/3 (33kton fiducial) detector, which should be enough to reach $\tau(p \rightarrow K^+\bar{\nu}) = 10^{34} years$ after several years of running, and gradually building up the rest.

## A PROTON DECAY DISCOVERY STRATEGY

In this section, we discuss a discovery strategy of the proton decay if it exists within the predicted range of $\tau(p \rightarrow K^+\bar{\nu}) < 10^{34} years$.

- Enough fiducial mass to observe the signal:
  The proton lifetime of $\tau(p \rightarrow K^+\bar{\nu}) = 10^{34} years$ gives 3 proton decay per year for a 100kton fiducial. Assuming an acceptance of 50%, $3 \times 0.5 \times 5 = 7.5$ proton decays after 5 years of running for $\tau(p \rightarrow K^+\bar{\nu}) = 10^{34} years$. In order to keep the cost within a few hundred million dollars, we cannot expect tens of events. This leads to the second point.

- Able to identify the signal with a couple of events:
  Discovering a signal is quite different from setting an upper limit particularly in the understanding of the background. In the case of an upper limit experiment, no candidate means no background. In the case of signal discovery, detailed understanding of the background is required to identify the signal from the backgrounds. It is particularly challenging when we want to identify the signal with a couple of events, in which case we cannot rely on observed signal distributions such as a signal peak. Because Monte Carlo simulations cannot represent very rare mechanisms nor hardware/software failures, it becomes crucial to measure the background levels from actual data using extra redundancies. A strategy for the $p \rightarrow K^+\bar{\nu}$ discovery, which follows the approach originally developed for a rare K decay experiment(9), is described in the following section.

- Compelling byproducts to justify the large cost:
  A large nucleon decay detector provides a rich program in neutrino physics addressing the neutrino mixing parameters including $U_{13}$ and leptonic CP violation phase as well as astrophysical neutrinos as described below.

- Prototyping/Calibration:
  Before building a large detector, we would want to test it and optimize it by prototyping and calibration (measuring cross sections). A scenario of prototyping/calibration discussed below provides rich physics outputs along the way.

## $P \rightarrow K^+\bar{\nu}$ BACKGROUND SUPPRESSION/MEASUREMENT

**(a)** Kinematics
A water Čerenkov detector like super-Kamiokande detector cannot see $K^+$ from proton decay since its velocity is below the water Čerenkov threshold. The Pb/scintillator calorimeter can observe the distinct topology with a heavily ionizing track of $K^+$ associated with minimum ionizing track of $\mu^+$ from $K^+ \rightarrow \mu^+\nu$ decay. Figure 3 shows the range and energy distributions of $K^+$ and $\mu^+$. $K^+$ distribution is smeared due to the Fermi motion of protons in the nucleus. The range and energy distributions of the $K^+ \rightarrow \mu^+\nu$ decay at rest show clear peaks. Compared to the SOUDAN2 detector, there is extra kinematic information of energy for this Pb/scintillator calorimeter. The range resolution, which is limited by range straggling, is also expected to be better than SOUDAN2 because it is packed with material without gas gaps. The range resolution from GEANT simulation is found to be about a factor of 2 better than SOUDAN2. The 90% CL upper limit set by SOUDAN2 experiment is $\tau_{p \rightarrow K^+\bar{\nu}} > 0.43 \times 10^{32} years$(7) with backgrounds appearing at $10^{32}$ years level. Additional energy measurement and better range resolution with the Pb/scintillator calorimeter would improve the kinematic rejection but is not enough to reach $10^{34}$ years and another handle is needed.

**(b)** Delayed coincidence
The background in the SOUDAN2 experiment originate from atmospheric neutrino and neutron interactions. The latter is produced by neutrino interaction with surrounding rocks ("rock" events). Figure 4 shows a typical track pattern for the signal and the background. The $\nu_\mu n \rightarrow \mu^- p$ reaction mimic the background when a heavily ionizing proton track fakes the $K^+$ track. In this case of background the time of the $\mu$ is the same time as proton track (prompt), whereas the time of $\mu$ is delayed by the $K^+$ decay time ($\sim \tau_K = 12 nsec$) for the signal. The "rock" event is also expected to be prompt. As a matter of fact, virtually all the background processes except for a single $K^+$ production are prompt and a delayed coincidence of the muon track provides a large rejection. A production of $K^+$ by atmo-

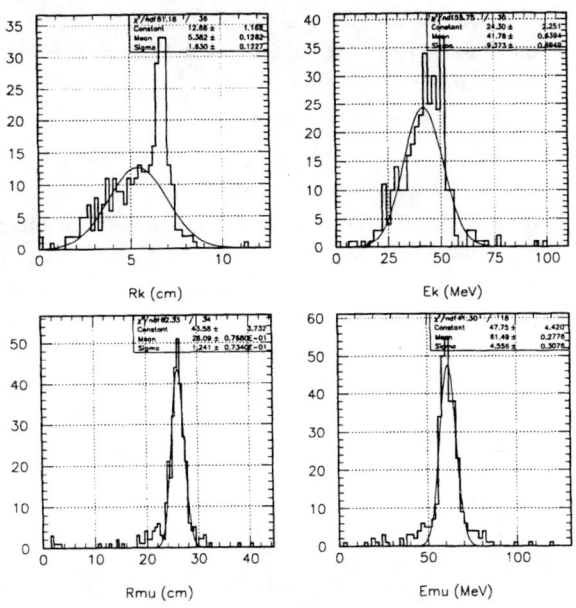

**FIGURE 3.** Expected range and energy distributions of $K^+$ and $\mu^+$.

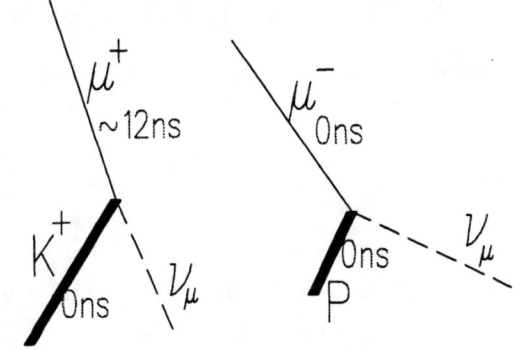

**FIGURE 4.** Decay topologies of the signal $p \to K^+\bar{\nu}; K^+ \to \mu^+\nu$ and the background $\nu_\mu n \to \mu^- p$.

spheric neutrinos or "rock" neutrons are associated with $K^-/\Lambda$ ($s\bar{s}$ production) and/or a charged lepton and can be highly suppressed by charged veto.

A similar prompt background in a similar geometry exists in a rare $K^+$ decay experiment (BNL-E787), which measures the branching ratio of $K^+ \to \pi^+\nu\bar{\nu}$. Figure 5 shows timing distributions of a prompt $\pi$ beam background (a prompt peak at zero) and the $K^+ \to \mu^+\nu$ decay signal. A delayed coincidence cut at around 2nsec gives a rejection of $R_{delco} \sim 10^4$. With the extra delayed coincidence rejection of $10^4$ on top of what is already achieved in the SOUDAN2

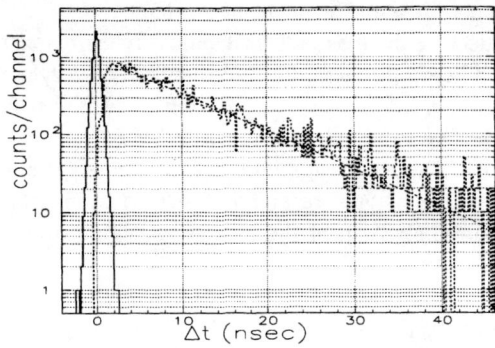

**FIGURE 5.** Decay time distribution for the signal $K^+ \to \mu^+\nu$ and the "prompt" beam background from the BNL-E787 experiment ($K^+ \to \pi^+\nu\bar{\nu}$).

experiment, the background is expected to be suppressed below $\tau_p = 10^{36}$ years level.

(c) Background measurement from the data:
Because the kinematic rejection ($R_{kin}$) and delayed coincidence $R_{delco}$ are independent in principle, we can measure each rejection using tagged samples with the other selection inverted. For example, one can select the prompt peak in Figure 5 and measure the kinematic rejection $R_{kin}$. By selecting the events outside the $R_\mu$ peak in Figure 3, we can measure the delayed coincidence rejection $R_{delco}$. Overall rejection is obtained as $R = R_{kin} \times R_{delco}$ from data without relying on Monte Carlo simulations, which is essential for very rare decay experiments such as proton decay and $K^+ \to \pi^+\nu\bar{\nu}$.

## NEUTRINO PHYSICS

The Pb/scintillator nucleon decay detector also provides a rich neutrino program as a general purpose neutrino detector. The large mass, good angular resolution, magnetic field, good $e/\pi^0$ identification, and a large enhancement in low energy $\nu_e Pb$ cross section makes this detector ideal for neutrino experiments.

1. Atmospheric neutrinos(10)
   The large mass and good angular resolution of the detector make it possible to observe the oscillation pattern of $\nu_\mu$ disappearance in the neutrino path length divided by energy (L/E) distribution. One could also look for $\nu_\tau$ appearance for example in $\tau^\pm \to \rho^\pm\nu$, taking advantage of the good $\pi^0$ reconstruction and $e/\pi^0$ separation(11). Matter oscillation (MSW effect) and $U_{e3}$ can also be studied in $\nu_e \to \nu_\mu$ oscillation(12).

2. Supernovae and solar neutrinos(13)

Due to enhancements in Gamov-Teller resonance and Fermi function, Pb is expected to be very sensitive to $\nu_e$(CC) and $\nu_\mu, \nu_\tau$(NC) (e.g. $\sim$100 times more sensitive to $\nu_e$ per gram than C for $E_{\nu_e} \sim 30 MeV$) (14). The event signature is a prompt electromagnetic(EM) signal (e and $\gamma$) followed by a delayed neutron signal near the location of the prompt signal:

$CC: \nu_e Pb \rightarrow Bi^* e^-; Bi^* \rightarrow Bi n \gamma$
$NC: \nu_e Pb \rightarrow Pb^* \nu_e; Pb^* \rightarrow Pb n \gamma$

Gd paper inserted between Pb/scintillator modules converts the delayed neutron into $\sim$8MeV $\gamma$. The EM energy and time corresponds to the energy and time of the neutrino. The Pb detector is sensitive to $\nu_e$ and complimentary to water Čerenkov which is sensitive to $\bar{\nu}_e$. Because of the large mass, large $\nu Pb$ cross section, and the clean signature, this detector would be very effective in detecting supernova neutrinos and higher energy (Hep) solar neutrinos.

The explosion of the supernova is triggered by a gravitational collapse of the core which first creates $\sim$10msec-long $\nu_e$ flash from shock breakout followed by seconds-long $\bar{\nu}_e, \nu_{\mu,\tau}$ signals. The Pb detector is ideal for detecting this $\nu_e$ (not $\bar{\nu}_e$) flash (CC) as well as $\nu_{\mu,\tau}$ (NC), providing detailed understanding of the explosion mechanism of supernovae. The neutrino oscillation converts $\nu_\mu$ and $\nu_\tau$, which are created at higher temperature, into $\nu_e$'s, enabling us to study the neutrino oscillation and its impact in supernovae. The clean signal signature may also make it possible to detect relic supernova neutrinos before the next supernova to explode(15). The 100kton detector would be sensitive beyond our galaxy to Andromeda galaxy, providing more chance of seeing the event. One may also study energy spectrum and seasonal and day-night effects of Hep solar neutrinos.

3. Neutrinos from accelerators(16)

Accelerator $\nu$ beams are created by $\pi$ decay-in-flight using neutrino horn or stopped $\pi/\mu$ decays in the beam dump. High intensity and high energy neutrinos may also be available from a $\mu$ storage ring (neutrino factory) in the future. A detector with large mass and charge/momentum measurement is ideal for the study of neutrino beam from the accelerator. The good electron(CC)/$\pi^0$(NC) separation of the detector becomes essential in some studies.

A precision measurement of $\nu_\mu$ oscillation can be performed by observing a detailed oscillation pattern in L/E, $\tau$ appearance through $\tau \rightarrow e\nu\bar{\nu}$, and NC($\pi^0$)/CC($l^+$) ratio. Searching for wrong sign $\mu$ provides $\nu_e \rightarrow \nu_\tau$ oscillation ($\theta_{13}$)(18). $\theta_{13}$ can also be measured in a matter enhanced $\nu_\mu \rightarrow \nu_e$ oscillation(19). The large mass and large cross section at lower energy may be ideal for T and CP measurement in neutrino oscillation(20).

A prototype detector can also be used for a short baseline experiment to study LSND type $\nu_\mu - \nu_e$ oscillation(17). The large enhancement and clean signature in $\nu_e Pb$ reaction enables us perform sensitive experiment using $\nu_\mu$ from stopped $\pi^+$ beam $\pi^+ \rightarrow \mu^+ \nu_\mu$; $\nu_\mu \rightarrow \nu_e$. Figure 6 shows the time and energy distribution of the signal and background from $\mu^+ \rightarrow e^+ \nu_e \bar{\nu}_\mu$ decay. The positive signatures of $\pi$ lifetime and monochromatic energy for the signal would help confirm/reject the signal in a clear way.

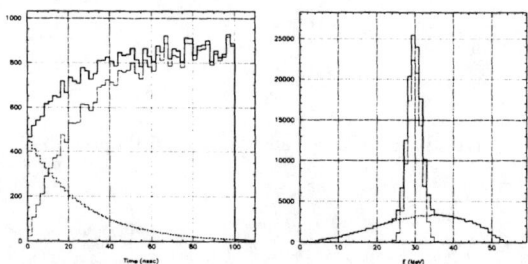

**FIGURE 6.** Time and the energy distributions of $\nu_e$ from the LSND type $\nu_\mu - \nu_e$ signal $\pi^+ \rightarrow \mu^+ \nu_\mu$; $\nu_\mu \rightarrow \nu_e$ and the background $\mu^+ \rightarrow e^+ \nu_e \bar{\nu}_\mu$. The signal has positive signatures of the pion lifetime and monochromatic energy.

# A SCENARIO FOR PROTOTYPING AND CALIBRATION

In building such a large detector, it is prudent to plan for prototyping and calibration of the detector. As we can see below, there will be rich physics opportunities along the way.

- 100ton (0.1%) detector:

  The barrel veto detector for KOPIO($K_L \rightarrow \pi^0 \nu\bar{\nu}$) experiment is about 100ton and has a same structure as the proton decay detector. By placing this detector at a meson factory, $\nu Pb$ cross section measurement below $E_\nu < 50 MeV$ can be performed. A study on the KARMEN timing anomaly can also be done. By placing the detector at the neutrino beam line at the Fermilab Booster parasitic to the mini-BooNE experiment, $\nu Pb$ cross section measurement between 0.2 to 2 GeV can be performed. Because of the fine tracking capability, $\nu_\mu p$ scattering cross sections could also be measured. The detector could also help monitoring the beam for mini-BooNE because of its good $e/\pi^0$ separation.

- 1kton (1%) detector:
  With a 1kton detector, one could make a definitive study on the LSND type neutrino oscillation, $\nu_\mu \to \nu_e$ at a meson factory as discussed in the previous section. By placing the detector underground such as the SNO cave, it will serve as an excellent galactic supernova and Hep solar neutrino detector due to a large $\nu - Pb$ cross section.

- 10kton (10%) detector:
  With a size of 10kton, the detector starts to be effective for the proton decay, eg. $p \to K^+ \bar{\nu}$ up to $10^{34}$ years. The finely segmented detector with this fiducial mass is large enough for long baseline $\nu$ experiments using accelerator neutrino beam. One could make a good atmospheric $\nu$ experiment.

- 100kton (100%) detector:
  The full 100kton detector works for proton decays particularly $p \to K^+ \bar{\nu}$ up to $\tau < 10^{35}$ years. The atmospheric neutrino oscillation can be studied in detail by L/E oscillation pattern, $\tau$ appearance, and matter effect. The detector would be even sensitive to the relic supernova neutrinos and extra galactic supernova neutrinos. Using long baseline neutrino beams from accelerators, one could study $\nu_e \to \nu_\tau (\theta_{13})$ and leptonic CP violation.

## SUMMARY

Nucleon decay is virtually the only direct search of GUT scale. Gauge unification with SUSY and $\nu$ oscillation independently suggests an energy scale of $10^{16}$ GeV. The favored decay mode is expected to be $p \to K^+ \bar{\nu}$ with a lifetime less than $\tau_{p \to K^+ \bar{\nu}} < 10^{34}$ years. Proton decay is waiting to be discovered.

A recent progress in scintillator and its readout technology made it possible to build a large scale fine grained calorimeter out of plastic scintillator and lead. A 100kton lead/extruded-scintillator detector would see 3 p-decays/yr for $\tau_{p \to K^+ \bar{\nu}} = 10^{34}$yr. A detailed kinematics(topology,range,energy) combined with delayed coincidence($T_K - T_\mu$) makes the measurement background free and reliable background measurement can be done using data.

The detector has a rich potential in neutrino physics. The large mass, fine segmentation (particularly good angular resolution and $e/\pi^0$ separation), and magnetic field makes it an ideal general purpose neutrino detector. One could consider studying all the neutrino mass/mixing parameters; matter oscillation, $\nu_e - \nu_\tau$, $\nu_e - \nu_\mu$, $\nu_{e,\mu} - \nu_s$, and CP violation. The large $\nu - Pb$ cross section at low energy also makes it an ideal supernova neutrino detector, detecting the shock breakout $\nu_e$, $\nu_{\mu,\tau}$(NC), and $\nu_{\mu,\tau} - \nu_e$ oscillation.

## REFERENCES

1. Super-Kamiokande collaboration, *Phys. Rev. Lett.* **81**, (1998) 3319.
2. F. Wilczek, *Nucl. Phys. Proc. Suppl.* **77**, (1999) 511, K. Babu, J. Pati, F. Wilczek, **hep-ph**/9812538, (1998).
3. See for example the MINOS experiment web page, http://www.hep.anl.gov/ndk/hypertext/numi.html.
4. Multipixel Hybrid Photodiode by DELFT Electronic Products BV, and Flat panel PM by Hamamatsu Photonics K.K.
5. http://sitka.triumf.ca/e926/index.html.
6. Super-Kamiokande collaboration, *Phys. Rev. Lett.* **81**, (1998) 2016, Super-Kamiokande collaboration, *Phys. Rev. Lett.* **83**, (1999) 1529, Brett Viren for the Super-Kamiokande Colloboration **hep-ex**/9903029, (1999).
7. SOUDAN2 collaboration, *Phys. Lett.* **B427**, (1998) 217.
8. Yu. Kudenko et. al., *BNL-E926* **TN008**, (1998).
9. A. Konaka, **hep-ex**/990316, (1999).
10. See the discussion of the Atmospheric and HE Astrophysical Neutrino Working Group in this workshop.
11. L. Hall and H. Murayama, *Phys. Lett.* **B463**, (1999) 241.
12. J. Pantaleone, *Phys. Rev. Lett.* **81**, (1998) 5060.
13. See the discussion of the Solar and Supernova Neutrino Working Group in this workshop.
14. C. K. Hargrove et.al., *Astroparticle Physics* **5**, (1996) 183, G. M. Fuller, W. C. Haxton, G. C. McLaughlin, *Phys. Rev.* **D59**, (1999) 085005.
15. M. Kaplinghat, G. Steigman, and T. Walker, **astro-ph**/9912391, (1999).
16. See the discussion of the Accelerator Neutrino Working Group in this workshop.
17. Cliff Hargrove, private communication.
18. S. Geer, *Phys. Rev.* **D57**, (1998) 6989.
19. I. Mochioiu and R. Shrock **hep-ph**/9910554, (1999), O. Yasuda, **hep-ph**/9910428, (1999).
20. M. Koike and J. Sato **hep-ph**/9909469, (1999).

# Nucleon Decay studies in a large Liquid Argon detector

A.Bueno[1], M.Campanelli[1],*, A.Ferrari[2], A.Rubbia[1]

[1] *Institut für Teilchenphysik, ETHZ, CH-8093 Zürich, Switzerland*
[2] *CERN, CH-1211 Geneve 23 Switzerland*

**Abstract.** Future nuclear decay experiments have to be able to combine a large mass, the capability of distinguishing between several possible decay channels and a good background discrimination, in order to increase their sensitivity linearly with the mass. We present the capabilities of the liquid Argon technology to fulfill these requirements.

## INTRODUCTION

Nucleon decay studies are presently in a "second generation" phase, after the enthusiasm following the developement of the first GUT theories. The minimal version of SU(5)(1), predicting the decay

$$p \to e^+ \pi^0$$

with a lifetime of about $10^{31}$ years. This model has been ruled out by the experimental limits ($\tau > 2.9 \times 10^{33}$ years), and also fails to predict the correct value of $\sin^2 \theta_W$.

Alternative models have been proposed, for instance SUSY GUTs, that predict values of $\sin^2 \theta_W$ closer to the measured ones, and the presence of an intermediate mass scale of $O(M_W)$, that seems to provide a better unification of the coupling constants, at higher values with respect to minimal SU(5)(2).

The higher unification mass pushes up the proton lifetime in the $p \to e^+ \pi^0$ channel, which has predicted lifetimes of $10^{36 \pm 1}$ years, compatible with the experimental limits.

However, in this scenario, other decay channels open up, where supersymmetric intermediate states are involved. In particular, s-quark production is favoured, and final states involving kaons (like $p \to K^+ \nu$) are present.

## EXPERIMENTAL SITUATION

In absence of background, the limit on nucleon lifetime in T years of observation is given by:

$\tau_p > 1.2 \times M \times T \times \eta$ ($10^{32}$ year) (90% C.L.) for the proton

---

* Presented by M.Campanelli

$\tau_n > 1.4 \times M \times T \times \eta$ ($10^{32}$ year) (90% C.L.) for the neutron

where M is the mass in kton and $\eta$ the detection efficiency.

If backround (mainly coming from atmospheric neutrinos) is present, the sensitivity increases only proportionally to $\sqrt{M}$. Given the variety of predicted decay modes, the ideal detectors should be as versatile as possible, very good in background rejection, and at the same time have the largest possible mass.

The ICARUS(3) collaboration has been developing for more than 10 years the technology to build a large Liquid Argon Time Projection Chamber, combining the imaging quality of a bubble chamber with the advantages of the electronic read-out.

The liquid argon technology allows the operation of large detectors with very good imaging, particle identification and energy resolution, all very important characteristics in the search for nucleon decays.

The detector consists of a large cryogenic pool equipped with wires and an electronic readout system. The main advantages of this detector are the following:

- it is uniform and continuously sensitive
- there is no need for external trigger
- it was shown to be safe for underground operation, even for large volumes
- particles can be easily identified via multiple dE/dx measurements along the tracks
- it can be considered an homogeneous calorimeter, with very good resolution (about $5\%/\sqrt{E}$ for electromagnetic and $15\%/\sqrt{E}$ for hadronic showers).

The ICARUS collaboration is presently building a 600 ton detector to be installed in Gran Sasso National Laboratories by the beginning of year 2001. The main physics goal with this kind of detector will be the study of atmospheric and solar neutrinos, possible thanks to the very low energy threshold ($\approx 5 MeV$) of this device.

In view of the CERN-Gran Sasso neutrino beam, a detector made of liquid Argon tanks separated by a tail-catcher calorimeter and a muon spectrometer has been proposed (ICANOE proposal, (4)). The total liquid Argon mass of this device will be more than 5 kton.

This detector will already have very good capabilities for nucleon decay searches ((5)), but the liquid argon technology can in principle push even further its limits. Last year, a 30 kton Super ICARUS was proposed, with the main goal to cover in $\tau$ appearance very small values of $\delta m^2$ for a long baseline neutrino experiment. This would also be an excellent detector for nucleon decay searches. An appropriate site is in this case needed, since for its dimensions it cannot be located in the Gran Sasso tunnel.

## NUCLEON DECAY ANALYSIS

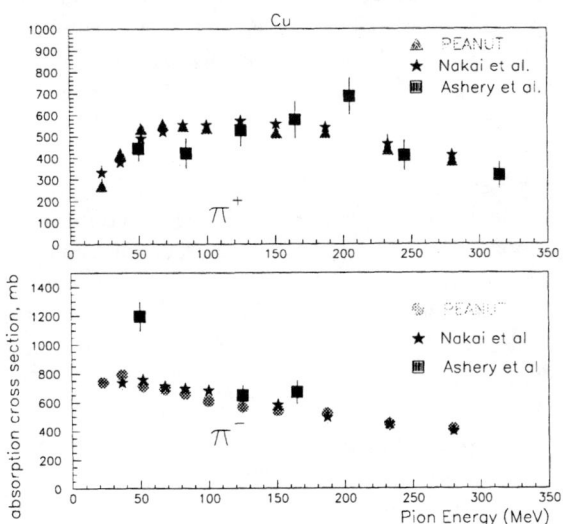

**FIGURE 1.** Low energy positive and negative pion absorption (copper target)

If nuclear reinteractions and Fermi motion are neglected, a proton decay event has the following kinematics:

$$\Sigma E_s = M_n c^2$$

$$\Sigma P_s = 0$$

where $\Sigma E_s$ and $\Sigma P_s$ are respectively the total energy and momenta of the nucleon decay products.

In the reality, the hadron produced in nucleon decays will reinteract with the nucleons of the atom, and

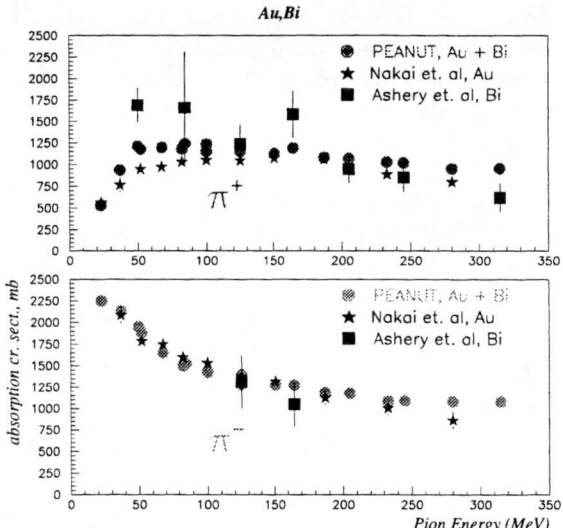

**FIGURE 2.** Low energy positive and negative pion absorption (gold target)

the clean scenario described above gets considerably distorted.

For the analysis presented in (5), only Fermi motion and a simple nucleon model were considered. The analysis was performed via an eye scan of a mixed sample of signal and background events, using large samples to extract background and efficiency.

In the following, an analysis based on a model with full nuclear reinteraction (FLUKA (6)) was used. The selection is based on cuts on the main kinematic variables, for the two most important channels

$$p \rightarrow e^+ \pi^0$$

$$p \rightarrow K \nu$$

The full simulation of nuclear processes is based on the PEANUT package. It gives a macroscopic description of the nuclear reaction, with a quasi-classical approach to the nuclear cascade. The nuclear thermalization acts in three stages: a preequilibrium, an evaporation/fission and a $\gamma$ deexcitation. The model includes:

- correct nuclear density and nuclear potential distributions
- quantum mechanical correction (Pauli blocking, formation time, nucleon hard core, antisymmetrization)
- multibody processes for pions

The model has been tested against a large variety of experimental data, taken with different types of particles and targets, at energy scales ranging from few MeV to many TeV. These tests have always given good agreement between the model and the experimental data, as can be seen for instance in the plots of figures 1 and 2, relative to low-energy pion interactions with copper and gold targets.

## $p \to e^+ \pi^0$ channel

This is the model favoured by minimal SU(5) theories, and has been extensively studied. The event is composed of three electromagnetic showers, one from the positron and the other two are photons from the $\pi^0$ decay. The total invariant mass is about 1 GeV. One example of such an event is shown in figure 3.

In our study the same cuts as the current Super Kamiokande analysis have been used (see table 1).

The idea behind these cuts is to have a balanced event, with all particles identified as such, and with a total invariant mass compatible with that of a proton. The accepted region in the total momentum-invariant mass plane for Argon or oxygen targets are shown in figures 4 and 5.

As expected, background rejection is extremely good (only 1 event has been found using the extreme exposure of 1000 Kton year) for an efficiency of about 18%. If Oxygen atoms are considered instead of Argon, the efficiency increases to 24%, still lower than the value quoted by SuperKamiokande for the same cuts. The reasons for this disagreement have to be further investigated, since they are likely to come from the different nuclear model used.

## $p \to \bar{\nu} K^+$ channel

This is one of the favourite SUSY decays, and is quite typical due to the presence of a strange meson in the final state. A liquid Argon detector can profit from its very good particle identification capabilities to tag the kaon and its decay products. We recall here that the kaon is typically below the Cerenkov threshold for water, therefore it can be seen in a large water detector only via its decay products.

An example of these capabilities is given in figure 6, where the energy loss of pions and kaons is plotted against the range of the track. The two bands are clearly separated. A cut around the kaon band gives a 77% kaon identification efficiency, which is almost 100% if we consider that about 20% of the kaons decay in flight. After this cut is performed, no pion background remains.

The two main kaon decay modes are $K^+ \to \mu^+ \nu_\mu$, with a branching ratio of 63.5%, and $K^+ \to \pi^+ \pi^0$, with a branching ratio of 21.2%. Both are two-body decays, and have quite clear signature. In the first case, a muon with fixed range of $\approx$ 55 cm is observed (see an example in figure 7), in the second a charged pion with a range of $\approx$ 40 cm, plus two electromagnetic showers from the decay $\pi^0 \to \gamma\gamma$.

The determination of the relative directions of the kaon and of the muon in the decay $K^+ \to \mu^+ \nu_\mu$ is essential to separate proton decay events from quasi-elastic $\nu_\mu$ charged current interactions with a recoiling proton, that could be misidentified as a kaon. This is achieved in via a ionization density measurement. Applying the cuts listed in table 2, good efficiency can be reached for a negligible background.

## CONCLUSIONS

Nucleon stability remains a fundamental open problem in particle physics. The ideal proton decay detector must be extremely versatile, to allow the study of several final states, and have a very good background rejection power. Furthermore, a detector with good imaging is probably essential in the case of a positive signal.

The ICARUS collaboration has been developing for 10 years the technology to have a reliable large mass liquid argon TPC for neutrino and nucleon decay physics. In principle, there are no limitations in reaching very high masses (O(30 kton)) and running for long data-taking periods (> 10 years) using this technology. Given the good imaging and particle identification capabilities of this kind of device, most of the nucleon decay channels are basically background-free, so the sensitivity grows linearly with exposure and mass, while in case background is present the sensitivity increases only like the square root of the exposure.

A correct understanding of the detection efficiency requires a careful study of nuclear reinteractions, and the present models seem to give a good description of these effects. Therefore, we think that the values of the efficiencies quoted are quite realistic, and that this kind of detector can have a very good discovery potential even with smaller mass with respect to other technologies.

## REFERENCES

1. H.Georgi and S.L.Glashow, Phys. Rev. Lett. **32** (1974) 438.

2. U.Amaldi, W. de Boer H.Füsternau, Phys. Lett. **B260** (1991) 447

3. P.Cennini *et al.* "ICARUS II. A second generation Proton Decay Experiment and Neutrino Observatory at the Gran

Sasso Laboratory". Experimental proposal LNGS 94/99-I and 94/99-II (1994)

4. ICANOE PROPOSAL LNGS-P21/99 INFN/AE-99-17 CERN/SPSC 99-25 SPSC/P314

5. see previous references, chapter 7 p. 149-158

6. A. Fassò, A. Ferrari, J. Ranft, P. R. Sala, G. R. Stevenson, and J. M. Zazula, Proc. of the workshop on *Simulating Accelerator Radiation Environment, SARE*, Santa Fè, 11-15 january (1993), (A. Palounek ed., Los Alamos LA-12835-C 1994), p. 134.

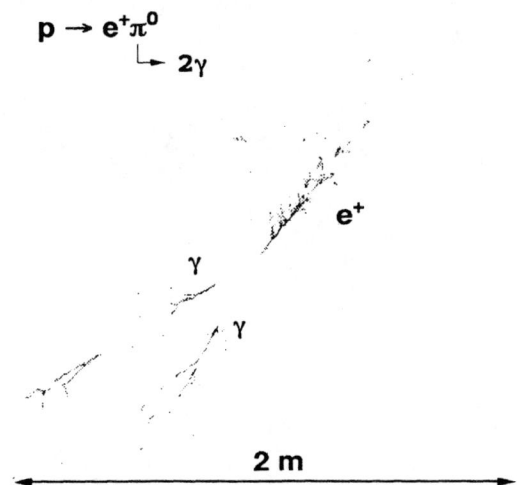

**FIGURE 3.** A typical $p \rightarrow e^+\pi^0$ candidate in ICARUS. Three electromagnetic showers coming from the positron and from the two photons from $\pi^0$ decay are clearly visible.

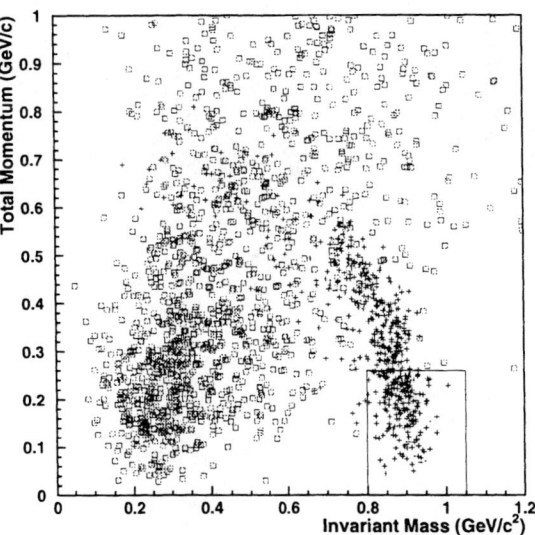

**FIGURE 4.** Kinematic cuts for $p \rightarrow e^+\pi^0$ decays in liquid Argon. Here, in the plane defined by total energy and momentum, crosses represent signal and squares background events. The box indicates the cut region, i.e. all events in the box are accepted

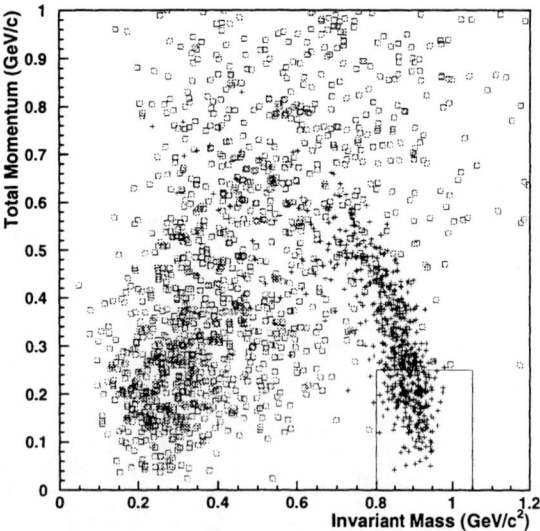

**FIGURE 5.** Kinematic cuts for $p \to e^+\pi^0$ decays in Oxygen. Like in the previous figure, signal and background are indicated by crosses and squares. The separation between the two classes is sligltly better due to the smaller size of Oxygen nuclei.

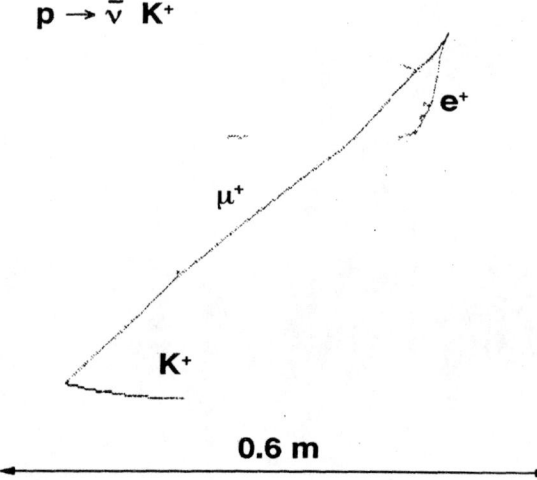

**FIGURE 7.** A typical $p \to \bar{\nu} K^+$ event in an imaging device like Icarus. The kaon track is visible, as well as a muon from its decay and an electron from the decay of the muon.

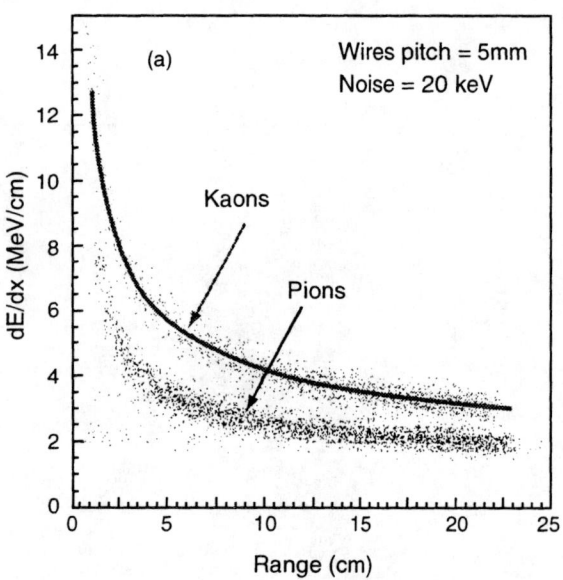

**FIGURE 6.** Particle identification capabilities of the Icarus detector. Left: dE/dx measurement vs the distance from the dtopping point. Right: distance from kaon hypothesis. All kaons not decaying in flight can be reconstructed as such, with no pion contamination

**Table 1.** Cuts for the $p \to e^+\pi^0$ analysis in Argon. Results are presented for nucleon decays in Argon, Oxygen and for the atmospheric neutrino background

| Cuts | $e+\pi^0$ Argon | $e+\pi^0$ Oxygen | $\nu_e$ CC | $\bar{\nu}_e$ CC | $\nu_\mu$ CC | $\bar{\nu}_\mu$ CC | $\nu$ NC | $\bar{\nu}$ NC |
|---|---|---|---|---|---|---|---|---|
| Initial | 100% | 100% | 59861 | 11707 | 106884 | 27273 | 64705 | 29612 |
| One $\pi^0$ | 54% | 70% | 5277 | 1696 | 11160 | 4388 | 6223 | 2278 |
| One $e$ | 54% | 70% | 5277 | 1696 | 7 | $<1$ | $<1$ | $<1$ |
| $T_p < 100$ MeV | 53% | 68% | 2505 | 1256 | $<1$ | $<1$ | $<1$ | $<1$ |
| $0.8 <$ Inv Mass $< 1.05$ GeV | 38% | 53% | 306 | 204 | $<1$ | $<1$ | $<1$ | $<1$ |
| Total Momentum $< 0.25$ GeV | 19% | 24% | 1 | $<1$ | $<1$ | $<1$ | $<1$ | $<1$ |

**Table 2.** Cuts for the $p \to \bar{\nu}K^+$ analysis in Argon. Results are presented for nucleon decays in Argon and for the atmospheric neutrino background

| Cuts | $K+\bar{\nu}$ | $\nu$ NC | $\bar{\nu}$ NC |
|---|---|---|---|
| Initial | 100% | 64705 | 29612 |
| No primary $\pi^\pm$ | 99.4% | 55481 | 26033 |
| No primary $\pi^0$ | 98.7% | 48397 | 23265 |
| Only one kaon | 98.5% | 108 | 22 |
| Total Energy $< 0.65$ GeV | 85% | $<1$ | $<1$ |

# A New Underground Laboratory in Finland

Juha T. Peltoniemi for the CUPP collaboration

*Centre for Underground Physics in Pyhäsalmi*
*University of Oulu, Finland*

**Abstract.** A new underground laboratory is being built in a deep mine in Pyhäsalmi in the middle of Finland. This site provides excellent opportunities for underground physics, with the equivalent depth being about 4 000 m.w.e., and no severe problems with background or instabilities. The first experiment in the mine consists of three small muon telescopes at different depths. New experiments are planned, the next one is probably a multipurpose muon detector at shallow depth.

## INTRODUCTION

The original idea for an underground laboratory came from quite an exceptional way. In 1993 the mayor of the city got worried that the mining in the Pyhäsalmi copper/zinc mine was supposed to stop around the year 2000. Consequently, around 300 people would lose their jobs and new activities had to be invented.

It occurred to someone that one could use the mine for underground experiments. A pre-feasiblity study on the possibility of using the mine for underground physics research was done, with the first background measurements and rock analyses. As a result Pyhäsalmi was found suitable for underground physics experiments.

Today the situation has changed completely. New ore has been found, and the mining in the Pyhäsalmi copper/zinc mine will probably go on till 2012 at least. No miner will lose his job, and there is no political or social pressure for scientists. Nevertheless, from the physics point of view, the site looks even more attractive.

After pre-feasibility studies a serious attempt for the underground physics programme began year 1999. With an approval for funding from the European Union, the Centre for Underground Physics in Pyhäsalmi (CUPP) is now planning new experiments and looking for establishing international collaboration.

## THE MAIN CHARACTERISTICS OF THE MINE

Pyhäsalmi is the biggest and oldest operating metal mine in Finland. It opened in 1962. During 1998, 1.5 million tonnes of ore was milled. Copper in concentrates totalled 9 500 tonnes, zinc in concentrates 30 700 tonnes and pyrites 770 000 tonnes.

The depth of the mine is 1 300 m at present. The final depth of the mine will probably be about 1500 m (4200 m.w.e.). The average rock density is 2.81–2.83 g/cm$^3$. There are no severe water or instability problems.

All the mine can be accessed by a truck. There is a spiral shaped road all the way down, with maximum load size 2.6 m x 2.8 m x 8 m. There is also a lift, and a new shaft down to the bottom is being constructed.

The mine provides modern and reliable infrastructure, including buildings, electricity, lifts, pumping and ventilation.

Presently there are several caverns with size 15 m x 5 m x 15 m at various depths. After 2001, caverns of sizes 10 m x 10 m x 100 m can be excavated and present caverns will be free. For bigger caverns, the typical excavation costs are about (order of magnitude estimate only):

- normal tunnel with a diameter of 5 m: 50 euros/cubic metre

- similar tunnel adequate for living: 100 euros/cubic metre

- giant tunnel with a diameter of 20 m: 150 euros/cubic metre

- a large cavern of a typical size of 25 m - 50 m: 200 euros/cubic metre

- a giant cavern with a diameter of 100 m: 300-400 euros/cubic metre

It would take some 5-7 years to excavate a spherical cavern of one million cubic metres.

## BACKGROUND

The rock above provides shielding against the cosmic rays. At present the equivalent depth is max 3 500 m.w.e., in the future about 4000 m.w.e. Because of a flat surface the shielding is better than in many mountain sites.

The concentration of dangerous isotopes in the rock has been measured multiple times. The results vary, depending on the rock (ore/granite etc). Standard measurements give the following results:

- U-238 = 0.8 ppm.
- Th-232 = 3.32 ppm
- K-40 = 1.17 %

The radon concentration in the air depends on the ventilation. In rooms with good ventilation one can reach Rn<30 Bq, and nowhere does it exceed the limits posed by the radiation protection authorities.

Better measurements of the background level in different depths and specific caverns are planned. Particularly, we are going to measure the fast neutron spectra by the SAGE spectrometer in winter 1999-2000 at 660 m.

## DEMONSTRATION EXPERIMENT

The experimental programme at Pyhäsalmi was started by a small cosmic ray experiment. The apparatus is based on a partial recycling of the old muon telescope 50x50x5 cm$^3$ scintillator elements. The phototubes, electronics, and data aquisition systems are, however, completely new.

The experiment starts at three levels: ground, 90 metres and 210 metres. At all levels there will be a data acquisition system connected to the surface.

This experiment will be expanded in the near future. So far only half of the old scintillators have been reused. The rest will be recycled when new light detectors will be installed.

## MULTI-MUON EXPERIMENT

The first new project will most probably be a multipurpose muon experiment at shallow depth (50 metres)(1). This experiment is designed for studying the cosmic rays, particularly in the knee region. It is also adapted for the studies of superhigh energy cosmic ray interactions as well as upward going atmospheric neutrino events, moon shadow, sun shadow, seasonal variations, to name a few.

The facility consists of seven identical detectors. Each detector is composed and two layers of two resistive plate chambers (RPC), separated by 3 m. The size of the detector is 10 m * 10 m * 3 m. The detectors are arranged as a hexagonal array, one detector in the middle and the other six in the corners.

The multi-muon experiment is reaching the phase of detailed design. If accepted, the construction could start within a couple of years, so that the measurements could begin after the CosmoLep experiment.

## OTHER PLANS

Another plan under consideration is a 2β-decay into excited states with a High Purity Germanium detector. This project is being developed together with the NEMO collaboration. This experiment would be placed deeper underground.

This site might also provide a target for long baseline neutrino oscillation experiments. The distance to the largest existing or planned accelerators are

- DESY: 1350 km
- CERN: 1900 km
- Fermilab: 6100 km
- Brookhaven: 5600 km

Particularly, getting a neutrino beam from a muon collider on the other side of the earth would open interesting possibilities for neutrino oscillation searches.

The site provides also excellent facilities for solar and supernova neutrino experiments. It would be particularly interesting for coincidence measurements for determining the direction of a newcoming supernova, because it is not located at the same latitude as most of the other detectors.

We are open for new ideas and projects.

## THE STRUCTURE OF THE PROJECT

The CUPP project belongs formally to the Sodankylä Geophysical Observatory which in turn is part of the University of Oulu. Other collaborators include Karlsruhe Forschungzentrum, NEMO group, Tsinghua University and the Institute for High Energy Physics of the Chinese Academy of Sciences. New collaborators are welcome.

The project is funded by European Union (ESF), Finnish Ministry of Education, Municipality of Pyhäsalmi, University of Oulu, Sodankylä Geophysical Observatory and Outokumpu Mining Ltd. The funding is secured for the muon test experiment and for establishing an international collaboration for a first large experiment by June 2001 (European Union funding limits).

The project includes also an expanding theory group. Its main interests are neutrino physics and astrophysics, as well as cosmic ray physics. The Centre is also active in Internet publishing, keeping the well known Ultimate Neutrino Page (http://cupp.oulu.fi/neutrino/).

## SUMMARY

Let us conclude by summarizing the advantages of the mine:

- Flat surface — simplifies simulations and gives better shielding

- Northern location — magnetic field quite vertical

- Many caverns available in different depths from 0 to 1100 m and below

- Lots of free caverns of different sizes available for big and small eperiments

- Possibilities to build very big caverns

- Very good accessibility

- Far enough from possible neutrino sources

- Local authorities very favorable

- No environmental problems

- Low rate of criminality

More actual info can be found on our web pages: http://cupp.oulu.fi/.

## REFERENCES

1. Ding, L. K., Zhu, Q., Shen, C., Z.G.Yao, T. C., and Vallinkoski, M., Expression of interest on multi-purpose underground muon detector, Report, 1999.

# Study of 1 Megaton water Cherenkov detectors for the future proton decay search

M. Shiozawa*

*Kamioka Observatory, ICRR, University of Tokyo
Higashi-mozumi, Kamioka-cho, Yoshiki-gun, Gifu 506-1205, JAPAN*

**Abstract.** The sensitivity of a possible future 1 Megaton water Cherenkov detector for proton decay searches was studied. For $p \rightarrow e^+\pi^0$ decay mode, the detection efficiency and the number of atmospheric neutrino backgrounds were estimated by using a detailed Monte Carlo simulation program. Moreover, their dependence on the number density of the photomultiplier tube (PMT) was investigated. With the PMT density of the Super-Kamiokande detector (2 PMT/m$^2$, 40% photocathode coverage), we will reach to $1.5 \times 10^{35}$ years partial lifetime limit at 90% confidence level by 10 years livetime of the detector ($10^4$ kton·year exposure). With a 1/4 (1/9) PMT density, the sensitivity for $p \rightarrow e^+\pi^0$ mode is decreased to $1 \times 10^{35}$ years ($7 \times 10^{34}$ years).

## INTRODUCTION

Baryon number violated nucleon decay search could be the direct test of Grand Unified Theories (1). In the past two decades, several large underground detector experiments have looked for the signal but no clear evidence has been reported (2, 3, 4, 5, 6, 7, 8). In table 1, proton decay search detectors and their results are summarized. One of the detector types is a ring imaging Cherenkov detector with a target of water; Kamiokande, IMB, and Super-Kamiokande (SK), and another is a fine grained tracking calorimeter with a main target of iron; Fréjus and Soudan 2. One of important advantages of water Cherenkov detectors is that we can obtain the large target volume with reasonable cost. Moreover, there are free protons in water which decay products are free from imperfectly known nuclear effect and total momentum invalance caused by Fermi motion of protons.

To push up the proton decay search, we would need a much larger detector. Therefore, feasibility of a water Cherenkov detector with fiducial volume of 1 Megaton, which is about 50 times larger than that of the SK detector, was studied. Among various proton decay modes, sensitivity for $p \rightarrow e^+\pi^0$ mode was studied in detail. Moreover, sensitivity dependence on the PMT density was studied in order to evaluate the proper PMT density.

---

* E-mail: masato@icrr.u-tokyo.ac.jp

## DETECTOR DESIGNS

As is shown in table 1, current lower limit on partial lifetime of protons via $p \rightarrow e^+\pi^0$ mode has been set as $3.3 \times 10^{33}$ years by the SK detector and the detector will reach to $10^{34}$ years in 10 years livetime. It seems natural to require that the next generation detector should have sensitivity over $10^{35}$ years for the decay mode. According to the requirement, I assumed the fiducial mass of the next generation detector is 1 Mton.

Figure 1 shows some detector designs. "A la Super-K" is a cylindrical tank measuring 100 m in height and 120 m in diameter. "Cubes" and "Doughnut" (9) comprise of 20 sub-detectors. I assume that all surfaces of each sub-detector in "Cubes" and "Doughnut" are instrumented with PMTs. Therefore, the geometry of these sub-detectors is close to that of the SK detector. I utilized the SK simulation program to estimate the performance of the 1 Mton detectors, "Cubes" or "Doughnut". However, due to much longer travel length of Cherenkov photons, results of my study presented here may not be applicable to 'a la Super-K'.

As is shown in figure 1, we need more than 300 thousands PMTs to obtain the same PMT density as the SK detector. Although these numbers may not be impossible, it is useful to consider the possibility to reduce them. I studied the following three detector configurations; detector-(A) which has same PMT density as the SK detector ($\sim 300,000$ PMTs/Mton in total), detector-(B) which has 1/4 PMT density of the SK detector ($\sim 80,000$ PMTs/Mton in total), and detector-(C) which has 1/9 PMT density of the SK detector ($\sim 35,000$

**Table 1.** Proton decay search detectors. Water Cherenkov detectors (Kamiokande, IMB-3, and Super-Kamiokande) and iron tracking detectors (Fréjus and Soudan 2) are listed. Partial lifetime limits have been set at 90% confidence level.

| detectors | fiducial mass [kt] | exposure [kt·yr] | limit on $p \to e^+\pi^0$ [$10^{31}$yrs] | limit on $p \to \bar{\nu}K^+$ [$10^{31}$yrs] |
|---|---|---|---|---|
| Kamiokande | 1.04 | 3.76 | 26 (2) | 10 (2) |
| IMB-3 | 3.3 | 7.6 | 54.0 (3)* | 15.1 (3) |
| Super-Kamiokande | 22.5 | 52.2 | 330 (4)† | |
|  |  | 33 |  | 67 (5) |
| Fréjus | 0.6 | 1.58 | 7.0 (6) |  |
|  |  | 1.3 |  | 1.5 (7) |
| Soudan 2 | 0.77 | 3.56 |  | 4.3 (8) |

* 85.5 by combining with IMB-1
† updated from the paper

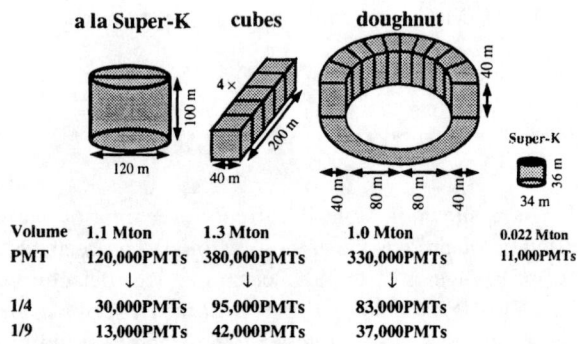

**FIGURE 1.** Some ideas for 1 Mton detector designs shown with the SK detector. Necessary PMT numbers are shown for each detector assuming the same PMT density with the SK detector. Below arrows, the number of PMTs are shown in case of reducing the PMT density by a factor of 4 or 9.

**Table 2.** Proton decay selection criteria. Criteria with circles are applied in each detector.

| detector | (A)* | (B) | (C) |
|---|---|---|---|
| PMT density | 1/1 | 1/4 | 1/9 |
| Number of rings (2 or 3 rings) | ○ | ○ | ○ |
| Particle ID (all showering) | ○ | — | — |
| $\pi^0$ invariant mass (85 – 185 MeV/c$^2$) | ○ | — | — |
| No muon decays | ○ | ○ | ○ |
| Total invariant mass & total momentum (800 < M < 1050 MeV/c$^2$, P < 250 MeV/c) | ○ | ○ | ○† |

* Super-Kamiokande
† Total invariant mass cut is loosened as 750 < M < 1050 MeV/c$^2$

PMTs/Mton in total). All Monte Carlo parameters were exactly same as those of the SK detector simulation program except the PMT density. Water transparency was set at about 100 m at wavelength of 420 nm and same to the three detectors.

## SENSITIVITY STUDY

### Detection Efficiency for $p \to e^+\pi^0$

Figure 2 shows event displays of a $p \to e^+\pi^0$ Monte Carlo event for detector-(A), (B), and (C). Three showering rings caused by a positron (lower left) and two γ (upper right) from the decay of $\pi^0$ are seen in each detector. Although Cherenkov rings become faint in detector-(B) and (C) due to the low PMT density, it seems possible to identify these rings.

Reconstruction of an event such as vertex position, Cherenkov rings, particle type, momentum, and the number of muon decays was automatically performed (see reference-(4) for details). Using these measured quantities, proton decay criteria were defined to reject atmospheric neutrino backgrounds but accept signal. Table 2 shows the criteria for each detector. By the number of ring criterion, one positron ring and one or two of γ rings were required. Particle ID criterion selected $e^\pm$ and γ. Although I did not require this criterion in detector-(B) and (C), one would improve the S/N ratio by using a tuned program to identify a particle type. For 3-ring events, at least one pair of rings must give $\pi^0$ invariant mass. This criterion was not required for detector-(B) and (C) because there was no clear $\pi^0$ mass peak in these detector. Finally, it was checked that the total invariant mass and total momentum correspond to the mass and momentum of the source proton, respectively.

Figure 3 shows the total invariant mass and total momentum distributions of $p \to e^+\pi^0$ events for the detector-(A), (B), and (C). Boxes are the selection cri-

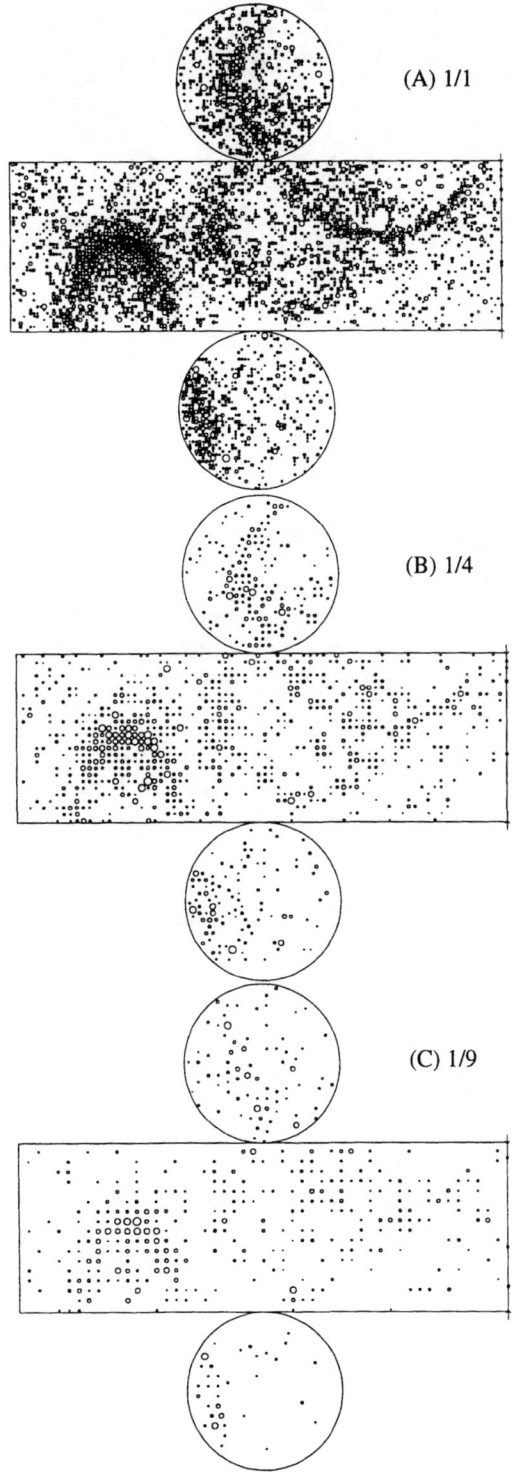

**FIGURE 2.** Event displays of a $p \to e^+\pi^0$ Monte Carlo event for detector-(A) with SK PMT density, (B) with 1/4 PMT density, and (C) with 1/9. Small circles indicate hit PMTs with the size proportional to detected photoelectrons. Positron ring (lower left) and two $\gamma$ rings (upper right) from the decay of $\pi^0$ are seen in each detector.

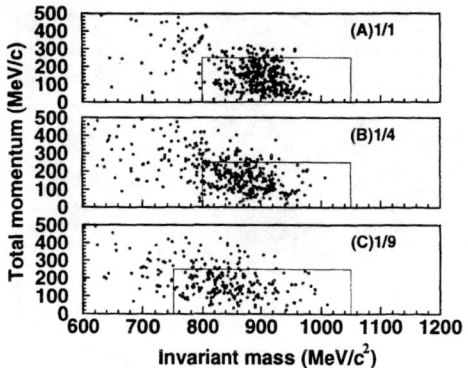

**FIGURE 3.** The total invariant mass and total momentum distributions for $p \to e^+\pi^0$ events in three detectors after the muon decay criterion (Table 2). The boxed region in each figure shows the selection criterion for the $p \to e^+\pi^0$ signal.

terion against the total mass and momentum. From the sample, detection efficiency in each detector was estimated as (A) 43%, (B) 32%, and (C) 21%. Dominant contribution of the inefficiency in detector-(A) comes from the $\pi^0$ interaction in the $^{16}$O nucleus; absorption, charge exchange, and scattering. In case of free protons which are free from the nuclear effect, detection efficiency was about 90% for the detector-(A). The difference of the efficiency between (A), (B), and (C) came mainly from the different performance of the ring identification. The fraction of events which were reconstructed as 2 or 3-ring events was 73% for (A) whereas 57% and 36% for (B) and (C), respectively.

## Atmospheric Neutrino Backgrounds

To estimate the number of background events, atmospheric neutrino Monte Carlo events were used. Here, due to limited CPUs for the reconstruction, I discarded all elastic scattering events ($\nu N \to \nu N, \nu N \to l^\pm N'$). And $\nu_\mu$ charged current (CC) events ($\nu_\mu N \to \mu^\pm X$) were also neglected because most of them will be rejected by the muon decay cut. When particle ID program becomes ready, particle type information will be also useful to reject them. I further reduced the Monte Carlo events by restricting the neutrino energy as $0.7 < E_\nu < 4$ GeV. Used Monte Carlo sample corresponded to 6.8 Mton·year exposure equivalent.

I applied the proton decay selection criteria to the background sample. Figure 4 shows the total invariant mass and total momentum distributions of the atmospheric neutrino events for the detector-(A), (B), and (C). The background level was same for these three detectors

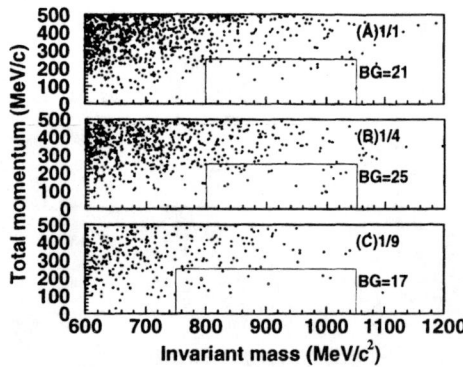

**FIGURE 4.** The total invariant mass and total momentum distributions for atmospheric neutrino Monte Carlo events after the muon decay criterion (Table 2). The boxed region in each figure shows the selection criterion for the $p \rightarrow e^+ \pi^0$ signal.

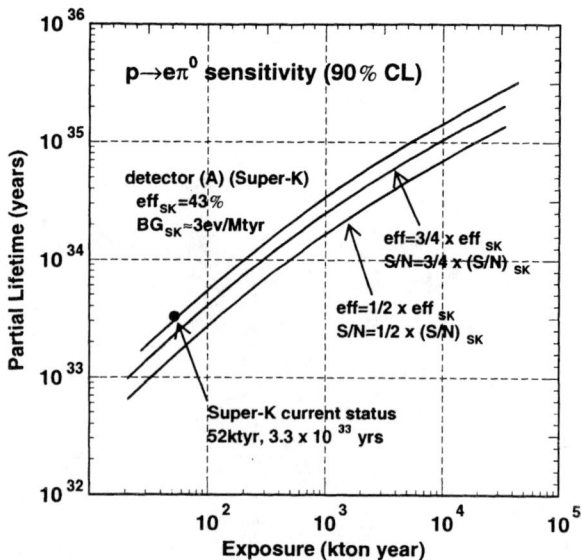

**FIGURE 5.** Expected sensitivity for the partial lifetime of protons. The sensitivity was calculated at 90% confidence level for the detector-(A) (upper line), detector-(B) (middle line), and detector-(C) (lower line).

within the statistical error. The number of background events was about 21/6.8 Mton·year $\simeq$ 3 Mton·year$^{-1}$.

## Sensitivity for Proton Lifetime

From the estimated detection efficiencies and the number of background events, expected sensitivity for the partial lifetime of proton was calculated in Figure 5. The sensitivity was culcated at 90% confidence level assuming simple Poisson processes with backgrounds (10). With 10 years livetime ($10^4$ kton·year exposure), we would reach to $1.5 \times 10^{35}$ years ($1 \times 10^{35}$ years) partial lifetime by the detector-(A) (detector-(B)). By the detector-(C), we would need 20 years livetime to reach to $1 \times 10^{35}$ years.

## DISCUSSIONS AND CONCLUSIONS

I have reported the sensitivity of 1 Mton water Cherenkov detector for $p \rightarrow e^+ \pi^0$ mode. With the same PMT density as the SK detector, the detection efficiency is 43% and the background level is 3 Mton·year$^{-1}$. The efficiency becomes 32% (21%) in case of 1/4 (1/9) PMT density while background level doesn't changed much. If we require the sensitivity of $10^{35}$ years lifetime, we could reduce the PMT density by a factor of 4 and probably up to 9. For a more precise results, I need optimized reconstruction program for each PMT density. Especially, optimization of the Cherenkov ring identification algorithm will improve the detection efficiency of the signal.

Finally, it should be noted that the proton decay search will be no longer background free and it will be crucial to precisely understand the tail of the atmospheric neutrino interactions and the detector responce.

## ACKNOWLEDGMENTS

I would like to appreciate Y. Suzuki, T. Kajita, and Y. Itow for useful discussions with them.

## REFERENCES

1. Jogesh C. Pati and Abdus Salam, *Phys. Rev. Lett.* **31**, 661 (1973); H. Georgi and S. L. Glashow, *Phys. Rev. Lett.* **32**, 438 (1974); P. Langacker, *Phys. Rep.* **72**, 185 (1981); G. G. Ross, *Grand Unified Theories* (ADDISON WESLEY, California, 1985).
2. K. S. Hirata *et al.*, *Phys. Lett.* **B220**, 308 (1989).
3. C. McGrew *et al.*, *Phys. Rev.* **D59**, 052004 (1999).
4. M. Shiozawa *et al.*, *Phys. Rev. Lett.* **81**, 3319 (1998).
5. Y. Hayato *et al.*, *Phys. Rev. Lett.* **83**, 1529 (1999).
6. C. Berger *et al.*, *Z. Phys.* **C50**, 385 (1991).
7. C. Berger *et al.*, *Nucl. Phys.* **B313**, 509 (1989).
8. W. W. M. Allison *et al.*, *Phys. Lett.* **B427**, 217 (1998).
9. M. Koshiba, *Phys. Rep.* **220**, 229 (1992).
10. Particle Data Group, *Phys. Rev.* **D50**, 1281 (1994).

# Comments on Future Proton Decay Experiments

Y. Suzuki

*Kamioka Observatory*
*Institute for Cosmic Ray Research, The University of Tokyo, Higashi-Mozumi, Kamioka, Gifu 506-1205 Japan*

## INTRODUCTION

Proton decay is recently revisited by many theorists (1) and experimentalists, partly because the discovery of the finite neutrino mass by Super-Kamiokande (2) leads to a suggestion that proton may decay within a reach of a current or a next generation proton decay detector. In this working group, the following presentations (related to experiments) were given by A.Konaka (TRIUMF) (Pb/scintillator calorimeter detector for proton decay), T.Ypsilantis (CERN) (Aqua-RICH), M.Campaneli (Zurich) (Liq.Ar detector: Super-ICARUS), K.Nakamura (KEK) (Super-Kamiokande and Hyper-Kamiokande), L.Peterson (CNA Engineering) (Rock engineering), C.K.Jung (SronyBrook) (Large water Cherenkov detector), M.Shiozawa (Kamioka,ICRR) (Monte Carlo calculation for the water Cherenkov detector: $e^+\pi^0$ mode) and C. McGraw (StonyBrook) (Monte Carlo calculation for water Cherenkov detectors: $\bar{\nu}K^+$ mode).

Here I will not intend to summarize what they have presented in the workshop session, since you can get all the information directly from their manuscripts in these proceedings. I would rather like to make some COMMENTS on the future proton decay experiments based partly on the discussion given at this workshop and partly on my personal opinion. Therefore the discussion is rather qualitative, not quantitative.

Unfortunately we do not have any 'good' proton decay candidates. This is the most important thing you should consider no matter what theorists' predictions are. There are no EXPERIMENTAL hints which would guide us to where we should attack.

Most stringent lower partial lifetime limits for typical decay modes are given by Super-Kamiokande: $3.3 \times 10^{33}$yrs for $e^+\pi^0$ (52ktyr exposure), $7.3 \times 10^{32}$yrs for $\bar{\nu}K$(33ktyr) and $1.0 \times 10^{33}$yrs for $\mu^+K^0$.

The Super-Kamiokande experiment will operate at least next 10 years and will reach to the sensitivity limits of $\sim 1.4 \times 10^{34}$yrs for $e\pi^0$, $\sim 1.8 \times 10^{33}$yrs for $\bar{\nu}K^+$ and $\sim 2 \times 10^{33}$yrs $\mu^+K^0$.

## SENSITIVITY

The magic numbers we have heard in this workshop for the next generation detectors are $\sim 1 \times 10^{35}$yrs for $e^+\pi^0$ and $\sim 5 \times 10^{34}$ yrs for $\bar{\nu}K^+$ and $\mu^+K^0$ and can be realized by 10 to 20 times larger water Cherenkov detector than SuperK or a few 100k tons calorimeter or other fine grained high efficiency type detectors. Note that in those sensitivity regions, even the $e\pi^0$ mode is limited by backgrounds. Sensitivity for $e^+\pi^0$ improves only a factor of approximately 7 comparing to the sensitivity that will be acheived by the 10 years of the Super-kamiokande running.

First question, then, is that the sensitivity of the next proton decay experiments of $10^{35}$ yrs for $e^+\pi^0$ and $5 \times 10^{34}$ yrs for $\mu^+K^0$ or $\bar{\nu}K^+$ is sufficient or not.

The theorists' best bets(3) are $\sim 1 \times 10^{35-36}$yrs for $e^+\pi^0$, $\sim 5 \times 10^{34}$yrs for $\bar{\nu}K^+$ and $\mu^+K^0$, but they may say that the DEFINITIVE ranges are $\sim 1 \times 10^{37-38}$yrs for $e^+\pi^0$ and $\sim 1 \times 10^{36}$yrs for $\bar{\nu}K^+$ and $\mu^+K^0$.

If there will be any good proton decay candidates to be found in Super-Kamiokande, we will be guided to the right direction (right decay mode) and right magnitudes for the next detectors. Then, either water Cherenkov of 10-20 times of SuperK (dedicated for $e^+\pi^0$ mode) or $\sim 100$kton calorimeter or Liq.-Ar detector (dedicated for $\bar{\nu}K^+$ and $\mu^+K^0$ mode) will be best suited.

Suppose we will not observe any good candidates within a reach of Super-K for next five years, we probably need a gigantic water Cherenkov detector to do a kind of general search (not a specific mode search). However, the water Cherenkov detector discussed here would—even if (we are lucky) in the right sensitivity next time—'find' AT MOST a few events (above backgrounds). Therefore, the sensitivity discussed here seems to be a little bit smaller than that probably needed for.

We should think of the sensitivity that completely covers the theorists best bets and approaches to the region of the 'definitive' discovery region. Namely, 50 to 100 times or more of the size of Super-Kamiokande should be considered. Cosequently, one of the important issue that is not discussed in this workshop is an EXPAND-ABILITY of the detector size.

Regardless of what we are discussing, the current important issue is to increase the experimental sensitivity of Super-Kamiokande not only by the natural statistical increase but also by efforts to improve the detection efficiency.

## BACKGROUNDS AND EFFICIENCY

Signals in the next generation water Cherenkov detector appear above the backgrounds even for the $e^+\pi^0$ mode. Therefore the trade off of the quality of the detector and the size of the detector should be avoided, although Shiozawa (4) pointed out that $e\pi^0$ could be detected with reasonable efficiency for the configuration with 1/4th of the PMT density comparing to the Super-Kamiokande. But the ability to detect other modes like $\bar{\nu}K^+$ was not discussed yet.

The backgrounds can be evaluated by the data that will be obtained by the front detectors of K2K experiments. Over the five years of running, for example at the 1kt Water Cherenkov detector located at KEK, we would expect more than 500k good muon neutrino interactions which will provide the precise information of the backgrounds to be produced by the atmospheric neutrinos for which we expect about 150k events for 1Mtonyr exposure.

The detection efficiency for the Water Cherenkov detector may be improved by the different choice of the material as is pointed out by M.Goldharber (5) in this workshop. The use of the liquid hydrogen, liquid $CH_4$, octane $C_8H_{18}$ and Ethanol $C_2H_5$ would increase the sensitivity due to the increase of the fractional ratio of the free protons. However, the size of the detector due to the lower material densities for some cases and the handling of those materials must be considered carefully.

## FEASIBILITY

### cavity or under sea/water

Where to place the detector—shallow or deep, under water (sea) or undergrounds—is not only related to the cost of the experiments, but also to the ability of the experiments.

We have heard from Peterson (6) that the cost of the shallow underground cavity would be equal or higher than those deep cavity around 1000m. Therefore if you place the detector underground, then deep detector would be cheap and have many other physics opportunities.

The argon detector or fine grained calorimeter detectors are only possible in the underground.

Underground cavity to house 1 Mton detector costs about $200M to $300M. We need to significantly reduce this cost. A collaboration with rock or mine engineering company would be necessary.

Those detectors placed shallow under water may be a particular interest. This is not addressed in this workshop, and should be studied in a serious way. The shallow under-water detector (especially those useing pure water in a tank) gives you an 'easy' expand-ability and a possible cost reduction due to no excavation costs.

The surface 1 Mton detector has 1~10 MHz of cosmic ray trigger. If you set the signal window of 500 nsec, then the detector has more than 50 to 100% dead time for proton decay search and not possible unless the detector is segmented into many pieces. If you go down to 100m water depth, then the muon rate become 10-100kHz for 1 Mton detector that creates about 0.5~5% dead time. The experiment for proton decay can be done with this depth. However, by those shallow detector, of course, the solar neutrino measurements are not possible, due to the spallation products from the high rate penetrating muons. Those spallation events produce about 20% dead time even for Super-Kamiokande at 2700m water equivalent depth.

The cost estimate for tanks in the shallow depth and the experimental feasibility should be studied. Those experiments need on-surface facility, and the accessibility and the easiness of the maintenance is also necessary.

The deep under water or under ice detector with VERY DENSE STRING of light sensors is another candidate, which we have not discussed at all in this workshop. We do not need either a cavity nor a container of the water. You can put all the cost to the light sensor. We should also study this possibility.

### detector

The cost of a PMT is $3000(/0.2m$^2$ coverage) for Super-Kamiokande. For the Water Cherenkov detector, it is absolutely necessary to develop new light sensor cheaper than the present one. Hybrid Photo-Diode (or Photo-Detector) was discussed by Ypsilantis (7) in this workshop. That is very attractive and very high quality detector. We wish that the HPD would be made cheaper than PMT in near future, The large flat panel PMT may be another candidate, which is particular interest of operating in the magnetic fields. But they may not sound great if you concern about the cost of those sensor. We may need entirely new idea to detect photons in very cheap way.

The calorimeter detector can be done by known technology, but the efforts to reduce costs are also necessary. Advantage of the detector is that it need: 1) small cavity, 2) lower threshold, 3) better detection efficiency for the $\nu K^0$ mode and 4) ability to in-cooperate with neutrino factory. However, obviously construction, calibration, monitor and maintenance are harder and take longer than uniform-detectors.

## VERSATILITY

It is a very persuasive argument that the next proton decay detectors should be multi-purpose. However, if a proton candidate is found in near future, the new detector may not need to be of versatile, since we have a strong and clear motivation to do that. But if there is no candidate at all, we must justify a gigantic detector not only for a proton decay search, but also for other physics.

Candidates for OTHER PHYSICS is the possible detection of solar neutrinos and the neutrino bursts from supernovae. Both are low energy events, of which the detection in large detectors is very difficult and challenging. Other possibility is a cooperation with neutrino factories. The energy of the events caused by neutrinos from the neutrino factories are high above GeV, therefore it is very natural to think of that as a feasible physics goal to be accomplished by a megaton detector.

### supernova neutrinos

The Mega-ton detector has an expanded sensitivity region for supernova bursts. By the Super-Kamiokande detector, we expect only 1 to 2 neutrino events for a supernova in Andromeda Galaxy, while you expect about a few tens of neutrino events in a Mega-ton detector. But you should note that there is no other spiral and irregular galaxies within the distance where Mega-ton detectors can search except Andromeda from which you only expect a few tens of events.

Remember that we have observed 11(Kamiokande)+8(IMB) supernova events for SN1987A. The correctness of the general supernova explosion theory has been demonstrated by the observation.

The expected events from Andromeda is small and we probably do not learn much about the detailed information beyond the knowledge from SN1987A. Therefore major aim of the supernova detection by megaton detectors is, I think, to provide high statistic and detailed information of the supernova neutrinos happenning in our galaxy (or LMC). Therefore it should provide the detailed information like the 'time structure' of the neutrino bursts, the precise energy spectrum, identification of the neutralization burst and the identification of $\nu_\mu$, $\nu_\tau$.

In this sense, the quality of the detector is important and we should study the possible supernova neutrino detection in Megato detectors in this aspects.

### solar neutrinos

The only meaningful measurements we can think of is the day night flux difference. The measurements related to the energy (for example distortion of the energy) are extremely hard since the difficulty of the energy calibration to the accuracy of 1% level in such a large detector both in the water Cherenkov and calorimeter detector. By the same reason, hep neutrino measurement is also difficult.

Note that solar neutrinos as well as supernova neutrinos can be cleanly measured in the liquid Ar detector.

### neutrino factory

It probably be possible the next generation proton decay detector serves as a far detector for the neutrino factory as SuperK is used as a far detector for the K2K experiment. However, the charge determination is one of the most important requirement for the oscillation experiments with a muon storage ring. Water Cherenkov detectors should have a magnetic fields inside of the water tank or place magnetic material in back. This may be possible, but very difficult. On the other hand, the magnetic fields are natural for the iron calorimeter, For water type detector, new cheap light sensors operative in strong magnetic fields have to be explored.

The physics most interested by using neutrino factories is a possible CP violation detection in the neutrino sector. This CP violation is only possible if the solution of the solar neutrinos is the large mixing angle. If the right solution is the small mixing angle, then there is no CP violation effect within the detectable range. We should know the solution of the solar neutrinos.

## CONCLUSION

If we find a good proton decay candidate in Super-Kamiokande in very near future, then a Megaton water Cherenkov detector (for $e^+\pi^0$ case), and a few 100k tons calorimeter or a gigantic liquid Ar detector (for $\bar{\nu}K$ case) discussed in this workshop, should be pushed forward as soon as possible. Therefore we need to prepare in advance by making a detailed feasibility study. We should also make efforts to reduce excavation cost. The cost of water Cherenkov detectors may be reduced by developing cheap light sensors.

If we do not find any good proton decay candidates in next five years or so, we need a general search for proton deccay. The water Cherenkov detector with the size of 10-20 times SuperK is not sufficient for the next step. We should think of much bigger detector (~100 times SuperK). A candidate detector, which should have an expand-ability, may probably be a gigantic water Cherenkov detector placed shallow WATER/SEA depth. This may be a dedicated proton decay detector. Significant R&D efforts are also necessary to realize such detectors.

Most important physics to be done by a combination with a muon storage ring is a study of the CP violation in the lepton sector. But the CP violation can not be observable if the solar neutrino solution is not the LARGE MIXING ANGLE.

In summary, 1) if the solar neutrino solution is the large mixing angle solution, the Pb-scintillator detector should get going; 2) if we find a candidate for $\nu K$ in Super-K, then the Pb-scintillator detector should get going; 3) if we find a candedate for $e\pi^0$ in Super-K, the 1 M ton Water Cherenkov should get going; and 4) if no candidates of proton decay are found in next several years in Super-K, then a general search for proton decay not by a 1 Mega ton, but by a several Mega-ton Water Chrenkov detector may be necessary, This can be constructed at shallow under water (100m). The detector , therefore is expand-able, which is, I think, very important.

Things probably have not matured yet to spend $500M or more. Meanwhile we should concentrate on the significant R&D efforts for various aspects.

## REFERENCES

1. See for example F.Wilczek, in Neutrino98, Proceedings of the XVIII International Conference on neutrino Physics and astrophysics, Takayama, Japan, 4-9, June, 1998, Y.Suzuki and Y. Totsuka eds.; and see also theoretical presentations in this workshop.

2. Y.Fukuda et al.,Phys.Rev.Lett.81,1562(1998).

3. W.Marciano, in these proceedings, and see other theoretical presentations in this workshop.

4. M.Shiozawa, in these proceedings.

5. M.Goldharber, in these proceedings.

6. L.Peterson, in these proceedings.

7. T.Ypsilantis, in these proceedings.

# Feasibility of a Next Generation Underground Water Cherenkov Detector: UNO

## Chang Kee Jung

*The State University of New York at Stony Brook, Stony Brook, New York 11794-3800, USA*

**Abstract.** The feasibility of a next generation underground water Cherenkov detector is examined and a conceptual design (UNO) is presented. The design has a linear detector configuration with a total volume of 650 kton which is 13 times the total volume of the Super-Kamiokande detector. It corresponds to a 20 times increase in fiducial volume for physics analysis. The physics goals of UNO are to increase the sensitivity of the search for nucleon decay by a factor of ten and to make precision measurements of the solar and atmospheric neutrino properties. In addition, the detection sensitivity for supernova neutrinos will reach as far as the Andromeda galaxy.

## INTRODUCTION

Large scale underground water Cherenkov detector experiments, Super-Kamiokande and its predecessors (IMB and Kamiokande), have been extremely successful in producing crucial physics results during the last two decades. Their accomplishments include: the first real time measurement of solar neutrinos, confirmation of solar neutrino flux deficit, observation of neutrino oscillations in the atmospheric neutrinos, observation of neutrinos from Supernova 1987A, and setting the world's best limit on nucleon decay.

These detectors were originally conceived for searches for nucleon decay predicted by various Grand Unification Theories (GUTs). While no positive observation of nucleon decay has been made to date with these detectors, the evidence for neutrino oscillations now firmly established by the Super-Kamiokande atmospheric neutrino analysis provides us with a breakthrough in particle physics beyond the Standard Model (1). This finding indicates that neutrino masses are indeed very small if we assume no degeneracy in mass eigenstates, which in turn indicates that there may be a new very high energy physics scale that facilitates small neutrino masses via the "See-saw" mechanism and allows protons to decay.

There are many theoretical models that predict proton decay and some of them were presented in this workshop (NNN99)(2). A specific example of such models can be found in Ref. (3) which presents a complete and detailed description of the interplay between neutrino masses, proton decay and other Standard Model observables in G(2,2,4) and SO(10) frameworks. The model predicts proton decay lifetime within the reach of Super-Kamiokande, especially in SUSY favored decay modes. These predictions along with many other predictions from other models encourage us to turn our attention back to proton decay searches with higher sensitivity.

If discovered, proton decay provides the only unambiguous direct evidence of GUT scale physics at low energies. It certainly will be remembered as one of the most revolutionary discoveries in particle physics history. Some say the discovery is around the corner.

In order to further our effort to search for nucleon decay, which I believe is in the category of "must-do" physics, and to do high statistics studies of neutrino physics including solar neutrinos, atmospheric neutrinos and supernova neutrinos, I propose a next generation large underground water Cherenkov detector which is named UNO (Ultra underground Nucleon decay and neutrino Observatory).

In the following sections, the design considerations, baseline configuration, physics capabilities and cost estimation of the UNO detector is described in some detail.

## ULTRA UNDERGROUND NUCLEON DECAY AND NEUTRINO OBSERVATORY DETECTOR

The design philosophy of UNO is to make a relatively simple extension of the well established water Cherenkov detector technology beyond Super-Kamiokande to achieve an order of magnitude better sensitivity in nucleon decay searches and to study various neutrino physics with higher precision. With water Cherenkov detector technology, we can utilize the tremendous amount of experience and expertise gained from the IMB, Kamioka and Super-Kamiokande detectors.

In order to establish reasonable detector parameters for physics capability studies and cost estimates, I set the benchmark fiducial volume of the UNO detector to be 20 times that of Super-Kamiokande. The design of the detector is kept simple and robust, and broad physics capabilities are required.

Several design options are considered keeping in mind the practical limitations in conventional water Cherenkov detector technique: namely, (1) the largest depth of water of a detector is limited by the current PMT pressure stress limit ( 8 atm for current 20″ Hamamatsu PMTs) which can be overcome if one is willing use an optical high pressure water tight container for each PMT and compromising the PMT efficiency; 2) The maximum dimension of the detector without active detection element is limited by the finite light attenuation length in pure water (~80 m @400 nm in Super-Kamiokande ).

Three different detector shapes are considered: Big, Torus and Linear. According to D. L. Petersen who is a rock engineer at the CNA Consulting Engineering at Chicago, Illinois, the excavation costs are approximately the same independent of the shapes of the detector (4). Then, if we want to keep the detector cost at minimum for a fixed fiducial volume (445 kton; ~20 times the Super-Kamiokande), we need to keep the ratio $rv$=(fiducial volume/total volume) large and keep $rs$=(PMT surface area/total volume) small. These requirements obviously prefer fat detectors.

The Big detector design option considered has a cube shape with 86x86x86m outer dimensions. It has a potential problem with the PMT pressure limit for the PMTs at the bottom of the detector where the water pressure will be about 9 atmospheric pressure. It also has a diagonal length of 150 m which is much longer than the light attenuation length currently measured in pure water.

The Torus detector design option considered turns out to be very inefficient in term of the $rv$ value and is physically not possible if the cross-section is 60x60 m for a benchmark detector size. Even with a 50x50 m cross-section a torus would be too tight, $i.e.$ the diameter of the central rock column will be too small to support the structure.

Thus, a Torus design will have to have a small cross-section which results in a small $rv$ and large $rs$ values. For example, a 40x40m cross-section torus has $rv$=0.6 compared to 0.7 for a Linear option considered below.

The Linear detector design option considered has a 60mx60mx180m outer dimensions and it appears to be the most optimal one. When compartmentalized as three 60mx60mx60m cubes in terms of active detection elements, it satisfies all of the requirements mentioned above and results in reasonable $rv$ and $rs$ values. The compartmentalization option naturally makes the detector cost more expensive but provides several significant benefits compared to an open geometry option. First, it minimizes so-called flasher background events, which commonly occur in all water Cherenkov detectors that use PMTs, by confining them in an isolated compartment. Second, it minimizes inefficiency and variation in efficiency due to finite light attenuation length in water. Third, it keeps the detector operational live time to close to 100%. The current detector operational live time for Super-Kamiokande is about 90%. The 10% inefficiency is mainly due to detector calibrations. With compartmentalization we can ensure that at least one compartment will be alive during the detector calibration times. Keeping the detector live time very high (virtually at 100%) is of course very important for physics programs such as a supernova watch.

Another issue that needs to be dealt with concerning the compartmentalization is whether we should make the divisions among the compartments rigid so that the water can be filled and drained independently. Certainly making the divisions rigid walls will make any future repairing of the detector easier. (It should be noted that it takes two months to drain the current Super-Kamiokande tank.) However, it is not clear whether this is an important enough reason for us to consider such an option which will incur substantial additional cost.

One could also argue about whether the next generation detector should be built underground or underwater. Indeed there are a few ideas presented in recent meetings that consider very large underwater detectors. In my opinion, however, one of the most serious disadvantage of an underwater detector will be its inaccessibility for calibration and repair. The experience gained from the Super-Kamiokande tells us that a well selected and maintained underground detector environment is a wonderful environment for us to work and to operate the detector. It provides us easy access to the detector for various calibrations and tests. On the other hand, I would assume that precision calibrations of a deep underwater detector will be extremely difficult if not impossible, especially if one is aiming to do a calibration for solar neutrino measurements using an electron linac or a DT generator etc. Thus, for an underwater detector design to become a viable option, a new set of elaborate remote calibration systems have to be developed and tested. In addition, location and accessibility of other service facilities such as a water purification system needs to be considered, and a practical and realistic solution must be found. Overall, I expect many more technical challenges for an underwater detector than for an underground detector.

Considering all of the above issues, I believe that, a large underground water Cherenkov detector with a linear configuration is the best option for a next generation nucleon decay and neutrino detector which can be built

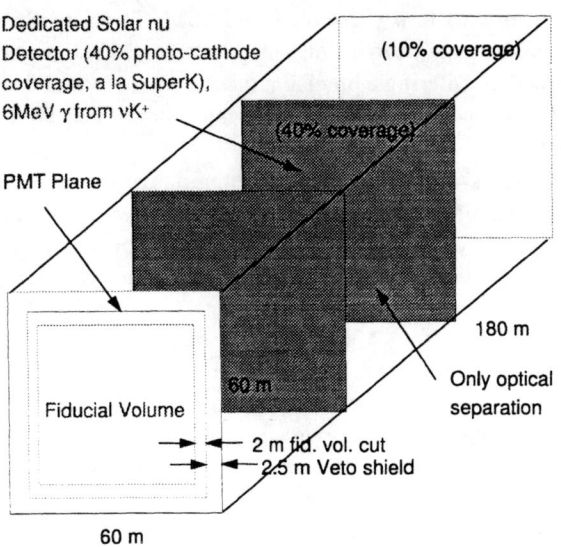

**FIGURE 1.** Baseline configuration of the UNO detector

within the next 10 years without requiring major R&D. A conceptual design of such a detector (UNO detector) is shown in figure 1.

The detector has three compartmentalized sections with 60m×60m×60m dimensions resulting in a total length of 180 m and a total volume of 648 kton. The outer detector region of the detector has 2.5 m depth of veto shield and is instrumented with 14,901 8″ PMTs with a PMT density of 0.33 PMTs/m$^2$. The inner detector region has a total fiducial volume of 445 kton which is defined as the water volume 2 m inside of the inner PMT planes. The inner detector region is instrumented with 20″ PMTs with a PMT density of 1.96 PMTs/m$^2$ (40% photocathode coverage a la Super-Kamiokande) for the central section and 0.49 PMTs/m$^2$ (10% photocathode coverage) for the two sections at the wings. The total number of 20″ PMTs is 56,650 for this configuration.

## UNO Physics Goals and Capabilities

The detector configuration proposed above presents excellent physics capabilities of UNO for broad ranges of nucleon decay and neutrino physics. The two wings of the detector with 10% photocathode coverage will have an energy threshold below 10 MeV which should be sufficient for nucleon decay searches, supernova neutrino detections, high statistics atmospheric neutrino studies and high statistics solar neutrino studies at the high energy end of the energy spectrum. The central compartment with 40% photocathode coverage will serve as a dedicated low energy solar neutrino detector with 5 MeV analysis threshold which will also provide UNO with a detection capability for 6 MeV $\gamma$'s from $p \to \nu K^+$ decays in the oxygen nucleus. It also provides the capability to observe low energy supernova neutrino events ($\sim 5$ MeV) which are important for core collapse modeling.

The primary physics goal of UNO is to obtain an order of magnitude better sensitivity in nucleon decay searches (especially in $p \to e^+\pi^0$ mode) than Super-Kamiokande. The exact amount of improvement in the sensitivity varies depending on the decay mode, particularly on the expected backgrounds. Obviously for the background free decay modes we expect more than a order of magnitude improvement but for the background limited decay modes we expect a factor of 4 or 5 improvement.

The amount of improvement also strongly depends on the optimization of the analysis for high statistics sample. For example, if we continue current standard Super-Kamiokande analysis for $p \to e^+\pi^0$, one will encounter a background limited analysis situation in ten years (0.5 years for UNO). However, if we make a tighter cut on total momentum in an event, say 150 MeV (rather than 250 MeV for current Super-Kamiokande standard analysis), the signal efficiency will go down to about 25-30% from current 44%, but we will be virtually background free for many UNO years. Since we gain a factor of 20 in total fiducial volume, by keeping the efficiency better than 22% we can insure a factor of 10 or more gain in the search sensitivity as long as we can keep the background under control.

In order to estimate the potential improvement reliably, we need to do much more work. First of all, we need to have better understanding of the characteristics of the atmospheric neutrino background events. This can be accomplished by studying the K2K 1 kton neutrino events which are similar in many aspects to the atmospheric neutrino background events relevant to nucleon decay searches. Second, we need to reduce the systematic uncertainties associated with the atmospheric neutrino flux and the neutrino interaction cross-section for water target. Some of this information can come from the K2K fine-grained detector data. Finally, we need to improve the search analysis by developing more sophisticated algorithms and optimizing the analysis for maximum sensitivity. I personally believe that there is much room for improvement in these areas, especially for the multi-ring event analysis. We also need to generate much larger MC event samples with detailed simulations for these purposes. While it is difficult to predict the exact sensitivity of the UNO detector for nucleon decay searches at this time, I am convinced that an order of magnitude improvement in the search sensitivity for various decay modes is attainable.

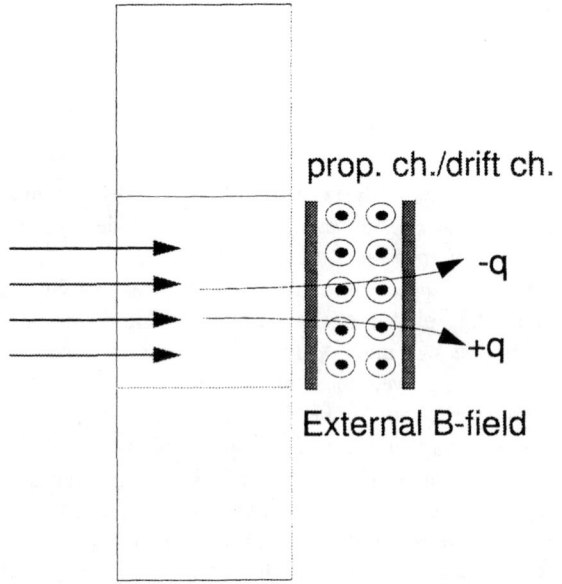

**FIGURE 2.** Conceptual layout of UNO as a far detector for a muon storage ring neutrino beam

Another equally important physics goal of UNO is supernova neutrino detection. With the proposed configuration UNO will be able to record 100k neutrino interaction events from a supernova explosion at 10 kpc. This will allow us to make a detailed map of the time structure of the neutrino flux providing valuable information for the theoretical modeling of the supernovae. The detection reach for neutrinos from supernova explosions will be about 1 Mpc from the earth which includes most of the local group of galaxies including Andromeda. We expect (optimistically) about one supernova explosion every ten years within this range.

Although current Super-Kamiokande measurements of the solar neutrino properties provide crucial information on solar neutrino problems with unprecedented statistics, the measurements on energy spectrum, day-night flux ratio, and seasonal variations are limited by the lack of statistics in providing an unambiguous resolution to the solar neutrino problems. Thus, it is strongly desirable to build a large scale detector that has a capability of precision studies of the solar neutrinos. The proposed configuration of UNO will provide about 20 times more statistics for high energy end of the solar neutrino events and at least 7 times more statistics for the low energy end events. Even if we are able to establish a solution to the solar neutrino problem with the results from Super-Kamiokande, SNO, KAMLAND and Borex-ino within this decade, we will need more precision measurements to firmly establish the SSM (Standard Solar Model) and neutrino oscillation parameters. I believe that we will be entering a precision measurement era for SSM in this decade as the two last decades have been for the Standard Model.

If we assume the neutrino oscillations observed in the Super-Kamiokande atmospheric neutrino events are due to $\nu_\mu$ to $\nu_\tau$ oscillations, there should be about 20 tau appearance events produced per year from the $\nu_\tau$ interactions in the Super-Kamiokande detector. For UNO the rate will be about 400 tau appearance events per year. While extracting a tau signal from background in a water Cherenkov detector is not expected to be easy, it will be a worthwhile project to pursue, especially with a reasonably high statistics event sample. (Currently, we are pursuing such an analysis with the Super-Kamiokande data and the initial results appear to be promising. We are hoping to get about three sigma effect.) If the $\Delta m^2$ turns out to be relative low, say $\sim 3 \times 10^{-3}$ eV$^2$ or lower, the long baseline neutrino experiments will have a difficult time in observing an unambiguous signal of tau appearance. And thus, establishing a direct observation of tau appearance coming from neutrino oscillations may take much longer than we currently hope for.

Finally, a possibility of using UNO as a far detector of a Muon Storage Ring produced neutrino beam experiment should be considered. A water Cherenkov detector is a good candidate for a far detector. It provides a relatively inexpensive large target volume, particle ID for electrons and muons, and reasonable calorimetry. It could also provide good charge separation if it is aided by an external spectrometer as shown in Figure 2.

Thus, if UNO is built, it will serve as a truly multipurpose general underground detector for variety of great physics. Especially for nucleon decay searches, it will serve as a robust multi-decay mode search detector.

## UNO Cost Estimates

In order to obtain a ball park figure cost estimate for UNO detector, I use the construction cost data for the Super-Kamiokande detector provided by K. Nakamura as a reference. I divide the major expense items into two categories according to their nature of scaling, i.e. volume-like scaling or surface-like scaling. And I apply my reasonable guesses to ultimately determine the scaling factor from Super-Kamiokande to UNO. Thus, these figures should not be taken too seriously. Table 1 shows my initial estimate of itemized costs for UNO along with the actual costs for Super-Kamiokande. The total cost for UNO with the baseline configuration proposed in this paper is then about $520M and the cost for a detector with full 40% photocathode coverage for all inner surfaces (a la Super-Kamiokande) is about $680M. And all of these figures do not include contingencies.

**Table 1.** UNO Cost Estimates (in thousands of US dollars): This is a complete list of expense items for all aspect of detector construction. The "v" and "s" symbols note the volume-like and surface-like nature of cost scalings used when the costs are extrapolated from the Super-Kamiokande (SuperK) to UNO. A conversion rate of 1$ = 100 yen is used.

| Item | SuperK | | UNO |
|---|---|---|---|
| Cavity Excavation | 27,640 | v | 168,000 |
| Water piping and pumps | 630 | v | 4,082 |
| Water Purification Sys | 1,850 | v | 11,988 |
| Power Station | 720 | v | 2,160 |
| Crane | 760 | v | 2,280 |
| Water Tank | 18,400 | s | 92,480 |
| PMT support structure | 4,580 | s | 23,019 |
| Counting Room | 330 | s | 990 |
| Computer Building | 1,860 | s | 2,232 |
| Main Building | 3,000 | s | 3,600 |
| 20″ PMT (including cables) | 34,670 | s | 173,664 |
| Electronics | 6,330 | s | 9,495 |
| DAQ | 1,090 | s | 1,635 |
| Air Conditioning | 210 | s | 315 |
| Veto instrumentation | 3,000 | s | 9,000 |
| 8″ PMT (including cables) | 2,262 | s | 17,881 |
| Total | 102,070 | | 522,822 |

For the major cost items of the detector, I use more reliable quotes and estimates for unit costs from vendors and experts. As shown in Table 2, the unit cost for excavation is taken from the estimates by D. L. Petersen, and the unit costs for PMTs are taken from Hamamatsu corporation. I take the liberty of reducing the per channel electronic cost by a factor of 5 and the water tank construction cost per unit surface by a factor of 2 from the Super-Kamiokande cost. The cost of these five items comprise the majority of the UNO detector cost.

## Cost Reduction Possibilities

I believe that there is quite a bit of room to reduce the total cost for UNO. First of all we can reduce the excavation cost by finding an existing cavity or underground facility. Or we could simply get a better quote. It is said that the actual excavation cost charged to the mining company in Kamioka, Japan is only $50/m$^3$. Sec-

**Table 2.** Estimated Unit Costs: The excavation cost is assuming a horizontal access tunnel and rock quality (Q value) of 100. The PMT unit cost including cable cost is based on a 50k PMT order. It is $2,850 if 100k PMTs are ordered. A conversion rate of 1$ = 100 yen is used.

| Item | Unit Cost | Source |
|---|---|---|
| Excavation | $260/m$^3$ | L. Petersen |
| 20″ PMTs | $3,100 | Hamamatsu |
| 8″ PMTs | $1,200 | Hamamatsu |
| Electronics | $170/channel | |
| Water Tank | $2,076/m$^2$ | |

ond, we could use Mineguard type of liner rather than a stainless-steel container for the container tank structure. Third, we should optimize the PMT size and granuality for our physics goals and look into a way of developing new cheaper photo-detectors. For example, we could use two 8″ PMTs instead of a 20″ PMT, which may result in overall better performance and cheaper cost. We could also look for other vendors (from Russia?) than Hamamatsu.

## Site Selection

Obviously, the issue of site selection is an important issue for the success of a project like UNO with this magnitude. In principle, a site near the Equator is desired if we are aiming for maximum information from UNO for the solar neutrino physics. However, the consensus among solar neutrino theorists indicate that it is not crucial unless the MSW small mixing angle solution looks very promising. Thus, I concentrate my attention to possible sites in the United States.

An extensive site search was conducted by R.R. Sharp, Jr. and A. Mann during the early 1980s for a site for potential National Underground Science Facility (NUSF). They surveyed mostly in the western states for existing tunnels, inactive mines, active mines and new locations. They found a couple of sites in Nevada particularly attractive (5). Although the size and depth requirements of the potential sites they searched for are quite different from the UNO requirements, the information gathered was found to be quite useful.

Another site search was conducted by W. Kropp *et. al.* concentrating on the site at San Jocinto, California (6). The results from this search also provide useful information for our current search.

The most attractive site so far presented to us is the WIPP (Waste Isolation Pilot Project) site at Carlsbad, New Mexico. The facility which is managed by DOE provides a hard salt rock geology, an already existing laboratory facility and tunneling machinery. With strong government interest to utilize this facility for scientific research, the site is indeed ideal for UNO or for any other large scale next generation underground detectors.

## CONCLUSIONS

Feasibility of a next generation very large underground water Cherenkov detector, UNO, is considered. The detector configuration of UNO is a linear compartmentalized detector (60mx60mx60mx3) with the middle section with 40% photocathode coverage and the two wing sections with 10% photocathode coverage. The middle section is dedicated for low energy solar neutrino and supernova neutrino studies and for detecting 6 MeV prompt $\gamma$'s from $p \rightarrow \nu K^+$ decays.

The detector which has a total volume of 650 kton and a corresponding fiducial volume of 445 kton (20 times larger than the fiducial volume of the Super-Kamiokande detector) can be built today at about $500M of construction cost. There are no known serious technical challenges in building such a detector. Although rigorous R&D is desired for cost reduction and improvement in detector performance, there are no critical path R&D items. If funding is available, we should be able to build this detector within next ten years.

Such a detector, if built, will provide us with a bonanza of exciting physics: Nucleon decay searches, precision solar and atmospheric neutrino studies, and supernova neutrino observations. The detector can also serve as a far detector for a possible future neutrino factory. It will also compliment any other major accelerator based initiative such as NLC or neutrino factories in terms of providing diverse but crucial physics programs to the US HEP community.

In my opinion, a physics program of searching for nucleon decay undoubtedly belongs to the "MUST-DO" physics category and it should be vigorously pursued. I also believe if such a large scale detector is built it should be a multi-purpose general detector with a robust multi-decay mode search capability for nucleon decay searches. It should not be a specialized single purpose detector with specific decay mode search capability.

## ACKNOWLEDGEMENTS

This work is supported by the funding from the U.S. Department of Energy under the contract No. DEFG0292ER40697. I wish to express my gratitude to Kenzo Nakamura who provided vital information on the Super-Kamiokande construction costs, D. Lee Petersen who provided information on the technical details concerning underground excavation, Al Mann who provided information on the underground site searches. I also like to thank John Bahcall, Bill Marciano, Jogesh Pati and Frank Wilczek for their theoretical input and encouragement, and Maurice Goldhaber and other workshop organizing committee members for their tireless work in making the workshop successful and support for this work.

## REFERENCES

1. Super-Kamiokande Collaboration, Y. Fukuda, *et al.*, Phys. Rev. Lett. **81**, 1562 (1998).

2. Next generation Nucleon decay and Neutrino detector Workshop (NNN99), Stony Brook, New York, September, 1999. See http://superk.physics.sunysb.edu/NNN99 for more information on the workshop and transparency copies of the talks.

3. K.S. Babu, J. C. Pati and F. Wilczek, hep-ph/9812538; also see the articles in this proceedings by Pati, by Wilczek and by Babu.

4. D. L. Petersen, See http://superk.physics.sunysb.edu/NNN99/talk-slides/petersen_rocks_101.pdf.

5. A. K. Mann and R. R. Sharp, Jr., unpublished documents in 1981, private communication.

6. W. Kropp *at. al.*, See http://www.ps.uci.edu/ nnn/day_2/kropp.pdf.

# THEORY

# DISCOVERY OF PROTON DECAY : A MUST FOR THEORY, A CHALLENGE FOR EXPERIMENT

Jogesh C. Pati

*Department of Physics, University of Maryland, College Park, MD 20742, USA.*

**Abstract.** It is noted that, but for one missing piece - proton decay - the evidence in support of grand unification is now strong. It includes: (i) the observed family-structure, (ii) the meeting of the gauge couplings, (iii) neutrino-oscillations, (iv) the intricate pattern of the masses and mixings of all fermions, including the neutrinos, and (v) the need for $B-L$ as a generator, to implement baryogenesis. Taken together, these not only favor grand unification but in fact select out a particular route to such unification, based on the ideas of supersymmetry, SU(4)-color and left-right symmetry. Thus they point to the relevance of an effective string-unified G(224) or SO(10)-symmetry.

A concrete proposal is presented, within a predictive SO(10)/G(224)-framework, that successfully describes the masses and mixings of all fermions, including the neutrinos - with eight predictions, all in agreement with observation. Within this framework, a systematic study of proton decay is carried out, which pays special attention to its dependence on the fermion masses, including the superheavy Majorana masses of the right-handed neutrinos. The study shows that a conservative upper limit on the proton lifetime is about $(1/2 - 1) \times 10^{34}$ yrs, with $\bar{\nu}K^+$ being the dominant decay mode, and as a distinctive feature, $\mu^+K^0$ being prominent. This in turn strongly suggests that an improvement in the current sensitivity by a factor of five to ten (compared to SuperK) ought to reveal proton decay. Otherwise some promising and remarkably successful ideas on unification would suffer a major setback.

## INTRODUCTION

It has been recognized since the early 1970's that the price one must pay to achieve a unification of quarks and leptons and simultaneously a unity of the three gauge forces, commonly called "grand unification", is proton decay (1, 2, 3, 4). This important process, which would provide the window for viewing physics at truly short distances ($< 10^{-30}$ cm), is yet to be seen. Nevertheless, as I will stress in this talk, there have appeared over the years an impressive set of facts, including the meeting of the gauge couplings and neutrino-oscillations, which not only favor the hypothesis of grand unification, but in fact select out a particular route to such unification, based on the ideas of supersymmetry (5) and SU(4)-color (2). These facts together provide a clear signal that the discovery of proton decay cannot be far behind.

To be specific, working within the framework of a unified theory (6), that incorporates the ideas mentioned above, I would argue that an improvement in the current sensitivity for detecting proton decay by a modest factor of five to ten should either produce real events, or else the framework would be excluded. In this sense, and as I will elaborate further, the discovery of proton decay is now crucial to the survival of some elegant ideas on unification, which are otherwise so successful. By the same token, proving or disproving their prediction on proton decay poses a fresh challenge to experiment.

The pioneering efforts by several physicists (7) in the mid 1950's through the early 70's had provided a lower limit on the proton lifetime of about $10^{26}$ yrs, independent of decay modes, and $10^{29}$-$10^{30}$ yrs for the $e^+\pi^0$-mode. Subsequent searches at the Kolar Goldfield and the NUSEX detectors in the early 80's (8) pushed this limit to about $10^{31}$ years in the $e^+\pi^0$-mode. Following the suggestion of proton decay in the context of grand unification, and thanks to the initiative of several experimenters, two relatively large-size detectors - IMB and Kamiokande - were built in the 80's, where dedicated searches for proton decay were carried out with higher sensitivity. These detectors helped to push the lower limit in the $e^+\pi^0$ channel to about $10^{32}$ yrs. This in turn clearly disfavored the minimal *non-supersymmetric* SU(5)-model of grand unification (3) - a conclusion that was strengthened subsequently by the measurements of the gauge couplings at LEP as well (see discussion later).

The searches for proton decay now continues with still greater sensitivity at the largest detector so far - at SuperKamiokande, completed in 1996. It is worth noting at this point that, although these detectors have not revealed proton decay yet, they did bring some major bonuses of monumental importance. These include : (a) the detection of the neutrinos from the supernova 1987a, (b) confirmation of the solar neutrino-deficit, and last but not least, (c) the discovery of atmospheric neutrino-oscillation. The SuperK water-Cerenkov detector with a fiducial volume of 22.5 kilotons currently provides (with three years of running) a lower limit on the inverse rate of proton decay of about $1.6 \times 10^{33}$ yrs for the theoretically favored ($\bar{\nu}K^+$)-channel (9) and of about $3.8 \times 10^{33}$ yrs for the ($e^+\pi^0$)-channel (10). It has the capability of improving these limits by a factor of two to three in each case within the next decade, unless of course it discovers real events for proton decay or strong candidate events in the meantime. I will return to this point and its relevance to theoretical expectations for proton decay in just a bit.

While proton decay is yet to be observed, it is worth stressing at this point, that the hypothesis of grand unification, especially that based on the ideas of SU(4)-color, left-right gauge symmetry, and supersymmetry, is now supported by several observations. As I will explain in sections 2-5, these include :

**(a) The observed family structure :** The five scattered multiplets of the standard model, belonging to a family, neatly become parts of a whole (*a single multiplet*), with their weak hypercharges precisely predicted by grand unification. Realization of this feature calls for an extension of the standard model symmetry $G(213) = SU(2)_L \times U(1)_Y \times SU(3)^C$ *minimally* to the symmetry group $G(224) = SU(2)_L \times SU(2)_R \times SU(4)^C$ (2), which can be extended further into the simple group SO(10) (11), but not SU(5) (3). The G(224) symmetry in turn introduces some additional attractive features (see Sec. 2), including especially the right-handed (RH) neutrinos ($\nu_R$'s) accompanying the left-handed ones ($\nu_L$'s), and B-L as a local symmetry. As we will see, both of these features, which are special to G(224), now seem to be needed on empirical grounds.

**(b) Meeting of the three gauge couplings :** Such a meeting is found to occur at a scale $M_X \approx 2 \times 10^{16}$ GeV, when the three couplings are extrapolated from their values measured at LEP to higher energies, in the context of supersymmetry (12). This dramatic phenomenon supports the ideas of both grand unification and supersymmetry. These in turn may have their origins within a string theory (13) or M-theory (14) (see discussion in Sec. 3).

**(c) Mass of $\nu_\tau \sim 1/20$ eV :** Subject to the well-motivated assumption of hierarchical neutrino masses, the recent discovery of atmospheric neutrino-oscillation at SuperKamiokande (15) suggests a value for $m(\nu_\tau) \sim 1/20 eV$. It has been argued (see e.g. Ref. (16) and discussion in Sec. 4) that a mass for $\nu_\tau$ of this magnitude points to the need for the RH neutrinos, and that it goes extremely well with the hypothesis of a supersymmetric unification, based on the symmetry G(224), and thus SO(10), though not SU(5). The SUSY unification-scale as well as the symmetry of SU(4)-color play crucial roles in making this argument.

**(d) Some intriguing features of fermion masses and mixings :** These include : (i) the "observed' near equality of the masses of the b-quark and the $\tau$-lepton at the unification-scale ($m_b^0 \approx m_\tau^0$), together with (ii) the empirical Georgi-Jarsklog relations: $m_s^0 \sim m_\mu^0/3$ and $m_d^0 \sim 3m_e^0$, and (iii) the observed largeness of the $\nu_\mu$-$\nu_\tau$ oscillation angle ($\sin^2 2\theta_{\nu_\mu\nu_\tau}^{osc} \geq 0.83$) (15), together with the smallness of the corresponding quark mixing parameter $V_{bc}(\approx 0.04)$ (17). As shown in recent work by Babu, Wilczek and me (6), it turns out that these features and more can be understood remarkably well (see discussion in Sec 5) within an economical and predictive SO(10)-framework based on a minimal Higgs system. The success of this framework is in large part due simply to the group-structure of SO(10). For most purposes, that of G(224) suffices.

**(e) Baryogenesis :** To implement baryogenesis (18) successfully, in the presence of electroweak sphaleron effects (19), which wipe out out any baryon excess generated at high temperatures in the (B-L)-conserving mode, it has become apparent that one would need B-L as a generator of the underlying symmetry, whose spontaneous violation at high temperatures would yield, for example, lepton asymmetry (leptogenesis). The latter in turn is converted to baryon-excess at lower temperatures by electroweak sphalerons. This mechanism, it turns out, yields even quantitatively the right magnitude for baryon excess (20). The need for B-L, which is a generator of SU(4)-color, again points to the need for G(224) or SO(10) as an effective symmetry near the unification-scale $M_X$.

The success of each of these five features (a)-(e) seems to be non-trivial. Together they make a strong case for both supersymmetric grand unification and simultaneously for the G(224)/SO(10)-route to such unification, as being relevant to nature. However, despite these successes, as long as proton decay remains undiscovered, the hallmark of grand unification - that is *quark-lepton transformability* - would remain unrevealed.

The relevant questions in this regard then are : What is the predicted range for the lifetime of the proton - in particular an upper limit - within the emperically favored route to unification mentioned above? What are the expected dominant decay modes within this route? Are these predictions compatible with current lower limits on proton lifetime mentioned above, and if so, can they still be tested at the existing or possible near-future detectors for proton decay?

Fortunately, we are in a much better position to answer these questions now, compared to a few years ago, because meanwhile we have learnt more about the nature of grand unification. As noted above (see also Secs.

2 and 4), the neutrino masses and the meeting of the gauge couplieings together seem to select the supersymmetric G(224)/SO(10)-route to higher unification. The main purpose of my talk here will therefore be to address the questions raised above, in the context of this route. For the sake of comparison, however, I will state the corresponding results for the case of supersymmetric SU(5) as well.

My discussion will be based on a recent study of proton decay by Babu, Wilczek and me (6), which, relative to previous ones, has three distinctive features :

**(a)** It systematically takes into account the link that exists between proton decay and the masses and mixings of all fermions, including the neutrinos.

**(b)** In particular, in adition to the contributions from the so-called "standard" $d = 5$ operators (22) (see Sec. 6), it includes those from a *new* set of $d = 5$ operators, related to the Majorana masses of the RH neutrinos (21). These latter are found to be as important as the standard ones.

**(c)** The work also incorporates GUT-scale threshold effects, which arise because of mass-splittings between the components of the SO(10)-multiplets, and lead to differences between the three gauge couplings.

Each of these features turn out to be *crucial* to gaining a reliable insight into the nature of proton decay. Our study shows that the inverse decay rate for the $\bar{\nu}K^+$-mode, which is dominant, is less than about $7 \times 10^{33}$ yrs. This upper bound is obtained by making generous allowance for uncertainties in the matrix elements and the SUSY-spectrum. Typically, the lifetime should of course be less than this bound. Furthermore, due to contributions from the new operators, the $\mu^+K^0$-mode is found to be prominent, with a branching ratio typically in the range of 10-50%. By contrast, minimal SUSY SU(5), for which the new operators are absent, would lead to branching ratios $\leq 10^{-3}$ for this mode. Thus our study of proton decay , correlated with fermion masses, strongly suggests that discovery of proton decay should be around the corner. In fact, one expects that at least candidate events should be observed in the near future already at SuperK. However, allowing for the possibility that the proton lifetime may well be closer to the upper bound stated above, a next-generation detector providing a net gain in sensitivity in proton decay-searches by a factor of 5-10, compared to SuperK, would certainly be needed not just to produce proton-decay events, but also to clearly distinguish them from the background. It would of course also be essential to study the branching ratios of certain sub-dominant but crucial decay modes, such as the $\mu^+K^0$. The importance of such improved sensitivity, in the light of the successes of supersymmetric grand unification, is emphasized at the end.

# ADVANTAGES OF THE SYMMETRY G(224) AS A STEP TO HIGHER UNIFICATION

The standard model (SM) based on the gauge symmetry $G(213) = SU(2)_L \times U(1)_Y \times SU(3)^C$ has turned out to be extremely successful empirically. It has however been recognized since the early 1970's that, judged on aesthetic merits, it has some major shortcomings. For example, it puts members of a family into five scattered multiplets, without providing a compelling reason for doing so. It also does not provide a fundamental reason for the quantization of electric charge. Nor does it explain the co-existence of quarks and leptons, and that of the three gauge forces, with their differing strengths. The idea of grand unification was postulated precisely to remove these shortcomings. That in turn calls for the existence of fundamentally new physics, far beyond that of the standard model. As mentioned before, recent experimental findings, including the meetings of the gauge couplings and neutrino-oscillations, seem to go extremely well with this line of thinking.

To illustrate the advantage of an early suggestion in this regard, consider the five standard model multiplets belonging to the electron-family as shown :

$$\begin{pmatrix} u_r & u_y & u_b \\ d_r & d_y & d_b \end{pmatrix}_L^{\frac{1}{3}} ;$$

$$\begin{pmatrix} u_r & u_y & u_b \end{pmatrix}_R^{\frac{4}{3}} ;$$

$$\begin{pmatrix} d_r & d_y & d_b \end{pmatrix}_R^{-\frac{2}{3}} ;$$

$$\begin{pmatrix} \nu_e \\ e^- \end{pmatrix}_L^{-1} ;$$

$$\begin{pmatrix} e^- \end{pmatrix}_R^{-2} . \quad (1)$$

Here the superscripts denote the respective weak hypercharges $Y_W$ (where $Q_{em} = I_{3L} + Y_W/2$) and the subscripts L and R denote the chiralities of the respective fields. If one asks : how one can put these five multiplets into just one multiplet, the answer turns out to be simple and unique. As mentioned in the introduction, the minimal extension of the SM symmetry G(213) needed, to achieve this goal, is given by the gauge symmetry (2) :

$$G(224) = SU(2)_L \times SU(2)_R \times SU(4)^C . \quad (2)$$

Subject to left-right discrete symmetry ($L \leftrightarrow R$), which is natural to G(224), all members of the electron family fall into the neat pattern :

$$F_{L,R}^e = \begin{bmatrix} u_r & u_y & u_b & \nu_e \\ d_r & d_y & d_b & e^- \end{bmatrix}_{L,R} \quad (3)$$

The multiplets $F_L^e$ and $F_R^e$ are left-right conjugates of each other and transform respectively as (2,1,4) and (1,2,4) of G(224); likewise for the muon and the tau families. Note that the symmetries $SU(2)_L$ and $SU(2)_R$ are just like the familiar isospin symmetry, except that they operate on quarks and well as leptons, and distinguish between left and right chiralities. The left weak-isospin $SU(2)_L$ treats each column of $F_L^e$ as a doublet; likewise $SU(2)_R$ for $F_R^e$; the symmetry SU(4)-color treats each row of $F_L^e$ and $F_R^e$ as a quartet, interpreting lepton number as the fourth color. Note also that postulating either SU(4)-color or $SU(2)_R$ forces one to introduce a right-handed neutrino ($\nu_R$) for each family as a singlet of the SM symmetry. *This requires that there be sixteen two-component fermions in each family, as opposed to fifteen for the SM*. The symmetry G(224) introduces an elegant charge formula :

$$Q_{em} = I_{3L} + I_{3R} + \frac{B-L}{2} \quad (4)$$

expressed in terms of familiar quantum numbers $I_{3L}$, $I_{3R}$ and $B$-$L$, which applies to all forms of matter (including quarks and leptons of all six flavors, gauge and Higgs bosons). Note that the weak hypercharge given by $Y_W/2 = I_{3R} + \frac{B-L}{2}$ is now completely determined for all members of the family. The values of $Y_W$ thus obtained precisely match the assignments shown in Eq. (1). Quite clearly, the charges $I_{3L}$, $I_{3R}$ and $B$-$L$, being generators respectively of $SU(2)_L$, $SU(2)_R$ and $SU(4)^c$, are quantized; so also then is the electric charge $Q_{em}$.

In brief, the symmetry G(224) brings some attractive features to particle physics. These include :
(i) Organization of all 16 members of a family into one left-right self-conjugate multiplet;
(ii) Quantization of electric charge;
(iii) Quark-lepton unification (through SU(4) color);
(iv) Conservation of parity at a fundamental level (2, 23);
(v) Right-handed neutrinos ($\nu'_R s$) as a compelling feature; and
(vi) $B$-$L$ as a local symmetry.
As mentioned in the introduction, the two distinguishing features of G(224) - i.e. the existence of the RH neutrinos and $B$-$L$ as a local symmetry - now seem to be needed on empirical grounds.

Believing in a complete unification, one is led to view the G(224) symmetry as part of a bigger symmetry, which itself may have its origin in an underlying theory, such as string theory. In this context, one might ask : Could the effective symmetry below the string scale in four dimensions (see sec.3) be as small as just the SM symmetry G(213), even though the latter may have its origin in a bigger symmetry, which lives however only in higher dimensions? I will argue in Sec. 4 that the data on neutrino masses and the need for baryogenesis provide an answer to the contrary, suggesting clearly that it is the effective symmetry in four dimensions, below the string scale, which must *minimally* contain either G(224) or a close relative G(214) = $SU(2)_L \times I_{3R} \times SU(4)^C$.

One may also ask : does the effective four dimensional symmetry have to be any bigger than G(224) near the string scale? In preparation for an answer to this question, let us recall that the smallest simple group that contains the SM symmetry G(213) is SU(5) (3). It has the virtue of demonstrating how the main ideas of grand unification, including unification of the gauge couplings, can be realized. However, SU(5) does not contain G(224) as a subgroup. As such, it does not possess some of the advantages listed above. In particular, it does not contain the RH neutrinos as a compelling feature, and $B$-$L$ as a local symmetry. Furthermore, it splits members of a family into two multiplets : $\bar{5}$ + 10.

By contrast, the symmetry SO(10) has the merit, relative to SU(5), that it contains G(224) as a subgroup, and thereby retains all the advantages of G(224) listed above. (As a historical note, it is worth mentioning that these advantages had been motivated on aesthetic grounds through the symmetry G(224) (2), and *all* the ideas of higher unification were in place (1, 2, 3), before it was noted that G(224)(isomorphic to SO(4)×SO(6)) embeds nicely into SO(10) (11)). Now, *SO(10) even preserves the 16-plet family-structure of G(224) without a need for any extension*. By contrast, if one extends G(224) to the still higher symmetry $E_6$ (24), the advantages (i)-(vi) are retained, but in this case, one must extend the family-structure from a 16 to a 27-plet, by postulating additional fermions. In this sense, there seems to be some advantage in having the effective symmetry below the string scale to be minimally G(224) (or G(214)) and maximally no more than SO(10). I will compare the relative advantage of having either a string-derived G(224) or a string-SO(10), in the next section. First, I discuss the implications of the data on coupling unification.

## THE NEED FOR SUPERSYMMETRY : MSSM VERSUS STRING UNIFICATIONS

It has been known for some time that the precision measurements of the standard model coupling constants (in particular $\sin^2\theta_W$) at LEP put severe constraints on the idea of grand unification. Owing to these constraints, the non-supersymmetric minimal SU(5), and for similar reasons, the one-step breaking minimal non-supersymmetric SO(10)-model as well, are now excluded (25). But the situation changes radically if one assumes that the standard model is replaced by the minimal supersymmetric standard model (MSSM), above a threshold of about 1 TeV. In this case, the three gauge couplings are found to meet (12), at least approximately, provided $\alpha_3(m_Z)$ is not too low (see Figs. in e.g. Refs. (23, 13)). Their scale of meeting is given by

$$M_X \approx 2 \times 10^{16} \text{GeV} \quad (\text{MSSM or SUSY SU(5)}) \quad (5)$$

This dramatic meeting of the three gauge couplings, or equivalently the agreement of the MSSM-based prediction of $\sin^2\theta_W(m_Z)_{\text{Th}} = 0.2315 \pm 0.003$ (26) with the observed value of $\sin^2\theta_W(m_Z) = 0.23124 \pm 0.00017$ (17), provides a strong support for the ideas of both grand unification and supersymmetry, as being relevant to physics at short distances.

The most straightforward interpretation of the observed meeting of the three couplings and of the scale $M_X$, is that a supersymmetric grand unification symmetry (often called GUT symmetry), like SU(5) or SO(10), breaks spontaneously at $M_X$ into the standard model symmetry G(213).

In the context of string or M theory, which seems to be needed to unify all the forces of nature including gravity and also to obtain a good quantum theory of gravity, an alternative interpretation is however possible. This is because, even if the effective symmetry in four dimensions emerging from a higher dimensional string theory is non-simple, like G(224) or G(213), string theory can still ensure familiar unification of the gauge couplings at the string scale. In this case, however, one needs to account for the small mismatch between the MSSM unification scale $M_X$ (given above), and the string unification scale, given by $M_{st} \approx g_{st} \times 5.2 \times 10^{17}$ GeV $\approx 3.6 \times 10^{17}$ GeV (Here we have put $\alpha_{st} = \alpha_{GUT}(\text{MSSM}) \approx 0.04$) (27). Possible resolutions of this mismatch have been proposed. These include : (i) utilizing the idea of *string-duality* (28) which allows a lowering of $M_{st}$ compared to the value shown above, or alternatively (ii) the idea of a *semi-perturbative* unification that assumes the existence of two vector-like families, transforming as $(16 + \overline{16})$, at the TeV-scale. The latter raises $\alpha_{GUT}$ to about 0.25-0.3 and simultaneously $M_X$, in two loop, to about $(1/2 - 2) \times 10^{17}$ GeV (29) (Other mechanisms resolving the mismatch are reviewed in Refs. (30) and (31)). In practice, a combination of the two mechanisms mentioned above may well be relevant [1]

---

[1] I have in mind the possibility of string-duality (28) lowering $M_{st}$ for the case of semi-perturbative unification (for which $\alpha_{st} \approx 0.25$, and thus, without the use of string-duality, $M_{st}$ would be about $10^{18}$ GeV) to a value of about $(1-2) \times 10^{17}$ GeV (say), and semi-perturbative unification (29) raising the MSSM value of $M_X$ to about $5 \times 10^{16}$ GeV $\approx M_{st}(1/2$ to $1/4)$ (say). In this case, an intermediate symmetry like G(224) emerging at $M_{st}$ would be effective only within the short gap between $M_{st}$ and $M_X$, where it would break into G(213). Despite this short gap, one would still have the benefits of SU(4)-color that are needed to understand neutrino masses (see sec.4). At the same time, Since the gap is so small, the

While the mismatch can thus quite plausibly be removed for a non-GUT string-derived symmetry like G(224) or G(213), a GUT symmetry like SU(5) or SO(10) would have an advantage in this regard because it would keep the gauge couplings together between $M_{st}$ and $M_X$ (even if $M_X \sim M_{st}/20$), and thus not even encounter the problem of a mismatch between the two scales. A supersymmetric GUT-solution (like SU(5) or SO(10)), however, has a possible disadvantage as well, because it needs certain color triplets to become superheavy by the so-called double-triplet splitting mechanism (see Sec. 6 and Appendix), in order to avoid the problem of rapid proton decay. However, no such mechanism has emerged yet, in string theory, for the GUT-like solutions (32).

Non-GUT string solutions, based on symmetries like G(224) or G(2113) for example, have a distinct advantage in this regard, in that the dangerous color triplets, which would induce rapid proton decay, are often naturally projected out for such solutions (33, 34). Furthermore, the non-GUT solutions invariably possess new "flavor" gauge symmetries, which distinguish between families. These symmetries are immensely helpful in explaining qualitatively the observed fermion mass-hierarchy (see e.g. Ref. (34)) and resolving the so-called naturalness problems of supersymmetry such as those pertaining to the issues of squark-degeneracy (35), CP violation (36) and quantum gravity-induced rapid proton decay (37).

Weighing the advantages and possible disadvantages of both, it seems hard at present to make a priori a clear choice between a GUT versus a non-GUT string-solution. As expressed elsewhere (31), it therefore seems prudent to keep both options open and pursue their phenomenological consequences. Given the advantages of G(224) or SO(10) in the light of the neutrino masses (see Secs. 2 and 4), I will thus proceed by assuming that either a suitable G(224)-solution with a mechanism of the sort mentioned above, or a realistic SO(10)-solution with the needed doublet-triplet mechanism, will emerge from string theory. We will see that with this broad assumption, an economical and predictive framework emerges that successfully accounts for a host of observed phenomena and makes some crucial testable predictions. I next discuss the implications of the mass of $\nu_\tau$ suggested by the SuperK data.

---

couplings of G(224), unified at $M_{st}$ would remain essentially so at $M_X$, so as to match with the "observed" coupling unification, of the type suggested in Ref. (29).

# MASS OF $\nu_\tau$: EVIDENCE IN FAVOR OF THE G(224) ROUTE

One can obtain an estimate for the mass of $\nu_L^\tau$ in the context of G(224) or SO(10) by using the following three steps (see e.g. Ref.(16)):

(i) Assume that B−L and $I_{3R}$, contained in a string-derived G(224) or SO(10), break near the unification-scale:

$$M_X \sim 2 \times 10^{16} \text{GeV}, \quad (6)$$

through VEVs of Higgs multiplets of the type suggested by string-solutions - i.e. $\langle(1,2,4)_H\rangle$ for G(224) or $\langle\overline{16}_H\rangle$ for SO(10), as opposed to $126_H$ (38). In the process, the RH neutrinos ($\nu_R^i$), which are singlets of the standard model, can and generically will acquire superheavy Majorana masses of the type $M_R^{ij} \nu_R^{iT} C^{-1} \nu_R^j$, by utilizing the VEV of $\langle\overline{16}_H\rangle$ and effective couplings of the form:

$$\mathcal{L}_M(SO(10)) = f_{ij}\, 16_i \cdot 16_j\, \overline{16}_H \cdot \overline{16}_H / M + h.c. \quad (7)$$

A similar expression holds for G(224). Here $i,j = 1,2,3$, correspond respectively to $e, \mu$ and $\tau$ families. Such gauge-invariant non-renormalizable couplings might be expected to be induced by Planck-scale physics, involving quantum gravity or stringy effects and/or tree-level exchange of superheavy states, such as those in the string tower. With $f_{ij}$ (at least the largest among them) being of order unity, we would thus expect M to lie between $M_{Planck} \approx 2 \times 10^{18}$ GeV and $M_{string} \approx 4 \times 10^{17}$ GeV. Ignoring for the present off-diagonal mixings (for simplicity), one thus obtains [2]:

$$M_{3R} \approx \frac{f_{33}\langle\overline{16}_H\rangle^2}{M} \approx f_{33}(2 \times 10^{14}\text{GeV})\eta^2 \times (M_{Planck}/M) \quad (8)$$

This is the Majorana mass of the RH tau neutrino. Guided by the value of $M_X$, we have substituted $\langle\overline{16}_H\rangle = (2 \times 10^{16}\text{GeV})\eta$, with $\eta \approx 1/2$ to 2 (say).

(ii) Now using SU(4)-color and the Higgs multiplet $(2,2,1)_H$ of G(224) or equivalently $10_H$ of SO(10), one obtains the relation $m_\tau(M_X) = m_b(M_X)$, which is known to be successful. Thus, there is a good reason to believe that the third family gets its masses primarily from the $10_H$ or equivalently $(2,2,1)_H$ (see sec.5). In turn, this implies:

$$m(\nu_{Dirac}^\tau) \approx m_{top}(M_X) \approx (100\text{-}120)\,\text{GeV} \quad (9)$$

Note that this relationship between the Dirac mass of the tau-neutrino and the top-mass is special to SU(4)-color. It does not emerge in SU(5).

(iii) Given the superheavy Majorana masses of the RH neutrinos as well as the Dirac masses as above, the see-saw mechanism (39) yields naturally light masses for the LH neutrinos. For $\nu_L^\tau$ (ignoring mixing), one thus obtains, using Eqs. (8) and (9),

$$m(\nu_L^\tau) \approx \frac{m(\nu_{Dirac}^\tau)^2}{M_{3R}} \approx$$
$$[(1/20)\,\text{eV}\,(1\text{-}1.44)/f_{33}\eta^2](M/M_{Planck}) \quad (10)$$

Now, assuming the hierarchical pattern $m(\nu_L^e) \ll m(\nu_L^\mu) \ll m(\nu_L^\tau)$, which is suggested by the see-saw mechanism, and further that the SuperK observation represents $\nu_L^\mu - \nu_L^\tau$ (rather than $\nu_L^\mu - \nu_X$) oscillation, the observed $\delta m^2 \approx 1/2(10^{-2}\text{-}10^{-3})\,\text{eV}^2$ corresponds to $m(\nu_L^\tau) \approx (1/15 - 1/40)$ eV. It seems *truly remarkable* that the expected magnitude of $m(\nu_L^\tau)$, given by Eq.(10), is just about what is suggested by the SuperK data, if $f_{33}\eta^2(M_{Planck}/M) \approx 1.3$ to $1/2$. Such a range for $f_{33}\eta^2(M_{Planck}/M)$ seems most plausible and natural (see discussion in Ref. (16)). Note that the estimate (10) crucially depends upon the supersymmetric unification scale, which provides a value for $M_{3R}$, as well as on SU(4)-color that yields $m(\nu_{Dirac}^\tau)$. *The agreement between the expected and the SuperK result thus clearly suggests that the effective symmetry below the string-scale should contain SU(4)-color. Thus, minimally it should be either G(214) or G(224), and maximally as big as SO(10), if not $E_6$.*

By contrast, if SU(5) is regarded as either a fundamental symmetry or as the effective symmetry below the string scale, there would be no compelling reason based on symmetry alone, to introduce a $\nu_R$, because it is a singlet of SU(5). Second, even if one did introduce $\nu_R^i$ by hand, their Dirac masses, arising from the coupling $h^i \bar{5}_i \langle 5_H\rangle \nu_R^i$, would be unrelated to the up-flavor masses and thus rather arbitrary (contrast with Eq. (9)). So also would be the Majorana masses of the $\nu_R^i$'s, which are SU(5)-invariant, and thus can be even of order string scale. This would give $m(\nu_L^\tau)$ in gross conflict with the observed value.

Before passing to the next section, it is worthnoting that the mass of $\nu_\tau$ suggested by SuperK, as well as the observed value of $\sin^2\theta_W$ (see Sec.3), provide valuable insight into the nature of GUT symmetry breaking. They both favor the case of a single-step breaking (SSB) of SO(10) or a string-unified G(224) symmetry at a scale of order $M_X$, into the standard model symmetry G(213), as opposed to that of a multi-step breaking (MSB). The latter would correspond, for example, to SO(10) (or G(224)) breaking at a scale $M_1$ into G(2213), which in turn breaks at a scale $M_2 \ll M_1$ into G(213). One reason why the

---

[2] The effects of neutrino-mixing and of possible choice of $M = M_{string} \approx 4 \times 10^{17}$ GeV (instead of $M = M_{Planck}$) on $M_{3R}$ are considered in Ref. (6).

case of single-step breaking is favored over that of multi-step breaking is that the latter can accomodate but not really predict $\sin^2\theta_W$, where as the former predicts the same successfully. Furthermore, since the Majorana mass of $\nu_R^\tau$ arises arises only after $B-L$ and $I_{3R}$ break, it would be given, for the case of MSB, by $M_{3R} \sim f_{33}(M_2^2/M)$, where $M \sim M_{st}$ (say). If $M_2 \ll M_X \sim 2 \times 10^{16}$ GeV, and $M > M_X$, one would obtain too low a value ($\ll 10^{14}$ GeV) for $M_{3R}$ (compare with Eq.(8)), and thereby too large a value for $m(\nu_L^\tau)$, compared to that suggested by SuperK. By contrast, the case of SSB yields the right magnitude for $m(\nu_\tau)$ (see Eq. (10)).

Thus the success of the result on $m(\nu_\tau)$ discussed above not only favors the symmetry G(224) or SO(10), but also clearly suggests that $B-L$ and $I_{3R}$ break near the conventional GUT scale $M_X \sim 2 \times 10^{16}$ GeV, rather than at an intermediate scale $\ll M_X$. In other words, the observed values of both $\sin^2\theta_W$ and $m(\nu_\tau)$ favor only the simplest pattern of symmetry-breaking, for which SO(10) or a string-derived G(224) symmetry breaks in one step to the standard model symmetry, rather than in multiple steps. It is of course only this simple pattern of symmetry breaking that would be rather restrictive as regards its predictions for proton decay (to be dicussed in Sec.6). I next dicuss the problem of understanding the masses and mixings of all fermions.

## UNDERSTANDING FERMION MASSES AND NEUTRINO OSCILLATIONS IN SO(10)

Understanding the masses and mixings of all quarks and charged leptons, in conjunction with those of the neutrinos, is a goal worth achieving by itself. It also turns out to be essential for the study of proton decay. I therefore present first a recent attempt in this direction, which seems most promising (6). A few guidelines would prove to be helpful in this regard. The first of these is motivated by the desire for economy and the rest by data.

**1) Hierarchy Through Off-diagonal Mixings** : Recall earlier attempts (40) that attribute hierarchical masses of the first two families to matrices of the form :

$$M = \begin{pmatrix} 0 & \varepsilon \\ \varepsilon & 1 \end{pmatrix} m_s^{(0)}, \qquad (11)$$

for the $(d,s)$ quarks, and likewise for the $(u,c)$ quarks. Here $\varepsilon \sim 1/10$. The hierarchical patterns in Eq. (11) can be ensured by imposing a suitable flavor symmetry which distinguishes between the two families (that in turn may have its origin in string theory (see e.g. Ref (34)). Such a pattern has the virtues that (a) it yields a hierarchy that is much larger than the input parameter $\varepsilon$ : $(m_d/m_s) \approx \varepsilon^2 \ll$ $\varepsilon$, and (b) it leads to an expression for the cabibbo angle :

$$\theta_c \approx \left| \sqrt{\frac{m_d}{m_s}} - e^{i\phi}\sqrt{\frac{m_u}{m_c}} \right|, \qquad (12)$$

which is rather successful. Using $\sqrt{m_d/m_s} \approx 0.22$ and $\sqrt{m_u/m_c} \approx 0.06$, we see that Eq. (12) works to within about 25% for any value of the phase $\phi$. Note that the square root formula (like $\sqrt{m_d/m_s}$) for the relevant mixing angle arises because of the symmetric form of $M$ in Eq. (11), which in turn is ensured if the contributing Higgs is a 10 of SO(10). A generalization of the pattern in Eq. (11) would suggest that the first two families (i.e. the $e$ and the $\mu$) receive masses primarily through their mixing with the third family ($\tau$), with (1,3) and (1,2) elements being smaller than the (2,3); while (2,3) is smaller than the (3,3). We will follow this guideline, except for the modification noted below.

**2) The Need for an Antisymmetric Component** : Although the symmetric hierarchical matrix in Eq. (11) works well for the first two families, a matrix of the same form fails altogether to reproduce $V_{cb}$, for which it yields :

$$V_{cb} \approx \left| \sqrt{\frac{m_s}{m_b}} - e^{i\chi}\sqrt{\frac{m_c}{m_t}} \right|. \qquad (13)$$

Given that $\sqrt{m_s/m_b} \approx 0.17$ and $\sqrt{m_c/m_t} \approx 0.0.06$, we see that Eq. (13) would yield $V_{cb}$ varying between 0.11 and 0.23, depending upon the phase $\chi$. This is too big, compared to the observed value of $V_{cb} \approx 0.04 \pm 0.003$, by at least a factor of 3. We interpret this failure as a *clue* to the presence of an antisymmetric component in $M$, together with symmetrical ones (thus $m_{ij} \neq m_{ji}$), which would modify the relevant mixing angle to $\sqrt{\frac{m_i}{m_j}}\sqrt{\frac{m_{ij}}{m_{ji}}}$, where $m_i$ and $m_j$ denote the respective eigenvalues.

**3) The Need for a Contribution Proportional to $B-L$** : The success of the relations $m_b^0 \approx m_\tau^0$, and $m_t^0 \approx m(\nu_\tau)_{Dirac}^0$ (see Sec. 4), suggests that the members of the third family get their masses primarily from the VEV of a SU(4)-color singlet Higgs field that is independent of $B-L$. This is in fact ensured if the Higgs is a 10 of SO(10). However, the empirical observations of $m_s^0 \sim m_\mu^0/3$ and $m_d^0 \sim 3m_e^0$ (41) clearly call for a contribution proportional to $B-L$ as well. Further, one can in fact argue that the suppression of $V_{bc}$ (in the quark-sector) together with an enhancement of $\theta_{\nu_\mu\nu_\tau}^{osc}$ (in the lepton sector) calls for a contribution that is not only proportional to $B-L$ but is also antisymmetric in the family space (as suggested above in item (2)). We note below how both of these requirements can be met, rather easily, in SO(10), even for a minimal Higgs system.

**4) Up-Down Asymmetry** : Finally, the up and the down-sector mass matrices must not be proportional to each other, as otherwise the CKM angles would all vanish.

Following Ref. (6), I now present a simple and predictive mass-matrix, based on SO(10), that satisfies *all three* requirements, (2), (3) and (4). The interesting point is that one can obtain such a mass-matrix for the fermions by utilizing only the minimal Higgs system, that is needed anyway to break the gauge symmetry SO(10). It consists of the set :

$$H_{minimal} = \{45_H, 16_H, \overline{16}_H, 10_H\}. \quad (14)$$

Of these, the VEV of $\langle 45_H \rangle \sim M_X$ breaks SO(10) into G(2213), and those of $\langle 16_H \rangle = \langle \overline{16}_H \rangle \sim M_X$ break G(2213) to G(213), at the unification-scale $M_X$. Now G(213) breaks at the electroweak scale by the VEV of $\langle 10_H \rangle$ to $U(1)_{em} \times SU(3)^c$.

One might have introduced large-dimensional tensorial multiplets of SO(10) like $126_H$ and $120_H$, both of which possess cubic level Yukawa couplings with the fermions. In particular, the coupling $16_i 16_j (120_H)$ would give the desired family-antisymmetric as well as (B-L)-dependent contribution. We do not however introduce these multiplets in part because they do not seem to arise in string solutions (38), and in part also because mass-splittings within such large-dimensional multiplets tend to give excessive threshold corrections to $\alpha_3(m_z)$ (typically exceeding 20%), rendering observed coupling unification fortuitous. By contrast, the multiplets in the minimal set (shown above) do arise in string solutions leading to SO(10). The added advantage of the minimal set is that the corresponding threshold corrections are found to be naturally small, and even to have the right sign, to go with the observed coupling unification (6).

The question is : does the minimal set meet all the requirements listed above? Now $10_H$ (even several 10's) can not meet the requirements of antisymmetry and (B-L)-dependence. Furthermore, a single $10_H$ cannot generate CKM-mixings. This impasse disappears, however, as soon as one allows for not only cubic, but also effective non-renormalizable quartic couplings of the minimal set of Higgs fields with the fermions. These latter couplings could of course well arise through exchanges of superheavy states (e.g. those in the string tower) involving renormalizable couplings, and/or through quantum gravity.

Allowing for such cubic and quartic couplings and adopting the guideline (1) of hierarchical Yukawa couplings, as well as that of economy, we are led to suggest the following effective lagrangian for generating Dirac masses and mixings of the three families (6) (for a related but different pattern, involving a non-minimal Higgs system, see Ref (42)).

$$\mathcal{L}_{Yuk} = h_{33} 16_3 16_3 10_H + [h_{23} 16_2 16_3 10_H$$
$$+ a_{23} 16_2 16_3 10_H 45_H/M + g_{23} 16_2 16_3 16_H 16_H/M]$$
$$+ \{a_{12} 16_1 16_2 10_H 45_H/M + g_{12} 16_1 16_2 16_H 16_H/M\}. \quad (15)$$

Here, $M$ could plausibly be of order string scale. Note that a mass matrix having essentially the form of Eq. (11) results if the first term $h_{33}\langle 10_H \rangle$ is dominant. This ensures $m_b^0 \approx m_\tau^0$ and $m_t^0 \approx m(\nu_{Dirac})^0$. Following the assumption of progressive hierarchy (equivalently appropriate flavor symmetries[3]), we presume that $h_{23} \sim h_{33}/10$, while $h_{22}$ and $h_{11}$, which are set to be zeros, are progressively much smaller than $h_{23}$ (see discussion in Ref. (31)). Since $\langle 45_H \rangle \sim \langle 16_H \rangle \sim M_X$, while $M \sim M_{st} \sim 10 M_X$, the terms $a_{23}\langle 45_H \rangle/M$ and $g_{23}\langle 16_H \rangle/M$ can quite plausibly be of order $h_{33}/10$, if $a_{23} \sim g_{23} \sim h_{33}$. By the assumption of hierarchy, we presume that $a_{12} \ll a_{23}$, and $g_{12} \ll g_{23}$

It is interesting to observe the symmetry properties of the $a_{23}$ and $g_{23}$-terms. Although $10_H \times 45_H = 10 + 120 + 320$, given that $\langle 45_H \rangle$ is along B-L, which is needed to implement doublet-triplet splitting (see Appendix), only 120 in the decomposition contributes to the mass-matrices. This contribution is, however, antisymmetric in the family-index and, at the same time, proportional to B-L. Thus the $a_{23}$ *term fulfills the requirements of both antisymmetry and (B-L)-dependence, simultaneously*[4]. With only $h_{ij}$ and $a_{ij}$-terms, however, the up and down quark mass-matrices will be proportional to each other, which would yield $V_{CKM} = 1$. This is remedied by the $g_{ij}$ coupling. Because, the $16_H$ can have a VEV not only along its SM singlet component (transforming as $\tilde{\bar{\nu}}_R$) which is of GUT-scale, but also along its electroweak doublet component – call it $16_d$ – of the electroweak scale. The latter can arise by the the mixing of $16_d$ with the corresponding doublet (call it $10_d$) in the $10_H$. The MSSM doublet $H_d$, which is light, is then a mixture of $10_d$ and $16_d$, while the orthogonal combination is superheavy (see Appendix). Since $\langle 16_d \rangle$ contributes only to the down-flavor mass matrices, but not to the up-flavor, the $g_{23}$ and $g_{12}$ couplings generate non-trivial CKM-mixings. *We thus see that the minimal Higgs system satisfies apriori all the qualitative requirements (2)-(4), including the con-*

---

[3] Although no explicit string solution with the hierarchy in $h_{ij}$ mentioned above, together with the $a_{ij}$ and $g_{ij}$ couplings of Eq. (15), exists as yet, flavor symmetries of the type alluded to, as well as SM singlets carrying flavor-charges and acquiring VEVs of order $M_X$ that can lead to effective hierarchical couplings, do emerge in string solutions. And, there exist solutions with top Yukawa coupling being leading (see e.g. Refs. (34) and (33)).

[4] The analog of $10_H \cdot 45_H$ for the case of G(224) would be $\chi_H \equiv (2,2,1)_H \cdot (1,1,15)_H$. Although in general, the coupling of $\chi_H$ to the fermions need not be antisymmetric, for a string-derived G(224), the multiplet $(1,1,15)_H$ is most likely to arise from an underlying 45 of SO(10) (rather than 210); in this case, the couplings of $\chi_H$ must be antisymmetric like that of $10_H \cdot 45_H$.

*dition of* $V_{CKM} \neq 1$. I now discuss that this system works well even quantitatively.

With these six effective Yukawa couplings, the Dirac mass matrices of quarks and leptons of the three families at the unification scale take the form:

$$U = \begin{pmatrix} 0 & \varepsilon' & 0 \\ -\varepsilon' & 0 & \varepsilon + \sigma \\ 0 & -\varepsilon + \sigma & 1 \end{pmatrix} m_U,$$

$$D = \begin{pmatrix} 0 & \varepsilon' + \eta' & 0 \\ -\varepsilon' + \eta' & 0 & \varepsilon + \eta \\ 0 & -\varepsilon + \eta & 1 \end{pmatrix} m_D,$$

$$N = \begin{pmatrix} 0 & -3\varepsilon' & 0 \\ 3\varepsilon' & 0 & -3\varepsilon + \sigma \\ 0 & 3\varepsilon + \sigma & 1 \end{pmatrix} m_U,$$

$$L = \begin{pmatrix} 0 & -3\varepsilon' + \eta' & 0 \\ 3\varepsilon' + \eta' & 0 & -3\varepsilon + \eta \\ 0 & 3\varepsilon + \eta & 1 \end{pmatrix} m_D. \quad (16)$$

Here the matrices are multiplied by left-handed fermion fields from the left and by anti–fermion fields from the right. $(U,D)$ stand for the mass matrices of up and down quarks, while $(N,L)$ are the Dirac mass matrices of the neutrinos and the charged leptons. The entries $1, \varepsilon,$ and $\sigma$ arise respectively from the $h_{33}, a_{23}$ and $h_{23}$ terms in Eq. (15), while $\eta$ entering into $D$ and $L$ receives contributions from both $g_{23}$ and $h_{23}$; thus $\eta \neq \sigma$. Similarly $\eta'$ and $\varepsilon'$ arise from $g_{12}$ and $a_{12}$ terms respectively. Note the quark-lepton correlations between $U$ and $N$ as well as $D$ and $L$, and the up-down correlations between $U$ and $D$ as well as $N$ and $L$. These correlations arise because of the symmetry property of $G(224)$. The relative factor of $-3$ between quarks and leptons involving the $\varepsilon$ entry reflects the fact that $\langle 45_H \rangle \propto (B-L)$, while the antisymmetry in this entry arises from the group structure of $SO(10)$, as explained above[4]. As we will see, this $\varepsilon$-entry helps to account for (a) the differences between $m_s$ and $m_\mu$, (b) that between $m_d$ and $m_e$, and also, (c) the suppression of $V_{cb}$ together with the enhancement of the $\nu_\mu$-$\nu_\tau$ oscillation angle.

The mass matrices in Eq. (16) contain 7 parameters[5]: $\varepsilon, \sigma, \eta, m_D = h_{33}\langle 10_d \rangle, m_U = h_{33}\langle 10_U \rangle, \eta'$ and $\varepsilon'$. These may be determined by using, for example, the following input values: $m_t^{phys} = 174\,\text{GeV}, m_c(m_c) = 1.37\,\text{GeV}, m_s(1\,\text{GeV}) = 110\text{-}116\,\text{MeV}$ (43), $m_u(1\,\text{GeV}) \approx 6\,\text{MeV}$ and the observed masses of $e, \mu$ and $\tau$, which lead to (see Ref. (6), for details):

$$\sigma \simeq 0.110, \eta \simeq 0.151, \varepsilon \simeq -0.095,$$
$$|\eta'| \approx 4.4 \times 10^{-3} \text{ and } \varepsilon' \approx 2 \times 10^{-4}$$

$$m_U \simeq m_t(M_U) \simeq (100\text{-}120)\,\text{GeV},$$
$$m_D \simeq m_b(M_U) \simeq 1.5\,\text{GeV}. \quad (17)$$

We have assumed for simplicity that the parameters are real, because a good fitting suggests that the relative phases of at least $\sigma, \eta$ and $\varepsilon$ are small ($< 10°$ say). Such fitting also fixes their relative signs. Note that in accord with our general expectations discussed above, each of the parameters $\sigma, \eta$ and $\varepsilon$ are found to be of order $1/10$, as opposed to being [6] $O(1)$ or $O(10^{-2})$, compared to the leading (3,3)-element in Eq. (16). Having determined these parameters, we are led to a total of five predictions involving only the quarks (those for the leptons are listed separately):

$$m_b^0 \approx m_\tau^0(1 - 8\varepsilon^2); \text{ thus } m_b(m_b) \simeq (4.6\text{-}4.9)\,\text{GeV} \quad (18)$$

$$|V_{cb}| \simeq |\sigma - \eta| \approx$$
$$\left| \sqrt{m_s/m_b} \left| \frac{\eta + \varepsilon}{\eta - \varepsilon} \right|^{1/2} - \sqrt{m_c/m_t} \left| \frac{\sigma + \varepsilon}{\sigma - \varepsilon} \right|^{1/2} \right| \simeq 0.045 \quad (19)$$

$$m_d(1\,\text{GeV}) \simeq 8\,\text{MeV} \quad (20)$$

$$\theta_C \simeq \left| \sqrt{m_d/m_s} - e^{i\phi}\sqrt{m_u/m_c} \right| \quad (21)$$

$$|V_{ub}/V_{cb}| \simeq \sqrt{m_u/m_c} \simeq 0.07. \quad (22)$$

In making these predictions, we have extrapolated the GUT-scale values down to low energies using $\alpha_3(m_Z) = 0.118$, a SUSY threshold of 500 GeV and $\tan\beta = 5$. The results depend weakly on these choices, assuming $\tan\beta \approx 2\text{-}30$. Further, the Dirac masses and mixings of the neutrinos and the mixings of the charged leptons also get determined. We obtain:

$$m_{\nu_\tau}^D(M_U) \approx 100\text{-}120\,\text{GeV}; \ m_{\nu_\mu}^D(M_U) \simeq 8\,\text{GeV}, \quad (23)$$

$$\theta_{\mu\tau}^\ell \approx -3\varepsilon + \eta \approx \sqrt{m_\mu/m_\tau} \left| \frac{-3\varepsilon + \eta}{3\varepsilon + \eta} \right|^{1/2} \simeq 0.437 \quad (24)$$

$$m_{\nu_e}^D \simeq [9\varepsilon'^2/(9\varepsilon^2 - \sigma^2)]m_U \simeq 0.4\,\text{MeV} \quad (25)$$

$$\theta_{e\mu}^\ell \simeq \left| \frac{\eta' - 3\varepsilon'}{\eta' + 3\varepsilon'} \right|^{1/2} \sqrt{m_e/m_\mu} \simeq 0.85$$
$$\sqrt{m_e/m_\mu} \simeq 0.06 \quad (26)$$

---

[5] Of these, $m_U^0 \approx m_t^0$ can in fact be estimated to within 20% accuracy by either using the argument of radiative electroweak symmetry breaking, or some promising string solutions (see e.g. Ref. (34)).

[6] This is one characteristic difference between our work and that of Ref. (38), where the (2,3)-element is even bigger than the (3,3).

$$\theta_{e\tau}^{\ell} \simeq \frac{1}{0.85}\sqrt{m_e/m_\tau}\,(m_\mu/m_\tau) \simeq 0.0012\,. \qquad (27)$$

In evaluating $\theta_{e\mu}^{\ell}$, we have assumed $\varepsilon'$ and $\eta'$ to be relatively positive.

Given the bizarre pattern of quark and lepton masses and mixings, it seems remarkable that the simple pattern of fermion mass-matrices, motivated by the group theory of G(224)/SO(10), gives an overall fit to all of them which is good to within 10%. This includes the two successful predictions on $m_b$ and $V_{cb}$ (Eqs.(18 and 19)). Note that in supersymmetric unified theories, the "observed" value of $m_b(m_b)$ and renormalization-group studies suggest that, for a wide range of the parameter $\tan\beta$, $m_b^0$ should in fact be about 10-20% *lower* than $m_\tau^0$ (44). This is neatly explained by the relation: $m_b^0 \approx m_\tau^0(1-8\varepsilon^2)$ (Eq. (18)), where exact equality holds in the limit $\varepsilon \to 0$ (due to SU(4)-color), while the decrease of $m_b^0$ compared to $m_\tau^0$ by $8\varepsilon^2 \sim 10\%$ is precisely because the off-diagonal $\varepsilon$-entry is proportional to $B$-$L$ (see Eq. (16)).

Specially intriguing is the result on $V_{cb} \approx 0.045$ which compares well with the observed value of $\simeq 0.04$. The suppression of $V_{cb}$, compared to the value of $0.17 \pm 0.06$ obtained from Eq. (13), is now possible because the mass matrices (Eq. (16)) contain an antisymmetric component $\propto \varepsilon$. That corrects the square-root formula $\theta_{sb} = \sqrt{m_s/m_b}$ (appropriate for symmetric matrices, see Eq. (11)) by the asymmetry factor $|(\eta+\varepsilon)/(\eta-\varepsilon)|^{1/2}$ (see Eq. (19)), and similarly for the angle $\theta_{ct}$. This factor suppresses $V_{cb}$ if $\eta$ and $\varepsilon$ have opposite signs. The interesting point is that, *the same feature necessarily enhances the corresponding mixing angle $\theta_{\mu\tau}^{\ell}$ in the leptonic sector*, since the asymmetry factor in this case is given by $[(-3\varepsilon+\eta)/(3\varepsilon+\eta)]^{1/2}$ (see Eq. (24)). This enhancement of $\theta_{\mu\tau}^{\ell}$ helps to account for the nearly maximal oscillation angle observed at SuperK (as discussed below). This intriguing correlation between the mixing angles in the quark versus leptonic sectors – *that is suppression of one implying enhancement of the other* – has become possible only because of the $\varepsilon$-contribution, which is simultaneously antisymmetric and is proportional to $B$-$L$. That in turn becomes possible because of the group-property of SO(10) or a string-derived G(224)[4].

Taking stock, we see an overwhelming set of evidences in favor of $B$-$L$ and in fact for the full SU(4)-color-symmetry. These include: (i) the suppression of $V_{cb}$, together with the enhancement of $\theta_{\mu\tau}^{\ell}$, just mentioned above, (ii) the successful relation $m_b^0 \approx m_\tau^0(1-8\varepsilon^2)$, (iii) the usefulness again of the SU(4)-color-relation $m(\nu_{Dirac}^\tau)^0 \approx m_t^0$ in accounting for $m(\nu_L^\tau)$ (see Sec. 4), and (iv) the agreement of the relation $|m_s^0/m_\mu^0| = |(\varepsilon^2 - \eta^2)/(9\varepsilon^2 - \eta^2)|$ with the data, in that the ratio is naturally *less than* 1, if $\eta \sim \varepsilon$. The presence of $9\varepsilon^2$ in the denominator is because the off-diagonal entry is proportional to $B$-$L$. Finally, the need for $(B$-$L)$- as a local symmetry, to implement baryogenesis, has been noted in Sec.1.

Turning to neutrino masses, while all the entries in the Dirac mass matrix $N$ are now fixed, to obtain the parameters for the light neutrinos, one needs to specify those of the Majorana mass matrix of the RH neutrinos ($\nu_R^{e,\mu,\tau}$). Guided by economy and the assumption of hierarchy, we consider the following pattern:

$$M_\nu^R = \begin{pmatrix} x & 0 & z \\ 0 & 0 & y \\ z & y & 1 \end{pmatrix} M_R\,. \qquad (28)$$

As discussed in Sec. 4, the magnitude of $M_R \approx (5\text{-}15) \times 10^{14}$ GeV can quite plausibly be justified in the context of supersymmetric unificaton[7] (e.g. by using $M \approx M_{st} \approx 4 \times 10^{17}$ GeV in Eq. (8)). To the same extent, the magnitude of $m(\nu_\tau) \approx (1/10\text{-}1/30)$ eV, which is consistent with the SuperK value, can also be anticipated. Thus there are effectively three new parameters: $x$, $y$, and $z$. Since there are six observables for the three light neutrinos, one can expect three predictions. These may be taken to be $\theta_{\nu_\mu\nu_\tau}^{osc}$, $m_{\nu_\tau}$ (see Eq. (10)), and for example $\theta_{\nu_e\nu_\mu}^{osc}$.

Assuming successively hierarchical entries as for the Dirac mass matrices, we presume that $|y| \sim 1/10, |z| \leq |y|/10$ and $|x| \leq z^2$. Now given that $m(\nu_\tau) \sim 1/20$ eV (as estimated in Eq. (10)), the MSW solution for the solar neutrino puzzle (45) suggests that $m(\nu_\mu)/m(\nu_\tau) \approx 1/10\text{-}1/30$. The latter in turn yields: $|y| \approx (1/18 \text{ to } 1/23.6)$, with $y$ having the same sign as $\varepsilon$ (see Eq. (17)). This solution for $y$ obtains only by assuming that $y$ is $O(1/10)$ rather than $O(1)$. Combining now with the mixing in the $\mu$-$\tau$ sector determined above (see Eq. (24)), one can then determine the $\nu_\mu$-$\nu_\tau$ oscillation angle. The two predictions of the model for the neutrino-system are then:

$$m(\nu_\tau) \approx (1/10\text{-}1/30)\,\text{eV} \qquad (29)$$

$$\theta_{\nu_\mu\nu_\tau}^{osc} \simeq \theta_{\mu\tau}^{\ell} - \theta_{\mu\tau}^{\nu} \simeq \left(0.437 + \sqrt{\frac{m_{\nu_2}}{m_{\nu_3}}}\right). \qquad (30)$$

Thus, $\sin^2 2\theta_{\nu_\mu\nu_\tau}^{osc} = (0.96, 0.91, 0.86, 0.83, 0.81)$ (31)

for $m_{\nu_2}/m_{\nu_3} = (1/10, 1/15, 1/20, 1/25, 1/30)$. (32)

Both of these predictions are extremely successful.

Note the interesting point that the MSW solution, together with the requirement that $|y|$ should have a natural hierarchical value (as mentioned above), lead to $y$ having the same sign as $\varepsilon$; that (it turns out) implies that the

---

[7] This estimate for $M_R$ is retained even if one allows for $\nu_\mu$-$\nu_\tau$ mixing (see Ref. (6)).

two contributions in Eq.(30) must *add* rather than subtract, leading to an *almost maximal oscillation angle* (6). The other factor contributing to the enhancement of $\theta^{osc}_{\nu_\mu\nu_\tau}$ is, of course, also the asymmetry-ratio which increases $|\theta^\ell_{\mu\tau}|$ from 0.25 to 0.437 (see Eq. (24)). We see that one can derive rather plausibly a large $\nu_\mu$-$\nu_\tau$ oscillation angle $\sin^2 2\theta^{osc}_{\nu_\mu\nu_\tau} \geq 0.8$, together with an understanding of hierarchical masses and mixings of the quarks and the charged leptons, while maintaining a large hierarchy in the seesaw derived masses $(m_{\nu_2}/m_{\nu_3} = 1/10\text{-}1/30)$, all within a unified framework including both quarks and leptons. In the example exhibited here, the mixing angles for the mass eigenstates of neither the neutrinos nor the charged leptons are really large, in that $\theta^\ell_{\mu\tau} \simeq 0.437 \simeq 23°$ and $\theta^\nu_{\mu\tau} \simeq (0.18\text{-}0.31) \approx (10\text{-}18)°$, *yet the oscillation angle obtained by combining the two is near-maximal.* This contrasts with most previous work, in which a large oscillation angle is obtained either entirely from the neutrino sector (with nearly degenerate neutrinos) or almost entirely from the charged lepton sector.

While $M_R \approx (5\text{-}15) \times 10^{14}$ GeV and $y \approx -1/20$ are better determined, the parameters $x$ and $z$ can not be obtained reliably at present because very little is known about observables involving $\nu_e$. Taking, for concreteness, $m_{\nu_e} \approx (10^{-5}\text{-}10^{-4}$ (1 to few)) eV and $\theta^{osc}_{e\tau} \approx \theta^\ell_{e\tau} - \theta^\nu_{e\tau} \approx 10^{-3} \pm 0.03$ as inputs, we obtain : $z \sim (1\text{-}5) \times 10^{-3}$ and $x \sim (1 \text{ to few})(10^{-6}\text{-}10^{-5})$, in accord with the guidelines of $|z| \sim |y|/10$ and $|x| \sim z^2$. This in turn yields : $\theta^{osc}_{e\mu} \approx \theta^\ell_{e\mu} - \theta^\nu_{e\mu} \approx 0.06 \pm 0.015$. Note that the mass of $m_{\nu_\mu} \sim 3 \times 10^{-3}$ eV, that follows from a natural hierarchical value for $y \sim -(1/20)$, and $\theta_{e\mu}$ as above, go well with the small angle MSW explanation[8] of the solar neutrinos puzzle.

In summary, we have proposed an economical and predictive pattern for the Dirac mass matrices, within the SO(10)/G(224)-framework, which is remarakbly successful in describing the observed masses and mixings of *all* the quarks and charged leptons. It leads to five predictions for just the quark- system, all of which agree with observation to within 10%. The same pattern, supplemented with a similar structure for the Majorana mass matrix, accounts for both the large $\nu_\mu$-$\nu_\tau$ oscillation angle and a mass of $\nu_\tau \sim 1/20$ eV, suggested by the SuperK data. It also accomodates a small $\nu_e$-$\nu_\mu$ oscillation angle relevant for theories of the solar neutrino deficit. Given this degree of success, it makes good sense to study proton decay concretely within this SO(10)/G(224)-framework. The results of this study (6) are presented in the next section.

Before turning to proton decay, it is worth noting that much of our discussion of fermion masses and mixings, including those of the neutrinos, is essentially unaltered if we go to the limit $\varepsilon' \to 0$ of Eq. (28). This limit clearly involves:

$$m_u = 0, \quad \theta_C \simeq \sqrt{m_d/m_s}, \quad m_{\nu_e} = 0, \quad \theta^\nu_{e\mu} = \theta^\nu_{e\tau} = 0.$$

$$|V_{ub}| \simeq \sqrt{\frac{\eta - \varepsilon}{\eta + \varepsilon}} \sqrt{m_d/m_b}(m_s/m_b) \simeq$$
$$(2.1)(0.039)(0.023) \simeq 0.0019 \quad (33)$$

All other predictions remain unaltered. Now, among the observed quantities in the list above, $\theta_C \simeq \sqrt{m_d/m_s}$ is a good result. Considering that $m_u/m_t \approx 10^{-5}$, $m_u = 0$ is also a pretty good result. There are of course plausible small corrections which could arise through Planck scale physics; these could induce a small value for $m_u$ through the (1,1)-entry $\delta \approx 10^{-5}$. For considerations of proton decay, it is worth distinguishing between these two variants, which we will refer to as cases I and II respectively.

$$\text{Case I}: \quad \varepsilon' \approx 2 \times 10^{-4}, \; \delta = 0$$
$$\text{Case II}: \quad \delta \approx 10^{-5}, \; \varepsilon' = 0. \quad (34)$$

## EXPECTATIONS FOR PROTON DECAY IN SUPERSYMMETRIC UNIFIED THEORIES

**6.1** Turning to the main purpose of this talk, I present now the reason why the unification framework based on SUSY SO(10) or G(224), together with the understanding of fermion masses and mixings discussed above, strongly suggest that proton decay should be imminent.

Recall that supersymmetric unfied theories (GUTs) introduce two new features to proton decay : (i) First, by raising $M_X$ to a higher value of about $2 \times 10^{16}$ GeV, they strongly suppress the gauge-boson-mediated $d = 6$ proton decay operators, for which $e^+\pi^0$ would have been the dominant mode (for this case, one typically obtains : $\Gamma^{-1}(p \to e^+\pi^0)|_{d=6} \approx 10^{36 \pm 1.5}$ yrs). (ii) Second, they generate $d = 5$ proton decay operators (22) of the form $Q_iQ_jQ_kQ_l/M$ in the superpotential, through the exchange of color triplet Higginos, which are the GUT partners of the standard Higgs(ino) doublets, such as those in the $5 + \bar{5}$ of SU(5) or the 10 of SO(10). Assuming that a suitable doublet-triplet splitting mecahnism provides heavy GUT-scale masses to these color triplets and light masses to the doublets, these "standard" $d = 5$ operators, suppressed by just one power of the heavy mass

---

[8] Although the small angle MSW solution appears to be more generic within the approach outlined above, we have found that the large angle solution can still plausibly emerge in a limited region of parameter space, without affecting our results on fermion masses.

and the small Yukawa couplings, are found to provide the dominant mechanism for proton decay in supersymmetric GUT (46, 47, 48, 49).

Now, owing to (a) Bose symmetry of the superfields in $QQQL/M$, (b) color antisymmetry, and especially (c) the hierarchical Yukawa couplings of the Higgs doublets, it turns out that these standard $d = 5$ operators lead to dominant $\bar{\nu}K^+$ and comparable $\bar{\nu}\pi^+$ modes, but in all cases to highly suppressed $e^+\pi^0$, $e^+K^0$ and even $\mu^+K^0$ modes. For instance, for minimal SUSY SU(5), one obtains (with $\tan\beta \leq 20$, say) :

$$[\Gamma(\mu^+K^0)/\Gamma(\bar{\nu}K^+)]_{std}^{SU(5)} \sim [m_u/m_c \sin^2\theta] R \approx 10^{-3}, \quad (35)$$

where $R \approx 0.1$ is the ratio of the relevant $|$matrix element$|^2 \times$(phase space), for the two modes.

It was recently pointed out that in SUSY unified theories based on SO(10) or G(224), which assign heavy Majorana masses to the RH neutrinos, there exists a new set of color triplets and thereby very likely a *new source* of $d = 5$ proton decay operators (21). For instance, in the context of the minimal set of Higgs multiplets[9] $\{45_H, 16_H, \overline{16}_H$ and $10_H\}$ (see Sec. 5), these new $d = 5$ operators arise by combining three effective couplings introduced before :– i.e., (a) the couplings $f_{ij}16_i16_j\overline{16}_H\overline{16}_H/M$ (see Eq. (7)) that are required to assign Majorana masses to the RH neutrinos, (b) the couplings $g_{ij}16_i16_j16_H16_H/M$, which are needed to generate non-trivial CKM mixings (see Eq. (15)), and (c) the mass term $M_{16}16_H\overline{16}_H$. For the $f_{ij}$ couplings, there are two possible SO(10)-contractions, and we assume both to have comparable strength[10]. In this case, the color-triplet Higgsinos in $\overline{16}_H$ and $16_H$ of mass $M_{16}$ can be exchanged between $\bar{q}_iq_j$ and $\bar{q}_kq_l$-pairs. This exchange generates a new set of $d = 5$ operators in the superpotential of the form

$$W_{new} = f_{ij}g_{kl}(16_i 16_j)(16_k 16_l)\langle\overline{16}_H\rangle\langle\overline{16}_H\rangle/M, \quad (36)$$

which induce proton decay. Note that these operators depend, through the couplings $f_{ij}$ and $g_{kl}$, both on the Majorana and on the Dirac masses of the respective fermions. *This is why within SUSY SO(10) or G(224), proton decay gets intimately linked to the masses and mixings of all fermions, including neutrinos.*

---

[9] The origin of the new $d = 5$ operators in the context of other Higgs multiplets, in particular in the cases where $126_H$ and $\overline{126}_H$ are used to break B-L, has been discussed in Ref. (21).

[10] One would expect such a general contraction to hold, especially if the $f_{ij}$ couplings are induced by non-perturbative quantum gravity. Furthermore, the $f_{ij}$ couplings with the contraction of the pair $(16_i \cdot \overline{16}_H)$, being effectively in 45 (rather than in 1) of SO(10), would be induced also by tree-level exchanges, if these pairs couple to the 45's in the string tower. Such a contraction would lead to proton decay.

## 6.2 Framework for Calculating Proton Decay Rate

To establish notations, consider the case of minimal SUSY SU(5) and, as an example, the process $\bar{c}\bar{d} \to \bar{s}\bar{\nu}_\mu$, which induces $p \to \bar{\nu}_\mu K^+$. Let the strength of the corresponding $d = 5$ operator, multiplied by the product of the CKM mixing elements entering into wino-exchange vertices, (which in this case is $\sin\theta_C \cos\theta_C$) be denoted by $\hat{A}$. Thus (putting $\cos\theta_C = 1$), one obtains:

$$\hat{A}_{\bar{c}\bar{d}}(SU(5)) = (h_{22}^u h_{12}^d/M_{H_C}) \sin\theta_c$$
$$\simeq (m_c m_s \sin^2\theta_C/v_u^2)(\tan\beta/M_{H_C})$$
$$\simeq (1.9 \times 10^{-8})(\tan\beta/M_{H_C}) \approx$$
$$(2 \times 10^{-24}\,\text{GeV}^{-1})(\tan\beta/2)(2 \times 10^{16}\,\text{GeV}/M_{H_C}), \quad (37)$$

where $\tan\beta \equiv v_u/v_d$, and we have put $v_u = 174$ GeV and the fermion masses extrapolated to the unification-scale – i.e. $m_c \simeq 300$ MeV and $m_s \simeq 40$ MeV. The amplitude for the associated four-fermion process $dus \to \bar{\nu}_\mu$ is given by:

$$A_5(dus \to \bar{\nu}_\mu) = \hat{A}_{\bar{c}\bar{d}} \times (2f) \quad (38)$$

where $f$ is the loop-factor associated with wino-dressing. Assuming $m_{\tilde{w}} \ll m_{\tilde{q}} \sim m_{\tilde{l}}$ one gets: $f \simeq (m_{\tilde{w}}/m_{\tilde{q}}^2)(\alpha_2/4\pi)$. Using the amplitude for $(du)(sv_\ell)$, as in Eq. (38), ($\ell = \mu$ or $\tau$), one then obtains (47, 48, 49, 6) :

$$\Gamma^{-1}(p \to \bar{\nu}_\tau K^+) \approx (2.2 \times 10^{31})\,\text{yrs} \times$$
$$\left(\frac{0.67}{A_S}\right)^2 \left[\frac{0.006\,\text{GeV}^3}{\beta_H}\right]^2 \times$$
$$\left[\frac{(1/6)}{(m_{\tilde{W}}/m_{\tilde{q}})}\right]^2 \left[\frac{m_{\tilde{q}}}{1\,\text{TeV}}\right]^2 \left[\frac{2 \times 10^{-24}\,\text{GeV}^{-1}}{\hat{A}(\bar{\nu})}\right]^2. \quad (39)$$

Here $\beta_H$ denotes the hadronic matrix element defined by $\beta_H u_L(\vec{k}) \equiv \varepsilon_{\alpha\beta\gamma}\langle 0|(d_L^\alpha u_L^\beta) u_L^\gamma|p,\vec{k}\rangle$. While the range $\beta_H = (0.003\text{-}0.03)$ GeV$^3$ has been used in the past (48), given that one lattice calculation yields $\beta_H = (5.6 \pm 0.5) \times 10^{-3}$ GeV$^3$ (50), we will take as a plausible range : $\beta_H = (0.006\,\text{GeV}^3)(1/2\text{-}2)$. Here, $A_S \approx 0.67$ stands for the short distance renormalization factor of the $d = 5$ operator. Note that the familiar factors that appear in the expression for proton lifetime – i.e., $M_{H_C}$, $(1 + y_{tK})$ representing the interference between the $\tilde{t}$ and $\tilde{c}$ contributions, and $\tan\beta$ (see e.g. Ref.(48)) – are all effectively contained in $\hat{A}(\bar{\nu})$. Allowing for plausible and rather generous uncertainties in the matrix element and the spectrum we take:

$$\beta_H = (0.006\,\text{GeV}^3)(1/2\text{-}2)$$
$$m_{\tilde{w}}/m_{\tilde{q}} = 1/6\,(1/2\text{-}2),$$
$$\text{and } m_{\tilde{q}} \approx m_{\tilde{l}} \approx 1\,\text{TeV}\,(1/\sqrt{2}\text{-}\sqrt{2}). \quad (40)$$

Using Eqs. (39-40), we get:

$$\Gamma^{-1}(p \to \bar{\nu}_\tau K^+) \approx (2.2 \times 10^{31} \, yrs)$$
$$[2.2 \times 10^{-24} GeV^{-1}/\hat{A}(\bar{\nu}_\ell)]^2 [32 - 1/32]. \quad (41)$$

This relation, as well as Eq. (39) are general, depending only on $\hat{A}(\bar{\nu}_\ell)$ and on the range of parameters given in Eq. (40). They can thus be used for both SU(5) and SO(10).

The experimental lower limit on the inverse rate for the $\bar{\nu}K^+$ modes is given by (9),

$$[\sum_\ell \Gamma(p \to \bar{\nu}_\ell K^+)]^{-1}_{expt} > 1.6 \times 10^{33} \, yrs. \quad (42)$$

Allowing for all the uncertainties to stretch in the same direction (in this case, the square bracket = 32), and assuming that just one neutrino flavor (e.g. $\nu_\mu$ for SU(5)) dominates, the observed limit Eq. (42) provides an upper bound on the amplitude[11]:

$$\hat{A}(\bar{\nu}_\ell) \leq \sqrt{2} \times 10^{-24} \, GeV^{-1} \quad (43)$$

which holds for both SU(5) and SO(10). For minimal SU(5), using Eq. (37) and $\tan\beta \geq 2$ (which is suggested on several grounds), one obtains a lower limit on $M_{HC}$ given by:

$$M_{HC} \geq 3 \times 10^{16} \, GeV \, (SU(5)) \quad (44)$$

At the same time, higher values of $M_{HC} > 3 \times 10^{16}$ GeV do not go very well with gauge coupling unification. Thus keeping $M_{HC} \leq 3 \times 10^{16}$ and $\tan\beta \geq 2$, we obtain from Eq. (37): $\hat{A}(SU(5)) \geq (4/3) \times 10^{-24} GeV^{-1}$. Using Eq. (41), this in turn implies that

$$\Gamma^{-1}(p \to \bar{\nu}K^+) \leq 1.5 \times 10^{33} \, yrs \, (SU(5)) \quad (45)$$

This is a conservative upper limit. In practise, it is unlikely that all the uncertainties, including that in $M_{HC}$, would stretch in the same direction to nearly extreme values so as to prolong proton lifetime. A more reasonable upper limit, for minimal SU(5), thus seems to be: $\Gamma^{-1}(p \to \bar{\nu}K^+)(SU(5)) \leq (0.7) \times 10^{33}$ yrs. Given the experimental lower limit (Eq. (42)), we see that minimal SUSY SU(5) is already or almost on the verge of being excluded by proton decay-searches. We have of course noted in Sec. 4 that SUSY SU(5) does not go well with neutrino oscillations observed at SuperK.

Now, to discuss proton decay in the context of supersymmetric SO(10), it is necessary to discuss first the mechanism for doublet-triplet splitting. Details of this discussion may be found in Ref. (6). A synopsis is presented in the appendix.

---

[11] If there are sub-dominant $\bar{\nu}_i K^+$ modes with branching ratio $R$, the right side of Eq. (43) should be divided by $\sqrt{1+R}$.

## 6.3. Proton Decay in Supersymmetric SO(10)

The calculation of the amplitudes $\hat{A}_{std}$ and $\hat{A}_{new}$ for the standard and the new operators for the SO(10) model, are given in detail in Ref. (6). Here, I will present only the vresults. It is found that the four amplitudes $\hat{A}_{std}(\bar{\nu}_\tau K^+)$, $\hat{A}_{std}(\bar{\nu}_\mu K^+)$, $\hat{A}_{new}(\bar{\nu}_\tau K^+)$ and $\hat{A}_{new}(\bar{\nu}_\mu K^+)$ are in fact very comparable to each other, within about a factor of two, either way. Since there is no reason to expect a near cancellation between the standard and the new operators, especially for both $\bar{\nu}_\tau K^+$ and $\bar{\nu}_\mu K^+$ modes, we expect the net amplitude (standard + new) to be in the range exhibited by either one. Following Ref. (6), I therefore present the contributions from the standard and the new operators separately. Using the upper limit on $M_{eff} \geq 3 \times 10^{18}$ GeV (see Appendix), we obtain a lower limit for the standard proton decay amplitude given by

$$\hat{A}(\bar{\nu}_\tau K^+)_{std} \geq$$
$$\begin{bmatrix} (7 \times 10^{-24} GeV^{-1})(1/6 - 1/4) & \text{case I} \\ (3 \times 10^{-24} GeV^{-1})(1/6 - 1/2) & \text{case II} \end{bmatrix} \quad (46)$$

Substituting into Eq. (41) and adding the contribution from the second competing mode $\bar{\nu}_\mu K^+$, with a typical branching ratio $R \approx 0.3$, we obtain

$$\Gamma^{-1}(\bar{\nu}K^+)_{std} \leq$$
$$\begin{bmatrix} (3 \times 10^{31} \, yrs.)(1.6 - 0.7) \\ (6.8 \times 10^{31} \, yrs.)(4 - 0.44) \end{bmatrix} (32 - 1/32) \quad (47)$$

The upper and lower entries in Eqs. (46) and (47) correspond to the cases I and II of the fermion mass-matrix - i.e. $\varepsilon' \neq 0$ and $\varepsilon' = 0$ - respectively, (see Eq. (34)). The uncertainty shown inside the square brackets correspond to that in the relative phases of the different contributions. The uncertainty of (32 to 1/32) arises from that in $\beta_H$, $(m_{\tilde{W}}/m_{\tilde{q}})$ and $m_{\tilde{q}}$ (see Eq. (40)). Thus we find that for MSSM embedded in SO(10), the inverse partial proton decay rate should satisfy:

$$\Gamma^{-1}(p \to \bar{\nu}K^+)_{std} \leq$$
$$\begin{bmatrix} 3 \times 10^{31 \pm 1.7} \, yrs. \\ 6.8 \times 10^{31^{+2.1}_{-1.5}} \, yrs. \end{bmatrix} \leq \begin{bmatrix} 1.5 \times 10^{33} \, yrs. \\ 7 \times 10^{33} \, yrs. \end{bmatrix}$$
$$\text{For SO(10)}. \quad (48)$$

The central value of the upper limit in Eq. (48) corresponds to taking the upper limit on $M_{eff}$. The uncertainties of matrix element and spectrum are reflected in the exponents. The uncertainty in the most sensitive entry of the fermion mass matrix - i.e. $\varepsilon'$ - is fully incorporated (as regards obtaining an upper limit on the lifetime) by going from case I to case II. Note that this increases the lifetime by almost a factor of five. Any non-vanishing value of $\varepsilon'$

would only shorten the lifetime compared to case II. In this sense, the larger of the two upper limits quoted above is rather conservative.

Evaluating similarly the contributions from only the new operators, we obtain:

$$\Gamma^{-1}(\bar{\nu}K^+)_{new} \approx (3 \times 10^{31} \, yrs) \times [16 \text{-} 1/1.7]\{32 \text{-} 1/32\}. \quad (49)$$

Note that this contribution is independent of $M_{eff}$. It turns out that it is also insensitive to $\varepsilon'$; thus it is nearly the same for cases I and II. Allowing for a net uncertainty at the upper end by as much as a factor of 20 to 200, arising jointly from the square and the curly brackets, i.e., without going to extreme ends of all parameters, the new operators related to neutrino masses, by themselves, lead to a proton decay lifetime bounded by:

$$\Gamma^{-1}(\bar{\nu}K^+)_{new}^{expected} \leq (0.6\text{-}6) \times 10^{33} \, yrs$$
$$(SO(10) \text{ or string } G(224)) \quad (50)$$

It should be stressed that while the standard $d = 5$ operators would be absent for a string-derived G(224)-model, the new $d = 5$ operators, related to the Majorana masses of the RH neutrinos and the CKM mixings, would still be present for such a model. *Thus our expectations for the proton decay lifetime (as shown in Eq. (50)) and the prominence of the $\mu^+K^0$ mode (see below) hold for a string-derived G(224)-model, just as they do for SO(10).*

### 6.4. The Charged Lepton Decay Mode $(p \to \mu^+ K^0)$

I now note a distinguishing feature of the SO(10) or the G(224) model presented here. Allowing for uncertainties in the way the standard and the new operators can combine with each other for the three leading modes i.e. $\bar{\nu}_\tau K^+$, $\bar{\nu}_\mu K^+$ and $\mu^+ K^0$, we obtain (see Ref. (6) for details):

$$B(\mu^+ K^0)_{std+new} \approx [1\% \text{ to } 50\%] \rho$$
$$(SO(10) \text{ or string } G(224)) \quad (51)$$

where $\rho$ denotes the ratio of the squares of relevant matrix elements for the $\mu^+K^0$ and $\bar{\nu}K^+$ modes. In the absence of a reliable lattice calculation for the $\bar{\nu}K^+$ mode (50), one should remain open to the possibility of $\rho \approx 1/2$ to 1 (say). We find that for a large range of parameters, the branching ratio $B(\mu^+K^0)$ can lie in the range of 20 to 40% (if $\rho \approx 1$). This prominence of the $\mu^+K^0$ mode for the SO(10)/G(224) model is primarily due to contributions from the new operators. This contrasts sharply with the minimal SU(5) model, in which the $\mu^+K^0$ mode is expected to have a branching ratio of only about $10^{-3}$. In short, prominence of the $\mu^+K^0$ mode, if seen, would clearly show the relevance of the new operators, and thereby reveal the proposed link between neutrino masses and proton decay (21).

### 6.5. Section Summary

In summary, our study of proton decay has been carried out within the SO(10) or the G(224)-framework[12], with special attention paid to its dependence on fermion masses and threshold effects. The study strongly suggests an upperlimit on proton lifetime, given by

$$\tau_{proton} \leq (1/2 \text{-} 1) \times 10^{34} \, yrs, \quad (52)$$

with $\bar{\nu}K^+$ being the dominant decay mode. Although there are uncertainties in the matrix element, in the SUSY-spectrum, and in certain sensitive elements of the fermion mass matrix, especially $\varepsilon'$ (see Eq. (48) for predictions in cases I versus II), this upper limit is obtained by allowing for a generous range in these parameters and stretching all of them in the same direction so as to extend proton lifetime. In this sense, while the predicted lifetime spans a wide range, the upper limit quoted above is quite conservative. In turn, it provides a clear reason to expect that the discovery of proton decay should be imminent. The implication of this prediction for a next-generation detector is emphasized in the next section.

## CONCLUDING REMARKS

The preceding sections show that one is now in possession of a set of facts, which may be viewed as the matching pieces of a puzzle; in that all of them can be resolved by just one idea - that of grand unification. These include : (i) the observed family-structure, (ii) meeting of the three gauge coulings, (iii) neutrino oscillations; in particular the mass of $\nu_\tau$ (suggested by SuperK), (iv) the intricate pattern of the masses and mixings of all the fermions, including the smallness of $V_{bc}$ and the largeness of $\theta^{osc}_{\nu_\mu\nu_\tau}$, and (v) the need for $B\text{-}L$ to implement baryogenesis. All these pieces fit beautifully together within a single puzzle board framed by supersymmetric unification, based on SO(10) or a string-unified G(224)-symmetry.

The one and the most notable piece of the puzzle still missing, however, is proton decay. Based on a systematic study of this process within the SO(10)/G(224)-framework (6), that is clearly supported by the pieces

---

[12] As described in Secs. 3 and 5.

listed above, I have argued here that proton lifetime, conservatively speaking, should not exceed about $(1/2 - 1) \times 10^{34}$ yrs. So, unless the fitting of all the pieces is a mere coincidence, and I believe that that is highly unlikely, discovery of proton decay should be around the corner. In particular, as mentioned in the Introduction, we expect that candidate events should be observed in the near future already at SuperK. However, allowing for the possibility that proton lifetime may well be near the upper limit stated above, a next-generation detector providing a net gain in sensitivity by a factor five to ten, compared to SuperK, would be needed to produce real events and distinguish them unambiguously from the background. Such an improved detector would of course be essential to study the branching ratios of certain crucial though subdominant decay modes such as the $\mu^+ K^0$.

The reason for pleading for such improved searches is that proton decay would provide us with a wealth of knowledge about physics at truly short distances ($< 10^{-30}$ cm), which cannot be gained by any other means. Specifically, the observation of proton decay, at a rate suggested above, with $\bar{\nu} K^+$ mode being dominant, would not only reveal the underlying unity of quarks and leptons but also the relevance of supersymmetry. It would also confirm a unification of the fundamental forces at a scale of order $2 \times 10^{16}$ GeV. Furthermore, prominence of the $\mu^+ K^0$ mode, if seen, would have even deeper significance, in that in addition to supporting the three features mentioned above, it would also reveal the link between neutrino masses and proton decay, as discussed in Sec. 6. *In this sense, the role of proton decay in probing into physics at the most fundamental level is unique*. In view of how valuable such a probe would be and the fact that the predicted upper limit on the proton lifetime is only a factor of three to six higher than the empirical lower limit, the argument in favor of building an improved detector seems compelling.

To conclude, the discovery of proton decay would undoubtedly constitute a landmark in the history of physics. It would provide the last, missing piece of gauge unification and would shed light on how such a unification may be extended to include gravity.

**Acknowledgements** : I would like to thank Kaladi S. Babu and Frank Wilczek for a most enjoyable collaboration, and Joseph Sucher for valuable discussions. I would also like to thank the organizers of the NNN99 workshop, especially Chang Kee Jung, Milind Diwan and Hank Sobel, for arranging a stimulating meeting and also for the kind hospitality. The research presented here is supported in part by DOE grant no. DE-FG02-96ER-41015.

# APPENDIX

In supersymmetric SO(10), a natural doublet–triplet splitting can be achieved by coupling the adjoint Higgs $\mathbf{45_H}$ to a $\mathbf{10_H}$ and a $\mathbf{10'_H}$, with $\mathbf{45_H}$ acquiring a unification–scale VEV in the $B$-$L$ direction (51): $\langle \mathbf{45_H} \rangle = (a, a, a, 0, 0) \times \tau_2$ with $a \sim M_U$. As discussed in Section 2, to generate CKM mixing for fermions we require $(\mathbf{16_H})_d$ to acquire a VEV of the electroweak scale. To ensure accurate gauge coupling unification, the effective low energy theory should not contain split multiplets beyond those of MSSM. Thus the MSSM Higgs doublets must be linear combinations of the $SU(2)_L$ doublets in $\mathbf{10_H}$ and $\mathbf{16_H}$. A simple set of superpotential terms that ensures this and incorporates doublet-triplet splitting is (6):

$$W_H = \lambda \mathbf{10_H} \mathbf{45_H} \mathbf{10'_H} + M_{10} \mathbf{10'_H}^2 + \lambda' \overline{\mathbf{16}}_H \overline{\mathbf{16}}_H \mathbf{10}_H + M_{16} \mathbf{16_H} \overline{\mathbf{16}}_H \,. \quad (A1)$$

A complete superpotential for $\mathbf{45_H}$, $\mathbf{16_H}$, $\overline{\mathbf{16}}_H$, $\mathbf{10_H}$, $\mathbf{10'_H}$ and possibly other fields, which ensure that $\mathbf{45_H}$, $\mathbf{16_H}$ and $\overline{\mathbf{16}}_H$ acquire unification scale VEVs with $\langle \mathbf{45_H} \rangle$ being along the $(B$-$L)$ direction, that exactly two Higgs doublets $(H_u, H_d)$ remain light, with $H_d$ being a linear combination of $(\mathbf{10_H})_d$ and $(\mathbf{16_H})_d$, and that there are no unwanted pseudoGoldstone bosons, can be constructed. With $\langle \mathbf{45_H} \rangle$ in the $B$-$L$ direction, it does not contribute to the Higgs doublet mass matrix, so one pair of Higgs doublet remains light, while all triplets acquire unification scale masses. The light MSSM Higgs doublets are

$$H_u = \mathbf{10}_u, \quad H_d = \cos\gamma \mathbf{10}_d + \sin\gamma \mathbf{16}_d, \quad (A2)$$

with $\tan\gamma \equiv \lambda' \langle \overline{\mathbf{16}}_H \rangle / M_{16}$. Consequently, $\langle \mathbf{10} \rangle_d = (\cos\gamma) v_d$, $\langle \mathbf{16}_d \rangle = (\sin\gamma) v_d$, with $\langle H_d \rangle = v_d$ and $\langle \mathbf{16}_d \rangle$ and $\langle \mathbf{10}_d \rangle$ denoting the electroweak VEVs of those multiplets. Note that $H_u$ is purely in $\mathbf{10_H}$ and that $\langle \mathbf{10}_d \rangle^2 + \langle \mathbf{16}_d \rangle^2 = v_d^2$. This mechanism of doublet-triplet (DT) splitting is rather unique for the minimal Higgs systems in that it meets the requirements of both D-T splitting and CKM-mixing. In turn, it has three special consequences:

(i) It modifies the familiar SO(10)-relation $\tan\beta \equiv v_u/v_d \approx m_t/m_b \approx 60$ to:

$$\tan\beta / \cos\gamma \approx m_t/m_b \approx 60 \quad (A3)$$

As a result, even low to moderate values of $\tan\beta \approx 3$ to 10 (say) are perfectly allowed in SO(10) (corresponding to $\cos\gamma \approx 1/20$ to $1/6$).

(ii) The most important consequence of the DT-splitting mechanism outlined above is this: In contrast to SU(5), for which the strengths of the standard d=5 operators are proportional to $(M_{H_C})^{-1}$ (where $M_{H_C} \sim few \times 10^{16}$ GeV (see Eq. (44)), for the SO(10)-model, they become proportional to $M_{eff}^{-1}$, where $M_{eff} = (\lambda a)^2/M_{10'} \sim$

$M_U^2/M_{10'}$. $M_{10'}$ can be naturally smaller (due to flavor symmetries) than $M_U$ and thus $M_{eff}$ correspondingly larger than $M_U$ by one to two orders of magnitude (see Ref. (6)). Now the proton decay amplitudes for SO(10) in fact possess an intrinsic enhancement compared to those for SU(5), owing primarily due to differences in their Yukawa couplings for the up sector (see Appendix C in Ref. (6)). As a result, these larger values of $M_{eff} \sim 10^{18}$ GeV are in fact needed for the SO(10)-model to be compatible with the observed limit on the proton lifetime. At the same time, being bounded above (see below), they allow optimism as regards future observation of proton decay.

(iii) $M_{eff}$ gets bounded above by considerations of coupling unification and GUT-scale threshold effects. Owing to mixing between $10_d$ and $16_d$ (see Eq. (A2)), the threshold correction to $\alpha_3(m_z)$ due to doublet-triplet splitting becomes proportional to $\ln(M_{eff}\cos\gamma/M_U)$. Inclusion of this correction and those due to splittings within the gauge and the Higgs multiplets (i.e. $45_H$, $16_H$, and $\overline{16}_H$)[13], together with the observed degree of coupling unification allows us to obtain a conservative upper limit on $M_{eff}$, given by (6):

$$M_{eff} \leq 3 \times 10^{18}\,\text{GeV}. \qquad (A4)$$

This in turn helps provide an upper limit on the expected proton decay lifetime (see text).

## REFERENCES

1. J. C. Pati and Abdus Salam; Proc. 15th High Energy Conference, Batavia, reported by J. D. Bjorken, Vol. 2, p. 301 (1972); Phys. Rev. **8**, 1240 (1973).

2. J. C. Pati and Abdus Salam, Phys. Rev. Lett. **31**, 661 (1973); Phys. Rev. **D10**, 275 (1974).

3. H. Georgi and S. L. Glashow, Phys. Rev. Lett. **32**, 438 (1974).

4. H. Georgi, H. Quinn and S. Weinberg, Phys. Rev. Lett. **33**, 451 (1974).

5. Y. A. Gelfand and E. S. Likhtman, JETP Lett. **13**, 323 (1971). J. Wess and B. Zumino, Nucl. Phys. **B70**, 139 (1974); D. Volkov and V. P. Akulov, JETP Lett. **16**, 438 (1972).

6. K. S. Babu, J. C. Pati and F. Wilczek, "*Fermion Masses, Neutrino Oscillations and Proton Decay in the Light of the SuperKamiokande*" hep-ph/981538V3; Nucl. Phys. B (to appear).

7. F. Reines, C. L. Cowan Jr. and M. Goldhaber, Phys. Rev. **96**, 1157 (1954); G. N. Flerov et al., Sov. Phys. Dokl. **3**, 79 (1958); W. R. Kropp Jr. and F. Reines, Phys. Rev. **B 137**, 740 (1965); F. Reines and M. F. Crouch, Phys. Rev. Lett. **32**, 493 (1974).

8. KGF: M. R. Krishnaswamy et al., Pramana, **19**, 525 (1982); NUSEX : G. Battisoni et al., Phys. Lett. **115 B**, 4, 349 (1982).

9. SuperK Collaboration : Y. Hayato, Proc. ICHEP, Vancouver (1998); M. Earl, NNN2000 Workshop, Irvine, Calif. (Feb. 25, 2000).

10. SuperK Collaboration : M. Shiozawa, Ph. D. Thesis, ICRR-Report-458-2000-2 (Nov. 26, 1999); M. Earl (see Ref (9)).

11. H. Georgi, in Particles and Fields, Ed. by C. Carlson (AIP, NY, 1975), p. 575; H. Fritzsch and P. Minkowski, Ann. Phys. **93**, 193 (1975).

12. For recent work, see P. Langacker and M. Luo, Phys. Rev. **D 44**, 817 (1991); U. Amaldi, W. de Boer and H. Furtenau, Phys. Rev. Lett. **B 260**, 131 (1991); F. Anselmo, L. Cifarelli, A. Peterman and A. Zichichi, Nuov. Cim. **A 104** 1817 (1991). The essential features pertaining to coupling unification in SUSY GUTS were noted earlier by S. Dimopoulos, S. Raby and F. Wilczek, Phys. Rev. **D 24**, 1681 (1981); W. Marciano and G. Senjanovic, Phys. Rev. **D 25**, 3092 (1982); M. Einhorn and D. R. T. Jones, Nucl. Phys. **B 196**, 475 (1982).

13. M. Green and J. H. Schwarz, Phys. Lett. **149B**, 117 (1984); D. J. Gross, J. A. Harvey, E. Martinec and R. Rohm, Phys. Rev. Lett. **54**, 502 (1985); P. Candelas, G. T. Horowitz, A. Strominger and E. Witten, Nucl. Phys. **B 258**, 46 (1985).

14. For a few pioneering papers on string-duality and M-theory, relevant to gauge-coupling unification, see E. Witten, Nucl. Phys. **B 443**, 85 (1995) and P. Horava and E. Witten, Nucl. Phys. **B 460**, 506 (1996). For reviews, see e.g. J. Polchinski, hep-th/9511157; and A. Sen, hep-th/9802051, and references therein.

15. SuperKamiokande Collaboration, Y. Fukuda et. al., Phys. Rev. Lett. **81**, 1562 (1998).

16. J. C. Pati, "Implications of the SuperKamiokande Result on the Nature of New Physics", in Neutrino 98, Takayama, Japan, June 98, hep-ph/9807315; Nucl. Phys. B (Proc. Suppl.) **77**, 299 (1999).

17. Caso et al., Particle Data Group, Review of Particle Physics, The European Physics Journal C, **3**, 1 (1998).

18. A. D. Sakharov, Pisma Zh. Eksp. Teor. Fiz. **5**, 32 (1967).

19. V. Kuzmin, Va. Rubakov and M. Shaposhnikov, Phys. Lett **BM155**, 36 (1985).

20. M. Fukugita and T. Yanagida, Phys. Lett. **B 174**, 45 (1986); M. A. Luty, Phys. Rev. **D 45**, 455 (1992); W. Buchmuller and M. Plumacher, hep-ph/9608308.

21. K. S. Babu, J. C. Pati and F. Wilczek, "*Suggested New Modes in Supersymmetric Proton Decay*", Phys. Lett. **B 423**, 337 (1998).

---

[13] The correction to $\alpha_3(m_z)$ due to Planck scale physics through the effective operator $F_{\mu\nu}F^{\nu\mu}45_H/M$ vanishes due to antisymmetry in the SO(10)-contraction.

22. N. Sakai and T. Yanagida, Nucl. Phys. **B 197**, 533 (1982); S. Weinberg, Phys. Rev. **D 26**, 287 (1982).

23. R. N. Mohapatra and J. C. Pati, Phys. Rev. **D 11**, 566, 2558 (1975); G. Senjanovic and R. N. Mohapatra, Phys. Rev. **D 12**, 1502 (1975).

24. F. Gürsey, P. Ramond and P. Sikivie, Phys. Lett. **B 60**, 177 (1976).

25. For recent reviews see e.g. P. Langacker and N. Polonsky, Phys. Rev. **D 47**, 4028 (1993) and references therein.

26. See e.g., Refs. (25) and (12).

27. P. Ginsparg, Phys. Lett. **B 197**, 139 (1987); V. S. Kaplunovsky, Nucl. Phys. **B 307**, 145 (1988); Erratum: *ibid.* **B 382**, 436 (1992).

28. E. Witten, hep-th/9602070.

29. J. C. Pati and K. S. Babu, "*The Problems of Unification – Mismatch and Low $\alpha_3$ : A Solution with Light Vector-Like Matter*", Phys. Lett. **B 384**, 140 (1996).

30. For a recent discussion, see K. Dienes, Phys. Rep. **287**, 447 (1997), and references therein.

31. J. C. Pati, hep-ph/9811442; Proc. Salam Memorial Meeting (1998), World Scientific; Int'l Journal of Modern Physics A, vol. 14, 2949 (1999); J. C. Pati; To appear in the Proc. of Dubna Conf. (July 1999) and Johns Hopkins Workshop (June, 1999), UMD-PP-0058.

32. See e.g. D. Lewellen, Nucl. Phys. **B 337**, 61 (1990); A. Font, L. Ibanez and F. Quevedo, Nucl. Phys. **B 345**, 389 (1990); S. Chaudhari, G. Hockney and J. Lykken, Nucl. Phys. **B 456**, 89 (1995), and Z. Kakushadze and S.H. Tye, hep-th/9605221, and hep-th/9609027; and references therein.

33. I. Antoniadis, G. Leontaris and J. Rizos, Phys. Lett **B245**, 161 (1990); G. Leontaris, Phys. Lett. **B 372**, 212 (1996)..

34. A. Faraggi, Phys. Lett. **B 278**, 131 (1992); Phys. Lett. **B 274**, 47 (1992); Nucl. Phys. **B 403**, 101 (1993); A. Faraggi and E. Halyo, Nucl. Phys. **B 416**, 63 (1994).

35. A. Faraggi and J.C. Pati, "*A Family Universal Anomalous U(1) in String Models as the Origin of Supersymmetry Breaking and Squark-degeneracy*", hep-ph/9712516v3, Nucl. Phys. **B 256**, 526 (1998).

36. K.S. Babu and J.C. Pati, "*Towards A Resolution of the Supersymmetric CP Problem Through Flavor and Left-Right Symmetries*", UMD-PP-0061 (To be submitted to Phys. Rev.).

37. J.C. Pati, "*The Essential Role of String Derived Symmetries in Ensuring Proton Stability and Light Neutrino Masses*", hep-ph/9607446, Phys. Lett. **B 388**, 532 (1996).

38. See e.g. K. R. Dienes and J. March-Russell, hep-th/9604112; K. R. Dienes, hep-ph/9606467.

39. M. Gell-Mann, P. Ramond and R. Slansky, in: *Supergravity*, eds. F. van Nieuwenhuizen and D. Freedman (Amsterdam, North Holland, 1979) p. 315; T. Yanagida, in: *Workshop on the Unified Theory and Baryon Number in the Universe*, eds. O. Sawada and A. Sugamoto (KEK, Tsukuba) 95 (1979); R. N. Mohapatra and G. Senjanovic, Phys. Rev. Lett. **44**, 912 (1980).

40. S. Weinberg, I.I. Rabi Festschrift (1977); F. Wilczek and A. Zee, Phys. Lett. **70 B**, 418 (1977); H. Fritzsch, Phys. Lett. **70 B**, 436 (1977).

41. H. Georgi and C. Jarlskog, Phys. Lett. **B 86**, 297 (1979).

42. For a related but different SO(10) model see C. Albright, K.S. Babu and S.M. Barr, Phys. Rev. Lett. **81**, 1167 (1998).

43. See e.g. R. Gupta and T. Bhattacharya, Nucl. Phys. Proc. Suppl. **53**, 292 (1997); and Nucl. Phys. Proc. Suppl. **63**, 45 (1998).

44. See e.g., D. M. Pierce, J. A. Bagger, K. Matchev and R. Zhang, Nucl. Phys. **B 491**, 3 (1997); K. S. Babu and C. Kolda, hep-ph/9811308, and references therein.

45. S. Mikhayev and A. Smirnov, Nuovo Cim. **9 C**, 17 (1986); L. Wolfenstein, Phys. Rev. **D 17**, 2369 (1978). For a comprehensive analysis of the solar neutrino data, see J. N. Bahcall, P. I. Krastev and A. Yu. Smirnov, Phys.Lett. B, 477 ,March 30, 2000;hep-ph/9911248 and references there in.

46. S. Dimopoulos, S. Raby and F. Wilczek, Phys. Lett. **B112**, 133 (1982); J. Ellis, D.V. Nanopoulos and S. Rudaz, Nucl. Phys. **B 202**, 43 (1982).

47. P. Nath, A.H. Chemseddine and R. Arnowitt, Phys. Rev. **D 32**, 2348 (1985);P. Nath and R. Arnowitt, hep-ph/9708469.

48. J. Hisano, H. Murayama and T. Yanagida, Nucl. Phys. **B 402**, 46 (1993).

49. K.S. Babu and S.M. Barr, Phys. Rev. **D 50**, 3529 (1994); **D 51**, 2463 (1995).

50. For a recent work, comparing the results of lattice and chiral lagrangian-calculations for the $p \to \pi^0, p \to \pi^+$ and $p \to K^0$ modes, see N. Tatsui et al (JLQCD collaboration), hep-lat/9809151.

51. S. Dimopoulos and F. Wilczek, Report No. NSF-ITP-82-07 (1981), in *The unity of fundamental interactions*, Proc. of the 19th Course of the International School on Subnuclear Physics, Erice, Italy, Erice, Italy, 1981, Plenum Press, New York (Ed. A. Zichichi); K.S. Babu and S.M. Barr, Phys. Rev. **D 48**, 5354 (1993).

# From Neutrino Masses to Proton Decay

P. Ramond

*Institute for Fundamental Theory*
*Department of Physics,*
*University of Florida*
*Gainesville, Fl 32611*

**Abstract.** Current theoretical and experimental issues are reviewed in the light of the recent SuperKamiokande discovery. By using quark-lepton symmetries, derived from Grand Unification and/or string theories, we show how to determine the necessary neutrino parameters. In addition, the seesaw neutrino masses set the scale for the proton decay operators by "measuring" the standard model cut-off. The SuperKamiokande values suggest that proton decay is likely to be observed early in the XXIst Century.

## NEUTRINO STORY

Once it became apparent that the spectrum of β electrons was continuous (1, 2), something drastic had to be done! In December 1930, in a letter that starts with typical panache, "*Dear Radioactive Ladies and Gentlemen...*", W. Pauli puts forward a "*desperate*" way out: there is a companion neutral particle to the β electron. Thus earthlings became aware of the *neutrino*, so named in 1933 by Fermi (Pauli's original name, *neutron*, superseded by Chadwick's discovery of a heavy neutral particle), implying that there is something small about it, specifically its mass, although nobody at that time thought it was *that* small.

Fifteen years later, B. Pontecorvo (3) proposes the unthinkable, that neutrinos can be detected: an electron neutrino that hits a $^{37}Cl$ atom will transform it into the inert radioactive gas $^{37}Ar$, which can be stored and then detected through radioactive decay. Pontecorvo did not publish the report, perhaps because of the times, or because Fermi thought the idea ingenious but not immediately relevant.

In 1956, using a scintillation counter experiment they had proposed three years earlier (4), Cowan and Reines (5) discover electron antineutrinos through the reaction $\bar{\nu}_e + p \to e^+ + n$. Cowan passed away before 1995, the year Fred Reines was awarded the Nobel Prize for their discovery. There emerge two lessons in neutrino physics: not only is patience required but also longevity: it took 26 years from birth to detection and then another 39 for the Nobel Committee to recognize the achievement! This should encourage physicists to train their children at the earliest age to follow their footsteps at the earliest possible age, in order to establish dynasties of neutrino physicists. Perhaps then Nobel prizes will be awarded to scientific families?

In 1956, it was rumored that Davis (6), following Pontecorvo's proposal, had found evidence for neutrinos coming from a pile, and Pontecorvo (7), influenced by the recent work of Gell-Mann and Pais, theorized that an antineutrino produced in the Savannah reactor could oscillate into a neutrino and be detected. The rumor went away, but the idea of neutrino oscillations was born; it has remained with us ever since.

Neutrinos give up their secrets very grudgingly: its helicity was measured in 1958 by M. Goldhaber (8), but it took 40 more years for experimentalists to produce convincing evidence for its mass. The second neutrino, the muon neutrino is detected (9) in 1962, (long anticipated by theorists Inouë and Sakata in 1943 (10)). This time things went a bit faster as it took only 19 years from theory (1943) to discovery (1962) and 26 years to Nobel recognition (1988).

That same year, Maki, Nakagawa and Sakata (11) introduce two crucial ideas: neutrino flavors can mix, and their mixing can cause one type of neutrino to oscillate into the other (called today flavor oscillation). This is possible only if the two neutrino flavors have different masses.

In 1964, using Bahcall's result (12) of an enhanced capture rate of $^8B$ neutrinos through an excited state of $^{37}Ar$, Davis (13) proposes to search for $^8B$ solar neutrinos using a 100,000 gallon tank of cleaning fluid deep underground. Soon after, R. Davis starts his epochal experiment at the Homestake mine, marking the beginning of the solar neutrino watch which continues to this day. In 1968, Davis et al reported (14) a deficit in the solar neutrino flux, a result that stands to this day as a truly remarkable experimental *tour de force*. Shortly after, Gribov and Pontecorvo (15) interpreted the deficit as evidence for neutrino oscillations.

In the early 1970's, with the idea of quark-lepton symmetries (16, 17) suggests that the proton could be unstable. This brings about the construction of underground detectors, large enough to monitor many protons, and instrumentalized to detect the Čerenkov light emitted by its decay products. By the middle 1980's, several such detectors are in place. They fail to detect proton decay, but in a remarkable serendipitous turn of events, 150,000 years earlier, a supernova erupted in the large Magellanic Cloud, and in 1987, its burst of neutrinos was detected in these detectors! All of a sudden, proton decay detectors turn their attention to neutrinos, while to this day still waiting for its protons to decay! Today, these detectors have shown great success in measuring the effects of solar and atmospheric neutrinos. They continue their unheralded watch for signs of proton decay, reassured in the knowledge that lepton number and baryon number violations are connected in most theories, leading to correlations between neutrino masses and proton decay rates.

## STANDARD MODEL NEUTRINOS

The standard model of electro-weak and strong interactions contains three left-handed neutrinos. The three neutrinos are represented by two-components Weyl spinors, $\nu_i$, $i = e, \mu, \tau$, each describing a left-handed fermion (right-handed antifermion). As the upper components of weak isodoublets $L_i$, they have $I_{3W} = 1/2$, and a unit of the global $i$th lepton number.

These standard model neutrinos are strictly massless. The only Lorentz scalar made out of these neutrinos is the Majorana mass, of the form $\nu_i^t \nu_j$; it has the quantum numbers of a weak isotriplet, with third component $I_{3W} = 1$, as well as two units of total lepton number. Higgs isotriplet with two units of lepton number could generate neutrino Majorana masses, but there is no such higgs in the Standard Model: there are no tree-level neutrino masses in the standard model.

Quantum corrections, however, are not limited to renormalizable couplings, and it is easy to make a weak isotriplet out of two isodoublets, yielding the $SU(2) \times U(1)$ invariant $L_i^t \vec{\tau} L_j \cdot H^t \vec{\tau} H$, where $H$ is the Higgs doublet. As this term is not invariant under lepton number, it is not be generated in perturbation theory. Thus the important conclusion: *The standard model neutrinos are kept massless by global chiral lepton number symmetry*. The detection of neutrino masses is therefore *a tangible indication of physics beyond the standard model*.

## EXPERIMENTAL ISSUES

From the solar neutrino deficit to the spectacular result from SuperKamiokande, experiments suggest that neutrinos have masses, providing the first credible evidence for physics beyond the standard model. As we stand at the end of this Century, there remains several burning issues in neutrino physics that can be settled by future experiments:

- The origin of the Solar Neutrino Deficit

  This is currently being addressed by SuperK, in their measurement of the shape of the $^8B$ spectrum, of day-night asymmetry and of the seasonal variation of the neutrino flux. Their reach will soon be improved by lowering their threshold energy.

  SNO is joining the hunt, and is expected to provide a more accurate measurement of the Boron flux. Its *raison d'être*, however, is the ability to measure neutral current interactions. If there are no sterile neutrinos, we might have a flavor independent measurement of the solar neutrino flux, while measuring at the same time the electron neutrino flux!

  This experiment will be joined by BOREXINO, designed to measure neutrinos from the $^7Be$ capture. These neutrinos are suppressed in the small angle MSW solution, which could explain the results from the $p-p$ solar neutrino experiments and those that measure the Boron neutrinos.

- Atmospheric Neutrinos

  Here, there are several long baseline experiments to monitor muon neutrino beams and corroborate the SuperK results. The first, called K2K, already in progress, sends a beam from KEK to SuperK. Another, called MINOS, will monitor a FermiLab neutrino beam at the Soudan mine, 730 km away. A Third experiment under consideration would send a CERN beam towards the Gran Sasso laboratory (also about 730 km away!). Eventually, these experiments hope to detect the appearance of a tau neutrino.

This brief survey of upcoming experiments in neutrino physics is intended to give a flavor of things to come. These experiments will not only measure neutrino parameters (masses and mixing angles), but will help us answer fundamental questions about the nature of neutrinos. But the future of neutrino detectors may be even brighter. Many of us expect them to detect proton decay, thus realizing the kinship between leptons and quarks. There is even increasing talk of producing intense neutrino beams in muon storage rings, and at this workshop of building a mammoth proton decay/neutrino detector!

# NEUTRINO MASSES

Neutrinos must be extraordinarily light: experiments indicate $m_{\nu_e} < 10$ eV, $m_{\nu_\mu} < 170$ keV, $m_{\nu_\tau} < 18$ MeV (18), and any model of neutrino masses must explain this suppression.

The natural way to generate neutrinos masses is to introduce for each one its electroweak singlet Dirac partner, $\overline{N}_i$. These appear naturally in the Grand Unified group $SO(10)$ where they complete each family into its spinor representation. Neutrino Dirac masses will then be generated by the couplings $L_i \overline{N}_j H$ after electroweak breaking. However, unless there are extraordinary suppressions, these couplings generate masses that are way too big, of the same order of magnitude as the masses of the charged elementary particles $m \sim \Delta I_w = 1/2$.

Based on recent ideas from string theory, it has been proposed (19) that the world of four dimensions is in fact a "brane" immersed in a higher dimensional space. In this view, all fields with electroweak quantum numbers live on the brane, while standard model singlet fields can live on the "bulk" as well. One such field is the graviton, others could be the right-handed neutrinos. Their couplings to the brane are reduced by geometrical factors, and the smallness of neutrino masses is due to the naturally small coupling between brane and bulk fields.

In the absence of any credible dynamics for the physics of the bulk, we think that *"one neutrino on the brane is worth two in the bulk"*. We take the more conservative approach where the bulk does opens up, but at much shorter scales. One indication of such a scale is that at which the gauge couplings unify, the other is given by the value of neutrino masses. this is achieved by introducing Majorana mass terms $\overline{N}_i \overline{N}_j$ for the right-handed neutrinos. The masses of these new degrees of freedom are arbitrary, as they have no electroweak quantum numbers, $M \sim \Delta I_w = 0$. If they are much larger than the electroweak scale, the neutrino masses are suppressed relative to that of their charged counterparts by the ratio of the electroweak scale to that new scale: the mass matrix (in $3 \times 3$ block form) is

$$\begin{pmatrix} 0 & m \\ m & M \end{pmatrix}, \quad (1)$$

leading, for each family, to one small and one large eigenvalue

$$m_\nu \sim m \cdot \frac{m}{M} \sim \left(\Delta I_w = \frac{1}{2}\right) \cdot \left(\frac{\Delta I_w = \frac{1}{2}}{\Delta I_w = 0}\right). \quad (2)$$

This seesaw mechanism (20) provides a natural explanation for small neutrino masses as long as lepton number is broken at a large scale $M$. With $M$ around the energy at which the gauge couplings unify, this yields neutrino masses at or below tenths of eVs, consistent with the SuperK results.

The lepton flavor mixing comes from the diagonalization of the charged lepton Yukawa couplings, and of the neutrino mass matrix. From the charged lepton Yukawas, we obtain $\mathcal{U}_e$, the unitary matrix that rotates the lepton doublets $L_i$. From the neutrino Majorana matrix, we obtain $\mathcal{U}_\nu$, the matrix that diagonalizes the Majorana mass matrix. The $6 \times 6$ seesaw Majorana matrix can be written in $3 \times 3$ block form

$$\mathcal{M} = \mathcal{V}_\nu^t \mathcal{D} \mathcal{V}_\nu \sim \begin{pmatrix} \mathcal{U}_{\nu\nu} & \varepsilon \mathcal{U}_{\nu N} \\ \varepsilon \mathcal{U}_{N\nu}^t & \mathcal{U}_{NN} \end{pmatrix}, \quad (3)$$

where $\varepsilon$ is the tiny ratio of the electroweak to lepton number violating scales, and $\mathcal{D} = \text{diag}(\varepsilon^2 \mathcal{D}_\nu, \mathcal{D}_N)$, is a diagonal matrix. $\mathcal{D}_\nu$ contains the three neutrino masses, and $\varepsilon^2$ is the seesaw suppression. The weak charged current is then given by

$$j_\mu^+ = e_i^\dagger \sigma_\mu \mathcal{U}_{MNS}^{ij} \nu_j, \quad (4)$$

where

$$\mathcal{U}_{MNS} = \mathcal{U}_e \mathcal{U}_\nu^\dagger, \quad (5)$$

is the Maki-Nakagawa-Sakata (11) (MNS) flavor mixing matrix, the analog of the CKM matrix in the quark sector.

In the seesaw-augmented standard model, this mixing matrix is totally arbitrary. It contains, as does the CKM matrix, three rotation angles, and one CP-violating phase. In the seesaw scenario, it also contains two additional CP-violating phases which cannot be absorbed in a redefinition of the neutrino fields, because of their Majorana masses (these extra phases can be measured only in $\Delta \mathcal{L} = 2$ processes).

Unfortunately, theoretical predictions of lepton hierarchies and mixings depend very much on hitherto untested theoretical assumptions. In the quark sector, where the bulk of the experimental data resides, the theoretical origin of quark hierarchies and mixings is a mystery, although there exits many theories, but none so convincing as to offer a definitive answer to the community's satisfaction. It is therefore no surprise that there are more theories of lepton masses and mixings than there are parameters to be measured. Nevertheless, one can present the issues as questions:

- Do the right handed neutrinos have quantum numbers beyond the standard model?

- Are quarks and leptons related by grand unified theories?

- Are quarks and leptons related by anomalies?

- Are there family symmetries for quarks and leptons?

The measured numerical value of the neutrino mass difference (barring any fortuitous degeneracies), suggests through the seesaw mechanism, a mass for the right-handed neutrinos that is consistent with the scale at which the gauge couplings unify. Is this just a numerical coincidence, or should we view this as a hint for grand unification?

Grand unified theories, originally proposed as a way to treat leptons and quarks on the same footing, imply symmetries much larger than the standard model's. Implementation of these ideas necessitates a desert and supersymmetry, but also a carefully designed contingent of Higgs particles to achieve the desired symmetry breaking. That such models can be built is perhaps more of a testimony to the cleverness of theorists rather than of Nature's. Indeed with the advent of string theory, we know that the best features of grand unified theories can be preserved, as most of the symmetry breaking is achieved by geometric compactification from higher dimensions (21).

An alternative point of view is that the vanishing of chiral anomalies is necessary for consistent theories, and their cancellation is most easily achieved by assembling matter in representations of anomaly-free groups. Perhaps anomaly cancellation is more important than group structure.

Below, we present two theoretical frameworks of our work, in which one deduces the lepton mixing parameters and masses. One is ancient (22), uses the standard techniques of grand unification, but it had the virtue of *predicting* the large $\nu_\mu - \nu_\tau$ mixing observed by SuperKamiokande. The other (23) is more recent, and uses extra Abelian family symmetries to explain both quark and lepton hierarchies. It also predicted large $\nu_\mu - \nu_\tau$ mixing, while both schemes predict small $\nu_e - \nu_\mu$ mixings.

## A Grand Unified Model

The seesaw mechanism was born in the context of the grand unified group $SO(10)$, which naturally contains electroweak neutral right-handed neutrinos. Each standard model family appears in two irreducible representations of $SU(5)$. However, the predictions of this theory for Yukawa couplings is not so clear cut, and to reproduce the known quark and charged lepton hierarchies, a special but simple set of Higgs particles had to be included. In the simple scheme proposed by Georgi and Jarlskog (24), the ratios between the charged leptons and quark masses is reproduced, albeit not naturally since two Yukawa couplings, not fixed by group theory, had to be set equal. This motivated us to generalize (22) their scheme to $SO(10)$, where it is (technically) natural, which meant that we had an automatic window into neutrino masses through the seesaw. The Yukawa couplings were of the Higgs-heavy, with **126** representations, but the attitude at the time was "damn the Higgs torpedoes, and see what happens". A modern treatment would include non-renormalizable operators (25), but with similar conclusion. The model yielded the mass relations

$$m_d - m_s = 3(m_e - m_\mu); \qquad m_d m_s = m_e m_\mu; \quad (6)$$

as well as

$$m_b = m_\tau, \quad (7)$$

and mixing angles

$$V_{us} = \tan\theta_c = \sqrt{\frac{m_d}{m_s}}; \qquad V_{cb} = \sqrt{\frac{m_c}{m_t}}. \quad (8)$$

While reproducing the well-known lepton and quark mass hierarchies, it predicted a long-lived $b$ quark, contrary to the lore of the time. It also made predictions in the lepton sector, namely **maximal** $\nu_\tau - \nu_\mu$ mixing, small $\nu_e - \nu_\mu$ mixing of the order of $(m_e/m_\mu)^{1/2}$, and no $\nu_e - \nu_\tau$ mixing.

The neutral lepton masses came out to be hierarchical, but heavily dependent on the masses of the right-handed neutrinos. The electron neutrino mass came out much lighter than those of $\nu_\mu$ and $\nu_\tau$. Their numerical values depended on the top quark mass, which was then supposed to be in the tens of GeVs!

Given the present knowledge, some of the features are remarkable, such as the long-lived $b$ quark and the maximal $\nu_\tau - \nu_\mu$ mixing. On the other hand, the actual numerical value of the $b$ lifetime was off a bit, and the $\nu_e - \nu_\mu$ mixing was too large to reproduce the small angle MSW solution of the solar neutrino problem.

The lesson should be that the simplest $SO(10)$ model that fits the observed quark and charged lepton hierarchies, reproduces, at least qualitatively, the maximal mixing found by SuperK, and predicts small mixing with the electron neutrino (26).

## A Non-grand-unified Model

There is another way to generate hierarchies, based on adding extra family symmetries to the standard model, without invoking grand unification. These types of models address only the Cabibbo suppression of the Yukawa couplings, and are not as predictive as specific grand unified models. Still, they predict no Cabibbo suppression between the muon and tau neutrinos. Below, we present a pre-SuperK model (23) with those features.

The Cabibbo supression is assumed to be an indication of extra family symmetries in the standard model. The idea is that any standard model-invariant operator, such as

$Q_i \bar{d}_j H_d$, cannot be present at tree-level if there are additional symmetries under which the operator is not invariant. Simplest is to assume an Abelian symmetry, with an electroweak singlet field θ, as its order parameter. Then the interaction

$$Q_i \bar{d}_j H_d \left(\frac{\theta}{M}\right)^{n_{ij}} \qquad (9)$$

can appear in the potential as long as the family charges balance under the new symmetry. As θ acquires a *vev*, this leads to a suppression of the Yukawa couplings of the order of $\lambda^{n_{ij}}$ for each matrix element, with $\lambda = \theta/M$ identified with the Cabibbo angle, and $M$ is the natural cut-off of the effective low energy theory. As a consequence of the charge balance equation

$$X^{[d]}_{if} + n_{ij} X_\theta = 0 , \qquad (10)$$

the exponents of the suppression are related to the charge of the standard model-invariant operator (27), the sum of the charges of the fields that make up the invariant.

This simple Ansatz, together with the seesaw mechanism, implies that the family structure of the neutrino mass matrix is determined by the charges of the left-handed lepton doublet fields.

Each charged lepton Yukawa coupling $L_i \bar{N}_j H_u$, has an extra charge $X_{L_i} + X_{N_j} + X_H$, which gives the Cabibbo suppression of the $ij$ matrix element. Hence, the orders of magnitude of these couplings can be expressed as

$$\begin{pmatrix} \lambda^{l_1} & 0 & 0 \\ 0 & \lambda^{l_2} & 0 \\ 0 & 0 & \lambda^{l_3} \end{pmatrix} \hat{Y} \begin{pmatrix} \lambda^{p_1} & 0 & 0 \\ 0 & \lambda^{p_2} & 0 \\ 0 & 0 & \lambda^{p_3} \end{pmatrix}, \qquad (11)$$

where $\hat{Y}$ is a Yukawa matrix with no Cabibbo suppressions, $l_i = X_{L_i}/X_\theta$ are the charges of the left-handed doublets, and $p_i = X_{N_i}/X_\theta$, those of the singlets. The first matrix forms half of the MNS matrix. Similarly, the mass matrix for the right-handed neutrinos, $\bar{N}_i \bar{N}_j$ will be written in the form

$$\begin{pmatrix} \lambda^{p_1} & 0 & 0 \\ 0 & \lambda^{p_2} & 0 \\ 0 & 0 & \lambda^{p_3} \end{pmatrix} \mathcal{M} \begin{pmatrix} \lambda^{p_1} & 0 & 0 \\ 0 & \lambda^{p_2} & 0 \\ 0 & 0 & \lambda^{p_3} \end{pmatrix} . \qquad (12)$$

The diagonalization of the seesaw matrix is of the form

$$L_i H_u \bar{N}_j \left(\frac{1}{\bar{N}\bar{N}}\right)_{jk} \bar{N}_k H_u L_l , \qquad (13)$$

from which the Cabibbo suppression matrix from the $\bar{N}_i$ fields *cancels*, leaving us with

$$\begin{pmatrix} \lambda^{l_1} & 0 & 0 \\ 0 & \lambda^{l_2} & 0 \\ 0 & 0 & \lambda^{l_3} \end{pmatrix} \hat{\mathcal{M}} \begin{pmatrix} \lambda^{l_1} & 0 & 0 \\ 0 & \lambda^{l_2} & 0 \\ 0 & 0 & \lambda^{l_3} \end{pmatrix}, \qquad (14)$$

where $\hat{\mathcal{M}}$ is a matrix with no Cabibbo suppressions. The Cabibbo structure of the seesaw neutrino matrix is determined solely by the charges of the lepton doublets! As a result, the Cabibbo structure of the MNS mixing matrix is also due entirely to the charges of the three lepton doublets. This general conclusion depends on the existence of at least one Abelian family symmetry, which we argue is implied by the observed structure in the quark sector.

The Wolfenstein parametrization of the CKM matrix (28),

$$\begin{pmatrix} 1 & \lambda & \lambda^3 \\ \lambda & 1 & \lambda^2 \\ \lambda^3 & \lambda^2 & 1 \end{pmatrix}, \qquad (15)$$

and the Cabibbo structure of the quark mass ratios

$$\frac{m_u}{m_t} \sim \lambda^8 \quad \frac{m_c}{m_t} \sim \lambda^4 \; ; \; \frac{m_d}{m_b} \sim \lambda^4 \quad \frac{m_s}{m_b} \sim \lambda^2 , \qquad (16)$$

can be reproduced (23, 29) by a simple *family-traceless* charge assignment for the three quark families, namely

$$X_{Q,\bar{u},\bar{d}} = \mathcal{B}(2,-1,-1) + \eta_{Q,\bar{u},\bar{d}}(1,0,-1) , \qquad (17)$$

where $\mathcal{B}$ is baryon number, $\eta_{\bar{d}} = 0$, and $\eta_Q = \eta_{\bar{u}} = 2$. Two striking facts are evident:

- the charges of the down quarks, $\bar{d}$, associated with the second and third families are the same,

- $Q$ and $\bar{u}$ have the same value for η.

To relate these quark charge assignments to those of the leptons, we need to inject some more theoretical prejudices. Assume these family-traceless charges are gauged, and not anomalous. Then to cancel anomalies, the leptons must themselves have family charges.

Anomaly cancellation generically implies group structure. In $SO(10)$, baryon number generalizes to $\mathcal{B} - \mathcal{L}$, where $\mathcal{L}$ is total lepton number, and in $SU(5)$ the fermion assignment is $\bar{5} = \bar{d} + L$, and $10 = Q + \bar{u} + \bar{e}$. Thus anomaly cancellation is easily achieved by assigning η = 0 to the lepton doublet $L_i$, and η = 2 to the electron singlet $\bar{e}_i$, and by generalizing baryon number to $\mathcal{B} - \mathcal{L}$, leading to the charges

$$X_{Q,\bar{u},\bar{d},L,\bar{e}} = (\mathcal{B} - \mathcal{L})(2,-1,-1) + \eta_{Q,\bar{u},\bar{d}}(1,0,-1) , \qquad (18)$$

where now $\eta_{\bar{d}} = \eta_L = 0$, and $\eta_Q = \eta_{\bar{u}} = \eta_{\bar{e}} = 2$.

The charges of the lepton doublets are simply $X_{L_i} = -(2,-1,-1)$. We have just argued that these charges determine the Cabibbo structure of the MNS lepton mixing matrix to be

$$\mathcal{U}_{MNS} \sim \begin{pmatrix} 1 & \lambda^3 & \lambda^3 \\ \lambda^3 & 1 & 1 \\ \lambda^3 & 1 & 1 \end{pmatrix}, \qquad (19)$$

implying *no Cabibbo suppression in the mixing between* $\nu_\mu$ *and* $\nu_\tau$. This is consistent with the SuperK discovery and with the small angle MSW (31) solution to the solar neutrino deficit. One also obtains a much lighter electron neutrino, and Cabibbo-comparable masses for the muon and tau neutrinos. Notice that these predictions are subtly different from those of grand unification, as they yield $\nu_e - \nu_\tau$ mixing. It also implies a much lighter electron neutrino, and Cabibbo-comparable masses for the muon and tau neutrinos.

On the other hand, the scale of the neutrino mass values depend on the family trace of the family charge(s). Here we simply quote the results our model (23). The masses of the right-handed neutrinos are found to be of the following orders of magnitude

$$m_{\overline{N}_e} \sim M\lambda^{13} \; ; \qquad m_{\overline{N}_\mu} \sim m_{\overline{N}_\tau} \sim M\lambda^7 \; , \qquad (20)$$

where $M$ is the scale of the right-handed neutrino mass terms, assumed to be the cut-off. The seesaw mass matrix for the three light neutrinos comes out to be

$$m_0 \begin{pmatrix} a\lambda^6 & b\lambda^3 & c\lambda^3 \\ b\lambda^3 & d & e \\ c\lambda^3 & e & f \end{pmatrix} \; , \qquad (21)$$

where we have added for future reference the prefactors $a, b, c, d, e, f$, all of order one, and

$$m_0 = \frac{v_u^2}{M\lambda^3} \; , \qquad (22)$$

where $v_u$ is the *vev* of the Higgs doublet. This matrix has one light eigenvalue

$$m_{\nu_e} \sim m_0 \lambda^6 \; . \qquad (23)$$

Without a detailed analysis of the prefactors, the masses of the other two neutrinos come out to be both of order $m_0$. The mass difference announced by superK (30) cannot be reproduced without going beyond the model, by taking into account the prefactors. The two heavier mass eigenstates and their mixing angle are written in terms of

$$x = \frac{df - e^2}{(d+f)^2} \; , \qquad y = \frac{d-f}{d+f} \; , \qquad (24)$$

as

$$\frac{m_{\nu_2}}{m_{\nu_3}} = \frac{1 - \sqrt{1-4x}}{1 + \sqrt{1-4x}} \; , \qquad \sin^2 2\theta_{\mu\tau} = 1 - \frac{y^2}{1-4x} \; . \qquad (25)$$

If $4x \sim 1$, the two heaviest neutrinos are nearly degenerate. If $4x \ll 1$, a condition easy to achieve if $d$ and $f$ have the same sign, we can obtain an adequate split between the two mass eigenstates. For illustrative purposes, when $0.03 < x < 0.15$, we find

$$4.4 \times 10^{-6} \leq \Delta m^2_{\nu_e - \nu_\mu} \leq 10^{-5} \text{ eV}^2 \; , \qquad (26)$$

which yields the correct non-adiabatic MSW (31) effect, and

$$5 \times 10^{-4} \leq \Delta m^2_{\nu_\mu - \nu_\tau} \leq 5 \times 10^{-3} \text{ eV}^2 \; , \qquad (27)$$

for the atmospheric neutrino effect. These were calculated with a cut-off, $10^{16}$ GeV $< M < 4 \times 10^{17}$ GeV, and a mixing angle, $0.9 < \sin^2 2\theta_{\mu-\tau} < 1$. This value of the cut-off is compatible not only with the data but also with the gauge coupling unification scale, a necessary condition for the consistency of our model, and more generally for the basic ideas of grand unification.

## Proton Decay

We have seen in the previous section that the ultraviolet cut-off $M$ appears directly in the seesaw masses. Now that it is determined by experiment, we can use it to estimate the strength of other interactions, in particular those that generate proton decay. In a supersymmetric theory with no R-parity violation, proton decay is caused by two types of operators that appear in the superpotential as

$$W = \frac{1}{M}[\kappa_{112i}\mathbf{Q}_1\mathbf{Q}_1\mathbf{Q}_2\mathbf{L}_i + \overline{\kappa}_{1jkl}\overline{\mathbf{u}}_1\overline{\mathbf{u}}_j\overline{\mathbf{d}}_k\overline{\mathbf{e}}_l] \qquad (28)$$

where for the first operator the flavor index $i = 1, 2$ if there is a charged lepton in the final state and $i = 1, 2, 3$ if there is a neutrino and $j = 2, 3, k, l = 1, 2$. Operators that involve only one family, such as $\mathbf{Q}_1\mathbf{Q}_1\mathbf{Q}_1\mathbf{L}_i$, and $\overline{\mathbf{u}}_1\overline{\mathbf{u}}_1\overline{\mathbf{d}}_1\overline{\mathbf{e}}_l$ are forbidden by symmetry. The reasons are that the combination $\mathbf{Q}_1\mathbf{Q}_1\mathbf{Q}_1$ vanishes identically in the color singlet channel, and the combination $\overline{\mathbf{u}}_1\overline{\mathbf{u}}_1$ transforms as a color sextet, and cannot make a color invariant with the addition of an extra antiquark. This is the well-known statement that in supersymmetric theories, proton decay products will necessarily involve strange particles. The conventional decay into first family members is still there but not dominant. It would be most amusing if the first experimental manifestation of supersymmetry were to be the detection of proton decay into kaons!

These interactions lead to dimension-five four-body interactions between two squarks and two sparticles (two squarks or two sleptons). After gaugino exchange, the two sparticles are turned into particles (32), leading to baryon number violating four fermion interactions, among them proton decay. The existing bounds on proton decay put severe constraints on the couplings $\kappa_{112i}$ and $\overline{\kappa}_{1jkl}$.

In theories where the Cabibbo suppression of operators is related to their charges, we expect these operators to be highly Cabibbo-suppressed. This is because of sum rules which relate their charges to those of standard model invariants.

Under the assumptions of tree-level top quark mass, zero $\mu$-term charge, and of the Green-Schwarz relation $C_{\text{color}} = C_{\text{weak}}$, the family-independent charges satisfy

$$X_{Q_1 Q_1 Q_2 L_i} = X_{\bar{u}_1 \bar{u}_j \bar{d}_k \bar{e}_l} = X_{Q_1 \bar{u}_1 H_u} . \quad (29)$$

Also, the branching ratios between different proton decay modes are determined by the $U(1)$ charges that are flavor dependent. In our model (23), the least suppressed operator is $Q_1 Q_1 Q_2 L_{2,3}$, with

$$\kappa_{1122} \sim \kappa_{1123} \sim \lambda^{11} , \quad (30)$$

leading to the estimate (with $M$ set by the neutrino mass values),

$$\Gamma(p \to K^0 + \mu^+) \sim 10^{32} \, \text{yr}^{-1} , \quad (31)$$

at the same level as the SuperK limits presented at this workshop by L. Sulak.

It is unfortunate that these models yield only orders of magnitude estimate, but it should be clear that those decay rates are tantalizingly close to the experimental bounds. Thus it is important to build a larger proton decay detector and improve the bounds by at least one order of magnitude.

## OUTLOOK

Theoretical predictions of neutrino masses and mixings depend on developing a credible theory of flavor. We have presented two flavor schemes, which predicted not only maximal $\nu_\mu - \nu_\tau$ mixing, but also small $\nu_e - \nu_\mu$ mixings. Neither scheme includes sterile neutrinos (33). The present experimental situation is somewhat unclear: the LSND results (34) imply the presence of a sterile neutrino; and superK favors $\nu_\mu - \nu_\tau$ oscillation over $\nu_\mu - \nu_{\text{sterile}}$. The origin of the solar neutrino deficit remains a puzzle, which several possible explanations. One is the non-adiabatic MSW effect in the Sun, which our theoretical ideas seem to favor, but it is an experimental question which is soon to be answered by the continuing monitoring of the $^8B$ spectrum by SuperK, and the advent of the SNO detector. If neutrino masses reflect (through the seesaw) the value of the ultraviolet cut-off, they set the scale for the strength of proton decay interactions, implying that observation may not be far in the future. Neutrino physics has given us a first glimpse of physics at very short distances, and proton decay cannot be too far behind.

## ACKNOWLEDGEMENTS

I wish to thank Professors C. K. Jung and M. V. Diwan for inviting me to this important and very stimulating workshop. This research was supported in part by the department of energy under grant DE-FG02-97ER41029.

## REFERENCES

1. J. Chadwick, Verh. d. D. Phys. Ges., **16**, 383(1914).
2. C. D. Ellis and W. A. Wooster, Proc. Royal Soc. **A117**, 109(1927).
3. B. Pontecorvo, Chalk river Report PD-205, November 1946, unpublished.
4. C. L. Cowan and F. Reines, Phys.Rev. **90**, 492(1953).
5. C.L. Cowan, F. Reines, F.B. Harrison, H.W. Kruse, A.D. McGuire, Science **124**, 103(1956).
6. Raymond Davis Jr., Phys Rev **97**, 766(1955).
7. B. Pontecorvo, JETP (USSR) **34**, 247(1958).
8. M. Goldhaber, L. Grodzins, A.W. Sunyar, Phys.Rev. **109**, 1015(1958).
9. G. Danby, J.M. Gaillard, K. Goulianos, L.M. Lederman, N. Mistry, M. Schwartz, J. Steinberger, Phys.Rev.Lett. **9**, 36(1962).
10. S. Sakata and T. Inouë, Prog. Theo. Physics, **1**, 143(1946).
11. Z. Maki, M. Nakagawa and S. Sakata, Prog. Theo. Physics, **28**, 247(1962). B. Pontecorvo, Zh. Eksp. Teor. Fiz. **53**, 1717(1967).
12. J. Bahcall, Phys. Rev. Lett. **12**, 300(1964).
13. Raymond Davis Jr., Phys. Rev. Lett. **12**, 303(1964).
14. Raymond Davis Jr., D. Harmer and K. Hoffman, Phys. Rev. Lett. **20**, 1205(1968).
15. V. Gribov and B. Pontecorvo, Phys. Lett. **B28**, 493(1969).
16. J. Pati and A. Salam, Phys. Rev. Lett. **31**, 661(1973)
17. H. Georgi and S. L. Glashow, Phys. Rev. Lett. **32**, ; H. Fritzsch and P. Minkowski, Annals Phys. **93**, 193(1975); H. Georgi, in AIP Conference Proceedings no 23, Williamsburg, Va, 1975; F. Gürsey, P. Ramond and P. Sikivie, Phys. Lett. **60B**, 177(1976)
18. Particle Data Group, R. M. Barnett *et al.*, Phys Rev **D54**, 1(1996).
19. N. Arkani-Hamed, S. Dimopoulos, Gia Dvali *Phys.Lett.* **B429**, 263(1998); E. Dudas, K. Dienes, and T. Gherghetta, *Phys.Lett.* **B436**, 55(1998)
20. M. Gell-Mann, P. Ramond, and R. Slansky in Sanibel Talk, CALT-68-709, Feb 1979 (unpublished), and in *Supergravity* (North Holland, Amsterdam 1979). T. Yanagida, in *Proceedings of the Workshop on Unified Theory and Baryon Number of the Universe*, KEK, Japan, 1979.
21. P. Candelas, G. Horowitz, A. Strominger and E. Witten, Nucl. Phys. **B258**, 46(1985)
22. J. A. Harvey, P. Ramond and D. B. Reiss, Nucl. Phys. **B199**, 223(1982)

23. N. Irges, S. Lavignac and P. Ramond, Phys. Rev. **D58**, 035003(1998).

24. H. Georgi and C. Jarlskog, Phys. Lett. **B86**, 297(1979)

25. K. S. Babu, J. Pati and F. Wilczek, hep-ph/9812538.

26. The Case for Neutrino Oscillations, P. Ramond, Proceedings of the Los Alamos Neutrino Workshop, LA-9358-C, June 1981.

27. C. Froggatt and H. B. Nielsen Nucl. Phys. B147 (1979) 277; P. Ramond, R.G. Roberts and G.G. Ross, Nucl. Phys. B406 (1993)

28. L. Wolfenstein, Phys. Rev. Lett. **51**, 1945(1983).

29. J. Elwood, N., Irges, and P. Ramond, Phys. Rev. Lett. **81**, 5064(1998)

30. Super-Kamiokande Collaboration, Phys. Rev. Lett. **81**, 1562(1998)

31. L. Wolfenstein, Phys. Rev. D17, 2369 (1978); S. Mikheyev and A. Yu Smirnov, Nuovo Cim. **9C**, 17 (1986).

32. H. Murayama, D.B. Kaplan, Phys. Lett. **B336** (1994), 221-228.

33. For an intriguing possibility see, E. Dudas, K. Dienes, and T. Gherghetta, hep-ph/9811428.

34. C. Athanassopoulos *et al.*, Phys. Rev. Lett. **75**, 2560(1995); **77**, 3082(1996); nucl-ex/9706006.

# Radical Conservatism And Nucleon Decay

Frank Wilczek

*Institute for Advanced Study,*
*School of Natural Sciences, Olden Lane,*
*Princeton, New Jersey 08540*

**Abstract.** Unification of couplings, observation of neutrino masses in the expected range, and several other considerations confirm central implications of straightforward gauge unification based on $SO(10)$ or a close relative and incorporating low-energy supersymmetry. The remaining outstanding consequence of this circle of ideas, yet to be observed, is nucleon instability. Clearly, we should aspire to be as specific as possible regarding the rate and form of such instability. I argue that not only esthetics, but also the observed precision of unification of couplings, favors an economical symmetry-breaking (Higgs) structure. Assuming this, one can exploit its constraints to build reasonably economical, overconstrained yet phenomenologically viable models of quark and lepton masses. Putting it all together, one arrives at reasonably concrete, hopeful expectations regarding nucleon decay. These expectations are neither ruled out by existing experiments, nor hopelessly inaccessible. Furthermore, the branching fractions can discriminate among different possibilities for physics at the unification scale.

Radical conservatism, in the sense of Wheeler, is the doctrine of taking good successful ideas seriously, and pressing them hard, to see if they break. It has a noble history, for example in quantum electrodynamics. Here I will follow this philosophy for straightforward gauge unification. In the recent literature many more exotic ideas about physics beyond the standard model have been explored (1), and there is nothing wrong with that, but one should not forget that the simplest possibilities, already broadly envisioned by the early 80s, have not been disproved. Quite the contrary. For reasons I will presently summarize, I believe that after years of marvelous precision work at LEP and elsewhere, the discovery of non-zero neutrino mass at SuperK, and the non-discovery of any among a plethora of suggested exotica, the early ideas look better than ever. Maybe it is a coincidence – excuse me, a series of coincidences – or a conspiracy. Maybe. But I doubt it, and so should you.

This is not to say that gauge theory unification is the end of all desire, or a Theory of Everything. It certainly is not. Even if true, it leaves many loose ends and unanswered questions. But if true it represents a worthy addition to the Standard Model, a major additional insight into Nature, and a foundation for further progress.

And, most fortunately, gauge theory unification is quite concretely a theory of Something. In many ways the crown jewel among its predictions is that nucleons should decay. The possibility of such decay directly reflects the unity of matter – interconvertibility of quarks and leptons – and connects to the cosmological asymmetry between matter and antimatter. The quest to observe nucleon decay has already inspired heroic, though so far fruitless, experimental efforts.

Actually, after a moment's reflection, I want to take back that 'so far fruitless'. Creative efforts to observe nucleon decay have led, through the great IMB, Kamiokande, and SuperK lineage of experiments, to technology that has proved immensely fruitful for neutrino physics. Highlights include observation of the supernova 1987a burst, observation of oscillations in neutrinos deriving from atmospheric cosmic rays (2), and observation of a non-zero but anomalous high-energy solar neutrino flux – each of these representing an achievement of historic proportions. And even the negative result of nucleon instability searches to date has been of genuine positive value. It provided an early motivation for supersymmetric unification, and continues to offer powerful guidance as to what proposals for physics beyond the standard model can be considered plausible.

In any case, we are gathered here to consider whether still more heroic, not to mention expensive, efforts in this direction are warranted. And I want to argue as forcefully as I can for what I believe, that they most certainly are. For upon putting together a number of elegant, successful ideas one arrives at reasonably concrete, hopeful expectations regarding nucleon decay, as I shall indicate. These expectations are neither hopelessly inaccessible, nor ruled out by existing experiments. Furthermore, the branching fractions can discriminate among different possibilities for physics at the unification scale. I will be drawing on results from a recent long, numerically dense analysis by K. Babu, J. Pati, and myself (3). This work in turn draws on an extensive previous literature; for details and references you should refer to our paper.

## THE CASE FOR UNIFICATION

The argument for gauge unification is powerful and many-faceted. I will review it in seven installments, starting with the strongest and working down:

1. the unification of quantum numbers and multiplets;

2. the unification of couplings, using supersymmetry;

3. the explanation of small neutrino masses, in the observed range;

4. the explanation of the $b/\tau$ mass ratio;

5. explaining why things that might otherwise happen, do not;

6. propinquity of the unification and quantum gravitational scales;

7. broad consistency with string/M theory.

### quantum numbers and multiplets

The standard model of particle physics is based upon the gauge groups $SU(3) \times SU(2) \times U(1)$ of strong, electromagnetic and weak interactions acting on the quark and lepton multiplets as shown in Figure 1.

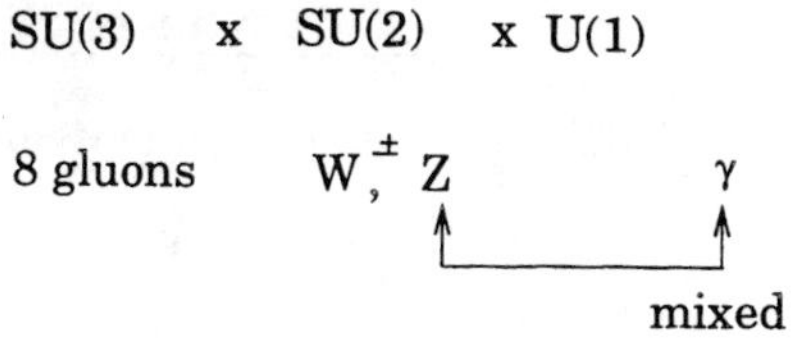

In this Figure I have depicted only one family $(u,d,e,\nu_e)$ of quarks and leptons; in reality there seem to be three families that are mere copies of one another as far as their interactions with the gauge bosons are concerned, but differ in mass. Actually in the Figure I have ignored masses altogether, and allowed myself the convenient fiction of pretending that the quarks and leptons have a definite chirality – right- or left-handed – as they would if they were massless. The more precise statement, of course, is that the gauge bosons couple to currents of definite chirality. The chirality is indicated by a subscript R or L. Finally the little number beside each multiplet is its assignment under the U(1) of hypercharge, which is the average of the electric charge of the multiplet.

While little doubt can remain that the Standard Model is essentially correct, a glance at Figure 1 is enough to reveal that it is not a complete or final theory. To remove its imperfections, while building upon its solid success, is a worthy challenge.

**FIGURE 1.** The gauge groups of the standard model, and the fermion multiplets with their hypercharges.

Given that the strong interactions are governed by transformations among three colors, and the weak by transformations between two others, what could be more natural than to embed both theories into a larger theory of transformations among all five colors (4)?

This idea has the additional attraction that an extra U(1) symmetry commuting with the strong SU(3) and weak SU(2) symmetries automatically appears, which we can attempt to identify with the remaining gauge symmetry of the standard model, that is hypercharge. For while in the separate SU(3) and SU(2) theories we must throw out the two gauge bosons which couple respectively to the color combinations R+W+B and G+P, in the SU(5) theory we only project out R+W+B+G+P, while the orthogonal, traceless combination $(R+W+B)-\frac{3}{2}(G+P)$ remains.

Finally, the possibility of unified gauge symmetry breaking is plausible by analogy; after all, we know for sure that gauge symmetry breaking occurs in the electroweak sector.

Georgi and Glashow (5) showed how these ideas can be used to bring some order to the quark and lepton sector, and in particular to supply a satisfying explanation of the weird hypercharge assignments in the standard model. As shown in Figure 2, the five scattered $SU(3) \times SU(2) \times U(1)$ multiplets get organized into just two representations of SU(5).

In making this unification it is necessary to allow transformations between (what were previously considered to be) particles and antiparticles, and also between quarks and leptons. It is convenient to work with left-handed fields only. Since the conjugate of a right-handed field is left-handed, we don't lose anything by doing so – though we must shed traditional prejudices about a rigorous distinction between matter and antimatter, since these get mixed up. Specifically, it will not be possible to de-

SU(5): 5 colors RWBGP

10: 2 different color labels (antisymmetric tensor)

$$\begin{array}{l} u_L: \quad RP, \quad WP, \quad BP \\ d_L: \quad RG, \quad WG, \quad BG \\ u_L^c: \quad RW, \quad WB, \quad BR \\ \phantom{u_L^c:} \quad (\bar{B}) \quad (\bar{R}) \quad (\bar{W}) \\ e_L^c: \quad GP \\ \phantom{e_L^c:} \quad () \end{array} \begin{pmatrix} 0 & u^c & u^c & u & d \\ & 0 & u^c & u & d \\ & & 0 & u & d \\ * & & & 0 & e \\ & & & & 0 \end{pmatrix}$$

$\bar{5}$: 1 anticolor label

$$\begin{array}{l} d_L^c: \quad \bar{R}, \quad \bar{W}, \quad \bar{B} \\ e_L: \quad \bar{P} \\ \nu_L: \quad \bar{G} \end{array} \quad (d^c \; d^c \; d^c \; e \; \nu)$$

$$\boxed{Y = -\tfrac{1}{3}(R+W+B) + \tfrac{1}{2}(G+P)}$$

**FIGURE 2.** Unification of fermions in SU(5) There is a beautiful extension of SU(5) to the slightly larger group SO(10). With this extension, one can unite all the observed fermions of a family, plus one more, into a *single* multiplet (6). The relevant representation for the fermions is a 16-dimensional spinor representation. Some of its features are depicted in Figure 3.

SO(10): 5 bit register

$(\pm\pm\pm\pm\pm)$ : <u>even</u> # *of* −

$$10: \begin{array}{l} (+\,+\,-|+\,-) \quad 6 \quad (u_L, d_L) \\ (+\,-\,-|+\,+) \quad 3 \quad u_L^c \\ (+\,+\,+|-\,-) \quad 1 \quad e_L^c \end{array}$$

$$\bar{5}: \begin{array}{l} (+\,-\,-|-\,-) \quad \bar{3} \quad d_L^c \\ (-\,-\,-|+\,-) \quad \bar{2} \quad (e_L, \nu_L) \end{array}$$

$1: (+\,+\,+|+\,+) \quad 1 \quad N_R$

**FIGURE 3.** Unification of fermions in SO(10). The rule is that all possible combinations of 5 + and − signs occur, subject to the constraint that the total number of − signs is even. The SU(5) gauge bosons within SO(10) do not change the numbers of signs, and one sees the SU(5) multiplets emerging. However there are additional transformations in SO(10) but not in SU(5), which allow any fermion to be transformed into any other.

clare that matter is what carries positive baryon and lepton number, since the unified theory does not conserve these quantum numbers.

As shown in Figure 2, there is one group of ten left-handed fermions that have all possible combinations of one unit of each of two different colors, and another group of five left-handed fermions that each carry just one negative unit of some color. These are the ten-dimensional antisymmetric tensor and the complex conjugate of the five-dimensional vector representation, commonly referred to as the five-bar. In this way, *the structure of the standard model, with the particle assignments gleaned from decades of experimental effort and theoretical interpretation, is perfectly reproduced by a simple abstract set of rules for manipulating symmetrical symbols.* Thus for example the object RB in this Figure has just the strong, electromagnetic, and weak interactions we expect of the complex conjugate of the right-handed up-quark, without our having to instruct the theory further.

A most impressive, though simple, exercise is to work out the hypercharges of the objects in Figure 2 and checking against what you need in the Standard Model. These ugly ducklings of the Standard Model have matured into quite lovely swans.

In addition to the conventional quarks and leptons the SO(10) spinor contains an additional particle, an SU(3)×SU(2)×U(1) singlet. (It is even an SU(5) singlet.) Usually when a theory predicts unobserved new particles they are an embarrassment. But these N particles – there are three of them, one for each family – are a notable exception. Indeed, they are central to the emerging connection between neutrino masses and unification, as I shall discuss below.

## unification of couplings using supersymmetry

We have just seen that simple unification schemes are spectacularly successful at the level of classification. New questions arise when we consider dynamics.

Part of the power of gauge symmetry is that it fully dictates the interactions of the gauge bosons, once an overall coupling constant is specified. Thus if SU(5) or some higher symmetry were exact, then the fundamental strengths of the different color-changing interactions would have to be equal, as would the (properly normalized) hypercharge coupling strength. In reality the coupling strengths of the gauge bosons in SU(3)×SU(2)×U(1) are not observed to be equal, but rather follow the pattern $g_3 \gg g_2 > g_1$.

Fortunately, experience with QCD emphasizes that couplings *run* (8). The physical mechanism of this effect is that in quantum field theory the vacuum must be regarded as a polarizable medium, since virtual particle-antiparticle pairs can screen charge. For charged gauge bosons, as arise in non-abelian theories, the paramagnetic (antiscreening) effect of their spin-spin interaction dominates, which leads to asymptotic freedom. As Georgi, Quinn, and Weinberg pointed out (9), if a gauge symmetry such as SU(5) is spontaneously broken at some very short distance then we should not expect that the effective couplings probed at much larger distances, such as are actually measured at practical accelerators, will be equal. Rather they will all have been affected to a greater or lesser extent by vacuum screening and antiscreening, starting from a common value at the unification scale but then diverging from one another. The pattern $g_3 \gg g_2 > g_1$ is just what one should expect, since the antiscreening effect of gauge bosons is more pronounced for larger gauge groups.

The running of the couplings gives us a truly quantitative handle on the ideas of unification. To specify the relevant aspects of unification, one basically needs only to fix two parameters: the scale at which the couplings unite, (which is essentially the scale at which the unified symmetry breaks), and their common value when they unite. Given these, one calculates three outputs, the three *a priori* independent couplings for the gauge groups in $SU(3) \times SU(2) \times U(1)$. Thus the framework is eminently falsifiable. The astonishing thing is, how close it comes to working (Figure 4).

The GQW calculation is remarkably successful in explaining the observed hierarchy $g_3 \gg g_2 > g_1$ of couplings and the approximate stability of the proton. In performing it, we assumed that the known and confidently expected particles of the standard model exhaust the spectrum up to the unification scale, and that the rules of quantum field theory could be extrapolated without alteration up to this mass scale – thirteen orders of magnitude beyond the domain they were designed to describe. It is a triumph for minimalism, both existential and conceptual.

On closer inspection, however, it is not quite good enough. Accurate modern measurements of the couplings show a small but definite discrepancy between the couplings, as appears in Figure 4. And heroic dedicated experiments to search for proton decay at the rate expected from exchange of the additional gauge bosons present in $SU(5)$ but not in the Standard Model did not find it (10). They currently exclude the minimal SU(5) prediction $\tau_p \sim 10^{31}$ yrs. by about two orders of magnitude.

If we just add particles in some haphazard way things will only get worse: minimal SU(5) nearly works, so a generic perturbation will be deleterious. Even if some *ad hoc* prescription could be made to work, that would be a disappointing outcome from what appeared to be one of our most precious, elegantly straightforward clues regarding physics well beyond the Standard Model.

Fortunately, there is a compelling escape from this impasse. That is the idea of supersymmetry (11). Supersymmetry is certainly not a symmetry in nature: for example, there is certainly no bosonic particle with the mass and charge of the electron. However there are several reasons for thinking that supersymmetry might be spontaneously, and only relatively mildly broken, so that the superpartners are no more massive than $\approx 1$ Tev. The most

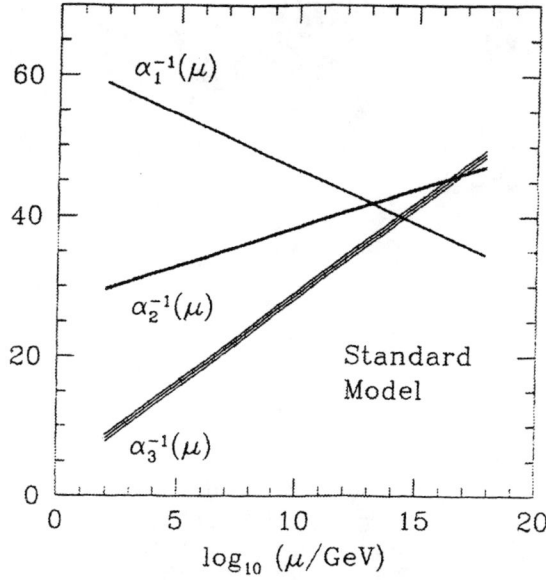

**FIGURE 4.** The failure of the running couplings, normalized according to SU(5) and extrapolated taking into account only the virtual exchange of the "known" particles of the standard model (including the Higgs boson) to meet. Note that only with fairly recent experiments (7), which greatly improved the precision of the determination of low-energy couplings, has the discrepancy become significant.

concrete arises in calculating radiative corrections to the (mass)$^2$ of the Higgs particle from diagrams of the type shown in Figure 5. One finds that they make an infinite, and also large, contribution. By this I mean that the divergence is quadratic in the ultraviolet cutoff. No ordinary symmetry will make its coefficient vanish. If we imagine that the unification scale provides the cutoff, we will find, generically, that the radiative correction to the (mass)$^2$ is much larger than the total value we need to match experiment. This is an ugly situation.

In a supersymmetric theory, if the supersymmetry is not too badly broken, it is possible to do better. For any set of virtual particles that might circulate in the loop there will be another graph with their supersymmetric partners circulating. If the partners were accurately degenerate, the contributions would cancel. Taking supersymmetry breaking into account, the threatened quadratic divergence will be cut off only at virtual momenta such that the difference in (mass)$^2$ between the virtual particle

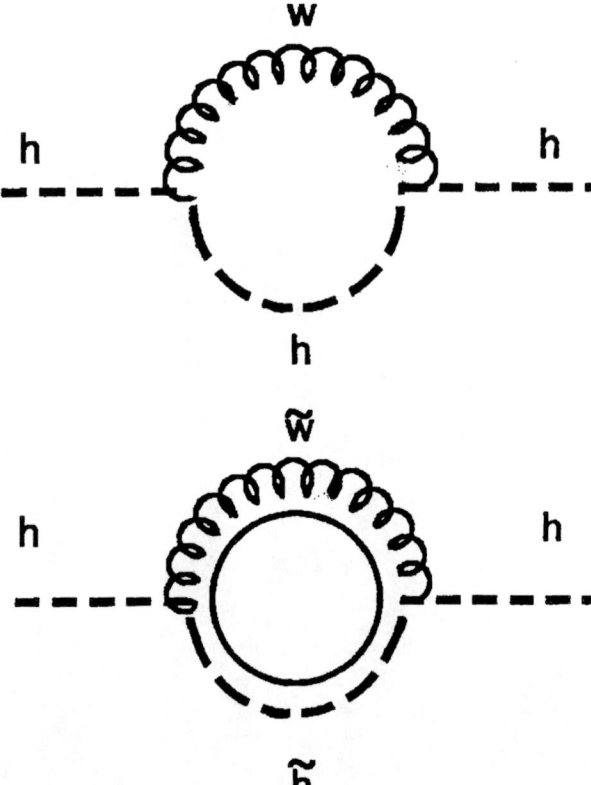

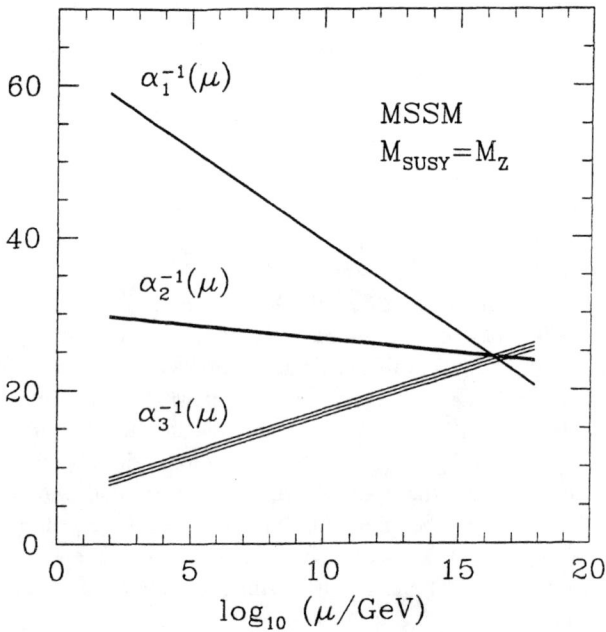

**FIGURE 5.** Contributions to the Higgs field self-energy. These graphs give contributions to the Higgs field self-energy which separately are formally quadratically divergent, but when both are included the divergence is removed. In models with broken supersymmetry a finite residual piece remains. If one is to obtain an adequately small finite contribution to the self-energy, the mass difference between Standard Model particles and their superpartners cannot be too great. This – and essentially only this – motivates the inclusion of virtual superpartner contributions in Figure 6 beginning at relatively low scales.

**FIGURE 6.** When the exchange of the virtual particles necessary to implement low-energy supersymmetry, a calculation along the lines of Figure 4 comes into adequate agreement with experiment.

and its supersymmetric partner is negligible. Notice that we will be assured adequate cancellation if and only if supersymmetric partners are not too far split in mass – in the present context, if the splitting times the square root of the fine structure constant is not much greater than the weak scale.

The effect of low-energy supersymmetry on the running of the couplings was first considered long ago (12), in advance of the precise measurements of low-energy couplings or of the modern limits on nucleon decay. One might have feared that such a huge expansion of the theory, which essentially doubles the spectrum, would utterly destroy the approximate success of the minimal SU(5) calculation. This is not true, however. To a first approximation since supersymmetry is a space-time rather than an internal symmetry it does not affect the group-theoretic structure of the calculation.

Thus to a first approximation the absolute rate at which the couplings run with momentum is affected, but not the relative rates. The main effect is that the supersymmetric partners of the color gluons, the gluinos, weaken the asymptotic freedom of the strong interaction. Thus they tend to make its effective coupling decrease and approach the others more slowly. Thus their merger requires a longer lever arm, and the scale at which the couplings meet increases by an order of magnitude or so, to about $10^{16}$ Gev. An immediate effect of raising the scale is to raise the mass of the gauge bosons that can mediate proton decay, so that the experimental bounds are no longer contradicted. (On nucleon stability, more below.)

I want to emphasize that this very large new mass scale has emerged unforced from the internal logic of the Standard Model itself. It will appear in several of our further considerations, and so for later reference let's give it a name, the unification scale, and the token $M_U$.

Since the running of the couplings with scale is logarithmic, the unification of couplings calculation is not terribly sensitive to the exact scale at which supersymmetry is broken, say between 100 Gev and 10 Tev. It is a result robust, at the few *per cent* level, against uncertainties of this sort. This robustness is fortunate (and virtually unique among the phenomenological signatures of supersymmetry), because at present the mechanism of supersymmetry breaking, and therefore the spectrum of sfermions and gauginos, is quite uncertain. The unification of couplings is also robust against addition of additional particles, so long as they come in complete, approximately degenerate $SU(5)$ multiplets. Additional uncertainties do arise from the details of the unified symmetry breaking at the high scale. I'll discuss these further below. The main conclusion is that these corrections are also at the few *per cent* level, so long as the symmetry breaking is implemented economically (*i.e.*, using the simplest Higgs field representations).

On the other hand, our successful unification of couplings calculation is most definitely *not* robust against radical changes in its embedding framework, such as abandoning low-energy supersymmetry, using radically different unification groups, or allowing the virtual particles to wander off into extra dimensions. If any of these ideas are correct, the spectacular existing agreement of theory and experiment, displayed in Figure 6, would seem to be a 'coincidence' – imputing to Mother Nature a rather sadistic propensity to tease.

## neutrino mass

It is important to realize that the degrees of freedom of the Standard Model permit neutrino masses. A minimal implementation of the construction requires an interaction of the type

$$\Delta \mathcal{L} = \kappa_{ij} L^{\alpha a i} L^{\beta b j} \varepsilon_{\alpha \beta} \phi_a^\dagger \phi_b^\dagger + \text{h.c.} , \qquad (1)$$

where $i$ and $j$ are family indices; $\kappa_{ij}$ is a symmetric matrix of coupling constants; the $L$ fields are the left-handed doublets of leptons, with Greek spinor indices, early Roman weak $SU(2)$ indices, and middle Roman flavor indices; and finally $\phi$ is the Higgs doublet, with its weak $SU(2)$ index. Two-component notation has been used for the spinors, to emphasize that this way of forming mass terms, although different from what we are used to for quarks and charged leptons, is in some sense more elementary mathematically. $\Delta \mathcal{L}$ becomes a neutrino mass term when the $\phi$ field is replaced by its vacuum expectation value $\langle \phi^a \rangle = v \delta_1^a$.

Although this Eq. (1) is a possible interaction for the degrees of freedom in the Standard Model, it is usually considered to be *beyond* the Standard Model, for a very good reason. The new term differs from the terms traditionally included in the Standard Model in that the product of fields has mass dimension 5, so that the coefficient $\kappa$ must have mass dimension -1. In the context of quantum field theory, it is a nonrenormalizable interaction. When one includes it in virtual particle loops, one will find amplitudes containing the dimensionless factors of the type $\kappa \Lambda$, where $\Lambda$ is an ultraviolet cutoff. In this framework, therefore, one cannot accept $\Delta \mathcal{L}$ as an elementary interaction. It can only be understood within a larger theoretical context.

Given a numerical value for the neutrino mass, we can infer a scale beyond which $\Delta \mathcal{L}$ cannot be accurate, and degrees of freedom beyond the Standard Model must open up. To get oriented, let us momentarily pretend that $\kappa$ is simply a number instead of a matrix, and that $m = 10^{-2}$ eV is the neutrino mass. Then, using $v = 250$ GeV for the vacuum expectation value, we calculate

$$1/M \equiv \kappa = m/v^2 = 1/(6 \times 10^{15} \text{ GeV}) . \qquad (2)$$

When energy and momenta of order $M$ begin to circulate in loops the form of the interaction must be modified. Otherwise the dangerous factor $\kappa \Lambda$ will become larger than unity, inducing large and uncontrolled radiative corrections to all processes, and rendering the success of the Standard Model accidental.

Thus we trace the "absurdly small" value of the observed neutrino mass scale to an "absurdly large" fundamental mass scale. You will not fail to notice that the new scale we infer here, directly from the observed value of the neutrino mass is, quantitatively, none other than $M_U$. This is most definitely not a coincidence, as I'll now explain.

Let us return to the question of the $N$ masses. Because the $N^i$ are singlets, mass terms of the type

$$\Delta \mathcal{L}_N = \eta_{ij} N^{\alpha i} N^{\beta j} \varepsilon_{\alpha \beta} \qquad (3)$$

with $\eta_{ij}$ a symmetric coupling matrix, are consistent with $SU(3) \times SU(2) \times U(1)$ symmetry. This term of course greatly resembles the effective interaction responsible for neutrino masses, Eq. (1), but conceptually the difference is crucial. Because the Ns are Standard Model singlets the Higgs doublets that occurred in Eq. (1) need not appear here. A consequence is that the operators appearing in Eq. (3) have mass dimension 3, so that the $\eta_{ij}$ must have mass dimension +1. This interaction therefore does not bring in any ultraviolet divergence problems.

What sets the scale for $\eta$? Although Eq. (3) is consistent with Standard Model gauge symmetries, or even $SU(5)$, it is not consistent with $SO(10)$. Indeed for the product of spinor 16 we have the decomposition $16 \times 16 = 10 + 120 + 126$, where only the 126 contains an $SU(5)$ singlet component. The most straightforward possibility for generating a term like Eq. (3) in the full theory is therefore to include a Higgs 126, and a Yukawa coupling of this to the 16s. If the appropriate components of the 126 acquire vacuum expectation values, Eq. (3) will emerge. The 126 is a five-index self-dual antisymmetric tensor under $SO(10)$, which may not be to everyone's taste. Alternatively, one can imagine that more complicated interactions, containing products of several simpler Higgs fields which condense, are responsible. These need not be fundamental interactions (they are, of course, non-renormalizable), but could arise through loop effects or by integrating out heavier particles even in a renormalizable field theory.

At this level there are certainly many more options than constraints, so that without putting the discussion of N masses in a broader context, and making some guesses, one can't be very specific or quantitatively precise. Nevertheless, I think it is fair to say that these general considerations strongly suggest that $\eta$ is associated with breaking of unified symmetries down to the Standard Model. Thus, if the general framework is correct, the expected scale for its entries is set by the one we met in the unification of couplings calculation, i.e. $\eta \sim M_U = 10^{16}$ Gev.

The Ns communicate with the familiar fermions through the Yukawa interactions

$$\Delta \mathcal{L}_{N-L} = g^i_j \bar{N}_i L^{aj} \phi^\dagger_a + \text{h.c.}, \qquad (4)$$

using the previous notations but now, in this more conventional term, suppressing the Dirac spinor indices. These interactions are of precisely the type that generate masses for the quarks and charged leptons in the Standard Model. If N were otherwise massless, the effect of Eq. (4) would be to generate neutrino masses, of the same order as ordinary quark and lepton masses. In $SO(10)$, indeed, these masses would be related by simple Clebsch-Gordon and renormalization factors of order unity. Fortunately, as we have seen, N is far from massless.

Indeed, N is so massive that for purposes of low-energy physics we can and should integrate it out. This is easy to do. The effect of combining Eq. (3) and Eq. (4) and integrating out N is to generate

$$\Delta \mathcal{L}_{\text{eff.}} = g^k_i g^l_j (\eta^{-1})_{kl} L^{\alpha a i} L^{\beta b j} \varepsilon_{\alpha\beta} \phi^\dagger_a \phi^\dagger_b + \text{h.c.} . \qquad (5)$$

Thus we arrive back at Eq. (1), with

$$\kappa_{ij} = g^k_i g^l_j (\eta^{-1})_{kl} . \qquad (6)$$

This so-called seesaw equation (13) provides a much more precise version of the loose connection between unification scale and neutrino mass we discussed at the outset. There is much uncertainty in the details, since there is no reliable detailed theory for the $g^k_i$ nor the $\eta$s. But if $g$ has an eigenvalue of order unity pointing toward the third family (this is suggested by symmetry and the value of the top quark mass, as discussed below), and if we set the scale for $\eta$ using the logic above, then we get close to $10^{-2}$ eV for the $\tau$ neutrino mass, as observed.

On the face of it, then, neutrino mass of the observed magnitude provide an additional confirmation of our well developed, straightforward, minimalist ideas for unification beyond the Standard Model. It also takes us in a pretty direction where we should be delighted to go: toward more complete symmetry, using $SO(10)$ (or perhaps, as Pati emphasizes, a smaller but still left-right symmetric variant). Within this circle of ideas, neutrino mass of the observed magnitude is a robust consequence. Outside that circle, it becomes another 'coincidence'.

## b/τ mass ratio

Within $SU(5)$, or any of its extensions, it is natural to expect certain kinds of regularities among their masses, since quarks and leptons are put together in common multiplets. Specifically, the right-handed $b$ quark (or, better, the left-handed $\bar{b}$) and the left-handed $\tau$ lepton can be found in a single $\bar{5}$ multiplet, whereas their oppositely-handed pairs can be found in a single 10 multiplet. If we assume that their masses are generated in the simplest possible way, using a Higgs field in the $\bar{5}$, we find a simple relation – in fact, equality – between the masses. Such equality does not hold, of course, of the observed physical masses. But we must remember that – again, in the circle of ideas around minimalist unification – the fundamental equality is between effective Yukawa couplings normalized at $M_U$. Just as for the gauge couplings, we must renormalize this prediction down to laboratory scales, taking into account the effect of virtual particles. When this is done – again, in the minimalist framework – one finds striking agreement between prediction and observation.

Within $SO(10)$, one obtains (with similar assumptions) in addition a similar relation between the top quark mass and the underlying Dirac mass of the $\tau$ neutrino (*i.e.*, the off-diagonal entry in the seesaw mass matrix). This reinforces the estimate of the heaviest neutrino mass presented above, and furthermore associates that mass with the $\tau$ neutrino.

The luster of these successes in correlating the masses of the heaviest fermion family is somewhat dimmed by the failure of the simplest hypotheses to explain the pattern of lighter fermion masses and mixings. Of course those masses, being smaller, are *a priori* more sensitive to quantitatively small complications, so that predictions for them are intrinsically less robust. The situation is far from desperate, and I'll say a bit more about it below.

## things that don't happen

One frequently encounters jeremiads about the danger of assuming lack of complications such as new strongly interacting sectors (technicolor), compositeness, and – recently popular – additional large dimensions, as one extrapolates from observed energy scales to $M_U$. Doubtless there are any number of ways that the radically conservative extrapolation of gauge field theory might go wrong. However, one should not discount the observation that similar jeremiads have been voiced for more than twenty years now, while so far no hint of any of the suggested deviations has in fact materialized.

Quite the contrary. As precision measurements of Standard Model parameters have converged on minimal supersymmetric unification of couplings, they have also put severe constraints on these picturesque and intuitively appealing, but speculative and phenomenologically gratuitous, possibilities. Likewise, searches for unconventional sources of CP violation and for effects of neutral flavor-changing interactions have come up empty, and put considerable pressure on any suggestion that the fundamental dynamics associated with non-universal flavor interactions, let alone with dynamics that connects quark and leptons, occurs below a scale of several Tev. Conversely, the idea that a large scale like $M_U$ characterizes these effects, if less tantalizing, is much safer.

There is a slightly naive, but not completely silly, general consideration worthy of mention here. Any ambitious extension of the Standard Model sufficient to unify quarks and leptons (including any incarnation of string/M theory) will almost certainly involve violation of baryon number, and therefore at some level nucleon instability. With a scale as large as $M_U$, we might – as discussed below – just squeeze by the experimental constraints. If the scale is significantly smaller, that becomes much more difficult.

## propinquity of the gravity scale

The value of $M_U$ is, on the appropriate logarithmic scale, remarkably close to the Planck scale $M_{Planck} \sim 10^{19}$ Gev. The Planck scale is the scale at which the classical Einstein description of gravity must break down; concretely it is the energy scale at which exchange of virtual gravitons competes quantitatively with the other interactions. Because $M_U$ is significantly smaller than the Planck mass, we need not be too nervous about the neglect of quantum gravity corrections to our unification of couplings calculation. Yet because it is not absurdly smaller, we can feel encouraged for the prospect of unification including both gravity and gauge forces, independent of any detailed model.

## broad consistency with string/M theory

String/M theory is at present the best candidate framework for incorporating quantum mechanics together with general relativity. Huge challenges remain for construction of scientific world-models on its basis. Specifically, for example, there is no generally accepted understanding of such basic questions as why the macroscopic world looks 3+1 dimensional (whereas the underlying theory is more naturally 9+1 or 10+1 dimensional), nor why the cosmological term is so small, nor even how to formulate either the basic equations or the initial-value problem. The rules of the game have changed over the years, and undoubtedly will continue to do so.

Nevertheless it is intriguing that, given a sense of humor and a bit of good will, one can descry most of the elements of reality utilized in gauge theory unification – the degrees of freedom of the Standard Model, the possibility of low-energy supersymmetry, and enough additional gauge symmetry for unification of couplings – within string/M theory. The classic (vintage 1984) weakly coupled heterotic phenomenology is mainly concerned with finding solutions to the equations of static classical string theory that reduce well below the Planck scale to effective theories resembling the supersymmetric standard model. (Of course, it is notorious that there are also zillions of apparently equally good solutions that look nothing like our world.) Within this class of models, there is a large subclass that also embodies something close to conventional gauge theory unification. Recent techniques support related constructions at strong coupling (14).

In any case, I think it is certainly fair to say that there is at present no clear contradiction between gauge theory unification as discussed here 'from the bottom up', based on straightforward extrapolation of established facts and principles, and string/M theory. This tends to reinforce the significance of our previous, model-independent result $M_U \sim M_{Planck}$.

# NUCLEON DECAY

## supersymmetry and the challenge of exotica

I have argued for the desirability of low-energy supersymmetry based on one major quantitative result (the unification of couplings) and one rather soft theoretical advantage (protection of the weak scale from radiative corrections). Other arguments can and have been made, but I think these two are by far the best, most concrete ones.

Against this less than overwhelming evidence we must weigh considerable complications and several embarrassments.

In the minimal version of the Standard Model (SM), without supersymmetry, one has the possibility of a clean, uniform explanation of the smallness of observed CP violating effects and of neutrino masses, and of the smallness of so far unobserved neutral flavor-changing effects in both the quark and lepton sectors, and of nucleon instability. For given the symmetries and matter content of the minimal SM, all these effects (except CP violation) arise only from higher-dimension, nonrenormalizable interactions. Thus they appear in the Lagrangian multiplied by coefficients inversely proportional to some mass scale, and if this mass scale is large (say approaching the Planck scale) they represent unobservable, or barely observable, small effects.

CP violation can arise through renormalizable interactions, but only in two special ways. One way is through complicated interference effects involving interference among all three families, as proposed by Kobayashi and Maskawa. The other is through the effects of the notorious θ term of QCD. Existing evidence is consistent with the idea that the first of these mechanisms is responsible for all CP violation so far observed; while the θ term is, for a reason presumably connected with Peccei-Quinn symmetry and the existence of axions, very small or zero. The adequacy of the minimal SM framework will be tested by future measurements of B-meson properties and searches for elementary electric dipole moments.

As one expands the SM to include supersymmetry this clean, uniform explanation of the absence or smallness of those many diverse species of possible exotica comes undone. Technically, this occurs because the accounting of possible 'relevant' (renormalizable, total mass dimension $\leq 4$) interactions is quite different in the supersymmetric case. The bosonic slepton and squark fields have mass dimension unity, as opposed to the fermionic lepton and quark fields, which have mass dimension 3/2 (and must appear in pairs within Lorentz invariant candidate interactions), which opens a considerably more capacious Pandora's box. For example, in the SM without supersymmetry possible baryon number violating interactions have mass dimension at least six, since to make a color singlet they must contain at least three quark fields, and then another fermion (lepton) to make a Lorentz singlet. Using the squark fields, baryon-number violating interactions with dimension 3 can be constructed. Supersymmetry forbids these particular terms (so they may be suppressed by the ratio of supersymmetry breaking to unification scales), but there are several possible supersymmetric dimension 4 and 5 terms. A complementary perspective, looking from the high scale down, is that exchange of heavy fermion partners of scalar or gauge fields brings in propagators with only one inverse power of the heavy scale, instead of two, and so is less suppressed at low energy. There are also many additional possible sources of CP violation, no longer necessarily involving all three families.

None of these problems appears insurmountable. Indeed, each presents opportunities for theoretical and experimental discovery, and each has generated its own sizable literature. The issue of nucleon instability, in view of its unique sensitivity and deep cosmological significance, may be the most critical and fundamental problem of all, and in the remainder of this talk I will focus on it exclusively. I will be brief, since my collaborators will be covering some of the same ground more thoroughly.

## from supersymmetry to Higgsino exchange

The analysis of nucleon instability in supersymmetric theories is difficult to discuss without introducing some technical machinery, since supersymmetry induces some special cancellations which are difficult to see without using superfields. A major result of the analysis is that the possible form of supersymmetric dimension 4 baryon number violating operators is quite restricted, and it is easily forbidden with an appropriate discrete symmetry. A second major result is that the main contribution to dimension 5 baryon number violation comes from Higgsino, not gaugino, exchange.

At first sight it might appear that the move from non-supersymmetric unification, where the leading contributions to nucleon instability arise from dimension 6 operators, suppressed by two inverse powers of the unification scale, to supersymmetric unification, which allows dimension 5 operators and nucleon instability suppressed by only one power of the unification scale, is catastrophic. A number of factors mitigate this crisis, however. The unification scale is somewhat larger, and the relevant Higgsino mass can be larger still; the bottom-line Higgs couplings to the light families are quite small; and one must at the end of the day dress the scalar (squark and slepton) fields appearing in the dimension 5 operators, by exchange of the standard model gauginos, into ordinary quarks and leptons.

Because of all this, in order to obtain a quantitative estimate of nucleon instability one must be quite concrete about masses of the superheavy Higgs fields and their couplings to ordinary fermions.

## doublet-triplet splitting

The Higgs doublet needed for electroweak symmetry breaking in the Standard Model can be embedded in various ways into a representation of the full unified gauge group. The simplest possibility, within $SU(5)$, is to embed it within a fundamental, *i.e.*, a **5**. The three extra components form a fractionally charged color triplet. The symmetry instructs us how this triplet couples, and we quickly discover that it is a very dangerous object, because its exchange violates baryon number and destabilizes nucleons. It must be extremely heavy, $M_{\text{triplet}} \gtrsim 10^{14}$ Gev, in order to be consistent with experimental limits. In particular, it must be very much heavier than its partner, the electroweak Higgs doublet. Theoretically, it is quite challenging to understand how such a large splitting could arise. This is the doublet-triplet splitting problem. Similar problems occur for other unification groups and embeddings.

A profound advantage of supersymmetric unification in $SO(10)$, which in my view forms an essential adjunct to its role in protecting the weak scale, is its ability to address the doublet-triplet splitting problem. Over and above its stability to radiative corrections, as mentioned above, there is the question of obtaining the splitting at the classical level in the first place. There are special constraints for the scalar potential due to supersymmetry, arising because it comes, roughly speaking, as the square of a simpler object, the superpotential. They make it possible – in $SO(10)$! – to assure the requisite classical splitting through a simple group-theoretic mechanism (15).

## fine structure of coupling unification

By persisting in the radically conservative hypothesis that the striking quantitative success of the unification of couplings calculation, as displayed in Figure 6, is not accidental, we are led to an important conclusion regarding the complexity of unified symmetry breaking.

In general, symmetry breaking effects will split the masses of different components of any Higgs field representation. These splittings lead to logarithmic changes in the running of couplings, as mentioned above. Let us see how they affect the fine structure of coupling constant unification. To lowest (one-loop) order the modifications to the predicted value of the couplings take the form

$$\alpha_i^{-1}(M_Z) = \alpha_U^{-1} - \frac{b_i}{2\pi}\ln(M_Z/M_U) - \Delta_i \quad (7)$$

where

$$\Delta_i = \sum_{\text{submultiplets } \kappa} \frac{-b_i^\kappa}{2\pi}\ln(M_U/m_\kappa). \quad (8)$$

Here $b_i^\kappa$ is the contribution to the $i^{\text{th}}$ gauge group $\beta$ function from the $\kappa$ submultiplet. If all the $m_\kappa$ are equal, the effect of the $\Delta_i$ is merely to renormalize $M_U$. In general, however, they will affect the predicted relation among the observed couplings. Suppose that we begin by taking all the $\Delta_i$ to vanish, which is known to lead to a successful result. Then if we accommodate the perturbations to $\alpha_1^{-1}$ and $\alpha_2^{-1}$ by adjusting the two free parameters $\alpha_U^{-1}$ and $M_U$, we are led to alter our prediction for the strong coupling $\alpha_3^{-1}$ according to

$$\frac{\delta\alpha_3(M_Z)}{\alpha_3(M_Z)^2} = \frac{5}{7}\Delta_1 - \frac{12}{7}\Delta_2 + \Delta_3. \quad (9)$$

Of course, before worrying about possible gratuitous corrections, we must make sure that the basic fields we use to break the unified symmetry down to the standard model give an acceptable zeroth-order answer to begin with. In particular, the contribution of the electroweak doublet, and its triplet partner must be handled carefully. It turns out that if we let the triplet become too heavy, say more than 100 times $M_U$, the successful zeroth-order prediction of $\alpha_3$ becomes endangered. Thus it is impossible to suppress nucleon stability due to this source down to arbitrarily low levels.

Now let us estimate the quantitative impact of different kinds of Higgs structure. If we take a $5 + \bar{5}$ of $SU(5)$, or a 10 of $SO(10)$, the result is

$$\frac{\delta\alpha_3(M_Z)}{\alpha_3(M_Z)^2} = \frac{1}{2\pi}\frac{9}{7}\ln(m_3/m_2). \quad (10)$$

If the logarithm is of order unity, this represents a few *per cent* correction to $\alpha_3(M_Z)$, which is tolerable.

On the other hand, consider the rank two symmetric traceless tensor 54 of $SO(10)$. This is still one of the simpler irreducible representations, but it contains a piece which goes as $(6,1,4/3) + (6,1,-4/3)$ under the standard model. If this piece is split from its brethren at $M_U$, the correction is

$$\frac{\delta\alpha_3(M_Z)}{\alpha_3(M_Z)^2} = \frac{1}{2\pi}\frac{51}{7}\ln(m_3/m_U), \quad (11)$$

which, for a logarithm of order unity, is in the neighborhood 10-20 %. Uncontrolled corrections of this sort could be expected to upset the applecart. Of course, for more complicated representations, containing more highly charged submultiplets, the situation only gets worse.

At face value, these considerations strongly suggest that the observed success of the unification of couplings can be construed as reassuring confirmation that Nature has good taste: She starts with lots of symmetry, and uses simple, minimalistic symmetry breaking patterns. Of course it's terribly dangerous to rely too heavily on a single number, but we're being radically conservative, and it's taking us where we want to go!

## a look toward fermion masses

As we've seen, in supersymmetric unification the leading source of nucleon instability is exchange of superheavy Higgsino fields. In order to pin this down, we must constrain which such fields are present, and how they couple to quarks and leptons. The immediately preceding considerations strongly encourage us to restrict ourselves to the simplest possible field content. For $SO(10)$, concretely, this means a small number of adjoints, fundamentals, and spinors.

Having chosen the Higgs content, we must address the coupling to quarks and leptons. There is, of course, a very large amount of data regarding the masses and mixing matrices of quarks and leptons that we should use for guidance. Let me briefly indicate the sorts of considerations that enter, sparing you the hairy details. (Actually, somewhat to my surprise, things work out rather elegantly, at least for the second and third families.)

In the supersymmetric standard model, which we want to recover at low energy, there are two electroweak doublets. These emerge as the dregs of a mass-generation process that gives superheavy masses $\sim M_U$ to all the other Higgs fields. In general, these dregs will be made up of bits and pieces coming from different irreducible $SO(10)$ multiplets, i.e. adjoints, fundamentals, and spinors.

Now a fundamental $10_H$ will couple to the matter 16s by a term of the form $g_{ij} 16_i \cdot 16_j \cdot 10_H$, where $i, j$ are family indices and the $g_{ij}$ are coupling constants. Group theory requires that $g_{ij}$ is symmetric in $i$ and $j$. When the $10_H$ acquires a vacuum expectation value, these couplings will contribute to the observable fermion mass matrices. The group theory also correlates the contributions to different quark and lepton mass matrices.

Similarly an effective coupling of the type $h_{ij} 16_i \cdot 16_j \cdot 10_H \cdot 45_H$, involving an adjoint field, can arise. Indeed, to implement a clean gauge symmetry breaking with doublet-triplet splitting we need a very specific form for the vacuum expectation value of $45_H$ (in the B-L direction). Group theory determines that $h_{ij}$ is antisymmetric in $i$ and $j$, and the required alignment of the $45_H$ introduces various factors of 3 (Georgi-Jarlskog (16) factors) into the relative contributions from this term to quark and lepton mass matrices.

By exploiting structures of this sort, and taking guidance from experiment, one can construct remarkably simple and overconstrained, yet not unrealistic, models of quark and lepton masses (3, 17, 18).

## numerical estimates; conclusion

In our long paper, we computed the numerical consequences of a complete model of this sort. In constructing the model we were forced to make several uncertain choices for the Higgs structure and couplings, and in getting to decay rates we were forced to make several further uncertain estimates of SUSY breaking parameters and strong matrix elements. I wish we could do better. With respect to the microscopic theory I'm afraid the situation is unlikely to improve dramatically any time soon. To make progress, we desperately need to open a dialogue with Nature, through experiment. On the other hand, given sufficient investment in numerical QCD one could improve the estimation of matrix elements. That direction should certainly be pursued, in order to insure that it will be possible to interpret the results properly when and if they do come in.

In any case, doing the best we know how, and with all our cards on the table, Babu, Pati and I by honest toil find

$$\Gamma^{-1}(p) \lesssim 10^{34} \text{yrs.} \tag{12}$$

within the circle of ideas here advocated. The dominant modes involve strange particles in the final state, and usually (though not necessarily) antineutrinos. The detailed branching ratios, and the nature of the subdominant modes, encode information on additional aspects of unification physics, which is very difficult to access otherwise.

If nucleon instability at these levels were observed it would constitute one of the greatest discoveries in the history of physics, and provide a unique window looking out into the deep structure of physical reality.

Research supported in part by DOE grant DE-FG02-90ER40542.

## REFERENCES

1. Reviewed in C. Bachas, hep-th/000193 (2000).
2. SuperKamiokande Collaboration, Y. Fukuda et. al., Phys. Rev. Lett. **81**, 1562 (1998).
3. J. Pati, K. Babu, and F. Wilczek, hep-ph/9812538, Nucl. Phys. **B**, in press.
4. J.C. Pati and A. Salam, Phys. Rev. Lett. **31**, 661 (1973) and Phys. Rev. **D10**, 275 (1974).
5. H. Georgi and S.L. Glashow, Phys. Rev. Lett. **32**, 438 (1974).
6. H. Georgi, in Particles and Fields, AIP, New York (1975), p. 575 (C.E. Carlson, Ed);
   H. Fritzsch and P. Minkowski, Ann. Phys. **93**, 193 (1975).
7. LEP Electroweak Working Group, preprint CERN-PPE/96-183 (Dec. 1996).

8. D. Gross and F. Wilczek, Phys. Rev. Lett. **30**, 1343 (1973); H. D. Politzer, Phys. Rev. Lett. **30**, 1346 (1973).

9. H. Georgi, H. Quinn, and S. Weinberg, Phys. Rev. Lett. **33**, 451 (1974)

10. See for example G. Blewitt, *et al*, Phys. Rev. Lett. **55**, 2114 (1985), and the latest Particle Data Group compilations.

11. A very useful introduction and collection of basic papers on supersymmetry is S. Ferrara, *Supersymmetry* (2 vols.) (World Scientific, Singapore 1986). Another excellent standard reference is N.-P. Nilles, Phys. Reports **110**, 1 (1984).

12. S. Dimopoulos, S. Raby, and F. Wilczek, Phys. Rev. D **24**, 1681 (1981).

13. M. Gell-Mann, P. Ramond, and R. Slansky, in *Supergravity*, ed. P. van Neiuwenhuizen and D. Freedman (North Holland, Amsterdam, 1979), p. 315; T. Yanagida, Proc. of the Workshop on Unified Theory and Baryon Number in the Universe, eds. O. Sawada and A. Sugamoto (KEK, 1979).

14. P. Horava and E. Witten, Nucl. Phys. **B460** 506 (1996); R. Donagi, B. Ovrut, T. Pantev, and D. Waldram hep-th/0001101 (2000).

15. S. Dimopoulos and F. Wilczek, Report No. NSF-ITP-82-07 (1981), in *The unity of fundamental interactions*, Proc. of the 19th Course of the International School on Subnuclear Physics, Erice, Italy, 1981, Plenum Press, New York (Ed. A. Zichichi);
K.S. Babu and S.M. Barr, Phys. Rev. **D48**, 5354 (1993).

16. H. Georgi and C. Jarlskog, Phys. Lett. **B 86**, 297 (1979).

17. S. Barr and C. Albright, hep-ph/0001052 (2000) and these proceedings.

18. S. Barr and S. Raby, Phys. Rev. Lett. **79**, 4748 (1997).

# Matter Effects on Long Baseline Neutrino Oscillation Experiments

Irina Mocioiu[a] and Robert Shrock[a,b]

*(a) C.N.Yang Institute for Theoretical Physics
State University of New York, Stony Brook, NY 11794
(b) Physics Department, Brookhaven National Laboratory
Upton, NY 11973*

**Abstract.**
We calculate matter effects on neutrino oscillations relevant for long baseline neutrino oscillation experiments. In particular, we compare the results obtained with simplifying approximations for the density profile in the Earth versus results obtained with actual density profiles. We study the dependence of the oscillation signals on both $E/\Delta m^2_{atm.}$ and on the angles in the leptonic mixing matrix. The results show quantitatively how matter effects can cause significant changes in the oscillation signals, relative to vacuum oscillations and can be useful in amplifying these signals and helping one to obtain measurements of mixing parameters and the magnitude and sign of $\Delta m^2$.

## INTRODUCTION

In a modern theoretical context, one generally expects nonzero neutrino masses and associated lepton mixing. Experimentally, there has been accumulating evidence for such masses and mixing. All solar neutrino experiments (Homestake, Kamiokande, SuperKamiokande, SAGE, and GALLEX) show a significant deficit in the neutrino fluxes coming from the Sun (1). This deficit can be explained by oscillations of the $\nu_e$'s into other weak eigenstate(s), with $\Delta m^2_{sol}$ of the order $10^{-5}$ eV$^2$ for MSW solutions (2) or of the order of $10^{-10}$ eV$^2$ for vacuum oscillations. Accounting for the data with vacuum oscillations requires almost maximal mixing. The MSW solutions include one for small mixing angle (SMA) and one with essentially maximal mixing (LMA).

Another piece of evidence for neutrino oscillations is the atmospheric neutrino anomaly, observed by Kamiokande, SuperKamiokande, IMB, MACRO, and Soudan-2 (3). Of these, the Superkamiokande data has especially high statistics - roughly 52 kton-years worth of data at present. This data can be well fit by the inference of $\nu_\mu \to \nu_x$ oscillations with $\Delta m^2_{atm} \sim 3.5 \times 10^{-3}$ eV$^2$ (3) and maximal mixing $\sin^2 2\theta_{atm} = 1$, where $\nu_x = \nu_\tau$ is favored. The possibility $\nu_x = \nu_{sterile}$ is disfavored at the $2\sigma$ level (5). (The possibility that $\nu_x$ is predominantly $\nu_e$ is ruled out by both the Superkamiokande data and the CHOOZ experiment (4)).

In addition, the LSND experiment has reported observing $\bar\nu_\mu \to \bar\nu_e$ and $\nu_\mu \to \nu_e$ oscillations with $\Delta m^2_{LSND} \sim$ $0.1 - 1$ eV$^2$ and moderately small mixing angle. This result is not confirmed by a similar experiment, KARMEN (6).

There are currently strong efforts to confirm and extend the evidence for neutrino oscillations in all of the various sectors – solar, atmospheric, and accelerator. Some of these are currently running: the Sudbury Neutrino Observatory, SNO, the K2K pioneering long baseline experiment between KEK and Kamioka. Others are in development and testing phases, such as Borexino, KamLAND, MINOS, mini-BOONE, and the CERN-Gran Sasso program. Among the long baseline neutrino oscillation (LBLNO) experiments, the distances are $L \simeq 250$ km for K2K, 730 km for both MINOS, from Fermilab to Soudan and the proposed CERN-Gran Sasso experiments. The sensitivity of these experiments should reach the region $\Delta m^2 \sim$ few $\times 10^{-3}$eV$^2$. Another generation of experiments, with even higher sensitivity will be required for precision measurements of oscillation parameters. One of the physics capabilities of the Next generation Nucleon decay and Neutrino detector discussed at this NNN99 workshop would be as part of a LBLNO experiment. An interesting possibility that is being studied intensively is a muon collider or storage ring that would serve as a source of quite high intensity, flavor-pure ($\nu_\mu + \bar\nu_e$ beams from $\mu^-$ and $\bar\nu_\mu + \nu_e$ beam from $\mu^+$) (anti)neutrino beams. Using these, one could perform LBLNO experiments with an existing deep underground detector, e.g., at Soudan, Gran Sasso, or Kamioka, the NNN detector, and/or a surface detector. Studies have shown that one can get hundreds of events per kiloton-year at distances of 7000-9000 km (7), (8). It is thus appropriate to begin planning for this next generation of very long baseline neutrino oscillation experiments.

An important effect that must be taken into account in such experiments is the matter-induced oscillations which neutrinos undergo along their flight path through the Earth from the source to the detector. In a hypothetical world in which there were only two neutrinos, $\nu_\mu$ and $\nu_\tau$, the $\nu_\mu \to \nu_\tau$ oscillations in matter would be the same as in vacuum, since both have the same forward scattering amplitude, via $Z$ exchange, with matter. However, in the realistic case of three generations, because of the indirect involvement of $\nu_e$ due to a nonzero $U_{13}$, and because of the fact that $\nu_e$ has a different forward scattering amplitude off of electrons, involving both $Z$ and $W$ exchange, there will be a matter-induced oscillation effect on $\nu_\mu \to \nu_\tau$ (as well as other channels). Early studies of matter effects are (9). Several recent studies are (8),(10).

Here we shall report on a study that we have carried out (11) of matter effects relevant to LBLNO experiments. We consider the usual three flavors of active neutrinos, with no light sterile (= electroweak-singlet) neutrinos. This is sufficient to describe the more established evidence from the solar and atmospheric neutrino deficit (if one tried to fit also the LSND experiment with a neutrino oscillation scenario, one would be led to include light sterile neutrinos). As suggested by the solar and atmospheric data, we consider that there is only one mass scale relevant for long baseline and atmospheric neutrino oscillations, $\Delta m^2_{atm} \sim$ few $\times 10^{-3}$ eV$^2$ and we work with the hierarchy

$$\Delta m^2_{21} = \Delta m^2_{sol} \ll \Delta m^2_{31} \approx \Delta m^2_{32} = \Delta m^2_{atm} \quad (1)$$

In our work we take into account the actual profile of the Earth, as given by geophysical seismic data (12) and compare the results with those calculated using the approximation of average density along the path of the neutrino. Further, we present the oscillation probabilities as functions of $E/\Delta m^2$ so one can determine which energies are best suited for precise measurements of $\Delta m^2$ in a given region. We study how these oscillation probabilities vary with the different input parameters and discuss the influence of the matter effects on the sensitivity to each of these parameters.

## MATTER EFFECTS

The evolution of the flavor eigenstates is given by

$$i\frac{d}{dx}\nu = \left(\frac{1}{2E}UM^2U^\dagger + V\right)\nu \quad (2)$$

where the flavor neutrino wavefunction is

$$\nu = U\nu_m \quad (3)$$

in terms of the mass eigenstates

$$\nu_m = \begin{pmatrix} \nu_1 \\ \nu_2 \\ \nu_3 \end{pmatrix} \quad (4)$$

and

$$M^2 = \begin{pmatrix} m_1^2 & 0 & 0 \\ 0 & m_2^2 & 0 \\ 0 & 0 & m_3^2 \end{pmatrix}, V = \begin{pmatrix} \sqrt{2}G_F N_e & 0 & 0 \\ 0 & 0 & 0 \\ 0 & 0 & 0 \end{pmatrix} \quad (5)$$

where $N_e$ is the electron number density and $\sqrt{2}G_F N_e [\text{eV}] = 7.6 \times 10^{-14} Y_e \rho$ [g/cm$^3$].

The leptonic mixing matrix $U$ can be written as

$$U = R_{23}KR_{13}K^*R_{12}K' \quad (6)$$

which is the standard CKM-type parametrization, with $R_{ij}$ being the rotation matrix in the $ij$ subspace, $c_{12} = \cos\theta_{12}$, $s_{12} = \sin\theta_{12}$, etc., $K = diag(e^{-i\delta}, 1, 1)$, and $K' = diag(e^{i\alpha_1}, e^{i\alpha_2}, 1)$ (the latter phases originate from the general presence of Majorana mass terms but will not be important here).

The atmospheric neutrino data suggests almost maximal mixing in the $2 - 3$ sector. However, a small but non-zero $s_{13}$ is still allowed, and this produces the matter effect in the traversal of neutrinos through the Earth. We use $\sin^2(2\theta_{13}) \leq 0.1$, consistent with the limits from the atmospheric neutrino data (3) and the CHOOZ experiment (4). We also assume the small mixing angle (SMA) MSW solution to the solar neutrino data. This assumption, together with the hierarchy of eq. (1), implies that, for the relevant energies $E \gtrsim 1$ GeV and pathlengths $L \sim 10^3 - 10^4$ km, only one squared mass scale, $\Delta m^2_{atm}$, is important for the oscillations, and the three-species neutrino oscillations can be described in terms of this quantity, $\Delta m^2_{atm}$, and the mixing parameters $\sin^2(2\theta_{23})$, and $\sin^2(2\theta_{13})$, with negligible dependence on $\sin^2(2\theta_{12})$ and $\delta$; hence also, CP violation effects would be negligibly small here, and $P(\nu_a \to \nu_b) = P(\nu_b \to \nu_a)$, $P(\bar\nu_a \to \bar\nu_b) = P(\bar\nu_b \to \bar\nu_a)$. Although, a priori, CP violation would lead to $P(\nu_a \to \nu_b) \neq P(\bar\nu_a \to \bar\nu_b)$ in vacuum, this inequality is true in matter even in the absence of CP violation.

For our purposes, we recall that the Earth is composed of crust, mantle, liquid outer core, and solid inner core, together with additional sublayers in the mantle, with particularly strong changes in density between the lower mantle and outer core. The density profile of the Earth is shown in fig. 1 from (12). The core has average density $\rho_{core} = 11.83$ g/cm$^3$ and electron fraction $Y_{e,core} = 0.466$, while the mantle has average density $\rho_{mantle} = 4.66$ g/cm$^3$

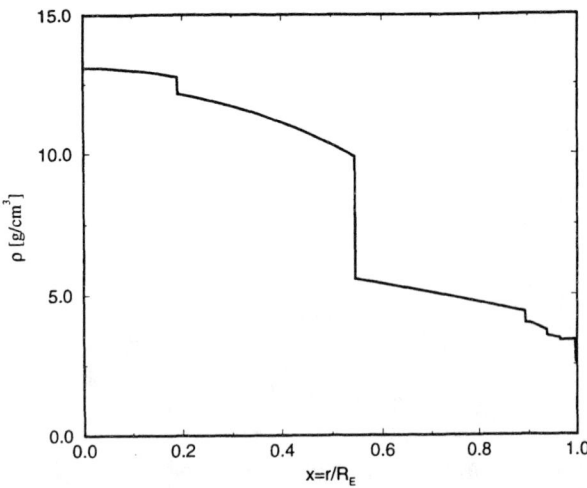

**FIGURE 1.** Density profile of the Earth

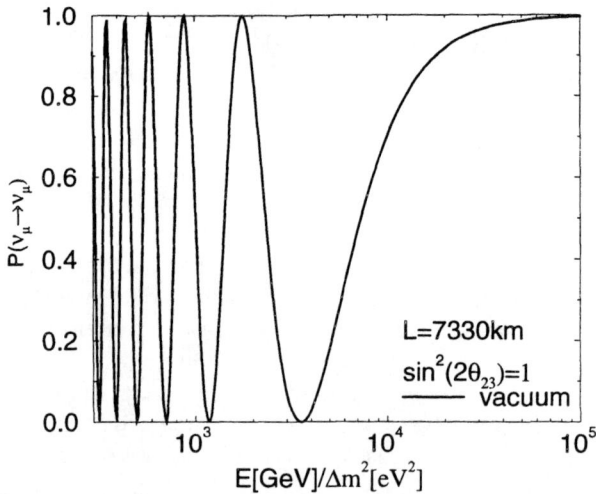

**FIGURE 2.**

and $Y_{e,mantle} = 0.494$. If one approximates the density as a constant along the neutrino flight path, the evolution equation can easily be solved, with well-known results. However, when one takes account of the actual variable-density situation in the earth, it is necessary to perform a numerical integration of the evolution equation, which we have done.

## RESULTS AND DISCUSSION

For long baseline experiments like K2K, Fermilab to Minos, and CERN to Gran-Sasso, the neutrino flight path only goes through the upper mantle. The density in this region is practically constant, and the oscillation probabilities can easily be calculated. The matter effects are small, but possibly detectable for the longer baselines. However, there are several motivations for very long baseline experiments, since, with sufficiently high-intensity sources, these can be sensitive to quite small values of $\Delta m^2$ and since the matter effects, being larger, can amplify certain oscillations and can, in principle, be used to get information on the sign of $\Delta m^2_{atm}$. Hence we concentrate here on these very long baseline experiments; for these, the neutrino flight path goes through several layers of the Earth with different densities, including the lower mantle. We show results for $L \simeq 7330$ km, the distance from Fermilab to Gran Sasso. We have also performed calculations for the Fermilab to SuperKamiokande and Fermilab to SLAC path lengths, $\sim 9120$ and $2880$ km, respectively. We calculate the probabilities of oscillation in long baseline experiments as a function of $E/\Delta m^2$, rather than using a particular value for $\Delta m^2$ or the energy. The relevant ranges are $\Delta m^2 \sim$ few $\times 10^{-3}$ eV$^2$ and energies

$E$ of the order of tens of GeV. This way of presenting the results can be useful in studying the optimization of the beam energy. In our work (11) we calcualte the oscillation probabilities for different values of the mixing angles $\theta_{13}$ and $\theta_{23}$ allowed by the atmospheric neutrino data and the CHOOZ experiment; for this workshop report we only show results for $\sin^2(2\theta_{23}) = 1$. We consider both neutrinos and antineutrinos. The matter effects change sign in these two cases; for antineutrinos, $V$ in (5) is replaced with $(-V)$. This implies that if $\Delta m^2$ is positive (as considered here), one can get a resonant enhancement of the oscillations for neutrinos, while for antineutrinos the matter effects would suppress the oscillations. The situation would be reversed if $\Delta m^2$ were negative.

We first study the survival probability of $\nu_\mu$. If the beam went through vacuum, the oscillation probability would look like the curve in fig. 2 for practically any value of $\sin^2(2\theta_{13})$. In matter, this probability becomes sensitive to all oscillation parameters, as can be seen from fig. 3 and fig. 4.

We also want to compare the solution in vacuum (fig. 2) with the solution in matter for neutrinos (fig. 3) and antineutrinos (fig. 5). In the legends for the figures with antineutrinos, "anti $\nu_a \to \nu_b$" means $\bar{\nu}_a \to \bar{\nu}_b$. One can see the opposite effects of matter on neutrinos and antineutrinos. The difference in the results for different mixing angles makes it possible in principle to use this probability for relatively precise measurements of the oscillation parameters. Measuring separately the probability for $\nu$ and $\bar{\nu}$ can be very useful in detecting the matter effects and using these to constrain the relevant mixings and squared mass difference. Clearly, if one could use two path lengths, as may be possible with a neutrino factory, this would provide more information and constraints.

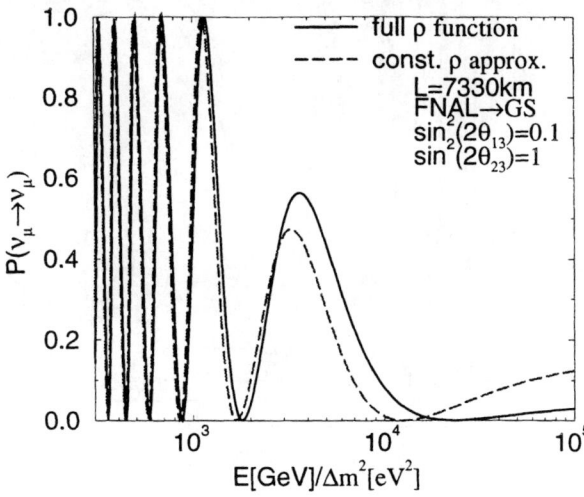

**FIGURE 3.**

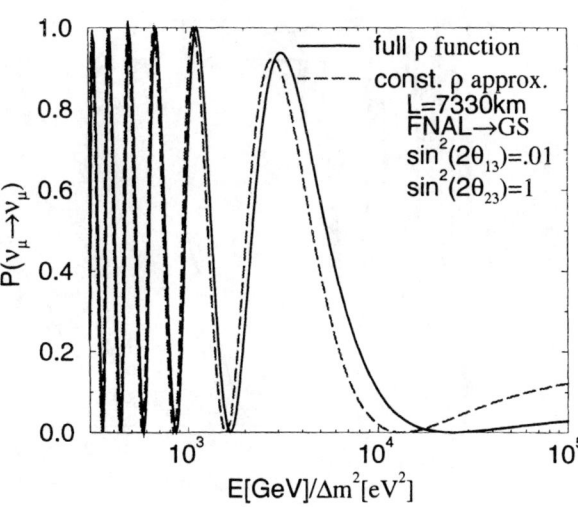

**FIGURE 4.**

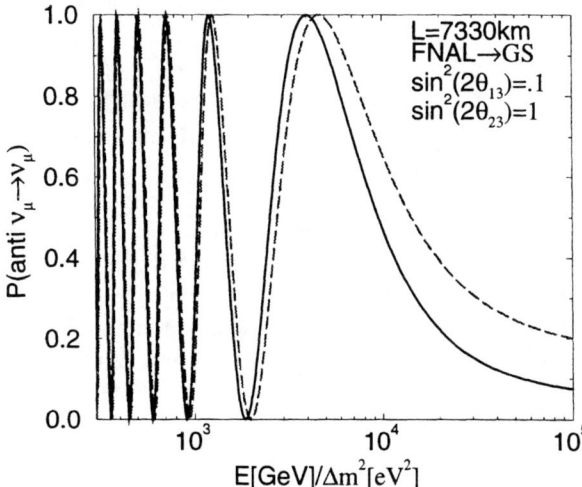

**FIGURE 5.**

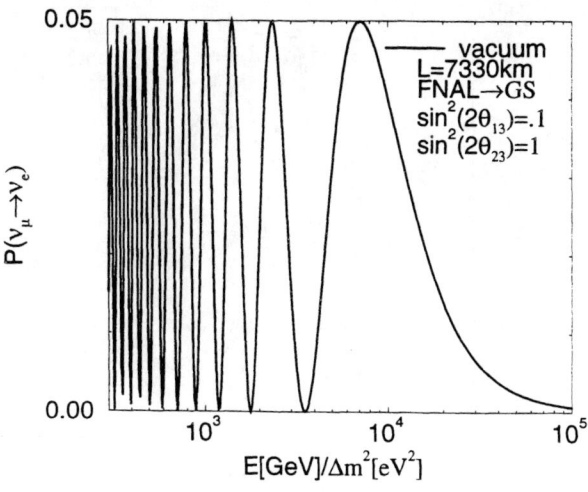

**FIGURE 6.**

The relative effects of matter can be especially dramatic in the oscillation probability $P(\nu_\mu \to \nu_e)$, since these directly involve $\nu_e$. If the beam were to go through the vacuum, $P(\nu_\mu \to \nu_e)$ would be probably too small to detect (fig. 6). Because of the matter effect however, this probability can be strongly enhanced, as is evident in fig. 7. The enhancement is largest for $E/\Delta m^2$ around 3000 GeV/eV$^2$. This is essentially equal to the ratio that one would get using a beam energy of $\sim 10$ GeV, given the indication from the data that $\Delta m_{atm} = 3.5 \times 10^{-3}$ eV$^2$. Hence the matter effect can amplify $P(\nu_\mu \to \nu_e)$ and enable this transition to be measured with reasonable accuracy, thereby yielding important information on the oscillation parameters. This probability is very sensitive to the value of $\theta_{13}$ (figs. 7, 8), so one could use it for a good determination of this angle. The sensitivity to $\Delta m^2$ is also quite strong, due to the pronounced peak given by the matter effect in the relevant region. Note that for antineutrinos, the oscillation is suppressed (fig. 9), so an independent measurement of the two channels ($\nu_\mu \to \nu_e$ and $\bar{\nu}_\mu \to \bar{\nu}_e$) would be very valuable.

The atmospheric neutrino data tells us that the dominant oscillation channel is actually $\nu_\mu \to \nu_\tau$. Consequently, it would be very useful to measure $P(\nu_\mu \to \nu_\tau)$; this would provide further confirmation of this oscillation and could also provide accurate determinations of $\Delta m^2$ and $\theta_{23}$. Fig. 10 shows $P(\nu_\mu \to \nu_\tau)$. Fig. 11 shows $P(\bar{\nu}_\mu \to \bar{\nu}_\tau)$.

Since with a muon collider or muon storage ring, $\nu_e$ ($\bar{\nu}_e$) beams would also be available, it would be interesting to study oscillation probabilities with these beams. We already have the results for $P(\nu_e \to \nu_\mu)$ since, as mentioned above, with our parameters, this is the same as $P(\nu_\mu \to \nu_e)$. We present here $P(\nu_e \to \nu_\tau)$ in fig. 12 and

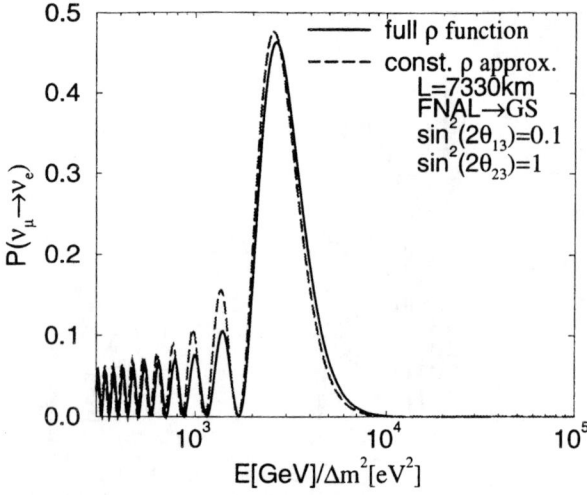

**FIGURE 7.**

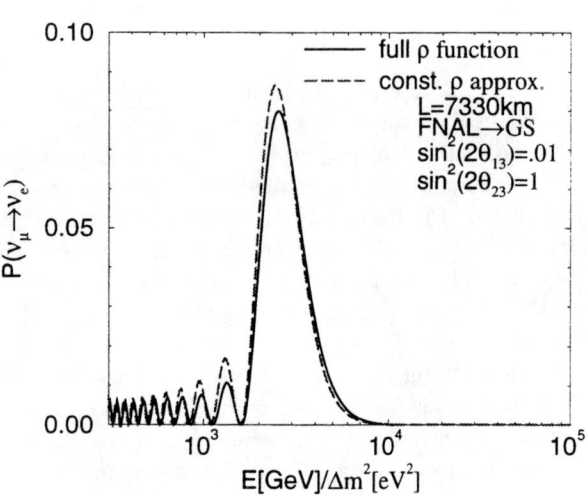

**FIGURE 8.**

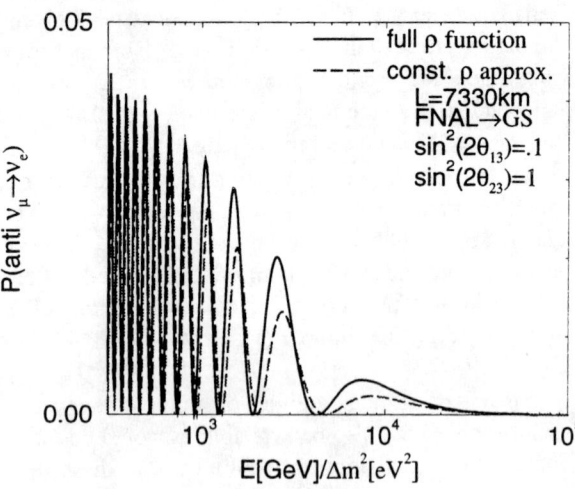

**FIGURE 9.**

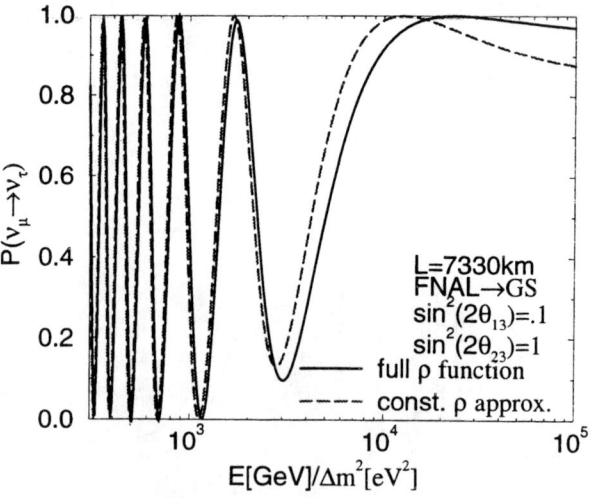

**FIGURE 10.**

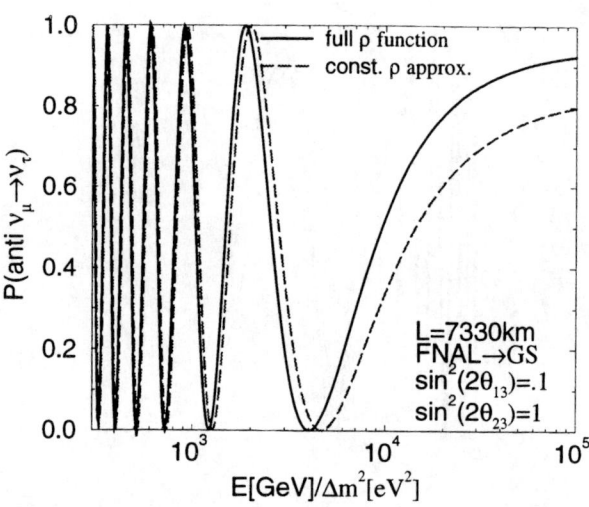

**FIGURE 11.**

$P(\bar{\nu}_e \to \bar{\nu}_\tau)$ in Fig. 13. These calculations show that matter effects are important and enhance oscillations of the neutrinos and suppress oscillations of antineutrinos in the relevant region of parameters.

To summarize, in planning for very long baseline neutrino oscillation experiments, it is important to take into account matter effects. We have performed a careful study of these, including realistic density profiles in the earth. Matter effects can be useful in amplifying neutrino oscillation signals and helping one to obtain measurements of mixing parameters and the magnitude and sign of $\Delta m^2$.

We thank Debbie Harris for a helpful comment. The research of R. S. was supported in part at Stony Brook by

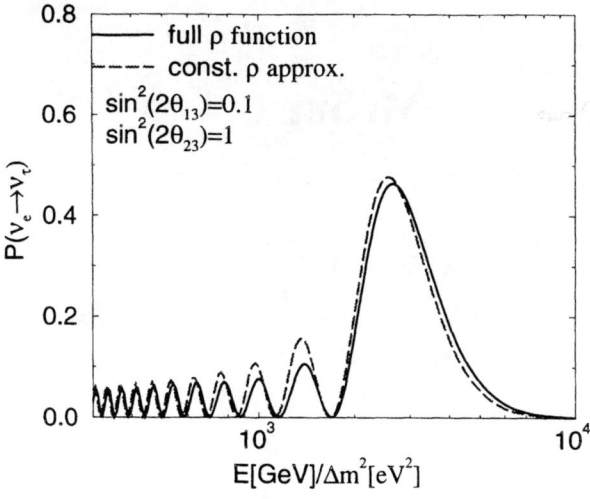

**FIGURE 12.**

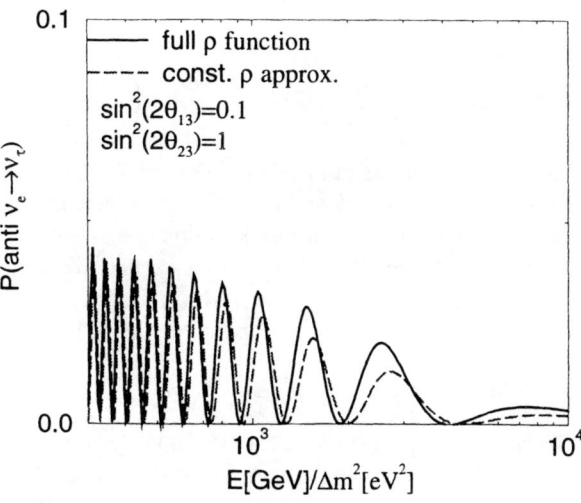

**FIGURE 13.**

the U. S. NSF grant PHY-97-22101 and at Brookhaven by the U.S. DOE contract DE-AC02-98CH10886. Accordingly, the U.S. government retains a non-exclusive royalty-free license to publish or reproduce the published form of this contribution or to allow others to do so for U.S. government purposes.

## REFERENCES

1. Fits and references to the data include J. Bahcall, P. Krastev, A. Smirnov, Phys. Rev. **D58**, 096016 (1998); J. Bahcall and P. Krastev, Phys. Lett. **B436**, 243 (1998). Recent Superkamiokande data is reported in SuperKamiokande Collab., Y.Fukuda et al., Phys. Rev. Lett. **82**, 1810, 2430 (1999).

2. L. Wolfenstein, Phys. Rev. **D17**, 2369 (1978); S. P. Mikheyev and A. Smirnov, Yad. Fiz. **42**, 1441 (1985) [Sov.J. Nucl. Phys. **42**, 913 (1986)], Nuovo Cim., **C9**, 17 (1986).

3. Kamiokande Coll., Y.Fukuda et al., Phys. Rev. Lett. **B335**, 237 (1994).

    IMB Coll., R.Becker-Szendy et al., Nucl. Phys.**B38**(Proc. Suppl.), 331 (1995).

    Soudan Coll., W.W.M. Allison et al, Phys. Lett. **B391**, 491 (1997).

    MACRO Coll., M. Ambrosio et al., Phys. Lett. **B434**, 451 (1998).

    Super-Kamiokande Coll., Y. Fukuda et al., Phys. Rev. Lett. **81**,1562 (1998); *ibid.*, **82**, 2644 (1999); hep-ex/9908049.

4. M. Apollonio et al., Phys. Lett. **B420**, 397 (1998); hep-ex/9907037.

5. J. Learned, this workshop.

6. LSND Coll., C. Athanassopoulous et al., Phys. Rev. Lett. **77**, 3082 (1996), LSND Coll., C. Athanassopoulous et al., Phys. Rev. Lett. **81**, 1774 (1998).

    KARMEN Coll., Nucl. Phys. Proc. Suppl. **70**, 210 (1999).

7. S. Geer, Phys.Rev.**D57**, 6989 (1998), erratum-ibid. **D59**, 039903 (1999).

8. V. Barger, S. Geer, K. Whisnant, hep-ph/9906487.

9. V. Barger, K. Whisnant, S. Pakvasa, R.J. Phillips, Phys Rev **D22**, 2718 (1980)

    A. J. Baltz, J. Weneser, Phys. Rev. **D37**, 3364 (1988).

    P. Krastev, S. Petcov, Phys. Lett. **B205**, 8 (1988).

    R. H. Bernstein, S.J. Parke, Phys Rev **D44**, 2069 (1991).

    G. Fogli, E. Lisi, D. Montanino, Phys.Rev. D49, 3626 (1994).

    E. Lisi, D. Montanino, Phys. Rev. **D56**, 1792 (1997).

10. E. Akhmedov, Nucl.Phys. **B538**, 25 (1999).

    E. Akhmedov, A. Dighe, P. Lipari, A. Smirnov, Nucl. Phys. **B542**, 3 (1999).

    De Rujula, M. B. Gavela, P. Hernandez, Nucl. Phys. **B547**, 21 (1999).

    M. Chiznov, M. Maris, S. Petcov, hep-ph/9810501.

    M. Chiznov, S. Petcov, hep-ph/9903424.

    P. Lipari, hep-ph/9903481.

    M. Campanelli, A. Bueno, A. Rubbia, hep-ph/9905240.

    D. Dooling, C. Giunti, K. Kang, C. W. Kim, hep-ph/9908513.

11. I. Mocioiu. R. Shrock, YITP-SB-99-13.

12. A.Dziewonski, Earth Structure, in: "The Encyclopedia of Solid Earth Geophysics", D.E.James (Ed.), (Van Nostrand Reinhold, New York, 1989), p.331.

# $SO(10)$ and Large $\nu_\mu - \nu_\tau$ Mixing

S.M. Barr[1] and C.H. Albright[2]

[1] *Bartol Research Institute, University of Delaware, Newark DE, 19716*
[2] *Fermi National Accelerator Laboratory, Batavia IL, 60510*

**Abstract.** A general approach to understanding the large mixing seen in atmospheric neutrinos is explained, as well as a highly predictive $SO(10)$ model which implements this approach. It is also seen how bimaximal mixing naturally arises in this scheme.

The problem of neutrino mixing should not be looked at in isolation, but as part of the larger "fermion mass problem", i.e. the problem of understanding the masses and mixings of the quarks and leptons. The fermion mass problem is very old; serious efforts at model building go back more than twenty-five years, and in that time hundreds of models of quark and lepton masses have been published. However, we are now in a new and more hopeful situation, for neutrino oscillations give us precious new clues in the search for the right theory as well as new ways of testing theories experimentally.

The most significant new clue, perhaps, is the largeness of the mixing of $\nu_\mu$ (presumably with $\nu_\tau$) seen at SuperK. What is this clue telling us? I will explain one idea for what it might be telling us which is based on certain features of grand unification. I will then mention and very briefly discuss a concrete model that incorporates this idea. (Actually, the model came first, and only then was it noticed that the model contains an attractive explanation of the large $\nu_\mu - \nu_\tau$ mixing!) At the end I will address the question whether large $\nu_e$ mixing is compatible with $SO(10)$ in a simple way. The answer is yes.

Large $\nu_\mu - \nu_\tau$ mixing came as a surprise. The puzzle is that the mixing of the second and third families is small for the quarks ($V_{cb} \cong 0.04$) and large for the leptons ($U_{\mu 3} \cong 1/\sqrt{2} \cong 0.7$). This is puzzling because both grand unification and flavor symmetry, the two most promising ideas for explaining fermion masses, tend to relate the quark and lepton parameters. Actually, there are *two* puzzles: the mixing of the second and third families is *too small for the quarks* and *too big for the leptons*, in a sense that I will now explain.

Many models are based on the old idea of Weinberg and Wilczek and Zee.[1] Looking at only the first two families, in the late 1970's they posited simple textures of the form

$$\overline{u_{iR}} U_{ij} u_{jL} = (\overline{u_R}, \overline{c_R}) \begin{pmatrix} 0 & a \\ a & b \end{pmatrix} \begin{pmatrix} u_L \\ c_L \end{pmatrix} \quad (1)$$

and

$$\overline{d_{iR}} D_{ij} d_{jL} = (\overline{d_R}, \overline{s_R}) \begin{pmatrix} 0 & a' \\ a' & b' \end{pmatrix} \begin{pmatrix} d_L \\ s_L \end{pmatrix}. \quad (2)$$

The crucial features of these matrices are that they are *hierarchical* (i.e. $a \ll b$ and $a' \ll b'$) and *symmetric*. The eigenvalues of the down quark matrix are $m_s \cong b'$, and $m_d \cong -a'^2/b'$, and the rotation angle needed to diagonalize it is $\theta_{ds} \cong a'/b' \cong \sqrt{m_d/m_s}$, with analogous formulas for the up quarks. Thus the Cabibbo angle is given by

$$V_{us} \cong \sqrt{m_d/m_s} - e^{i\phi_{12}}\sqrt{m_u/m_c}. \quad (3)$$

Since $V_{us} \cong 0.21$, $\sqrt{m_d/m_s} \cong 0.21$, and $\sqrt{m_u/m_c} \cong 0.07$, this relation works well for $\phi_{12} \sim \pi/2$. If one chooses analogous hierarchical and symmetric forms for the lepton mass matrices one obtains

$$U_{e2} \cong \sqrt{m_e/m_\mu} - e^{i\phi'_{12}}\sqrt{m_{\nu_e}/m_{\nu_\mu}}. \quad (4)$$

This also can work reasonably well if the small angle MSW solution to the solar neutrino problem is correct, as then $U_{e2} \sim 0.04$, and $\sqrt{m_e/m_\mu} \cong 0.07$.

If one supposes that the full $3 \times 3$ mass matrices are symmetrical and hierarchical, one obtains similar predictions for the mixing of the second and third families, For example, extending the Weinberg-Wilczek-Zee pattern to the third family, as Fritzsch did,[2] one finds

$$V_{cb} \cong \sqrt{m_s/m_b} - e^{i\phi_{23}}\sqrt{m_c/m_t}, \quad (5)$$

and

$$U_{\mu 3} \cong \sqrt{m_\mu/m_\tau} - e^{i\phi'_{23}}\sqrt{m_{\nu_\mu}/m_{\nu_\tau}}. \quad (6)$$

Since $\sqrt{m_s/m_b} \cong 0.14$, and $\sqrt{m_c/m_t} \cong 0.04$, one sees that the experimental value of the quark mixing $V_{cb}$ ($\cong 0.04$) is about a factor of 3 smaller than the Fritzschian expectation. On the other hand, since $\sqrt{m_\mu/m_\tau} \cong 0.24$, and $\sqrt{m_{\nu_\mu}/m_{\nu_\tau}}$ may be presumed to be small, one sees that the experimental value of the lepton mixing $U_{\mu 3}$ ($\cong 0.7$) is about a factor of 3 larger than the Fritzschian expectation.

What we have argued in several papers[3,4,5] is that the trouble with such Fritzsch-like textures when applied to the heavier two families is that they are based on symmetric matrices. Let us see what happens if there are instead highly asymmetric or, as we have called them, "lopsided" textures. Consider a toy example with matrices

$$\overline{d_{iR}} D_{ij} d_{jL} = m(\overline{d_R}, \overline{s_R}, \overline{b_R}) \begin{pmatrix} - & - & - \\ - & 0 & \sigma \\ - & \varepsilon & 1 \end{pmatrix} \begin{pmatrix} d_L \\ s_L \\ b_L \end{pmatrix}, \quad (7)$$

$$\overline{\ell_{iR}} L_{ij} \ell_{jL} = m(\overline{e_R}, \overline{\mu_R}, \overline{\tau_R}) \begin{pmatrix} - & - & - \\ - & 0 & \varepsilon \\ - & \sigma & 1 \end{pmatrix} \begin{pmatrix} e_L \\ \mu_L \\ \tau_L \end{pmatrix}, \quad (8)$$

where $\varepsilon \ll \sigma \sim 1$. We are interested at the moment in the heavier two families, so we do not write the entries of the first row and column, which are assumed to be small. The important feature of these matrices is the large, lopsided off-diagonal entry ($\sigma$). For the mass matrices of the up quarks and neutrinos we assume that there is no such large off-diagonal element. Note another very important feature of these matrices, which is that the charged lepton mass matrix $L$ is the *transpose* of the down quark mass matrix $D$. $L = D^T$ is a "minimal $SU(5)$" relation. The realistic model we will discuss later is based on $SO(10)$, nevertheless the $SU(5)$ subgroup of $SO(10)$ in that model relates $L$ to $D^T$. The reason $L$ is related to $D^T$ is simple: $SU(5)$ unifies the $\ell_L^-$ with the $d_R$ in $\bar{5}$ multiplets and the $\ell_R^-$ with the $d_L$ in $10$ multiplets. Thus the mass matrix of the charged leptons is related to that of the down quarks only up to a left-right transposition. As we shall see, this feature allows a simple explanation of the double puzzle of $U_{\mu 3}$ and $V_{cb}$.

The point is that the observed mixings are of *left-handed* fermions. Specifically, we may write $U_{\mu 3} \cong \theta_{\mu\tau}^{left} - \theta_{\nu_\mu\nu_\tau}^{left}$, and $V_{cb} \cong \theta_{sb}^{left} - \theta_{ct}^{left}$. Thus $V_{cb}$ and $U_{\mu 3}$ are really not directly related to each other by $SU(5)$; rather each is related to some *right-handed* mixing angle. Specifically, $SU(5)$ relates $\theta_{sb}^{left}$ to $\theta_{\mu\tau}^{right}$, and in the toy example both are of order $\varepsilon$, as Eqs. (7) and (8) show. And $SU(5)$ relates $\theta_{\mu\tau}^{left}$ to $\theta_{sb}^{right}$, and in the toy example both are of order $\sigma$. Moreover, from the form of the matrices in Eqs. (7) and (8) one sees that the ratio of masses of the fermions of the second and third families are given by $m_2/m_3 \sim \sigma\varepsilon$. Thus, given the experimental values of the quark and lepton mass ratios, the values of $\varepsilon$ and $\sigma$ are inversely related, so that the smallness of $V_{cb}$ and the largeness of $U_{\mu 3}$ are two sides of the same coin. The Fritzschian expectation would be that the mixing angles $\theta_{23}^{Fritzschian}$ are of order $\sqrt{m_2/m_3}$, which as we just have seen is to say that they are of order $\sqrt{\sigma\varepsilon}$. However, the form of Eqs. (7) and (8) tells us that actually $V_{cb} \sim \varepsilon$ and $U_{\mu 3} \sim \sigma$. Thus, as observed, $V_{cb}/\theta_{23}^{Fritzschian} \sim \theta_{23}^{Fritzschian}/U_{\mu 3} \sim \sqrt{\varepsilon/\sigma} < 1$.

A striking fact about the mechanism just described is that the large mixing seen in atmospheric neutrino data is coming from a large off-diagonal entry in $L$, the mass matrix of the *charged* leptons. One is accustomed to speak of "neutrino mixing angles", but this is obviously a misnomer, since the leptonic angles are really the mismatch between the charged and neutral leptons, just as the KM angles are the mismatch between the up and down quarks. The mechanism I have just described has three ingredients: (1) There is large mixing in $L$; (2) $L$ is highly asymmetric in its 23 block; and (3) $L$ is related by $SU(5)$ to the transpose of $D$.

Although I have described it in a toy example based on $SU(5)$, this mechanism first emerged in a realistic model based on $SO(10)$.[3,4] Several papers[6] have appeared that consider this idea in models with the unified group $SU(5)$, while an $SO(10)$ model quite similar in some respects to the one I am about to discuss has been proposed by Babu, Pati and Wilczek.[7] In a short talk I can not go into the details of our model. The point I wish to emphasize is that this model was first constructed without any thought to the pattern of neutrino masses and mixings that would arise. The goal in our first paper[3] was rather to construct a realistic $SO(10)$ model with as simple a Higgs structure as possible. The "minimal Higgs structure" in $SO(10)$ involves the use of only the following non-singlet Higgs representations to break $SO(10)$ down to the standard model group $G_{SM}$: $\mathbf{45}_H + \mathbf{16}_H + \overline{\mathbf{16}}_H$. That the breaking down to $G_{SM}$ can be done with these fields was shown in Ref. 8. Starting with this minimal Higgs structure for $SO(10)$ breaking, we looked for the simplest effective Yukawa operators that can reproduce the pattern of quark and lepton masses that is seen. By "simple" operators we mean operators of low dimension, which can arise from simple tree-level diagrams where small multiplets are integrated out.

We were led, practically uniquely, to a set of six Yukawa operators that give the following Dirac mass matrices for the up quarks, down quarks, neutrinos, and charged leptons at the GUT scale.[3,4] (The Majorana mass matrix $M_R$ of the right-handed neutrinos comes from different terms.)

$$U = \begin{pmatrix} \eta & 0 & 0 \\ 0 & 0 & \varepsilon/3 \\ 0 & -\varepsilon/3 & 1 \end{pmatrix} m_U, \quad (9)$$

$$D = \begin{pmatrix} 0 & \delta & \delta' \\ \delta & 0 & \sigma+\varepsilon/3 \\ \delta' & -\varepsilon/3 & 1 \end{pmatrix} m_D, \quad (10)$$

$$N = \begin{pmatrix} \eta & 0 & 0 \\ 0 & 0 & -\varepsilon \\ 0 & \varepsilon & 1 \end{pmatrix} m_U, \quad (11)$$

$$L = \begin{pmatrix} 0 & \delta & \delta' \\ \delta & 0 & -\varepsilon \\ \delta' & \sigma+\varepsilon & 1 \end{pmatrix} m_D. \quad (12)$$

Much about the pattern of entries in these matrices can be understood in purely group-theoretic terms. The "1" entries come from the usual minimal Yukawa term $\mathbf{16_3 16_3 10}_H$. As is well-known, such a term gives equal contributions to the up quark matrix $U$ and neutrino Dirac matrix $N$, and it also gives equal contributions to $D$ and $L$. That is why we can write the matrices in terms of only two overall scales $m_U$ and $m_D$. The "$\varepsilon$" entries arise from the lowest dimension Yukawa operator that involves the $\mathbf{45}_H$, namely $\mathbf{16}_i \mathbf{16}_j \mathbf{10}_H \mathbf{45}_H$. Because the vacuum expectation value of the $\mathbf{45}_H$ must be proportional to the generator $B-L$ in order to do the breaking of $SO(10)$ (in particular the doublet-triplet splitting), there arises from this term a relative factor of $-1/3$ between the quark matrices and the lepton matrices, which can be seen in Eqs. (9) — (12). It can also be shown that with $\langle \mathbf{45}_H \rangle \propto B-L$ the quartic Yukawa operator involving this VEV gives a flavor-antisymmetric contribution, as also seen in Eqs. (9) — (12). The lopsided $\sigma$ entries come from a quartic term $\mathbf{16}_2 \mathbf{16}_3 \mathbf{16}'_H \mathbf{16}_H$, where the $SO(10)$ indices are contracted in a certain way. (No effective Yukawa of lower dimension can be written down that involves the $\mathbf{16}_H$.) It is well-know that such a four-16 operator contributes only to the down quark and charged lepton mass matrices. Moreover, since the expectation value of the spinor $\mathbf{16}_H$ breaks $SO(10)$ only down to $SU(5)$, this operator respects the minimal $SU(5)$-relation $L = D^T$. The entries $\delta$ and $\delta'$ also arise from four-16 operators, though ones where the $SO(10)$ indices are contracted differently in a way that gives flavor-symmetric contributions. However, for the same reason as in the case of $\sigma$, these only appear in $D$ and $L$.

One cannot in a short talk explain in detail the structure of the model. Suffice it to say that the mass matrices arise from very simple Yukawa structures that in turn arise from very simple particle content. The model is thus not only simple at the level of the mass matrix "textures", but also at the level of the underlying unified model.

Although it has very few parameters, this model gives a remarkably good fit to all the quark and lepton masses and mixings.[4] There are altogether *nine* predictions. Three of them are well-known relations that arise in many models because of the group theory of $SU(5)$ and $SO(10)$: **(1)** $m_b^0 \cong m_\tau^0$ (the superscript zero refers throughout to quantities evaluated at the unification scale); **(2)** $m_s^0 \cong \frac{1}{3} m_\mu^0$; **(3)** $m_d^0 \cong 3 m_e^0$ (these last two relations are the well-known Georgi-Jarlskog relations).

The fourth prediction is that $m_u$ is relatively small. **(4)** $m_u/m_t \ll m_d/m_b, m_e/m_\tau$. The point is that the Yukawa terms (involving the parameters $\delta$ and $\delta'$) that generate masses for the other fermions of the first family ($d$ and $e$) leave the $u$ quark massless. (We have seen the group-theoretical reason for this.) This accords well with the fact that $m_d^0/m_b^0 \cong 10^{-3}$, and $m_e^0/m_\tau^0 \cong 0.3 \times 10^{-3}$, whereas (assuming $m_u \approx 4$ MeV) the comparable ratio $m_u^0/m_t^0$ is only about $0.6 \times 10^{-5}$. Thus a small Yukawa term ($\eta$) must be introduced in this model to give a non-vanishing mass for the $u$ quark. This $\eta$ term has interesting consequences for neutrino masses, as we shall see.

A remarkable postdiction of the model is the charm quark mass[4]: **(5)** $m_c(m_c) = (1.1 \pm 0.1)$ GeV. This is remarkable for two reasons. First, the charm quark mass is generally a severe problem for $SO(10)$, since the simplest $SO(10)$ schemes predict that $m_c^0/m_t^0 = m_s^0/m_b^0$, which is an order of magnitude too large. Second, this model not only gives a postdiction of $m_c$ that is of the right order of magnitude, but even predicts it correctly to within about 15% accuracy, which is quite acceptable given the uncertainties.

Another remarkable success of the model is a prediction of $V_{ub}$: **(6)** $V_{ub} = 0.0052 e^{i\theta} - 0.0028$. The phase $e^{i\theta}$ is the single non-trivial physical phase angle contained in the parameters appearing in Eqs. (9) — (12). This phase is not fixed by other measured masses or mixings. So the prediction of the model is that $V_{ub}$ lies on a certain circle in the complex plane. As it happens, this circle slices precisely through the middle of the presently allowed region in the $(Re(V_{ub}), Im(V_{ub}))$ plane.

Finally, there are predictions for the three neutrino mixing angles. **(7)** The most significant prediction is that the $\nu_\mu - \nu_\tau$ mixing is large. This stems from the large entry $\sigma$ in $L$. A fit to the known quark and lepton masses and mixings gives $\sigma \cong 1.8$, $\varepsilon \cong 0.14$, $\delta \cong 0.008$, $|\delta'| \cong 0.008$, and $\eta \cong 0.6 \times 10^{-5}$. Thus the angle $\theta_{\mu\tau}^{left}$ appearing in $U_{\mu 3} = \sin(\theta_{\mu\tau}^{left} - \theta_{\nu_\mu \nu_\tau}^{left})$ is given by $\theta_{\mu\tau}^{left} \cong \tan^{-1} \sigma \cong \pi/3$. The angle $\theta_{\nu_\mu \nu_\tau}^{left}$ is not exactly predicted by the model, since it depends on the unknown Majorana mass matrix $M_R$ of the right-handed neutrinos. But we can say from the form of Eq. (11) that it is of order $\varepsilon$ and thus small.

**(8)** The prediction for the mixing of the electron neutrino is quite interesting because, depending on what one assumes for $M_R$, it comes out quite naturally to give *either* the small-angle MSW solution to the solar neutrino problem *or* the vacuum "just-so" solution. If one supposes that the matrix $M_R$ does not have large mixing between the first family and the other families, i.e. that the 12, 21, 13, and 31 elements of $M_R$ are negligible, then the mass matrix of the light neutrinos, which has the usual "see-saw" form $M_\nu = -N^T M_R^{-1} N$, also gives very little mixing between the first family and the others. (See Eq. (11).) This is also the case no matter what the form of $M_R$ if the parameter $\eta = 0$ (meaning that $m_u = 0$). In these cases, the mixing of the electron neutrino comes entirely, or almost entirely, from diagonalizing $L$. In that case one gets a sharp prediction that $\sin^2 2\theta_{e\mu} \cong 16 \times 10^{-3} \cos^2 \theta_{\mu\tau}$. For maximal mixing of $\nu_\mu - \nu_\tau$, as needed to fit the atmospheric neutrino data, this gives $\sin^2 2\theta_{e\mu} \cong 8 \times 10^{-3}$, which is in the allowed range for small-angle MSW.

On the other hand, if one assumes that $M_R$ has large 12, 21 and/or 13, 31 elements, something quite remarkable happens. To illustrate, consider the form

$$M_R = \begin{pmatrix} 0 & D\varepsilon^3 & 0 \\ D\varepsilon^3 & B\varepsilon^2 & 0 \\ 0 & 0 & A \end{pmatrix} m_R, \quad (13)$$

where we parameterize using powers of $\varepsilon$ merely for convenience. The light neutrino mass matrix comes out to be

$$M_\nu = -N^T M_R^{-1} N = \begin{pmatrix} \frac{\eta^2}{\varepsilon^4}\frac{AB}{D^2} & 0 & -\frac{\eta}{\varepsilon^2}\frac{A}{D} \\ 0 & \varepsilon^2 & \varepsilon \\ -\frac{\eta}{\varepsilon^2}\frac{A}{D} & \varepsilon & 1 \end{pmatrix} \frac{m_U^2}{A m_R}. \quad (14)$$

One sees that the 2-3 block has vanishing determinant, so that a rotation in the 2-3 plane by an angle $\theta_{23}^\nu \cong \varepsilon$ brings $M_\nu$ to the form

$$M_\nu' \cong \begin{pmatrix} \frac{\eta^2}{\varepsilon^4}\frac{AB}{D^2} & \frac{\eta}{\varepsilon}\frac{A}{D} & -\frac{\eta}{\varepsilon^2}\frac{A}{D} \\ \frac{\eta}{\varepsilon}\frac{A}{D} & 0 & 0 \\ -\frac{\eta}{\varepsilon^2}\frac{A}{D} & 0 & 1 \end{pmatrix} \frac{m_U^2}{A m_R}. \quad (15)$$

The important thing to notice is that the 1-2 block has a pseudo-Dirac form. What happens, as a result of this, is that there is almost exactly maximal mixing between the electron neutrino and the muon neutrino. In fact, one has "bimaximal" mixing. Interestingly, the large value of $U_{\mu 3}$ arises, as we have said, from the *charged* lepton mass matrix, whereas the large value of $U_{e2}$ is coming, as we have just seen, from the *neutrino* mass matrix. When the parameter values are looked at more closely, it is found that the vacuum solution is easily obtained, but the large-angle MSW solution requires some fine-tuning of parameters.

**(9)** Finally, there is a prediction for the $\nu_e - \nu_\tau$ mixing angle. It is easy to show that in *both* cases, the small-angle MSW case and the bimaximal vacuum oscillation case, the mixing $U_{e3}$ comes out the same: $U_{e3} \cong 0.07 \sin\theta_{\mu\tau}$.

In conclusion, we have shown that a very simple explanation exists based on the group theory of $SU(5)$ that accounts for the fact that the mixing of the left-handed fermions of the second and third families is small for quarks ($V_{cb} \cong 0.04$) and large for leptons ($\sin^2 2\theta_{\mu\tau} \cong 1$). We also showed how this explanation arose from a particular $SO(10)$ model of fermion masses. This model was seen to be very simple, both at the level of the underlying $SO(10)$ structures, and at the level of the resulting mass matrices. Because these matrices involve few parameters, no fewer than nine predictions result. Several of these will provide very non-trivial tests of the model, in particular the predictions for $V_{ub}$ and the mixings $U_{e2}$ and $U_{e3}$. For further details the reader can consult the series of papers in Refs 4 and 5.

## REFERENCES

1. Weinberg, S. *Trans. N.Y. Acad. Sci.* **38**, 185 (1977). Wilczek, F. and Zee, A. *Phys. Rev. Lett.* **42**, 421 (1979).
2. Fritzsch, H. *Phys. Lett.* **73B**, 317 (1979).
3. Albright, C.H. and Barr, S.M. *Phys. Rev.* **D58**, 013002 (1998).
4. Albright, C.H., Babu, K.S. and Barr, S.M. *Phys. Rev. Lett.* **81**, 1167 (1998) (hep-ph/9802314). Albright, C.H. and Barr, S.M. *Phys. Lett.* **B452**, 287 (1999).
5. Albright, C.H. and Barr, S.M. *Phys. Lett.* **B461**, 218 (1999).
6. Hagiwara, K. and Okamura, N. *Nucl. Phys.* **B548**, 60 (1999) (hep-ph/9811495). Altarelli, G. and Feruglio, F. *Phys. Lett.* **B451**, 388 (1999) (hep-ph/9812475). Berezhiani, Z. and Rossi, A. hep-ph/9907397.
7. Babu, K.S., Pati, J. and Wilczek, F. hep-ph/9812538.
8. Barr, S.M. and Raby, S. *Phys. Rev. Lett.* **79**, 4748 (1997).

# Nucleon instability and (B-L) non-conservation

Yu. Kamyshkov

*Department of Physics, University of Tennessee, Knoxville, TN 37996-1200*

**Abstract.**
Proton decay with the modes and rates predicted by the original SU(5) grand unification scheme conserving (B-L) is not observed experimentally. There are reasons to expect that (B-L) might not be conserved in nature. Among possible observable manifestations of (B-L) non-conservation are Majorana masses of neutrinos, neutrinoless double-beta decay, decay of protons to lepton + (X), and neutron to anti-neutron transitions. Feasible progress in experimental search for some (B-L) non-conserving processes is discussed here.

## INTRODUCTION

Searches for nucleon instability (1) are motivated by two outstanding concepts of contemporary physics beyond the Standard Model of particle physics: the interpretation of baryon asymmetry of the universe (BAU) (2, 3) and the idea of Unification of particles and their forces (4, 5). However, even within the scope of the Standard Model at the non-perturbative level (6) baryon number is not conserved (the latter non-conservation is so weak that it does not lead to directly observable nucleon decay effects).

So far, nucleon instability has not been experimentally observed (7). Conservation of angular momentum in nucleon decay (nucleon spin 1/2) requires leptons (fermions) to appear in the final state. Two possibilities can be realized here: with $\Delta B = \Delta L$ or $\Delta B = -\Delta L$ ($B$ and $L$ are baryon and lepton numbers respectively). The first leads to the conservation of $(B-L)$ and the second to processes which violate $(B-L)$ conservation by two units. Stringent nucleon decay limits are experimentally established (8, 7) for the nucleon decay modes where $(B-L)$ is conserved (e.g. $p \to e^+\pi^0$) ruling out the original SU(5) Unification model (5). It is important to notice that in the original SU(5) model, as well as in the Standard Model, $(B-L)$ is strictly conserved. New generation of experiments with huge-mass detectors, such as those discussed at this Workshop, will continue to test the stability of nucleons with the respect to the $(B-L)$ conservation. That is particularly important in a view of the new theoretical predictions of the super-symmetric models (9). In this paper we discuss the prospects for experimental searches for the processes which do not conserve $(B-L)$.

## IS (B-L) CONSERVED?

Naively we would expect that $(B-L)$ number is violated (the number of neutrons in our laboratory samples is in excess of equal number of protons and electrons). However, most of the leptons in the universe likely exist as, yet undetected, relic $\nu$ and $\bar{\nu}$ radiation. Thus, the conservation of $(B-L)$ on a scale of the whole universe is an open question.

Can $(B-L)$ be conserved in a way similar to the conservation of electrical charge? From tests of Equivalence Principle (10) one can exclude the existence of massless long-range gauge field of "$B-L$ photons" at a level of strength $\sim 10^{-12}$ of the gravitational strength. It is interesting to notice that "baryonic photons" responsible for conservation of baryon charge are excluded from the same tests only on the level $\sim 10^{-10}$ (11). Unless $(B-L)$ is globally conserved in nature, it is very natural to expect that $(B-L)$ is violated.

In nucleon decay processes (with $\Delta B = -1$) the non-conservation of $(B-L)$ implies the existence of transitions of the type *nucleon* $\to$ *lepton* $+ ...$ (the conservation of $(B-L)$ corresponds to *nucleon* $\to$ *anti-lepton* $+ ...$ transitions). If $(B-L)$ is violated by two units, it is natural to assume (and it follows from Unification models (4, 12)) that processes with $\Delta L = 2$ and $\Delta B = 2$ are also the components of the physics of $(B-L)$ non-conservation. Examples of these are heavy Majorana neutrinos with $\Delta L = 2$ transitions $\nu_M \to \bar{\nu}_M$ and oscillations of neutrons to anti-neutrons $n \to \bar{n}$.

Since 1973, when $(B-L)$ non-conservation was first considered in theory (4), it was discussed within the framework of Unification models in a number of theoretical papers (12, 13, 14, 15, 16). In Unification models, like SO(10) (9), massive Majorana neutrinos with $\Delta L = 2$ transitions violating $(B-L)$ by two units are used in a "see-saw" mechanism to generate the masses of conventional neutrinos. In the left-right symmetric Unification models the violation of $(B-L)$ arises at the same energy scale where the left-right symmetry is restored (4, 13, 12). Probably the most compelling reason for the existence of $(B-L)$ non-conservation in nature follows from the fact (17) that electroweak non-perturbative mechanisms ("sphalerons") erase the baryon asymmetry of the universe if $(B-L)$ is globally conserved. Thus, the most natural explanation of BAU would require non-conservation of $(B-L)$ at an energy scale above the electro-weak scale. In this sense, experimental discovery of the nucleon decay into "standard" decay modes (like $p \to e^+ + \pi^0$ or $p \to \mu^+ + K^0$) with conservation of $(B-L)$ would leave BAU unexplained.

## NUCLEON INSTABILITY WITH $\Delta(B-L) = -2$

For some nucleon decay modes (see complete list of experimental limits in (7)) the experimental limits can be interpreted as limits for both $(B-L)$ conserving or $(B-L)$ violating processes. This is due to the presence of undetectable anti-neutrinos or neutrinos in the final states (for example, mode $N \to \nu + \pi$). In this sense only processes with $\Delta B = 2$ or $\Delta L = 2$ would unambiguously indicate in experiment the conservation or violation of $(B-L)$.

Let us discuss several $(B-L)$ violating processes where we believe significant progress can be made in the near future. These are: (a) the neutron to anti-neutron transition or the intranuclear disappearance of two nucleons; (b) proton decay $p \to e^+ \nu\nu$; and (c) intranuclear neutron decay $n \to \nu\nu\bar{\nu}$. The latter two processes might be enhanced by a large phase space factor as compared to many other modes of nucleon decay (if all of them are originated by the same mechanism).

### $n \to \bar{n}$ or $NN \to pions$

Transitions $n \to \bar{n}$ can be searched for with free neutrons from reactors or in the intranuclear disappearance of two nucleons (neutron to anti-neutron transition followed by two-nucleon annihilation into pions inside nuclei). Probability of coherent transition of free neutrons to anti-neutrons as function of observation time $t$ is given by $(t/\tau_{n\bar{n}})^2$ (12), where $\tau_{n\bar{n}}$ is a characteristic oscillation time experimentally limited to $\tau_{n\bar{n}} \geq 8.6 \cdot 10^7$ sec (18). The intranuclear $n \to \bar{n}$ transition is strongly suppressed by the difference of nuclear potential for neutrons and anti-neutrons (see most recent theoretical paper (19) and references therein). This suppression leads to the regular exponential probability of decay with the lifetime $\tau_A$ related to $\tau_{n\bar{n}}$ by $\tau_A = R \cdot \tau_{n\bar{n}}^2$, where $R$ is a dimensional suppression factor predictable from nuclear theoretical models to an accuracy of $\sim \pm(20-25)\%$ (19).

Experimental limits for $\tau_A$ in intranuclear search of $n \to \bar{n}$ transitions were set by IMB, Kamiokande, and Fréjus experiments (7) at a level $2.4$-$6.5 \cdot 10^{31}$ years for oxygen and iron nuclei, corresponding to a limit on oscillation time $\tau_{n\bar{n}} \geq 1.2 \cdot 10^8$ sec (7). The Super-Kamiokande detector is expected after several years of running to improve this limit to $\tau_A \geq 1-2 \cdot 10^{33}$ years (20) or $\tau_{n\bar{n}} \geq 5 \cdot 10^8$ sec.

Future prospects of $n \to \bar{n}$ transition search with free neutrons in reactor experiments are discussed elsewhere (21). With existing research reactor facilities it is possible to extend the search limit for $n \to \bar{n}$ beyond $\tau_{n\bar{n}} \geq \sim 3 \cdot 10^9$ sec and to explore the stability of matter (in this intranuclear nucleon instability mode with $\Delta B = -2$) beyond the limit of $\sim 7 \cdot 10^{34}$ years.

As was pointed out in (22), the existence of $n \to \bar{n}$ transitions would provide a unique opportunity to test the CPT-theorem with unprecedented accuracy by looking at the mass difference of neutron and anti-neutron. Such a mass difference (or small gravitational non-equivalence of neutron and anti-neutron, or small non-compensated magnetic field on the neutron flight path) can suppress the $n \to \bar{n}$ transition for free neutrons but is too small to produce a sizable additional effect in intranuclear transitions where very large nuclear suppression is already taking place. Therefore, searches in both directions with free neutrons and with the neutrons bound inside nuclei are desirable (22).

### $p \to e^+ \nu\nu$

The present experimental lifetime limit $\tau \geq 1.1 \cdot 10^{31}$ years for this process was set in Fréjus experiment. Difficulty with this decay mode is the fact that the observable final state (with existing experimental techniques a single positron with energy of few hundred MeV is indistinguishable from an electron) overlaps with the final state of atmospheric neutrino ($\nu_e$ and $\bar{\nu}_e$) interactions in the detectors. In a 1992 paper (23) an attempt was made to attribute the entire atmospheric neutrino anomaly to this mode of proton decay by interpreting the Sub-GeV data sample of Kamiokande detector (within neutrino flux normalization uncertainties) as an excess in the electron spectrum with a characteristic Michel-type energy shape rather than as a deficit in the muon spectrum.

New Super-Kamiokande results (24) provided a new evidence of the zenith angle dependence of muon events rendering the neutrino oscillation hypothesis a more viable explanation of the atmospheric neutrino anomaly. However, Super-K Collaboration has so far not ruled out the possibility of excess in the Sub-GeV electron-type spectrum. In Super-K analysis of atmospheric neutrino events the normalization factor of electron-type neutrino flux is used as a free parameter (within flux prediction uncertainties) to reduce the uncertainties of the absolute muon flux. This procedure will be certainly absolutely correct if it is known a priori that electron-type events are pure atmospheric neutrino interactions without possible admixture of any other effects. One can hope that new analysis and increased statistics in Super-K experiment as well as new more precise calculations and measurements of the absolute atmospheric neutrino fluxes will permit isolation of the possible contribution of proton decay in the electron-type Sub-GeV data or set a new higher lifetime limit for the process $p \to e^+ \nu \nu$.

$$n \to \nu \nu \bar{\nu}$$

Surprisingly this elusive decay mode was experimentally explored by two different methods. First method, used in IMB and Fréjus experiment (25) treated the Earth as a source of neutrinos and the detector selection criteria were optimized for an energy range typical for neutrinos emitted from the decay process. The IMB limit for muon neutrinos was $\tau \geq 5 \cdot 10^{26}$ years; Fréjus limits for electron and muon type of neutrinos were $3 \cdot 10^{25}$ and $1.2 \cdot 10^{26}$ years respectively.

The second method used by Kamiokande II Collaboration (26) was based on the detection of nuclear de-excitation produced by the hole left by the neutron decay in the $S_{1/2}$ nuclear state of $^{16}O$. De-excitations of $S_{1/2}$ hole would typically occur via the emission of proton or neutron; and with small probability $Br \simeq (2.7 - 10.4) \cdot 10^{-5}$ (estimated theoretically in (26)) it can proceed via the emission of an energetic photon with the energy above 19 MeV. (The entire solar neutrino spectrum is below 19 MeV threshold). The lifetime limit obtained by Kamiokande Collaboration from the observation of two background events was $\tau \geq 4.9 \cdot 10^{26}$ years and independent on the type of neutrino produced in the decay. More recent theoretical re-evaluation of the probability of de-excitation of $S_{1/2}$ hole with emission of energetic photon (27) suggested for the same data an improved lifetime limit of $\tau \geq 2.3 \cdot 10^{27}$ years.

The two methods mentioned above can be hopefully used in Super-Kamiokande detector to search for $n \to \nu \nu \bar{\nu}$ decay. In the presence of background the search limits here can be extended by factor of square root of detector mass ratio, i.e. approximately by an order of magnitude from the present limit and reach $\sim 10^{28}$ years.

A more sensitive approach in exploration of this decay mode channel should be possible with the new low-threshold large scintillating detector, KamLAND. Although the total fiducial mass of KamLAND will be $\sim 1$ Kton, it should detect with full efficiency the de-excitation of nuclear state holes left by the disappearing neutron. Consider as an example the following process. In liquid scintillator ($\sim CH_2$) 1/3 of all neutrons are in $S_{1/2}$ state of carbon nuclei. Hole in this state will de-excite mostly by proton emission (since $^{11}C$ is a proton-rich nucleus), but with a branching of several percent (28) neutron emission is also possible, leaving an excited $^{10}C^*$ state. Detection of such event in KamLAND detector will start with detection of $\gamma$ from de-excitation of $^{10}C^*$ (detection threshold is $\sim 0.2$ MeV) followed be the detection of neutron (by capture on hydrogen in the liquid scintillator: capture lifetime is $\sim 180$ $\mu$sec with a detected signal of 2.2 MeV). Following this pair of events after $\sim 19.2$ sec $^{10}C_{gs}$ will $\beta^+$-decay with maximum energy release of 3.65 MeV. All three events in the sequence must be reconstructed to the same point in the detector within reconstruction inaccuracies. Preliminary estimates show that random background for triple-coincidence events will be negligible and it will be possible to explore the intranuclear neutron stability for $3\nu$ final state up to a lifetime limit of $\sim 3 \cdot 10^{30}$ years. Possible sources of background events with similar signature arising from the atmospheric neutrino interactions with carbon nuclei must be carefully accounted for through nuclear model calculations.

The experimental method of detection of nuclear final states created as a result of nucleon disappearance (into neutrinos or any other invisible or undetectable particles) is very important as a complement to the exclusive modes of nucleon decay search. Together they provide an experimental basis for establishing the decay-mode-independent limit for nucleon instability which is presently, according to PDG (7), is at the level of only $1.6 \cdot 10^{25}$ years.

## CONCLUSIONS

The experimental search for $\Delta(B-L) \neq 0$ processes is at least as important as the search for conventional baryon number violating processes with $\Delta(B-L) = 0$. For several of $\Delta(B-L) \neq 0$ processes discussed above it is possible to improve the discovery potential by a significant factor. These improvements can be made with the existing or currently constructed detectors (Super-Kamiokande,

KamLAND) and with existing reactor facilities (for the case of free neutron to anti-neutron transitions).

I would like to thank Professor W.M. Bugg for useful discussions.

# REFERENCES

1. M. Goldhaber, "Search for Nucleon Instability (Origin and History)", *in Proceedings of International Workshop on Future Prospects of Baryon Instability Search in p-Decay and $n \to \bar{n}$ Oscillation Experiments*, Oak Ridge, 1996, p.1. and these Proceedings

2. A.D. Sakharov, *JETP Lett.* **5** (1967) 24

3. V.A. Kuzmin, *JETP Lett.* **12** (1970) 228; *ZhETF Pis. Red.* **12**, No. 6 (1970) 335

4. J. Pati and A. Salam, *Phys.Rev.* **D8** (1973) 1240; *Phys. Rev.* **D10** (1974) 275

5. H. Georgi and S. Glashow, *Phys. Rev. Lett.* **32** (1974) 438

6. G. 't Hooft, *Phys. Rev. Lett.* **37** (1976) 8-11

7. Particle Data Group, "Review of Particle Physics", *Eur. Phys. J.* **C3** (1998) 1-794

8. See contributions of L. Sulak and Y. Suzuki in these Proceedings

9. See contributions of K. Babu, H. Murayama, J. Pati, P. Ramond, and F. Wilczek in these Proceedings

10. E. v. Eötvös, D. Pekar, and E. Fekete, *Ann. d. Phys.* **68** (1922) 11; P. G. Roll, R. Krotkov, and R. H. Dicke, *Ann. Phys. (U.S.A)* **26** (1964) 442; V. G. Braginsky and V. I. Panov, *Sov. Phys. JETP* **34** (1971) 464

11. T.D. Lee, C.N. Yang, *Phys.Rev.* **98** (1955) 1501; J.C. Pati and A. Salam, *Phys.Rev.Lett.* **31** (1973) 661; A.D. Sakharov, "Baryon Asymmetry of the Universe", Reprint of the review paper presented at the A.A. Friedmann Centenary Conference, Leningrad, USSR, June 22-26, 1988, published in *Sov.Phys.Usp.* **34**, No.5 (1991) 417

12. R. Mohapatra and R. Marshak, *Phys. Lett.* **91B** (1980) 222; *Phys. Rev. Lett.* **44** (1980) 1316; *Phys. Lett.* **94B** (1980) 183

13. A. Davidson, *Phys.Rev.* **D20** (1979) 776

14. F. Wilczek and A. Zee, *Phys. Lett.* **88B** (1979) 311

15. L.N. Chang and N.P. Chang, *Phys. Lett.* **B92** (1980) 103

16. F. Buccella, G. Mangano, A. Masiero, L. Rosa, *Phys. Lett.* **B320** (1994) 313

17. V. Kuzmin, V. Rubakov, and M. Shaposhnikov, *Phys. Lett.* **155B** (1985) 36; V. Rubakov and M. Shaposhnikov, *Phys. Usp.* **39** (1996) 461

18. M. Baldo-Ceolin et al., *Z. Phys* **C63** (1994) 409

19. J. Hufner and B.Z. Kopeliovich, *Mod.Phys.Lett.* **A13** (1998) 2385

20. J. Stone, 1996, Private Communication

21. Yu.A. Kamyshkov, in the Proceedings of CIPANP 97, "Big Sky 1997, Intersections between particle and nuclear physics", pp 335-341; in the Proceedings of Workshop on Physics Beyond the Standard Model, "Tegernsee 1997, Beyond the desert", pp. 542-553

22. Yu. Abov, F. Djeparov, and L. Okun, *JETP Lett.* **39** (1984) 493; L. Okun, "Test of CPT", preprint ITEP-TH-55/96; hep-ph/9612247 (1996)

23. W.A. Mann, T. Kafka, and W. Leeson, *Phys. Lett.* **B291** (1992) 200; W.A. Mann, W. Leeson, and D. Wall, *in Proceedings of International Workshop on Future Prospects of Baryon Instability Search in p-Decay and $n \to \bar{n}$ Oscillation Experiments*, Oak Ridge, 1996, p.175)

24. Y. Fukuda et al., *Phys. Rev. Lett.* **81** (1998) 1562

25. J.Learned et al., *Phys. Rev. Lett.* **43** (1979) 907; C. Berger et al., *Phys. Lett.* **B269** (1991) 227

26. Y. Suzuki et al., *Phys. Lett.* **B311** (1993) 357

27. J.-F. Glicenstein, *Phys. Lett.* **B411** (1997) 326

28. D. Dean, ORNL, 1999, Private Communication

# ASTROPHYSICAL

# NEUTRINOS

# SOLAR NEUTRINOS

## J. N. BAHCALL

*Institute for Advanced Study, Princeton, NJ 08540, USA*
*E-mail: jnb@sns.ias.edu*

**Abstract.** I summarize the current state of solar neutrino research and then answer the question: What should we do next?

## INTRODUCTION

The reader who is familiar with solar neutrino research may wish to skip directly to the last section entitled: What Next?

Solar neutrinos have been detected experimentally with fluxes and energies that are qualitatively consistent with solar models that are constructed assuming that the sun shines by nuclear fusion reactions. The first experimental result, obtained by Ray Davis and his collaborators in 1968, (1, 2) has now been confirmed by four other beautiful experiments, Kamiokande, (3) SAGE, (4) GALLEX, (5) and SuperKamiokande. (6) The observation of solar neutrinos with approximately the predicted energies and fluxes establishes empirically the theory (7) that main sequence stars derive their energy from nuclear fusion reactions in their interiors and has inaugurated what we all hope will be a flourishing field of observational neutrino astronomy.

Although the calculated neutrino fluxes depend upon high powers of the central temperature of the solar model, the experiments and the solar model theory are so precise that persistent quantitative discrepancies have existed between the model predictions and the solar model calculations for over thirty years (8, 9, 10)

Important experiments are underway that will provide diagnostic information about the physical properties of neutrinos that are created in the center of the sun and detected on earth in really long baseline experiments. At this workshop, we will hear discussions of the Super-Kamiokande, SNO, BOREXINO, HELLAZ, HERON, ICARUS, LENS, and KamLAND experiments.

I will discuss predictions of the combined standard model in the main part of this review. By 'combined' standard model, I mean the predictions of the standard solar model and the predictions of the standard electroweak model. We need a solar model to tell us how many neutrinos of what energy are produced in the sun and we need an electroweak theory to tell us how the number and flavor content of the neutrinos are changed as they make their way from the center of the sun to detectors on earth. For all practical purposes, the standard electroweak model states that nothing happens to solar neutrinos after they are created in the deep interior of the sun. Using standard electroweak theory and fluxes from the standard solar model, one can calculate the rates of neutrino interactions in different terrestrial detectors with a variety of energy sensitivities. The combined standard model also predicts that the energy spectrum from a given neutrino source should be the same for neutrinos produced in terrestrial laboratories and in the sun and that there should not be measurable time-dependences (other than the seasonal dependence caused by the earth's orbit around the sun). The spectral and temporal departures from standard model expectations are expected to be small in all currently operating experiments (11) and have not yet yielded definitive results. Therefore, I will concentrate here on inferences that can be drawn by comparing the total rates observed in solar neutrino experiments with the combined standard model predictions.

I will begin by reviewing in Section the quantitative predictions of the combined standard solar model and then describe in Section the three solar neutrino problems that are established by the chlorine, Kamiokande, SAGE, GALLEX, and SuperKamiokande experiments. In Section , I detail the uncertainties in the standard model predictions and then show in Section that helioseismological measurements indicate that the standard solar model predictions are accurate for our purposes. In Section , I discuss the implications for solar neutrino research of the precise agreement between helioseismological measurements and the predictions of standard solar models. Next, ignoring all knowledge of the sun, I cite analyses in Section that show that one cannot fit the existing experimental data with neutrino fluxes that are arbitrary parameters, unless one invokes new physics to change the shape or flavor content of the neutrino energy spectrum. I summarize in Section the characteristics of the best-fitting neutrino

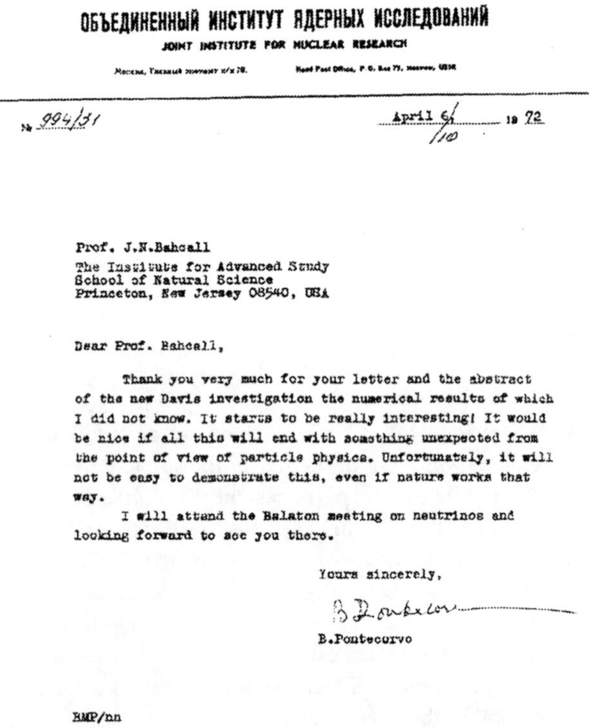

**FIGURE 1.** Letter from Bruno Pontecorvo in 1972.

oscillation descriptions of the experimental data. Finally, I will discuss and summarize the results in Section .

If you want to obtain numerical data or subroutines that are discussed in this talk, or to see relevant background information, you can copy them from my Web site: http://www.sns.ias.edu/~jnb.

Before we begin the detailed discussion, I want to make just a brief historical diversion. Nearly all of the current interest in solar neutrinos centers around the opportunity to use the sun as a neutrino source in a very long baseline oscillation experiment. In preparing for a talk in honor of Fred Reines a few months ago, I ran across a long forgotten 1972 letter from Bruno Pontecorvo, the originator of the hypothesis that oscillations may be observed in solar neutrino experiments. For your interest, I enclose a reproduction of this letter in Fig. 1.

For the benefit of neutrino pioneers of today, it is perhaps worth remarking that Ray Davis and I never considered the possibility that solar neutrinos could be used to learn more about neutrinos when, in the early 1960's, we were first analyzing the potentialities of a practical chlorine experiment. We sold the experiment as a fundamental test of the hypothesis that the sun shines by nuclear fusion reactions in its interior. Only after the first results of the chlorine experiment showed in 1968 a conflict with the solar model calculations and Gribov and Pontecorvo published their epochal 1969 paper on vacuum oscillations of solar neutrinos did we begin to consider the possibility that solar neutrinos might tell us something new about particle physics. Maybe there are previously unimagined physics treasures to be discovered in future neutrino experiments.

**Table 1.** Standard Model Predictions (BP98): solar neutrino fluxes and neutrino capture rates, with $1\sigma$ uncertainties from all sources (combined quadratically).

| Source | Flux ($10^{10}$ cm$^{-2}$s$^{-1}$) | Cl (SNU) | Ga (SNU) |
|---|---|---|---|
| pp | $5.94 \left(1.00^{+0.01}_{-0.01}\right)$ | 0.0 | 69.6 |
| pep | $1.39 \times 10^{-2} \left(1.00^{+0.01}_{-0.01}\right)$ | 0.2 | 2.8 |
| hep | $2.10 \times 10^{-7}$ | 0.0 | 0.0 |
| $^7$Be | $4.80 \times 10^{-1} \left(1.00^{+0.09}_{-0.09}\right)$ | 1.15 | 34.4 |
| $^8$B | $5.15 \times 10^{-4} \left(1.00^{+0.19}_{-0.14}\right)$ | 5.9 | 12.4 |
| $^{13}$N | $6.05 \times 10^{-2} \left(1.00^{+0.19}_{-0.13}\right)$ | 0.1 | 3.7 |
| $^{15}$O | $5.32 \times 10^{-2} \left(1.00^{+0.22}_{-0.15}\right)$ | 0.4 | 6.0 |
| $^{17}$F | $6.33 \times 10^{-4} \left(1.00^{+0.12}_{-0.11}\right)$ | 0.0 | 0.1 |
| Total | | $7.7^{+1.2}_{-1.0}$ | $129^{+8}_{-6}$ |

## STANDARD MODEL PREDICTIONS

Table 1 gives the neutrino fluxes and their uncertainties for our best standard solar model, hereafter BP98. (10) Figure 2 shows the predicted neutrino fluxes from the dominant $p$-$p$ fusion chain.

The BP98 solar model includes diffusion of heavy elements and helium, makes use of the nuclear reaction rates recommended by the expert workshop held at the Institute of Nuclear Theory, (12) recent (1996) Livermore OPAL radiative opacities, (13) the OPAL equation of state, (14) and electron and ion screening as determined by the recent density matrix calculation. (15, 16) The neutrino absorption cross sections that are used in constructing Table 1 are the most accurate values available (17, 18) and include, where appropriate, the thermal energy of fusing solar ions and improved nuclear and atomic data. The validity of the absorption cross sections has recently been confirmed experimentally using intense radioactive sources of $^{51}$Cr. The ratio, $R$, of the capture rate measured (in GALLEX and SAGE) to the calculated $^{51}$Cr capture rate is $R = 0.95 \pm 0.07$ (exp) $+^{+0.04}_{-0.03}$ (theory)

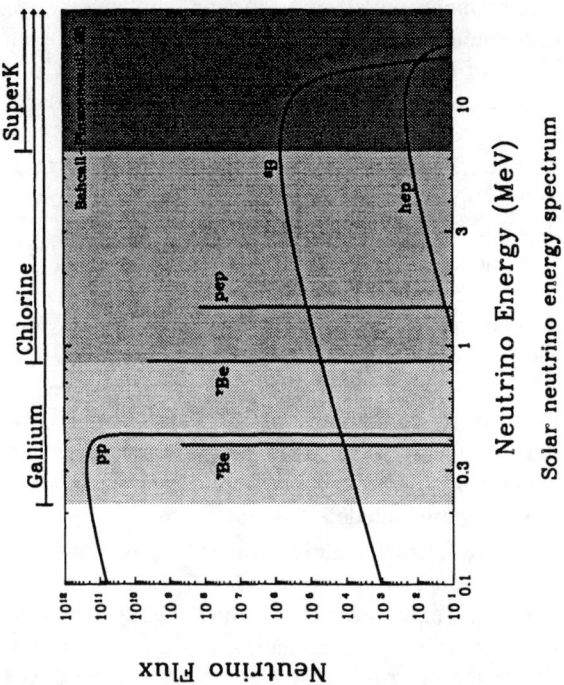

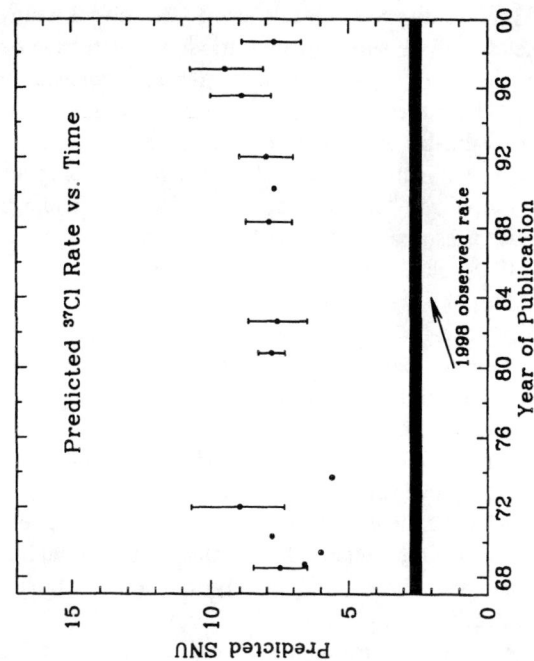

**FIGURE 2.** The energy Spectrum of neutrinos from the pp chain of interactions in the Sun, as predicted by the standard solar model. Neutrino fluxes from continuum sources (such as pp and 8B) are given in the units of counts per cm2 per second. The pp chain is responsible for more than 98% of the energy generation in the standard solar model. Neutrinos produced in the carbon-nitrogen-oxygen CNO chain are not important energetically and are difficult to detect experimentally. The arrows at the top of the figure indicate the energy thresholds for the ongoing neutrino experiments.

**FIGURE 3.** The predictions of John Bahcall and his collaborators of neutrino capture rates in the $^{37}$Cl experiment are shown as a function of the date of publication (since the first experimental report in 1968. (1)) The event rate SNU is a convenient product of neutrino flux times interaction cross section, $10^{-36}$ interactions per target atom per sec. The format is from Figure 1.2 of the book Neutrino Astrophysics. (9) The predictions have been updated through 1998.

and was discussed extensively at Neutrino 98 by Gavrin and by Kirsten. The neutrino-electron scattering cross sections, used in interpreting the Kamiokande and Super-Kamiokande experiments, now include electroweak radiative corrections. (19)

Figure 3 shows for the chlorine experiment all the predicted rates and the estimated uncertainties (1σ) published by my colleagues and myself since the first measurement by Ray Davis and his colleagues in 1968. This figure should give you some feeling for the robustness of the solar model calculations. Many hundreds and probably thousands of researchers have, over three decades, made great improvements in the input data for the solar models, including nuclear cross sections, neutrino cross sections, measured element abundances on the surface of the sun, the solar luminosity, the stellar radiative opacity, and the stellar equation of state. Nevertheless, the most accurate predictions of today are essentially the same as they were in 1968 (although now they can be made with much greater confidence). For the gallium experiments, the neutrino fluxes predicted by standard solar models, corrected for diffusion, have been in the range 120 SNU to 141 SNU since 1968. (17) A SNU is a convenient unit with which to describe the measured rates of solar neutrino experiments: $10^{-36}$ interactions per target atom per second.

There are three reasons that the theoretical calculations of neutrino fluxes are robust: 1) the availability of precision measurements and precision calculations of input data; 2) the connection between neutrino fluxes and the measured solar luminosity; and 3) the measurement of the helioseismological frequencies of the solar pressure-mode ($p$-mode) eigenfrequencies. I have discussed these reasons in detail in another talk. (20)

Figure 4 displays the calculated $^7$Be and $^8$B neutrino fluxes for all 19 standard solar models which have been published in the last 10 years in refereed science journals. The fluxes are normalized by dividing each published value by the flux from the BP98 solar model; (10) the abscissa is the normalized $^8$B flux and the ordinate is the normalized $^7$Be neutrino flux. The rectangular box shows the estimated $3\sigma$ uncertainties in the predictions of the BP98 solar model.

All of the solar model results from different groups fall within the estimated $3\sigma$ uncertainties in the BP98 analysis (with the exception of the Dar-Shaviv model whose results have not been reproduced by other groups). This agreement demonstrates the robustness of the predictions since the calculations use different computer codes (which achieve varying degrees of precision) and involve a variety of choices for the nuclear parameters, the equation of state, the stellar radiative opacity, the initial heavy element abundances, and the physical processes that are included.

The largest contributions to the dispersion in values in Figure 4 are due to the choice of the normalization for $S_{17}$ (the production cross-section factor for $^8$B neutrinos) and the inclusion, or non-inclusion, of element diffusion in the stellar evolution codes. The effect in the plane of Fig. 4 of the normalization of $S_{17}$ is shown by the difference between the point for BP98 (1.0,1.0), which was computed using the most recent recommended normalization, (12) and the point at (1.18,1.0) which corresponds to the BP98 result with the earlier (CalTech) normalization. (22)

Helioseismological-observations have shown (10, 23) that element diffusion is occurring and must be included in solar models, so that the most recent models shown in Fig. 4 now all include helium and heavy element diffusion. By comparing a large number of earlier models, it was shown that all published standard solar models give the same results for solar neutrino fluxes to an accuracy of better than 10% if the same input parameters and physical processes are included. (24, 25)

Bahcall, Krastev, and Smirnov (11) have compared the observed rates with the calculated, standard model values, combining quadratically the theoretical solar model and experimental uncertainties, as well as the uncertainties in the neutrino cross sections. Since the GALLEX and SAGE experiments measure the same quantity, we treat the weighted average rate in gallium as one experimental number. We adopt the SuperKamiokande measurement as the most precise direct determination of the higher-energy $^8$B neutrino flux.

Using the predicted fluxes from the BP98 model, the $\chi^2$ for the fit to the three experimental rates (chlorine, gallium, and SuperKamiokande, see Fig. 5) is

$$\chi^2_{SSM}(3 \text{ experimental rates}) = 61. \quad (1)$$

The result given in Eq. (1), which is approximately equivalent to a $20\sigma$ discrepancy, is a quantitative expression of the fact that the standard model predictions do not fit the observed solar neutrino measurements.

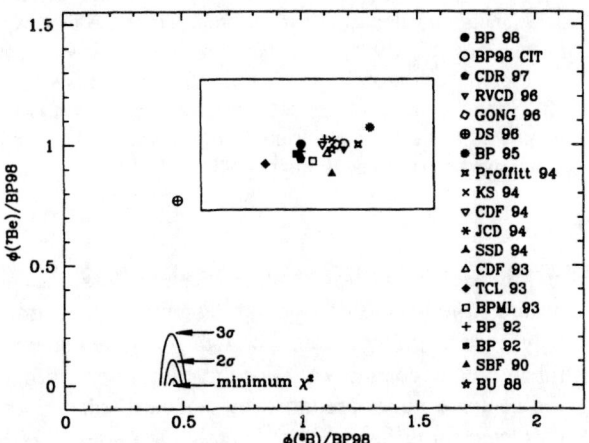

**FIGURE 4.** Predictions of standard solar models since 1988. This figure, which is Fig. 1 of Bahcall, Krastev and Smirnov (1998)(11), shows the predictions of 19 standard solar models in the plane defined by the $^7$Be and $^8$B neutrino fluxes. The abbreviations that are used in the figure to identify different solar models are defined in the bibliographical item, Ref. (21). The figure includes all standard solar models with which I am familiar that were published in refereed journals in the decade 1988-1998. All of the fluxes are normalized to the predictions of the Bahcall-Pinsonneault 1998 solar model, BP98. (10) The rectangular error box defines the $3\sigma$ error range of the BP98 fluxes. The best-fit $^7$Be neutrino flux is negative. At the 99% C.L., there is no solution (11) with all positive neutrino fluxes (see discussion in Section ). All of the standard model solutions lie far from the best-fit solution, even far from the $3\sigma$ contour.

## THREE SOLAR NEUTRINO PROBLEMS

I will now compare the predictions of the combined standard model with the results of the operating solar neutrino experiments.

We will see that this comparison leads to three different discrepancies between the calculations and the observations, which I will refer to as the three solar neutrino problems.

Figure 5 shows the measured and the calculated event rates in the five ongoing solar neutrino experiments. This figure reveals three discrepancies between the experimental results and the expectations based upon the combined standard model. As we shall see, only the first of these

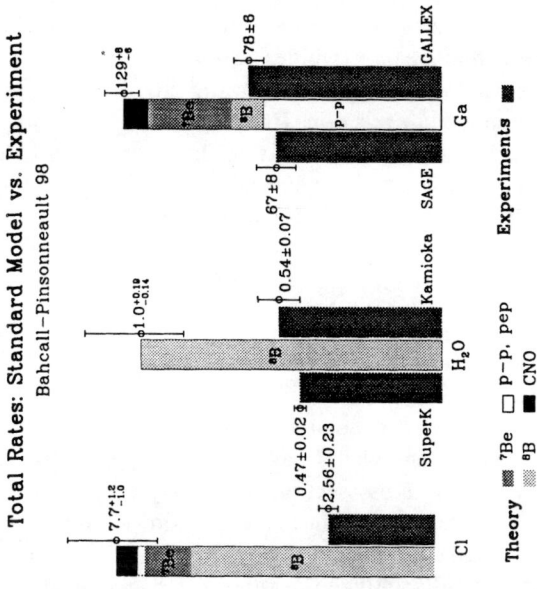

**FIGURE 5.** Comparison of measured rates and standard-model predictions for five solar neutrino experiments. (2, 3, 4, 5, 6) The unit for the radiochemical experiments (chlorine and gallium) is SNU (see Fig. 3 for a definition); the unit for the water-Cerenkov experiments (Kamiokande and SuperKamiokande) is the rate predicted by the standard solar model plus standard electroweak theory. (10)

discrepancies depends in an important way upon the predictions of the standard solar model.

## Calculated versus Observed Absolute Rate

The first solar neutrino experiment to be performed was the chlorine radiochemical experiment, (2) which detects electron-type neutrinos that are more energetic than 0.81 MeV. After more than a quarter of a century of operation of this experiment, the measured event rate is $2.56 \pm 0.23$ SNU, which is a factor of three less than is predicted by the most detailed theoretical calculations, $7.7^{+1.2}_{-1.0}$ SNU. (10) Most of the predicted rate in the chlorine experiment is from the rare, high-energy $^8$B neutrinos, although the $^7$Be neutrinos are also expected to contribute significantly. According to standard model calculations, the *pep* neutrinos and the CNO neutrinos (for simplicity not discussed here) are expected to contribute less than 1 SNU to the total event rate.

This discrepancy between the calculations and the observations for the chlorine experiment was, for more than two decades, the only solar neutrino problem. I shall refer to the chlorine disagreement as the "first" solar neutrino problem.

## Incompatibility of Chlorine and Water Experiments

The second solar neutrino problem results from a comparison of the measured event rates in the chlorine experiment and in the Japanese pure-water experiments, Kamiokande (3) and SuperKamiokande. (6) The water experiments detect higher-energy neutrinos, most easily above 7 MeV, by by observing the Cerenkov radiation from neutrino-electron scattering: $\nu + e \longrightarrow \nu' + e'$. According to the standard solar model, $^8$B beta decay, and possibly the *hep* reaction, (26) are the only important source of these higher-energy neutrinos.

The Kamiokande and SuperKamiokande experiments show that the observed neutrinos come from the sun. The electrons that are scattered by the incoming neutrinos recoil predominantly in the direction of the sun-earth vector; the relativistic electrons are observed by the Cerenkov radiation they produce in the water detector. In addition, the water Cerenkov experiments measure the energies of individual scattered electrons and therefore provide information about the energy spectrum of the incident solar neutrinos.

The total event rate in the water experiments, about 0.5 the standard model value (see Fig. 5), is determined by the same high-energy $^8$B neutrinos that are expected, on the basis of the combined standard model, to dominate the event rate in the chlorine experiment. I have shown elsewhere (27) that solar physics changes the shape of the $^8$B neutrino spectrum by less than 1 part in $10^5$. Therefore, we can calculate the rate in the chlorine experiment (threshold 0.8 MeV) that is produced by the $^8$B neutrinos observed in the Kamiokande and SuperKamiokande experiments at an order of magnitude higher energy threshold.

If no new physics changes the shape of the $^8$B neutrino energy spectrum, the chlorine rate from $^8$B alone is $2.8 \pm 0.1$ SNU for the SuperKamiokande normalization ($3.2 \pm 0.4$ SNU for the Kamiokande normalization), which exceeds the total observed chlorine rate of $2.56 \pm 0.23$ SNU.

Comparing the rates of the SuperKamiokande and the chlorine experiments, one finds–assuming that the shape of the energy spectrum of $^8$B $\nu_e$'s is not changed by new physics–that the net contribution to the chlorine experiment from the *pep*, $^7$Be, and CNO neutrino sources is negative: $-0.2 \pm 0.3$ SNU. The contributions from the *pep*, $^7$Be, and CNO neutrinos would appear to be completely missing; the standard model prediction for the combined contribution of *pep*, $^7$Be, and CNO neutrinos is a relatively large 1.8 SNU (see Table 1). On the other hand, we know that the $^7$Be neutrinos must be created in the sun since they are produced by electron capture on the same isotope ($^7$Be) which gives rise to the $^8$B neutrinos by proton capture.

Hans Bethe and I pointed out (28) that this apparent incompatibility of the chlorine and water-Cerenkov experiments constitutes a "second" solar neutrino problem that is almost independent of the absolute rates predicted by solar models. The inference that is usually made from this comparison is that the energy spectrum of $^8$B neutrinos is changed from the standard shape by physics not included in the simplest version of the standard electroweak model.

## Gallium Experiments: No Room for $^7$Be Neutrinos

The results of the gallium experiments, GALLEX and SAGE, constitute the third solar neutrino problem. The average observed rate in these two experiments is $73 \pm 5$ SNU, which is accounted for in the standard model by the theoretical rate of 72.4 SNU that is calculated to come from the basic *p-p* and *pep* neutrinos (with only a 1% uncertainty in the standard solar model *p-p* flux). The $^8$B neutrinos, which are observed above 6.5 MeV in the Kamiokande experiment, must also contribute to the gallium event rate. Using the standard shape for the spectrum of $^8$B neutrinos and normalizing to the rate observed in Kamiokande, $^8$B contributes another 6 SNU. (The contribution predicted by the standard model is 12 SNU, see Table 1.) Given the measured rates in the gallium experiments, there is no room for the additional $34 \pm 3$ SNU that is expected from $^7$Be neutrinos on the basis of standard solar models (see Table 1).

The seeming exclusion of everything but *p-p* neutrinos in the gallium experiments is the "third" solar neutrino problem. This problem is essentially independent of the previously-discussed solar neutrino problems, since it depends strongly upon the *p-p* neutrinos that are not observed in the other experiments and whose theoretical flux can be calculated accurately.

The missing $^7$Be neutrinos cannot be explained away by a change in solar physics. The $^8$B neutrinos that are observed in the Kamiokande experiment are produced in competition with the missing $^7$Be neutrinos; the competition is between electron capture on $^7$Be versus proton capture on $^7$Be. Solar model explanations that reduce the predicted $^7$Be flux generically reduce much more (too much) the predictions for the observed $^8$B flux.

The flux of $^7$Be neutrinos, $\phi(^7\text{Be})$, is independent of measurement uncertainties in the cross section for the nuclear reaction $^7\text{Be}(p,\gamma)^8\text{B}$; the cross section for this proton-capture reaction is the most uncertain quantity that enters in an important way in the solar model calculations. The flux of $^7$Be neutrinos depends upon the proton-capture reaction only through the ratio

$$\phi(^7\text{Be}) \propto \frac{R(e)}{R(e) + R(p)}, \quad (2)$$

where $R(e)$ is the rate of electron capture by $^7$Be nuclei and $R(p)$ is the rate of proton capture by $^7$Be. With standard parameters, solar models yield $R(p) \approx 10^{-3} R(e)$. Therefore, one would have to increase the value of the $^7\text{Be}(p,\gamma)^8\text{B}$ cross section by more than two orders of magnitude over the current best-estimate (which has an estimated experimental uncertainty of $\sim 10\%$) in order to affect significantly the calculated $^7$Be solar neutrino flux. The required change in the nuclear physics cross section would also increase the predicted neutrino event rate by more than 100 in the Kamiokande experiment, making that prediction completely inconsistent with what is observed.

I conclude that either: 1) at least three of the five operating solar neutrino experiments (the two gallium experiments plus either chlorine or the two water Cerenkov experiments, Kamiokande and SuperKamiokande) have yielded misleading results, or 2) physics beyond the standard electroweak model is required to change the energy spectrum of $\nu_e$ after the neutrinos are produced in the center of the sun.

## UNCERTAINTIES IN THE FLUX CALCULATIONS

I will now discuss uncertainties in the solar model flux calculations.

Table 2 summarizes the uncertainties in the most important solar neutrino fluxes and in the Cl and Ga event rates due to different nuclear fusion reactions (the first four entries), the heavy element to hydrogen mass ratio (Z/X), the radiative opacity, the solar luminosity, the assumed solar age, and the helium and heavy element diffusion coefficients. The $^{14}\text{N} + p$ reaction causes a 0.2% uncertainty in the predicted pp flux and a 0.1 SNU uncertainty in the Cl (Ga) event rates.

The predicted event rates for the chlorine and gallium experiments use recent improved calculations of neutrino

**Table 2.** Average uncertainties in neutrino fluxes and event rates due to different input data. The flux uncertainties are expressed in fractions of the total flux and the event rate uncertainties are expressed in SNU. The $^7$Be electron capture rate causes an uncertainty of ±2% (29) that affects only the $^7$Be neutrino flux. The average fractional uncertainties for individual parameters are shown.

| <Fractional uncertainty> | pp 0.017 | $^3$He$^3$He 0.060 | $^3$He$^4$He 0.094 | $^7$Be+p 0.106 | Z/X 0.033 | opac see text | lum 0.004 | age 0.004 | diffuse 0.15 |
|---|---|---|---|---|---|---|---|---|---|
| Flux | | | | | | | | | |
| pp | 0.002 | 0.002 | 0.005 | 0.000 | 0.002 | 0.003 | 0.003 | 0.0 | 0.003 |
| $^7$Be | 0.0155 | 0.023 | 0.080 | 0.000 | 0.019 | 0.028 | 0.014 | 0.003 | 0.018 |
| $^8$B | 0.040 | 0.021 | 0.075 | 0.105 | 0.042 | 0.052 | 0.028 | 0.006 | 0.040 |
| SNUs | | | | | | | | | |
| Cl | 0.3 | 0.2 | 0.5 | 0.6 | 0.3 | 0.4 | 0.2 | 0.04 | 0.3 |
| Ga | 1.3 | 0.9 | 3.3 | 1.3 | 1.6 | 1.8 | 1.3 | 0.20 | 1.5 |

absorption cross sections. (17, 18) The uncertainty in the prediction for the gallium rate is dominated by uncertainties in the neutrino absorption cross sections, +6.7 SNU (7% of the predicted rate) and −3.8 SNU (3% of the predicted rate). The uncertainties in the chlorine absorption cross sections cause an error, ±0.2 SNU (3% of the predicted rate), that is relatively small compared to other uncertainties in predicting the rate for this experiment. For non-standard neutrino energy spectra that result from new neutrino physics, the uncertainties in the predictions for currently favored solutions (which reduce the contributions from the least well-determined $^8$B neutrinos) will in general be less than the values quoted here for standard spectra and must be calculated using the appropriate cross section uncertainty for each neutrino energy. (17, 18)

The nuclear fusion uncertainties in Table 2 were taken from Adelberger et al., (12) the neutrino cross section uncertainties from Bahcall (1997)(17) and Bahcall et al. (1996),(18) the heavy element uncertainty was taken from helioseismological measurements, (30) the luminosity and age uncertainties were adopted from BP95, (25) the 1σ fractional uncertainty in the diffusion rate was taken to be 15%, (31) which is supported by helioseismological evidence, (23) and the opacity uncertainty was determined by comparing the results of fluxes computed using the older Los Alamos opacities with fluxes computed using the modern Livermore opacities. (24) To include the effects of asymmetric errors, the now publicly-available code for calculating rates and uncertainties (see discussion in previous section) was run with different input uncertainties and the results averaged. The software contains a description of how each of the uncertainties listed in Table 2 were determined and used.

The low energy cross section of the $^7$Be + p reaction is the most important quantity that must be determined more accurately in order to decrease the error in the predicted event rates in solar neutrino experiments. The $^8$B neutrino flux that is measured by the Kamiokande, (3) Super-Kamiokande, (6) and SNO (32) experiments is, in all standard solar model calculations, directly proportional to the $^7$Be + p cross section. If the 1σ uncertainty in this cross section can be reduced by a factor of two to 5%, then it will no longer be the limiting uncertainty in predicting the crucial $^8$B neutrino flux (cf. Table 2).

## HOW LARGE AN UNCERTAINTY DOES HELIOSEISMOLOGY SUGGEST?

Could the solar model calculations be wrong by enough to explain the discrepancies between predictions and measurements for solar neutrino experiments? Helioseismology, which confirms predictions of the standard solar model to high precision, suggests that the answer is probably "No."

Figure 6 shows the fractional differences between the most accurate available sound speeds measured by helioseismology (33) and sound speeds calculated with our best solar model (with no free parameters). The horizontal line corresponds to the hypothetical case in which the model predictions exactly match the observed values. The rms fractional difference between the calculated and the measured sound speeds is $1.1 \times 10^{-3}$ for the entire region over which the sound speeds are measured, $0.05 R_\odot < R < 0.95 R_\odot$. In the solar core, $0.05 R_\odot < R < 0.25 R_\odot$ (in which about 95% of the solar energy and neutrino flux is produced in a standard model), the rms fractional difference between measured and calculated sound speeds is $0.7 \times 10^{-3}$.

Helioseismological measurements also determine two other parameters that help characterize the outer part of the sun (far from the inner region in which neutrinos are

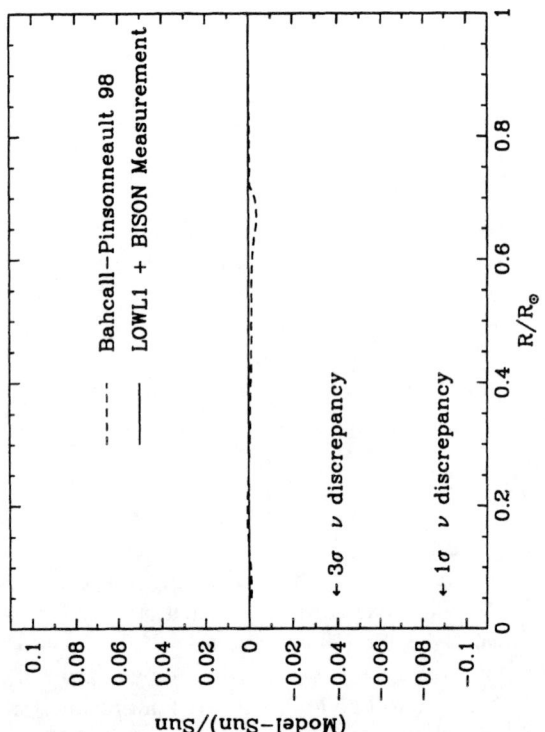

**FIGURE 6.** Predicted versus Measured Sound Speeds. This figure shows the excellent agreement between the calculated (solar model BP98, Model) and the measured (Sun) sound speeds, a fractional difference of 0.001 rms for all speeds measured between $0.05R_\odot$ and $0.95R_\odot$. The vertical scale is chosen so as to emphasize that the fractional error is much smaller than generic changes in the model, 0.04 to 0.09, that might significantly affect the solar neutrino predictions.

produced): the depth of the solar convective zone (CZ), the region in the outer part of the sun that is fully convective, and the present-day surface abundance by mass of helium ($Y_{surf}$). The measured values, $R_{CZ} = (0.713 \pm 0.001)R_\odot$, (34) and $Y_{surf} = 0.249 \pm 0.003$, (30) are in satisfactory agreement with the values predicted by the solar model BP98, namely, $R_{CZ} = 0.714R_\odot$, and $Y_{surf} = 0.243$. However, we shall see below that precision measurements of the sound speed near the transition between the radiative interior (in which energy is transported by radiation) and the outer convective zone (in which energy is transported by convection) reveal small discrepancies between the model predictions and the observations in this region.

If solar physics were responsible for the solar neutrino problems, how large would one expect the discrepancies to be between solar model predictions and helioseismological observations? The characteristic size of the discrepancies can be estimated using the results of the neutrino experiments and scaling laws for neutrino fluxes and sound speeds.

All recently published solar models predict essentially the same fluxes from the fundamental pp and pep reactions (amounting to 72.4 SNU in gallium experiments, cf. Table 1), which are closely related to the solar luminosity. Comparing the measured gallium rates and the standard predicted rate for the gallium experiments, the $^7$Be flux must be reduced by a factor $N$ if the disagreement is not to exceed $n$ standard deviations, where $N$ and $n$ satisfy $72.4 + (34.4)/N = 72.2 + n\sigma$. For a $1\sigma$ ($3\sigma$) disagreement, $N = 6.1(2.05)$. Sound speeds scale like the square root of the local temperature divided by the mean molecular weight and the $^7$Be neutrino flux scales approximately as the 10th power of the temperature. (35) Assuming that the temperature changes are dominant, agreement to within $1\sigma$ would require fractional changes of order 0.09 in sound speeds ($3\sigma$ could be reached with 0.04 changes), if all model changes were in the temperature[1]. This argument is conservative because it ignores the $^8$B and CNO neutrinos which contribute to the observed counting rate (cf. Table 1) and which, if included, would require an even larger reduction of the $^7$Be flux.

I have chosen the vertical scale in Fig. 6 to be appropriate for fractional differences between measured and predicted sound speeds that are of order 0.04 to 0.09 and that might therefore affect solar neutrino calculations. Fig. 6 shows that the characteristic agreement between solar model predictions and helioseismological measurements is more than a factor of 40 better than would be expected if there were a solar model explanation of the solar neutrino problems.

## FITS WITHOUT SOLAR MODELS

Suppose (following the precepts of Hata et al., (36) Parke, (37) and Heeger and Robertson (38)) we now ignore everything we have learned about solar models over the last 35 years and allow the important $pp$, $^7$Be, and $^8$B fluxes to take on any non-negative values. What is the best fit that one can obtain to the solar neutrino measurements assuming only that the luminosity of the sun is supplied by nuclear fusion reactions among light elements (the so-called 'luminosity constraint')? (39)

---

[1] I have used in this calculation the GALLEX and SAGE measured rates reported by Kirsten and Gavrin at Neutrino 98. The experimental rates used in BP98 were not as precise and therefore resulted in slightly less stringent constraints than those imposed here. In BP98, we found that agreement to within $1\sigma$ with the then available experimental numbers would require fractional changes of order 0.08 in sound speeds ($3\sigma$ could be reached with 0.03 changes.)

**Table 3.** Neutrino Oscillation Solutions.

| Solution | $\Delta m^2$ | $\sin^2 2\theta$ |
|---|---|---|
| SMA | $5 \times 10^{-6}$ eV$^2$ | $5 \times 10^{-3}$ |
| LMA | $2 \times 10^{-5}$ eV$^2$ | 0.8 |
| LOW | $8 \times 10^{-8}$ eV$^2$ | 0.96 |
| VAC | $8 \times 10^{-11}$ eV$^2$ | 0.7 |

The answer is that the fits are bad, even if we completely ignore what we know about the sun. I quote the results from Bahcall, Krastev and Smirnov (1998).

If the CNO neutrino fluxes are set equal to zero, there are no acceptable solutions at the 99% C. L. ($\sim 3\sigma$ result). The best-fit is worse if the CNO fluxes are not set equal to zero. All so-called 'solutions' of the solar neutrino problems in which the astrophysical model is changed arbitrarily (ignoring helioseismology and other constraints) are inconsistent with the observations at much more than a $3\sigma$ level of significance. No fiddling of the physical conditions in the model can yield the minimum value, quoted above, that was found by varying the fluxes independently and arbitrarily.

Figure 4 shows, in the lower left-hand corner, the best-fit solution and the $1\sigma$–$3\sigma$ contours. The $1\sigma$ and $3\sigma$ limits were obtained by requiring that $\chi^2 = \chi^2_{min} + \delta\chi^2$, where for $1\sigma$ $\delta\chi^2 = 1$ and for $3\sigma$ $\delta\chi^2 = 9$. All of the standard model solutions lie far from the best-fit solution and even lie far from the $3\sigma$ contour.

Since standard model descriptions do not fit the solar neutrino data, we will now consider models in which neutrino oscillations change the shape of the neutrino energy spectra.

## NEUTRINO OSCILLATIONS

The experimental results from all five of the operating solar neutrino experiments (chlorine, Kamiokande, SAGE, GALLEX, and SuperKamiokande) can be fit well by descriptions involving neutrino oscillations, either vacuum oscillations (as originally suggested by Gribov and Pontecorvo (40)) or resonant matter oscillations (as originally discussed by Mikheyev, Smirnov, and Wolfenstein (MSW) (41)).

Table 3 summarizes the four best-fit solutions that are found in the two-neutrino approximation. (11, 26) Only the SMA and vacuum oscillation solutions fit well the recoil electron energy spectrum measured in the SuperKamiokande experiment–if the standard value for the *hep* production reaction cross section ($^3$He + $p \to ^4$He + $e^+ + \nu_e$) is used. (11) However, for over a decade I have not given an estimated uncertainty for this cross section. (9) The transition matrix element is essentially forbidden and the actual quoted value for the production cross section depends upon a delicate cancellation between two comparably sized terms that arise from very different and hard to evaluate nuclear physics. I do not see anyway at present to determine from experiment or from first principles theoretical calculations a relevant, robust upper limit to the *hep* production cross section (and therefore the *hep* solar neutrino flux).

The possible role of *hep* neutrinos in solar neutrino experiments is discussed extensively in Bahcall and Krastev (1998) (26) The most important unsolved problem in theoretical nuclear physics related to solar neutrinos is the range of values allowed by fundamental physics for the *hep* production cross section.

## DISCUSSION

When the chlorine solar neutrino experiment was first proposed, (42) the only stated motivation was "...to see into the interior of a star and thus verify directly the hypothesis of nuclear energy generation in stars." This goal has now been achieved,

The focus has shifted to using solar neutrino experiments as a tool for learning more about the fundamental characteristics of neutrinos as particles. Experimental effort is now concentrated on answering the question: What are the probabilities for transforming a solar $\nu_e$ of a definite energy into the other possible neutrino states? Once this question is answered, we can calculate what happens to $\nu_e$'s that are created in the interior of the sun. Armed with this information from weak interaction physics, we can return again to the original motivation of using neutrinos to make detailed, quantitative tests of nuclear fusion rates in the solar interior. Measurements of the flavor content of the dominant low energy neutrino sources, $p$-$p$ and $^7$Be neutrinos, will be crucial in this endeavor and will require another generation of superb solar neutrino experiments (see the comments in Section ).

Three decades of refining the input data and the solar model calculations has led to a predicted standard model event rate for the chlorine experiment, 7.7 SNU, which is very close to 7.5 SNU, the best-estimate value obtained in 1968. (8) The situation regarding solar neutrinos is, however, completely different now, thirty years later. Four experiments have confirmed the original chlorine detection of solar neutrinos. Helioseismological measurements are in excellent agreement with the standard solar model predictions and very strongly disfavor (by a factor of 40 or more) hypothetical deviations from the standard model that are require to fits the neutrino data (cf. Fig. 6). Just in the last two years, improvements in the helioseismological measurements have resulted in a five-fold improvement in the agreement between the calculated standard solar model sound speeds and the measured solar velocities (cf. Figure 2 of the Neutrino 96 talk (43) with Figure 6 of this talk).

## WHAT NEXT?

More than 98% of the calculated standard model solar neutrino flux lies below 1 MeV. The rare $^8$B neutrino flux is the only solar neutrino source for which measurements of the energy have been made, but $^8$B neutrinos constitute a fraction of less than $10^{-4}$ of the total solar neutrino flux.

The next goal of solar neutrino astronomy is to measure neutrino fluxes below 1 MeV. We should begin today preparing for experiments that will measure the $^7$Be neutrinos (energy of 0.86 MeV) and the fundamental $p$-$p$ neutrinos ($< 0.43$ MeV). Indeed, we have heard at this workshop some marvelously exciting descriptions of how such low energy experiments could be carried out. The BOREXINO observatory, which can detect $\nu - e$ scattering, is the only approved solar neutrino experiment which can measure energies less than 1 MeV.

The $p$-$p$ neutrinos are overwhelmingly the most abundant source of solar neutrinos, carrying about 91% of the total flux according to the standard solar model. The $^7$Be neutrinos constitute about 7% of the total standard model flux.

If we want to test and to understand neutrino oscillations with high precision using solar neutrino sources, then we have to measure the neutrino-electron scattering rate with $^7$Be neutrinos, as will be done with the BOREXINO experiment, and also the CC (neutrino-absorption) rate with $^7$Be neutrinos (no approved experiment). With a neutrino line as provided by $^7$Be electron-capture in the sun, unique and unambiguous tests of neutrino oscillation models can be carried out if one knows both the charged-current and the neutral current reaction rates (44).

I believe we have calculated the flux of $p$-$p$ neutrinos produced in the sun to an accuracy of $\pm 1\%$. Unfortunately, we do not yet have a direct measurement of this flux. The gallium experiments only tell us the rate of capture of all neutrinos with energies above 0.23 MeV.

The most urgent need for solar neutrino research is to develop a practical experiment to measure directly the $p$-$p$ neutrino flux and the energy spectrum of electrons produced by target interactions with $p$-$p$ neutrinos. Such an experiment can be used to test the precise and fundamental standard solar model prediction of the $p$-$p$ neutrino flux. Moreover, the currently favored neutrino oscillation solutions all predict a strong influence of oscillations on the low-energy flux of $\nu_e$.

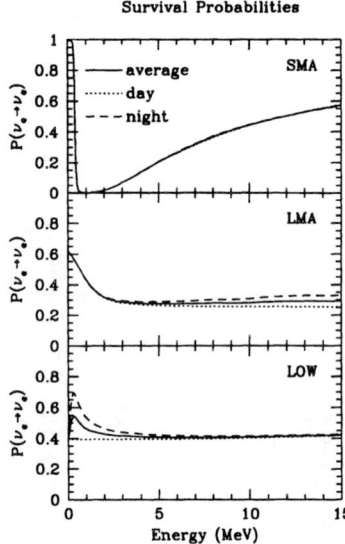

**FIGURE 7.** Survival probabilities for MSW solutions. The figure presents the yearly-averaged survival probabilities for an electron neutrino that is created in the sun to remain an electron neutrino upon arrival at the SuperKamiokande detector. There are only slight differences between the computed regeneration probabilities for the detectors located at the positions of Super-Kamiokande, SNO and the Gran Sasso Underground Laboratory. The full line refers to the average survival probabilities computed taking into account regeneration in the earth and the dotted line refers to calculations for the daytime that do not include regeneration. The dashed line includes regeneration at night. This is Fig. 9 of the 1998 analysis by Bahcall, Krastev, and Smirnov (11).

Figure 7 shows the calculated neutrino survival probability as a function of energy for three global best-fit MSW oscillation solutions. You can see directly from this figure why we have to have accurate measurements for the $p$-$p$ and $^7$Be neutrinos: the currently favored solutions exhibit their most characteristic and strongly energy dependent features below 1 MeV. In all of these solutions, the survival probability shows a dramatic increase with energy below 1 MeV, whereas in the region above 5 MeV (accessible to SuperKamiokande and to SNO) the energy dependence of the survival probability is at best modest.

The $p$-$p$ neutrinos are the gold ring of solar neutrino astronomy. Their measurement will constitute a simultaneous and critical test of stellar evolution theory and of neutrino oscillation solutions.

The most exciting result of this workshop for me has been the possibility discussed here of a synergistic experiment involving a huge (megaton?) nucleon decay detector in which an inner region is reserved for solar neutrino experiments(see, for example, the talks by Jung, Nakahata, and Ypsilantis at this workshop). The most straightforward solar neutrino experiments that could be carried out with this detector would be precision measurements of the temporal dependences of the relatively high-energy $^8$B neutrinos. One could measure with such a detector the zenith-angle dependence of the solar neutrino-event rate (the generalization of the day-night difference) and the seasonal dependence (generalization of the winter-summer difference). The design could relax somewhat the precise requirements for energy calibration and for energy resolution used for the SuperKamiokande and SNO experiments and concentrate instead on limiting the systematic uncertainties in the detector that could contribute to the error budget in the day-night or seasonal dependences. After three years of very careful measurements, the SuperKamiokande experiment has, as we have heard at this conference, about a $2\sigma$ result for the day-night difference. They do not yet have the statistics to report a meaningful measurement of the full zenith-angle dependence or the seasonal dependence. The predicted temporal effects are small, generally of order a percent, with the currently favored neutrino oscillation solutions.

Nature has provided us with many different baselines and with many different matter column densities with which to do Very Long Baseline (VLB) studies of neutrino oscillations. The earth-sun distance varies continuously during the year between $1.496(1.0\pm0.017)10^{13}$ cm and the column density through the earth to a terrestrial detector varies from 0 gm cm$^{-2}$ during the day to more than $10^9$ gm cm$^{-2}$ at night.

A solar neutrino detector ten or more times the volume of the current SuperKamiokande experiment, as discussed in concept at this workshop, could measure precisely the results of many different VLB neutrino oscillation experiments. This would be a hell of a 'Smoking Gun' detector. I have a hard time sitting down when imagining such an exciting possibility.

## ACKNOWLEDGMENTS

I acknowledge support from NSF grant #PHY95-13835.

## REFERENCES

## REFERENCES

1. R. Davis, Jr., D.S. Harmer and K.C. Hoffman, Phys. Rev. Lett. 20 (1968) 1205.
2. R. Davis, Jr., Prog. Part. Nucl. Phys. 32 (1994) 13; B.T. Cleveland, T. Daily, R. Davis, Jr., J.R. Distel, K. Lande, C.K. Lee, P.S. Wildenhain, and J. Ullman, Astrophys. J. 496 (1998) 505.
3. KAMIOKANDE Collaboration, Y. Fukuda et al., Phys. Rev. Lett. 77 (1996) 1683.
4. SAGE Collaboration, V. Gavrin et al., in: K. Enqvist, K. Huitu, J. Maalampi (Eds.), Neutrino '96, Proceedings of the 17th International Conference on Neutrino Physics and Astrophysics (Helsinki), World Scientific, Singapore, 1997, p. 14; SAGE collaboration, J.N. Abdurashitov et al., Phys. Rev. Lett. 77 (1996) 4708.
5. GALLEX Collaboration, P. Anselmann et al., Phys. Lett. B 342 (1995) 440; GALLEX Collaboration, W. Hampel et al., Phys. Lett. B 388 (1996) 364.
6. SuperKamiokande Collaboration, Y. Suzuki, in: Y. Suzuki, Y. Totsuka (Eds.), Neutrino 98, Proceedings of the XVIII International Conference on neutrino Physics and Astrophysics, Takayama, Japan, 4-9 June 1998. To be published in Nucl. Phys. B (Proc. Suppl.); Super-Kamiokande Collaboration, Y. Fukuda et al., Phys. Rev. Lett. 81 (1998) 1562; Y. Fukuda et al., hep-ex/9812009; Y. Totsuka, in: A. Olinto, J. Frieman and D. Schramm (Eds.), Proceedings of the 18th Texas Symposium on Relativistic Astrophysics and Cosmology, December 15–20, 1996, Chicago, Illinois, World Scientific, Singapore, 1998, p. 114.
7. H.A. Bethe, Phys. Rev. 55 (1939) 434.
8. J.N. Bahcall, N.A. Bahcall and G. Shaviv, Phys. Rev. Lett. 20 (1968) 1209.
9. J.N. Bahcall, Neutrino Astrophysics, Cambridge University Press, Cambridge, 1989.
10. J.N. Bahcall, S. Basu and M.H. Pinsonneault, Phys. Lett. B 433 (1998) 1.
11. J.N. Bahcall, P.I. Krastev and A.Yu. Smirnov, Phys. Rev. D 58 (1998) 096016-1; J.N. Bahcall, P.I. Krastev and A.Yu. Smirnov, Phys. Rev. D (November 1999), hep-ph/9905220.
12. E. Adelberger et al., Rev. Mod. Phys. 70 (1998) 1265.
13. C.A. Iglesias and F.J. Rogers, Astrophys. J. 464 (1996) 943; D.R. Alexander and J.W. Ferguson, Astrophys. J. 437 (1994) 879. These references describe the different versions of the OPAL opacities.
14. F.J. Rogers, F.J. Swenson and C.A. Iglesias, Astrophys. J. 456 (1996) 902.
15. A.V. Gruzinov and J.N. Bahcall, Astrophys. J. 504 (1998) 996.
16. E.E. Salpeter, Australian J. Phys. 7 (1954) 373.
17. J.N. Bahcall, Phys. Rev. C 56 (1997) 3391.
18. J.N. Bahcall, E. Lisi, D.E. Alburger, L. De Braeckeleer, S.J. Freedman and J. Napolitano, Phys. Rev. C 54 (1996) 411.
19. J.N. Bahcall, M. Kamionkowski and A. Sirlin, Phys. Rev. D 51 (1995) 6146.
20. J.N. Bahcall, Astrophys. J. 467 (1996) 475.

21. (GONG) J. Christensen-Dalsgaard et al., GONG Collaboration, Science 272 (1996) 1286; (BP95) J.N. Bahcall and M.H. Pinsonneault, Rev. Mod. Phys. 67 (1995) 781; (KS94) A. Kovetz and G. Shaviv, Astrophys. J. 426 (1994) 787; (CDF94) V. Castellani, S. Degl'Innocenti, G. Fiorentini, L.M. Lissia and B. Ricci, Phys. Lett. B 324 (1994) 425; (JCD94) J. Christensen-Dalsgaard, Europhys. News 25 (1994) 71; (SSD94) X. Shi, D.N. Schramm and D.S.P. Dearborn, Phys. Rev. D 50 (1994) 2414; (DS96) A. Dar and G. Shaviv, Astrophys. J. 468 (1996) 933; (CDF93) V. Castellani, S. Degl'Innocenti and G. Fiorentini, Astron. Astrophys. 271 (1993) 601; (TCL93) S. Turck-Chièze and I. Lopes, Astrophys. J. 408 (1993) 347; (BPML93) G. Berthomieu, J. Provost, P. Morel and Y. Lebreton, Astron. Astrophys. 268 (1993) 775; (BP92) J.N. Bahcall and M.H. Pinsonneault, Rev. Mod. Phys. 64 (1992) 885; (SBF90) I.-J. Sackman, A.I. Boothroyd and W.A. Fowler, Astrophys. J. 360 (1990) 727; (BU88) J.N. Bahcall and R.K. Ulrich, Rev. Mod. Phys. 60 (1988) 297; (RVCD96) O. Richard, S. Vauclair, C. Charbonnel and W.A. Dziembowski, Astron. Astrophys 312 (1996) 1000; (CDR97) F. Ciacio, S. Degl'Innocenti and B. Ricci, Astron. Astrophys. Suppl. Ser. 123 (1997) 449.
22. C.W. Johnson, E. Kolbe, S.E. Koonin and K. Langanke, Astrophys. J. 392 (1992) 320.
23. J.N. Bahcall, M.H. Pinsonneault, S. Basu and J. Christensen-Dalsgaard, Phys. Rev. Lett. 78 (1997) 171.
24. J.N. Bahcall and M.H. Pinsonneault, Rev. Mod. Phys. 64 (1992) 885.
25. J.N. Bahcall and M.H. Pinsonneault, Rev. Mod. Phys. 67 (1995) 781.
26. J.N. Bahcall and P.I. Krastev, Phys. Lett. B 436 (1998) 243.
27. J.N. Bahcall, Phys. Rev. D 44 (1991) 1644.
28. J.N. Bahcall and H.A. Bethe, Phys. Rev. Lett. 65 (1990) 2233.
29. A.V. Gruzinov and J.N. Bahcall, Astrophys. J. 490 (1997) 437.
30. S. Basu and H.M. Antia, Mon. Not. R. Astron. Soc. 287 (1997) 189.
31. A.A. Thoul, J.N. Bahcall, and A. Loeb, Astrophys. J. 421 (1994) 828.
32. A.B. McDonald, in: A. Astbury et al. (Eds.), Proceedings of the 9th Lake Louise Winter Institute, World Scientific, Singapore, 1994, p. 1.
33. S. Basu et al., Mon. Not. R. Astron. Soc. 292 (1997) 234.
34. S. Basu and H.M. Antia, Mon. Not. R. Astron. Soc. 276 (1995) 1402.
35. J.N. Bahcall and A. Ulmer, Phys. Rev. D 53 (1996) 4202.
36. N. Hata, S. Bludman and P. Langacker, Phys. Rev. D 49 (1994) 3622.
37. S. Parke, Phys. Rev. Lett. 74 (1995) 839.
38. K.M. Heeger and R.G.H. Robertson, Phys. Rev. Lett. 77 (1996) 3720.
39. J.N. Bahcall and P.I. Krastev, Phys. Rev. D 53 (1996) 4211.
40. V.N. Gribov and B.M. Pontecorvo, Phys. Lett. B 28 (1969) 493; B. Pontecorvo, Sov. Phys. JETP 26 (1968) 984.
41. L. Wolfenstein, Phys. Rev. D 17 (1978) 2369; S.P. Mikheyev and A.Yu. Smirnov, Yad. Fiz. 42 (1985) 1441 [Sov. J. Nucl. Phys. 42 (1985) 913]; Nuovo Cimento C 9 (1986) 17.
42. J.N. Bahcall, Phys. Rev. Lett. 12 (1964) 300; R. Davis, Jr., Phys. Rev. Lett. 12 (1964) 303; J.N. Bahcall, R. Davis Jr., in: R.F. Stein, A.G. Cameron (Eds.), Stellar Evolution, Plenum Press, New York, 1966, p. 241 [proposal first made in 1963 at this conference].
43. J.N. Bahcall, M.H. Pinsonneault, in: K. Enqvist, K. Huitu, J. Maalampi (Eds.), Neutrino '96, Proceedings of the 17th International Conference on Neutrino Physics and Astrophysics (Helsinki), World Scientific, Singapore, 1997, p. 56.
44. J. N. Bahcall, and P. I. Krastev, Phys. Rev. C 55 (1997) 929.

# HELLAZ - The new generation solar neutrino experiment to measure the spectrum of $\nu_{pp}$ and $\nu_{Be}$

## Th. Patzak

*Tufts University, Department of Physics and Astronomy, 4 Colby Street, Medford, MA 02155, USA, e-mail: patzak@tuhepf.phy.tufts.edu, on behalf of the HELLAZ collaboration*

**Abstract.** The HELLAZ experiment is dedicated to measure the spectrum of neutrinos originating in the sun. PP neutrinos are the most important since they are at the origin of the nuclear fusion chain of the sun. Moreover, the flux of pp-neutrinos is well constrained by measurements of the solar luminosity and helio seismology. Hellaz is a real-time experiment measuring the neutrino energy in elastic neutrino-electron scattering reactions. The target consists of cooled helium contained in a 2000 m³ TPC.

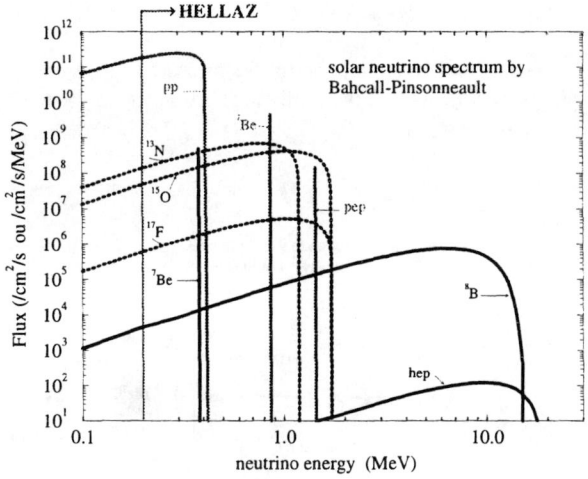

**FIGURE 1.** Solar neutrino spectrum.

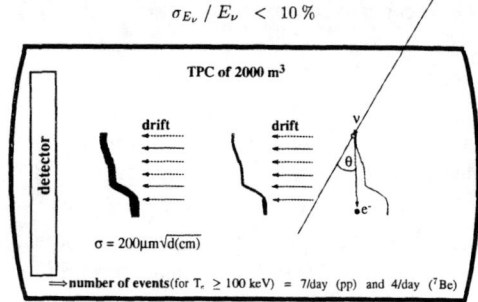

**FIGURE 2.** Schematic illustration of the experimental concept.

## THE EXPERIMENT

Our main objective with the HELLAZ (**HEL**ium at Liquid **AZ**ote - nitrogen - temperature) experiment is to measure the spectrum of pp - neutrinos and ⁷Be neutrinos (1). The solar neutrino spectrum as predicted by the standard solar model is depicted in figure 1.

The concept of the experiment is based on the measurement of the solar neutrino energy, $E_\nu$, by measuring the kinetic energy, $T_e$, and the scattering angle, θ, of recoil electrons from elastic neutrino - electron scattering. The kinetic energy of recoil electrons is measured by counting the individual electrons in an ionisation cloud generated by the energy loss of the recoil electron due to ionisation in the helium gas. A schematic view of the principle of the experiment is shown in figure 2. A large time projection chamber (TPC), filled with cooled helium will serve as active target. The apparatus with proper shielding will be situated in the Gran Sasso underground laboratory in Italy. To obtain a reasonable event rate the detector

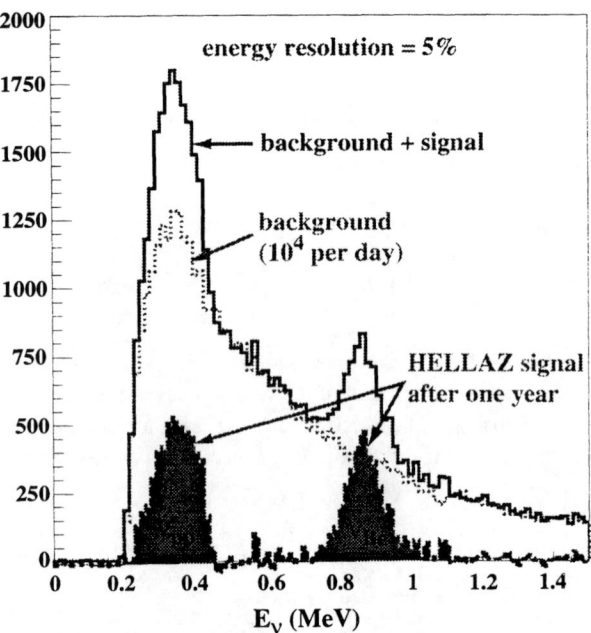

**FIGURE 3.** MC simulation for the solar neutrino spectrum measured with the HELLAZ detector for one year data taking. The number of background events is $10^4$ per day.

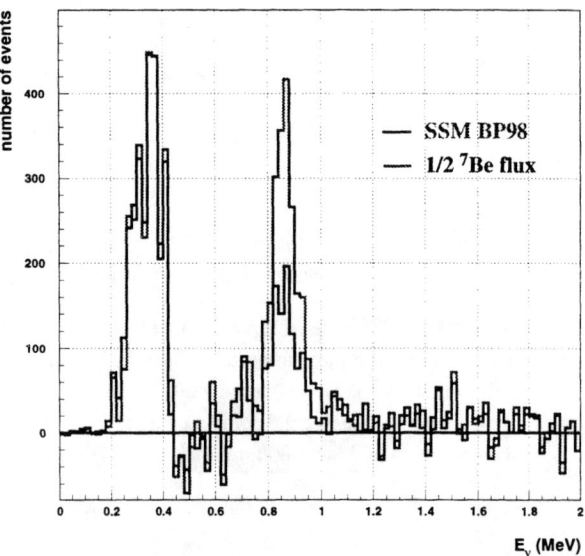

**FIGURE 4.** MC simulation for the measured spectrum assuming a model with half the Be-flux. This test was suggested by S. Turck-Chieze.

is operated at low temperature and high pressure such as to produce $2 \times 10^{30}$ target electrons in a 2000 m$^3$ TPC.

Using standard solar model calculations the expected event rate is 7/day and 4/day for pp-neutrinos and $^7$Be-neutrinos, respectively. Figure 3 shows a simulation of the spectrum as it will be measured by the HELLAZ experiment overlayed with the simulated spectrum of background and the curve of the sum of signal and background.

After having measured with high precision the spectrum one can test different hypothesis if a derivation from the standard solar model is observed. A neutrino-energy resolution of 10 % allows to detect deformations of the pp spectrum as predicted for the MSW and vacuum oscillation solutions. As example for the discriminating power of the HELLAZ experiment the result for a solution with half of the Be-flux is shown in figure 4.

It is clear that such an experiment needs a fundamental R&D program to develop a suitable detector and to demonstrate the feasibility of the experiment. To study the response of the detector to single electrons and more important to two single electrons which follow each other on the same strip of the chamber we constructed an generator of single electrons where the time between two electrons can be tuned precisely between 0 and 100 ns. The

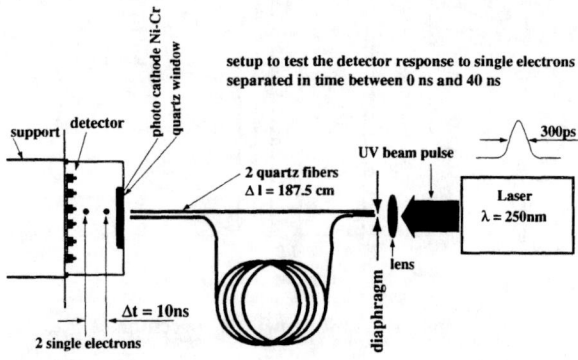

**FIGURE 5.** Generator for two single electrons following each other by a tunable time difference precisely between 0 and 100 ns. In this figure an example for 10 ns is shown.

layout of the generator is shown in figure 5. For more detailed information on this device see (2).

Different detectors of type "micro structure" have been tested. A MGWC (3) has been tested with this device and very good single electron efficiency was obtained. Two electrons, separated by 20 ns could be unambiguously be identified. The chamber suffers from its fragility and needs a extremely clean environment.
The second option tested was a Micromegas (4). This chamber gave very good results for the single electron sensitivity and the time resolution. We can distinguish two electrons which are separated by 10 ns. A typical

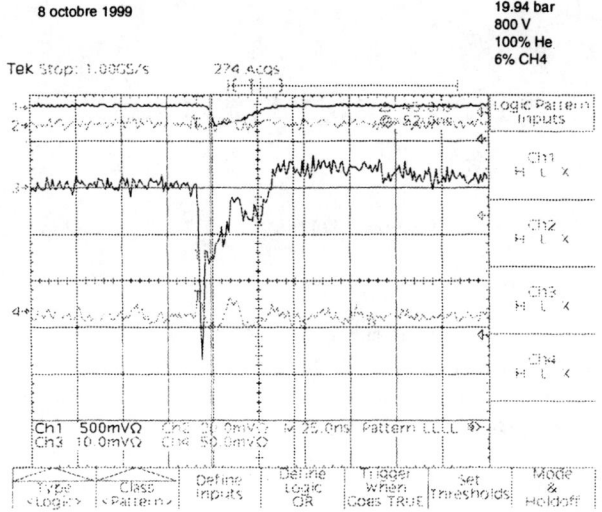

**FIGURE 6.** Typical pulse of a Micromegas chamber at a pressure of 20 bars and room temperature in Helium/Methane.

pulse from the Micomegas chamber is shown in figure 6. The sharp peak due to the movement of electrons is well distinguishable from the tail, originating from the movement of positive ions.

An example of the detection of two electrons using a Micromegas chamber is shown in figure 7.

## FUTURE

The HELLAZ experiment is a true new generation experiment to measure the spectrum of solar neutrinos at low energy. This precision measurement will allow an important contribution to solve the long standing solar neutrino problem. Furthermore, the experiment will provide a handle to distinguish different solutions such as MSW and vacuum oscillations and determine the flux and flavor of interacting neutrinos.
To realize such new experiment a large effort of detector R&D is ongoing to solve the problem of fast electron detection. In parallel studies begun on low radioactivity materials.
The prove of feasibility of the experiment will be provided by a small prototype, called Hellaz 1. It consists of a TPC with maximal drift length of 50 cm and a detector surface of 100 cm$^2$. With this setup we will generate electron tracks inside the TPC volume using $\gamma$ Compton scattering. A $^{22}$Na source emitting two 511 keV $\gamma$'s back-to-back is employed. One $\gamma$ is used to create a trigger signal, the second $\gamma$ will undergo Compton scattering in

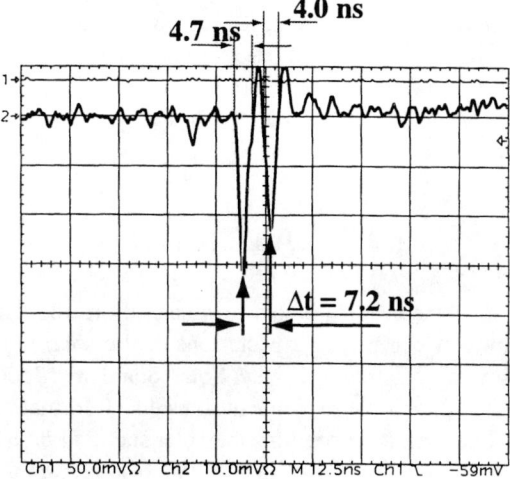

**Micromegas**

- 100% He + 6% Isobutan ($P_{abs}$ = 1 bar, T = 310 K)
- distance photo cathode - mesh = 8.5 mm
- distance mesh - cathode = 100 $\mu$m
- ampli = ORTEC 120 (+ filter C= 50pF, R=50$\Omega$)
- source = laser + 1 quartz fiber
- HT(pc) = -1900 V
- HT(mesh) = -570 V

- signals of two single electrons
- $\Delta$t between the 2 signals = 7.2 ns
- $\approx$ 4.5 ns on the base of each signal

**FIGURE 7.** Signal from two electrons, separated by 7 ns observed with Micromegas. The experimental conditions are given inside the figure.

the chamber. The angle and energy of the scattered $\gamma$ is measured with outside detectors. Thus, the kinematics of the recoil electron is fully determined allowing to test the single electron efficiency and tracking capability of the chamber under same conditions as in the final HELLAZ experiment.

## REFERENCES

1. F. Arzarello et al., CERN-LAA/94-19.
2. Th. Patzak et al., CERN-99:014, accepted for publication in NIM A.
3. E. Christophel and M. Dracos, Centre de Recherches Nucleaires Strasbourg, CRN 97-05.
4. Y. Giomataris et al., Nucl. Instr. and Meth. A 376 (1996) 29-35.

# Status of the BOREXINO Solar Neutrino Experiment

L. Oberauer

*Technische Universität München, James-Franck Str., D-85747 Garching*
*SFB 375 Astro- Particle Physics*
*on behalf of the Borexino collaboration*

**Abstract.** Aim of BOREXINO is to measure in real time solar neutrinos with an energy threshold of about 250 keV via pure leptonic neutrino electron scattering $\nu + e \rightarrow \nu + e$. Special interest in BOREXINO is the first determination of the low energy $^7$Be solar neutrino flux. In the MSW-scenario a strong depletion of this flux should be found. In case of vacuum oscillations a clear time fluctuation in the measured flux is to be expected. Further physics goals are described in the text.

## INTRODUCTION

The solar neutrino problem originates from the discrepancy between the $\nu_e$ expectations of the solar neutrino flux, as calculated by the Solar Standard Model (see e.g. (1)), and the experimental results. Information from helioseismology confirms the solar standard model to an accuracy within <1%. In addition, results from the Luna experiment (2) at low energies disfavour explanations based on discrepancies in this nuclear cross section.

Data analysis of the existing experiments leads to the assumption of severe suppression of the charged current solar $^7$Be-branch. Fig. 1 shows a model independent constraint on this flux (4). The only inputs are the integral flux of the SuperKamiokande experiment, which is taken as the value for the $^8B-$neutrinos, and the known luminosity of the sun, whereas contributions from other branches of the pp-cycle and the CNO-cycle are even neglected. Within the 90% confidence limits a flux of only 20% of the expected charged current $^7$Be neutrinos is consistent with the data, revealing a strong hint for physics beyond the standard model.

If neutrinos have non-zero masses and if they mix in analogy to the quark sector, neutrino oscillations arise. It can be shown that certain parameter on neutrino masses and mixing angles provide a solution to the result of all solar neutrino experiments. There exist three parameter areas: the MSW[1] small mixing angle solution with values typically for neutrino mass differences $\Delta m^2 \approx 10^{-6} \rightarrow 10^{-5}\ eV^2$ and mixing strength in the region $10^{-6} < \sin^2 2\theta < 2 \cdot 10^{-5}$, the MSW large mixing angle solution for the same mass region but almost full mixing strength, and the vacuum oscillation solution with strong mixing at $\Delta m^2 \approx 10^{-10}$ eV$^2$.

Fig. 2 shows the deformation of the solar neutrino spectrum as expected in case of the MSW solution. The $^7$Be neutrino line at 861 keV is almost fully suppressed in the small mixing angle solution as it is indicated by experimental data. The spectral shape of the high energy $^8$B neutrino branch is deformed. The measurement of this shape is aim of SuperKamiokande and SNO, whereas BOREXINO will deliver direct information about the $^7$Be neutrino flux.

## PHYSICS WITH BOREXINO

BOREXINO is going to be build up in the underground laboratory at Gran Sasso, Italy. The collaboration consists of several european and american institutes and is listed at the end of this article.

The aim of BOREXINO is to measure in real time the solar neutrino flux at a low energy threshold with high statistics and energy resolving via pure leptonic neutrino electron scattering $\nu + e \rightarrow \nu + e$. Liquid scintillator (300t total mass) will be target and detector material. Monoenergetic $^7$Be-neutrinos give rise to a compton like recoil spectrum in BOREXINO with a edge at 660 keV. Assuming validity of the standard model a counting rate for $^7$Be-neutrinos, which would consist in this case purely as $\nu_e$, of roughly 55/day in BOREXINO is expected. Here charged current as well as neutral current interaction in the neutrino electron scattering takes place.

In scenarios of total neutrino flavour conversion, i.e. for neutrino mass differences $\Delta m^2 \approx 10^{-6} \rightarrow 10^{-5}\ eV^2$, a reduced flux of approximately 12/day would be measured

---

[1] Mikhejev, Smirnov, Wolfenstein effect: resonant flavour conversion inside the sun, e.g. ref (5).

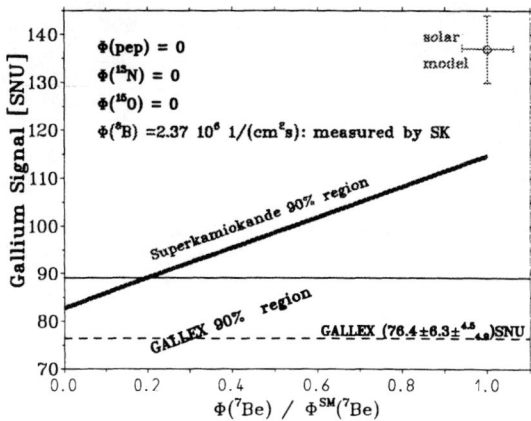

**FIGURE 1.** Model independent constraint on the charged current solar 7-Be neutrino flux derived by the experimental results from SuperKamiokande, Gallex, and the value of the solar luminosity. The signal strength (1 SNU = 1 capture in $10^{36}$ target atoms per second) of a gallium experiment is depicted as function of the 7-Be neutrino flux, resulting in the black line. The width of the line represents the experimental uncertainty of SuperKamiokande. Within 90% CL only about 20% of the expected 7-Be neutrino flux is compatible with experimental results.

due to the lower cross section of $\nu_{\mu,\tau}$ scattering, which occurs only via neutral current interaction.

In case of vacuum oscillations, i.e. for neutrino mass differences $\Delta m^2 \approx 10^{-10} \, eV^2$, BOREXINO would see a distinct time dependent periodical neutrino signal due to the seasonal eccentricity of the earths orbit around the sun. In fig. 3 the counting rate of the 7-Be solar neutrino flux in BOREXINO is depicted in case of vacuum oscillations for parameters favoured by the recent SuperKamiokande results. A very distinct seasonal dependence should be observed. As a dashed line the oscillation probability is indicated with the scale on the right side. But also for other parameters in the range of $\Delta m^2 \approx 10^{-10} \, eV^2$ a very clear seasonal fluctuation should be expected.

For neutrino mass differences in the range of $\Delta m^2 \approx 10^{-7} \, eV^2$ and for large mixing BOREXINO should see a 'day/night' effect due to electron neutrino recovery during the path through the earth.

BOREXINO also can serve for additional projects in neutrino physics. Search for a magnetic moment can be performed by means of terrestial neutrino sources by investigating the electron recoil shape at low momentum transfer. Via the inverse beta-decay $\bar{\nu}_e + p \to e^+ + n$ BOREXINO can look for signals from geophysical neutrinos (6) as well as for neutrinos emitted by european nuclear power plants (7). The latter would serve as a long baseline neutrino oscillation experiment probing the large mixing angle solution for the solar neutrino problem.

## The BOREXINO Detector

BOREXINO is placed in hall C of the underground laboratory at Gran Sasso, Italy. An overburden of about 3500 m.w.e. suppresses the cosmic muon flux to $1.1 \, /hm^2$. The detector is shielded successively from outer radioactivity. The adjacent inner layer serves as shielding and has to provide an increased purity in terms of internal radioactivity. Fig. 4 shows the experimental setup.

Inside the 'external' tank a stainless steel sphere will support about 2200 phototubes on the inside and 200 tubes at the outside. Latter provide the muon veto Cherenkov system. Tubes inside the sphere will be equipped with light guides in order to increase the geometrical coverage and hence the energy resolution.

The steel sphere will be filled with a transparent, high purity buffer liquid which itself holds a nylon sphere, filled with organic scintillator. Between the steel sphere and the scintillator region an additional nylon shroud hinders radon convection from the outer area towards the critical region. The active scintillator mass will be around 300t. As our first option serves pseudocumene (PC) with a wavelengthshifter at concentrations of about 0.15%.

Time information in each channel provides reconstruction of the event position. A fiducial volume of about 100t for solar neutrino interaction will be defined, establishing a counting rate of 55 neutrinos per day according to the standard solar model. The outer part of the scintillator sphere serves as an active additional shielding against external background.

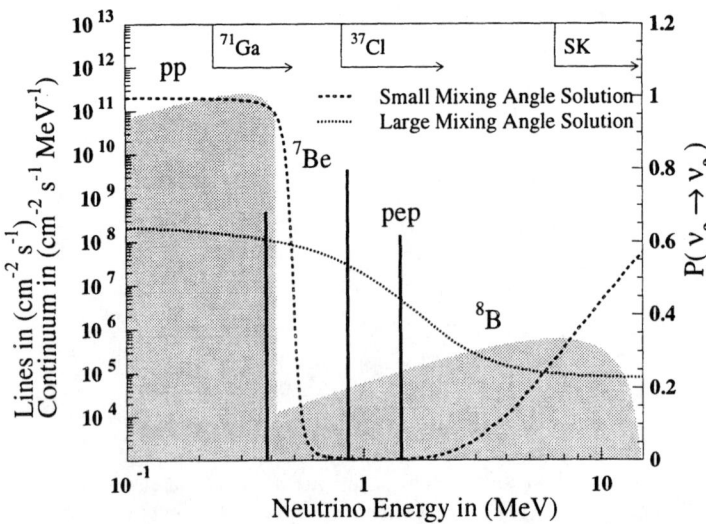

**FIGURE 2.** The solar neutrino spectrum as calculated in the standard model and the electron neutrino survival curve due to resonant neutrino flavour transversion. The 7-Be neutrino line at 861 keV is almost completely suppressed in the small mixing angle solution, as it is indicated by experimental data. The spectral shape of the high energy 8-B neutrino branch should be deformed. The measurement of this shape is aim of SuperKamiokande and SNO, BOREXINO will deliver information about the 7-Be neutrino flux.

## RESULTS OF CTF AND NAA

The demands on purity in terms of radioactivity in BOREXINO, especially for the scintillator itself, are challenging. In order to be able to extract a clear signal from background events also in case of total flavour conversion, an intrinsic concentration in Uranium and Thorium of about $10^{-16}$ should not be exceeded significantly. The amount of $^{14}C/^{12}C$ must not be higher than $\approx 10^{-18}$. In order to test scintillating materials a large Counting Test Facility (CTF) has been built up in hall C of the underground laboratory at Gran Sasso, which resembles a small prototype (ca. 5t of scintillator) of BOREXINO. In addition high sensitive neutron activation analysis (NAA) has been developed at the Technical University Munich in Garching, Germany.

### CTF results

From beginning of 1995 until summer 1997 several CTF-tests about the feasibility of BOREXINO have been performed. This includes procedures to maintain or to improve the purity of the scintillator and encouraging results have been obtained: $^{14}C/^{12}C = 1.94 \cdot 10^{-18}$, $^{238}U = (3.5 \pm 1.3) \cdot 10^{-16}$ g/g, $^{232}Th = (4.4 \pm 1.5) \cdot 10^{-16}$ g/g. A complete discussion of the CTF results including experimental techniques for further background suppression are given in ref. (8) and ref. (9). Details about the experimental setup of the CTF can be found in ref. (10).

Due to the three-dimensional array of phototubes the position of events in CTF could be reconstructed. Careful data analysis including source tests demonstrated, that the observed single rate was dominated by external background, mainly by radon present in the water shielding.

Several purification tests on the scintillator compounds have been performed, including $N_2$-bubbling, water extraction, distillation as well as column separation. Removal of radioactive gases present in the environment like Krypton and improvements in the concentration levels of $^{210}Pb$-daughters could be demonstrated (8). Column separation has been tested with an alternative scintillator and very promising results have been obtained.

The CTF response on intersecting cosmic muons has been studied and the necessity of an outer veto system in BOREXINO became clear. According to the requirements of the experiment a design for the outer detector has been finished. Measurements done at the high energy muon beam at SPS in CERN allowed for the careful study of cross sections of the generation of cosmogenic radionuclides in the scintillator (11).

Experience with the CTF helped to understand technical problems. Some of them concern deterioration of detector materials, the sealing of the tubes, radon diffusion

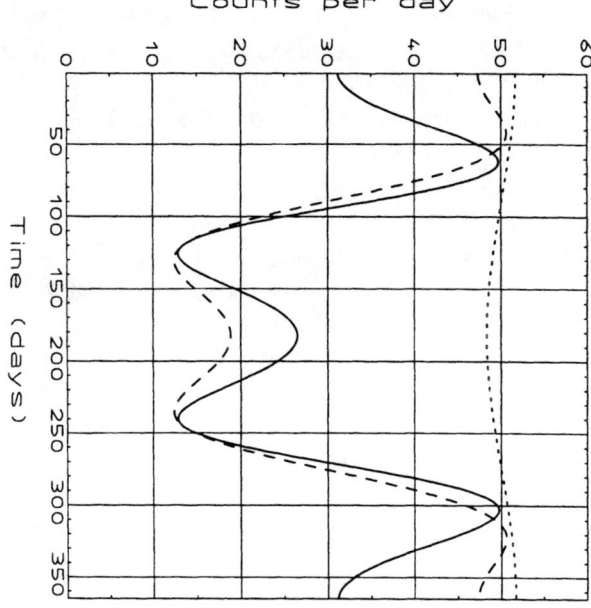

**FIGURE 3.** Daily counting rate in BOREXINO for vacuum oscillations. The dotted curve demonstrates the expected signal in case of the standard solar model and no oscillation. The depletion in summer is due to the earth's eccentricity and accounts here for a ≈7% effect. Neutrino oscillations appear in a very distinct manner for the monoenergetic $^7$Be-line. Two oscillation parameter has been used: $\Delta m^2 = 4.2 \cdot 10^{-10} \, eV^2$ (best fit to Superkamiokande data) and $\Delta m^2 = 3.2 \cdot 10^{-10} \, eV^2$ to demonstrate that this parameter can be determined with high accuracy. In both cases full mixing was assumed. The broadening of the Be-line due to the temperature in the solar center as well as the source position distribution in the sun has been taken into account.

through wet nylon, and radon tightness of liquid handling systems.

## NAA results

High sensitive neutron activation analysis (NAA) of scintillation samples performed in Garching, Germany, provides an independent test on the radiopurity and allows important tests on the secular equilibrium of the decay chains. With NAA an upper limit for uranium in PC/PPO of $^{238}U < 2 \cdot 10^{-16}$ g/g (90% CL) has been obtained. In addition concentration values or limits have been measured by this method for various isotopes, including man-made nuclei for different detector materials. For details, see ref. (12).

In Garching a new low background laboratory with shielding against the cosmic hadronic and soft electromagnetic component has been built up and completed in 1998. Now, obtained sensitivities in concentrations of Thorium and Uranium in liquid scintillators are $1 \cdot 10^{-16}$ and $1 \cdot 10^{-17}$, respectively. Indeed these limits have been reached at 90% CL for phenylxylylethane (PXE) after a multi-step purification using a silica-gel column combined with water extraction (13).

## CONCLUSION

BOREXINO is dedicated to detect in real time low energy solar neutrinos. The demands on the purity of detector materials are challanging. However, results from the CTF and NAA show very encouraging results on the radiopurity of liquid scintillators. The CTF will be reinstalled and finds further use for testing purposes. NAA in Garching has been developed to concentration levels far below the $10^{-16}$ range. In hall C of the underground laboratory at Gran Sasso the external tank and the inner steel sphere of BOREXINO has been finished. Start of the filling procedure of BOREXINO is expected at the end of the next year.

## COLLABORATION LIST

Germany: Max-Planck-Institut für Kernphysik Heidelberg: B. Freudiger, W. Hampel, J.Handt, G. Heusser, J. Kiko, T. Kirsten, H. Neder, W. Rau, M.Wojcik, Y. Zakharov. Technische Universität München: F. von Feilitzsch, C. Hagner, T. Hagner, R. von Hentig, G. Korschinek, L. Oberauer, J. Jochum, S. Schönert, K.H. Schuhbeck.

Hungary: KFKI-RMKI Budapest: L. Cser, D. Kiss, I. Manno, G. Marx.

Italy: Universita e INFN di Genova: F. Gatti, V. Lagomarsino, G. Manuzio, P. Musico, A. Nostro, A. Razeto, E. Resconi, C. Salvo, G. Testera, S. Vitale. LNGS, Gran Sasso: C. Arpesella, M. Balata, A. Falgiani, A. Goretti, A. Ianni, M. Laubenstein, M. Neff, S. Nisi, R. Tartaglia. Universita e INFN di Milano: G. Alimonti, G. Bellini, S. Bonetti, A. Brigatti, B. Caccianiga, R. Dossi, C. Galbiati, A. Garagiola, M.G. Giammarchi, D. Giugni, A. Golubchikov, F.X. Hartmann, G. Korga, P. Lombardi, S. Magni, S. Malvezzi, J. Maneira, E. Meroni, L. Perasso, G. Pieri, G. Ranucci, P. Saggese, R. Scardaoni. Universita e INFN di Pavia: G. Cecchet, A. De Bari, A. Perotti, G. Sau. Universita e INFN di Perugia: F. Elisei, F. Masetti, U. Mazzucato.

Russia: J.I.N.R. Dubna: O. Smirnov, A. Sotnikov, O. Zaimidoroga.

USA: AT&T Bell Laboratories: R.S. Raghavan. Massachusetts Institute of Technology: M. Deutsch. Princeton University: J. Benziger, M. Johnson, L. Cadonati, F. Calaprice, M. Chen, R. Eisenstein, R. Fernholz, F. Loeser, R. Parsells, R.B. Vogelaar, R. Walls.

## REFERENCES

1. J.N. Bahcall, M. Pinsonneault, Rev. Mod. Phys. 67, (1995) 781.
2. M. Junker et al., Nucl. Phys. B 70, (Proc. Suppl.), (1999), 382.
3. Gallex collaboration, Phys. Lett. B 388, (1996), 384.
4. M. Altmann, Naturwissenschaften 84, (1997), 105.
5. S.P. Mikheyev, A. Yu. Smirnov, Sov. J. Nucl. Phys. 42, (1985), 913. L. Wolfenstein, Phys. Rev. D20, (1979), 2634.
6. R.S.Raghavan, S. Schönert, S. Enomoto, J. Shirai, F. Suekane and A. Suzuki, Phys. Rev. Letters 80, (1998) 635
7. S. Schönert, Nucl. Phys. B 70, (Proc. Suppl.), (1999), 195.
8. G. Alimonti et al., BOREXINO collaboration, Astr. Part. Phys. 8 (1998), 141.
9. G. Alimonti et al., BOREXINO collaboration, Phys. Lett. B 422, (1998), 349.
10. G. Alimonti et al., BOREXINO collaboration, Nucl. Instr. Meth. (1998), accepted for publication.
11. T. Hagner et al., PhD thesis, 'Myoninduzierter Untergrund im solaren Neutrinoexperiment Borexino', Technische Universität München, Fakultät für Physik, (1999).
12. T.Goldbrunner et al., Journ. of Rad. Nucl. Chem. 216, (1997) 293.
13. R. von Hentig, Proc. of 'International Conference on Advanced Technology and Particle Physics', Como, Italy, October 5-9, (1998).

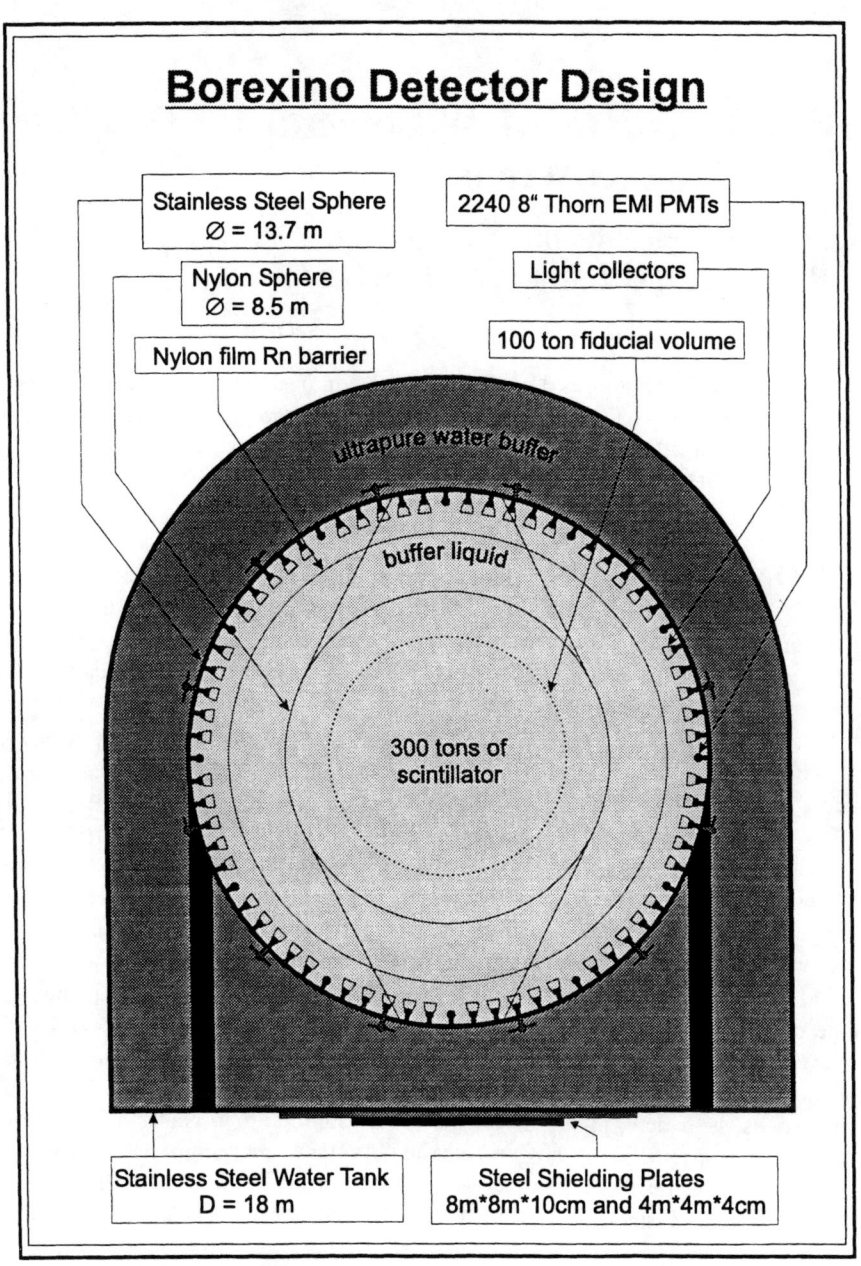

**FIGURE 4.** Schematic view of BOREXINO. The ultrapure scintillator (300t) inside a nylon vessel is shielded differentially by means of a liquid buffer (1040t), a steel sphere on which the tubes are mounted, and the outer water buffer which is contained in an external steel tank with dimensions of about 18m. A transparent nylon shroud hinders radon convection in the liquid buffer region. Additional tubes mounted on the outer surface of the steel sphere allow detection of penetrating muons via the Cherenkov effect.

# Progress on HERON: A Real-time Detector for P-P Solar Neutrinos

J.S. Adams, A. Fleischmann, Y.H. Huang, Y.H. Kim,
R.E. Lanou, H.J. Maris & G.M. Seidel

*Department of Physics, Brown University, Providence, RI 02912*

**Abstract.** The HERON project is an R&D effort to create a detector suitable for real-time, high rate measurement of neutrinos from both the p-p and $^7$Be reactions in the Sun using superfluid helium as the target medium. Progress on studies of particle detection processes in superfluid, on development of sensors, on backgrounds, and on event energy and position measurement which are related to this goal are discussed.

## INTRODUCTION

The study of solar neutrinos continues to play a major role in the effort to measure the properties of the neutrino families. The need to measure the total low energy neutrino flux and its composition is emerging as the next goal of solar neutrino research. The entire neutrino energy spectrum from all the reactions in the Sun extends from 0 MeV up to nearly 19 MeV. Measuring as much of this spectrum as possible is a principal part of the program to establish whether or not neutrino oscillations are the cause of the observed discrepancies in fluxes measured at the Earth, and if that is the case, to measure as precisely as possible their mass and mixing parameters.

No single detection technique can suffice for all of the measurements needed to cover the total energy range. Flavor composition of the flux adds another dimension to the measurements. For the latter case, use is made of inverse beta-decay reactions which are flavor specific to $\nu_e$ but they must be complemented by flavor-mixed reactions such as elastic scattering from electrons or purely neutral current reactions to test for $\nu_{\mu,\tau,s}$. At lower energies, the radiochemical experiments (Homestake, GALLEX and SAGE) measure the integrated $\nu_e$ flux above a threshold (0.8 for the former and 0.23 MeV, for the latter two). The current, real-time experiments (Kamiokande, SuperKamiokande and SNO) measure energies for a variety of reactions from the neutrinos of the $^8$B (> 5 MeV) flux using the Cherenkov effect in water and heavy water (SNO). The combination of these experiments is expected to give the first "smoking gun" evidence for or against oscillation of solar neutrinos. The maximum recoiling electron energy from elastic scattering of the principal $^7$Be line is 665 keV; while that from the p-p continuum is only 261 keV. The upcoming BOREXINO detector, based on the use of a highly purified liquid scintillator, will give the first direct window on the $^7$Be flux in real-time using the elastic scattering reaction but is not expected to access the p-p flux because of internal backgrounds at the lowest energies.

Extending the measurements of the p-p neutrinos in the near future is important for several reasons. Even after the completion of all of the above mentioned experiments we still will have a direct measure for < 10% of the total solar neutrino flux or of its flavor composition. More than 90% of the solar neutrino flux, namely that from the p-p reaction, will remain to be understood. The region in the Sun where this reaction occurs coincides with that of maximum solar energy production. In standard solar models, the p-p neutrino rate can be precisely ($\pm 1\%$) (1) related to solar luminosity. The solar luminosity is well measured; if complemented by experiments designed to measure the total neutrino flux at these low energies an important direct comparison of two measured quantities can be made. Additionally, all of the currently favored, model-dependent solutions (e.g., MSW and vacuum oscillations) make quite different predictions for the flux composition at p-p energies. Even if one solution remains strongly favored after the present generation of experiments, a direct measurement of this flux composition can be an important ingredient in establishing the mass-mixing parameters accurately. Except for the large mixing angle MSW solution, these parameters are not likely to be amenable to entirely terrestrially-based experiments.

Detection reactions with accurately known cross sections for the p-p and $^7$Be neutrinos are available — for the ground state transitions in Ga and for elastic electron scattering. The continuing gallium-based detectors (GNO and SAGE), with increased statistics and improved systematic errors, will provide more precise measures of the $\nu_e$ flux for energies > 0.23 MeV. Already this flux is known to be at or below the amount allowed by solar models from the p-p reaction alone. A real-time experiment utilizing spectral measurements for the flavor-mixed and precisely known elastic cross-section is needed to provide the essential, but presently missing, sensitivity to the components $\nu_{\mu,\tau}$ of the low energy flux. An experiment with this capability would also be sensitive to the $^7$Be neutrinos and, if done by a technique with very different systematics from BOREXINO, could provide an important corroboration of that important measurement.

There are a number of reasons why a real-time, p-p experiment is very difficult and no technique has yet been shown to be fully feasible. Principal among the difficulties are the residual radioactivity in the target and the target vessel and the challenge of retrieving energy information for depositions as low as 50-100 keV in a massive target. We have been conducting a series of experimental studies which suggest that a superfluid-based detector may be feasible. We present a synopsis of these studies and their current status.

## GENERAL PRINCIPLES OF HERON

In the HERON detector, the detection reaction is

$$\nu_{e,\mu,\tau} + e^- \rightarrow \nu_{e,\mu,\tau} + e^-.$$

Fortunately, although its cross-section is low ($\sim 10^{-45}$ cm$^2$) the expected flux is high ($\sim 6 \times 10^{10}$ cm$^{-2} \cdot$s$^{-1}$ for p-p neutrinos in the standard solar model) so that $\sim 10$ tons of target would yield about 20 p-p events and 7 $^7$Be events per day with recoil energies above 50 keV. HERON would use superfluid helium as the target material. Helium in the superfluid state is self-cleaning and retains no other atomic species in solution. It has a density of 0.145 g·cm$^{-3}$ at 30 mK. Consequently it can be made into a very compact target completely free of radioactivity. It has no long lived isotopes and its first excited nuclear state is high, $\sim 20$ MeV. The technology for handling large volumes of liquid helium is well established and is regularly used in the industrial and research communities. It is inexpensive; at $30K per ton.

Two copious forms of radiation are produced in the liquid by an ionizing particle: UV photons and phonons/rotons (quasi-particles). The photons (15 per keV) are in a narrow range centered at 16 eV; an energy at which helium is completely transparent. About $10^5$ phonons/rotons per keV are also produced and propagate ballistically (2) in the superfluid at speeds of $\sim 200$ m·s$^{-1}$. The signals from both single photons and multiple phonons/rotons will be used to determine the position and energy of each event. These signals are detected on an array of cryogenic wafer-calorimeters placed above the surface of the liquid. The prompt photons are absorbed directly by the wafers providing a fast trigger and can be used to find the event position. The wafers of silicon or sapphire are arranged in two patterned planes to form a coded aperture array. The distribution of photons hitting the array is used to determine the location of the event in a manner similar to that used in x-ray astronomy and tomography. The delayed phonons/rotons are detected through the secondary process of quantum evaporation (3) and adsorption (4); this signal is combined with that of the photons for energy determination and the relative timing among the wafers is used as a check on event position.

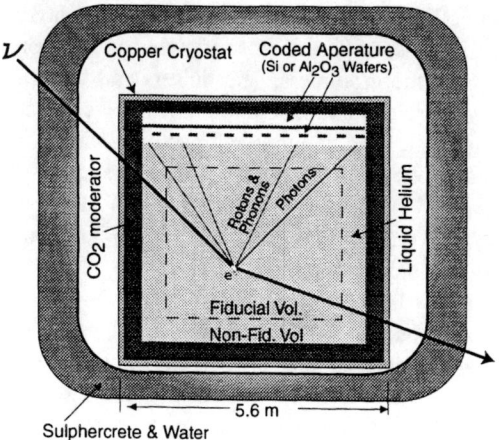

**FIGURE 1.** A schematic of the HERON configuration.

While the target itself can be made completely free of background producing materials, backgrounds emanating from the containment cryostat must be tagged and subtracted. This is to be done by logically dividing the helium volume into a fiducial and non-fiducial region and making a statistical subtraction. Figure 1 represents a generalized view of a detector based upon these principles.

## EXPERIMENTS ON SINGLE PARTICLE DETECTION

We have carried out a series of experiments using stopping electrons and alpha particles in order to acquire a quantitative understanding of the energy loss process in superfluid. The experiments have been carried out in a test cell which contains $\sim 3$ liters of superfluid typically held at 30 mK by a dilution refrigerator. The interior of the cell is instrumented with a variety of devices which can be re-configured appropriately for each experiment. Besides thermometers, heat pulsers and liquid level sensors, there are small, superconducting stepper motors for positioning and moving radioactive sources within the liquid volume. The sources were low activity $^{241}$Am for 3-5 MeV alphas and $^{113}$Sn for 364 keV electrons. X-rays from $^{55}$Fe as well as from the Sn and Am sources were used for calibration.

Thin silicon or sapphire wafers ranging in size from 5 cm dia. discs to $1 \times 2$ cm rectangles, were mounted above the liquid surface to be used as calorimeters. The temperature rise in these wafers was measured by Ir-Au thin-film superconducting transition edge sensors (TES)(5) deposited on the surface of the wafers. The TES devices were run in a self-biasing mode and read out with conventional SQUID electronics. The wafers were kept free of superfluid film by a "film burning" device (6). The signals from the phonons/rotons and photons are detected as follows.

The ballistically propagating phonons/rotons which reach the free surface of the liquid and whose energy is greater than the helium-helium binding energy can cause quantum evaporation. This is a process in which one phonon or roton can evaporate one helium atom and does so about 30% of the time (3). Energy and momentum conservation at the liquid surface place constraints on the incident roton angles which can cause evaporation (7). The helium binding energy is 0.7 meV and the typical phonon/roton energies involved are $\geq 0.8$ meV. Thus in the case of a charged particle in the 50 keV range (ie a recoil electron from a neutrino event) a very large number of atoms are involved, and are subsequently evaporated. The evaporated atoms are adsorbed onto the wafer and their deposited binding energy produces a heat pulse in the TES. The binding energy of the helium to the substrate is approximately a factor of ten higher than that of the helium-helium binding energy, thus there is an effective amplification in the evaporation-adsorption process.

The scintillation (UV photons) result from the radiative decay of helium dimers formed along the track and have energies in the range of 14-20 eV, but peaked at 16 eV. Since the first excited state of atomic helium is at 20 eV the liquid is transparent to lower energies. The absorption of photons in the silicon and sapphire wafers create a prompt heat pulse in them. The two contributions, photons and adsorbed helium atoms, can be distinguished by their different times of arrival on the wafer. In the results described here, the energy threshold of the wafers was 300 eV, so single photons were not detectable. The principal results for alpha particles, including evidence for particle directionality, are discussed elsewhere (8).

The $^{113}$Sn had an activity of a few nano-curie so that single electrons could be distinguished. A 364 keV elec-

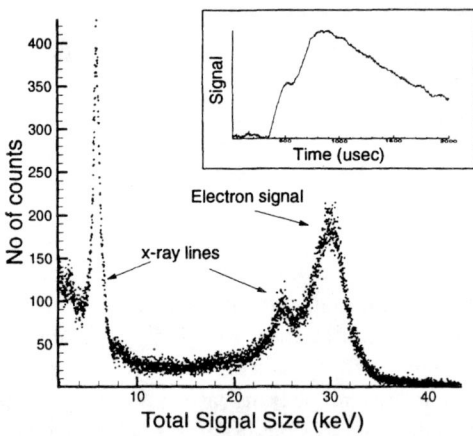

**FIGURE 2.** Inset: a typical 364 keV electron pulse from superfluid helium; the initial rise is due to photons and the later one to evaporation by phonons/rotons. The main figure is the energy spectrum for 364 keV electrons in superfluid helium; also shown are the 6 and 25 keV x-ray calibrations.

tron is produced by isomeric transition in $^{113}$In following electron capture; a 25 keV x-ray is also produced and is used as one of the calibration points for the wafers. For the data illustrated in Figure 2, the source was $\sim$5 cm below the wafer and the time delay difference can be clearly seen in the pulse shape even for that very short distance. Also shown in the figure is the measured single electron spectrum.

We find that the division of the electron's initial energy is 25% into UV photons (or 15 photons/keV) and 10% into detectable phonons/rotons (or $10^5$ phonons/keV). This contrasts with the 8% and 40%, respectively, for alpha particles. It should be noted that the energy fraction represents the energy deposited directly into the helium into $4\pi$ sr and has been corrected for geometrical factors. However in the case of the phonon/roton component, this energy also differs from what is measured on the bolometers due to the effective gain from binding energy differences. These results agree well with a model in which the formation of excited helium dimers in the close vicinity of the track, and their subsequent de-excitation by radiation or collision, plays a central role in determining the above energy fraction. (9).

## WAFER SENSITIVITY

For the coded aperture array to be useful for recoil energies as low as 50 keV in a large detector, the wafers must be able to detect with high probability a single 16 eV photon. In our prototype cell so far using transition edge sensors on our 5 cm dia.. wafers our threshold was

~20 photons. Thus to ensure that our modified detection technique is successful we must develop somewhat larger wafer/calorimeters with a threshold sensitivity of at least 10 eV.

We have taken as a benchmark event for establishing the performance requirements of a large-scale neutrino detector, the detection and measurement of an approximately 50 keV electron, 4 meters below the surface of a container of superfluid helium. Wafers of about 150 cm$^2$ with a 10 eV threshold are needed to meet the performance requirements of a full-scale neutrino detector; however, the successful development of a neutrino detector does not depend upon any anticipated improvements in the performance of transition edge sensors. Rather, we have shifted to a very promising and rapidly developing different technology which should allow wafers with the necessary increase in size and improvement in sensitivity. The technology is that of the magnetic micro-calorimeter which would replace the TES as the sensor on the wafers. A magnet sensor for calorimetry consists of a material with a strong temperature-dependent magnetization and a means to measure that magnetization. Small amounts of energy deposited in the calorimeter cause a magnetic flux change which is measured with dc SQUID electronics. For low temperature operation we have used a system of dilute paramagnetic rare earth ions in metallic gold. The initial development experiments have been carried out in Heidelberg and at Brown using Er as the paramagnetic ions.

To see the power of the technique it is useful to cite an early result of our collaborating colleagues in Heidelberg in which they had achieved at 25 mK a resolution of 135 eV at 6 keV and excellent linearity (11). The resolution of 135 eV is, by itself not exceptional, but the fact that the calorimeter in the measurement had a heat capacity of about $4 \times 10^{-9}$ J/K makes the result important. This heat capacity is comparable to that of 15 cm$^3$ (the volume of a 20 cm diameter, 0.5 mm thick wafer) of silicon at 50 mK. There are a large number of experimental parameters that have not been optimized for our particular application: physical size, spin concentration, magnetic field, temperature, bandwidth and filtering, flux coupling to the SQUID, etc. We base our confidence in being able to predict the performance achievable with a magnetic calorimeter on the fact that it is a well defined and well characterized thermodynamic system. The response of the system to an energy input can be calculated. Measurements so far are in good agreement with theory. Our expectation is that we shall eventually come close to the thermodynamic noise limit set by the temperature, heat capacity, coupling to the thermal reservoir and bandwidth of measurement.

Significant progress continues and just recently a major step forward was made. A resolution of 13 eV at 6 keV has now been achieved by the Brown-Heidelberg collaboration with a magnetic calorimeter having a heat capacity of about $1 \times 10^{-12}$ J/K (10). Since the sensitivity varies roughly as $C^{-1/3}$ for magnetic calorimeters, we require a factor of 5 improvement over present performance. Such an improvement is well within that predicted to be achievable with better calorimeter design. Based on the experience gained so far, we believe that calorimeter performance adequate for large-area wafers with a 10 eV threshold is feasible.

## EVENT LOCATION AND BACKGROUNDS

In liquid helium the range of p-p and $^7$Be recoil electrons is less than 2 cm. On the scale of the detector (volume ~60 m$^3$), these events are effectively point sources of photons and phonons/rotons. In contrast, most of the background gamma-rays, which originate in the cryostat, make spatially extended energy depositions by multiple Compton scattering. Although there is always a small number of single Compton scatterings, the major contribution to background is from multiple depositions originating from an initial gamma-ray. To suppress the background, multiple Compton scattering events must be identified and the spatial distribution of point energy depositions obtained. The background from point depositions can be determined statistically using the measured energy and spatial distribution of such events in the non-fiducial region. A knowledge of the position of the event in a large detector is essential not only for background rejection and diagnostics but also for the determination of the energy of the event.

Coded aperture arrays, mentioned above, are often used to determine the position of a source in circumstances where the radiation is difficult or impossible to focus or is of low intensity, such as for x-rays. In simple terms, the original (12) principle of operation is based on the concept of decoding an image formed on an image plane by the "shadow" of a mask constituting a multiple hole pin-hole camera where the mask pattern and plane separations are known. A large modern literature exists, and a large variety of mask patterns are employed (13).

In our application there are important differences and simplifications in implementation. There are, of course, image and mask planes but here the "pixels" are wafer-calorimeters, with the lower mask plane having roughly 50% or more transmission. Unlike the usual method of coded apertures, the wafers forming the mask for HERON will also be active calorimeters and be used in both position finding and energy determination. Thus, no signal is lost by the presence of the mask. Depending on photon statistics, the distribution of their signals on the image plane uniquely determines the source position. The accuracy with which the array is able to measure event location can be directly tested in-situ for a large detector by using a movable radioactive source.

To evaluate potential performance of this application, we have initiated Monte Carlo simulations in which the distribution of photons incident on the wafers from a source anywhere in the detector is used to determine the most probable position of an event based on a maximum likelihood method. In the calculations we have assumed the wafers have an energy threshold sufficiently low to detect single 16 eV photons. We have simulated a HERON-sized array using a model in which the image plane consists of 1600, 150 cm$^2$, densely packed wafers, separated by 50 cm from the mask plane of 800 wafers. These early simulations suggest that it may be possible to determine the position of an energy deposition by an electron over most of a 60 m$^3$ cube of liquid with a resolution of 10-15 cm FWHM provided more than about 75 photons hit the array. The 75 corresponds to an energy of 30 keV for an event near the top and 75 keV near the bottom of the fiducial volume. We have not yet optimized the parameters of the array (among them: pixel size, mask pattern, plane spacing, etc.) nor the search algorithms. Additionally, a thorough study of the background involving multiple Compton scattering events, which should be distinguished by the poor quality of the determination of event location, must be made.

Devising a strategy for the control of backgrounds is also a key issue. For this purpose we have carried out Monte Carlo simulations based on assumptions of detector performance and knowledge of expected sources. Because the signal signature is a single recoil electron appearing in the detector volume, Compton scatterings of gamma-rays are the most dangerous. (The relative pulse heights of photon vs. phonon initiated signals is a discriminant against heavily ionizing deposition.) At the energies from likely sources, Compton scatters constitute more than 90% of $\gamma$-ray conversions in liquid helium. The major sources are from cosmogenically produced isotopes in the cryostat materials while on the surface of the Earth and from heavy elements (U and Th) in the cryostat or any material placed in immediate contact with the helium. Decays from such activities located in the wafer planes or from slow neutron captures elsewhere are minor sources. Several steps can be taken to control the signal to background. Those we have included in our simulations are: minimization of cosmogenics by control of exposure and storage time (two weeks above and two years below ground), minimizing U and Th ($<10^{-12}$g/g) in the copper of the cryostat by electroforming, providing a cryostat liner ("moderator") of low U/Th content ($<10^{-14}$g/g) and providing a non-fiducial volume of helium sufficient to allow a statistical background subtraction. Liquid helium is not an efficient absorber of high energy ($\sim$2 MeV) $\gamma$-rays; however, while it is possible to use an excess volume of liquid helium itself as an attenuator of these external $\gamma$'s to do so makes the size impractically large. Conversely, the detector dimensions are large enough that lower energy ($\leq 1$ MeV) $\gamma$'s produce a multiplicity of spatially well separated Compton recoils. The main purpose of the "moderator" then is energy degradation of the higher energy $\gamma$'s to produce a more distinctive background signature as well as some attenuation.

In a typical simulation to study background, a 10 ton cylindrical copper cryostat (5.6m dia.x5.6m ht.) with a 25 cm thick moderator of $CO_2$ (or frozen non-polar liquid) and containing a total ( fiducial plus non-fiducial volume ) of $\sim$18 tons of superfluid helium was used. The cosmogenic spectrum and rates in the material were taken from double-beta decay and dark matter experiments (14) as per the above mentioned storage conditions (45$\mu$Bq/kg). The levels of U and Th in the materials was chosen to make their overall contribution to the background 10-20% of that from the cosmogenics. Single photon sensitivity and the ability of the coded aperture array to perform as described was assumed. High statistics runs were made to be the equivalent of six month's data taking. A variety of different relative sizes and shapes of fiducial and non-fiducial volumes were used. Typical cuts on events required energy depositions $\leq$800 keV and single or multiple Comptons within the assumed spatial resolution from the coded aperture array. The resulting deposited energy spectra in the fiducial and non-fiducial regions are remarkably similar down to <100 keV thus facilitating a direct bin-by-bin subtraction. The normalization is taken from the Monte Carlo. Since the division between fiducial and non-fiducial volumes is done in software, in a real detector that division could be varied to check for consistency of subtraction. As examples from the simulation with a fixed total volume, a fiducial to non-fiducial volume ratio of 1:1 gave an unsubtracted signal to background of 3.5:1 while a volume ratio of 1.5:1 gave signal to background of 0.8:1. We believe we have taken realistic assumptions as Monte Carlo input so if the conditions of performance and activity assumed can be achieved then this background strategy should be adequate for obtaining the electron recoil rate and spectrum from low energy solar neutrinos.

## SUMMARY AND CONCLUSIONS

At each stage in the history of solar neutrino research imaginative and unique new experimental tools have been successfully developed to achieve the task then at hand.

There is now increasing interest to study the lowest energy part of the neutrino spectrum which constitutes >90% of the total flux. However, real-time detection of p-p and $^7$Be neutrino interactions presents severe experimental challenges to developing a detector because of the backgrounds inherent in the target material and the material surrounding the target. Another concern is the very low energy signal that must be extracted from a massive target. Once again, a new technique is needed.

We have presented results on the progress made so far toward developing a new type of detector, HERON, to have this capability. Its essential feature is the use of superfluid helium as the target material. Only at these very low temperatures can helium be made entirely free of internal backgrounds. The goal of our experiments and studies so far has been to gain sufficient knowledge of the conditions needed to establish feasibility of the new technique. Several of these conditions have either been met or appear to be achievable. We now have a thorough understanding of the processes by which low energy electrons can be detected using both photons and quasi-particles (phonons/rotons). We have developed calorimeter sensors able to detect both phonon and quasi-particle induced signals. The sensitivity of these devices is showing rapid improvement. We have carried out detailed studies of strategies for controlling externally produced backgrounds. The initial simulations of position finding within a detector by the new coded aperture method shows considerable promise. Recently, we have upgraded our simulation capability for optimizing the array and testing with what precision an event can be found with no prior information as to its approximate location in the entire detector and with what signal to background selectivity. For neutrino experiments at these low energies, we believe that the ability to locate an event position inside a detector is an essential control on backgrounds from sources outside the target material itself. The requirements for radioactivity in the cryostat and moderator are stringent but appear possible. Presently, we are concentrating on two topics which are fundamental to the success of the superfluid technique: achieving single, 16 eV photon sensitivity on wafer sensors and demonstrating the efficacy of the coded aperture for event location and background signature recognition.

## REFERENCES

1. Bahcall, J.N. and Krastev, P.I., *Phys. Rev.* **D53**, 4211 (1996).
2. Hendry, P.C. and McClintock, P.V.E., *Cryogenics* **27**, 131 (1987) Lowering the natural concentration of $^3$He reduces roton scattering. Using a single pass with the Hendry-McClintock heat flush technique during detector filling we have reduced the $^3$He content of the superfluid to $1:10^{-9}$.
3. Brown, M and Wyatt, A.F.G.,*J. Phys.: Condens Matt.*, **2**, 5025 (1989).
4. Adams, J.S. et al, *Phys. Lett.* **B341**, 431 (1995).
5. Bühler, M. et al, Proceedings of 7th International Workshop on Low Temperature Detectors, 149 (1997), Editor: S. Cooper (Max Planck Institut, Munich). We wish to thank the von Feilitzsch Group for the deposition of these thin films on our wafers.
6. Torii, R. et al, *Rev. Sci. Instr.* **63**, 230 (1992).
7. Bandler, S.R. et al, *Phys. Rev. Lett.* **68**, 2429 (1992).
8. Bandler, S.R. et al, *Phys. Rev. Lett.* **74**, 3169 (1997); Bandler, S.R. et al, *Nucl. Instr. & Meth.* **A370**, 578 (1996).
9. Adams, J.S., Kim, Y.H., Lanou, R.E., Maris, H.J. and Seidel, G.M., *J. Low Temp Phys.*, **113**, 1121 (1998). See also Adams, J.S. et al, (To appear in *Nucl. Phys.B*: Proceedings of Workshop on Low Temperature Detectors ("LTD8") Dalfsen, Holland; in press (1999)).
10. Schönefeld, J. et al, (To appear in *Nucl. Phys.B*: Proceedings of Workshop on Low Temperature Detectors ("LTD8") Dalfsen, Holland; in press (1999)).
11. Fleischmann, A., et al, (To appear in *Nucl. Phys.B*: Proceedings of Workshop on Low Temperature Detectors ("LTD8") Dalfsen, Holland; in press (1999)).
12. See for example, Dicke, R.H., *Astroph. J.* **153**, L101 (1968).
13. See for example, Fenimore, K.E. and Cannon, T.M., *Applied Optics* **17**, 337 (1978).
14. Private communication: Zerle, L. (CRESST collaboration); Heuser, G. *Low-Level Measurements of Radioactivity in the Environment (Proc. of 3rd Intl. Sch.*, 69 (1994), Editor: M. Garcia-Leon (World Sci. Publ.).

Supported in part by DoE DE-FG02-88ER40452 & NSF PHY-9870276

# The Sudbury Neutrino Observatory

J. Heise (on behalf of the SNO Collaboration*)

*Department of Physics and Astronomy, University of British Columbia, Vancouver, BC, Canada V6T 1Z1*
*E-mail: jaret@physics.ubc.ca*

**Abstract.** The Sudbury Neutrino Observatory (SNO) has been in operation since April 1998. Light and heavy water fill was completed in April, 1999, and since May, 1999 the detector has been taking physics data. The SNO detector is a 1000 tonne heavy water Čerenkov detector situated 2070 m (5900 m.w.e.) underground in INCO's Creighton Mine near Sudbury, Ontario, Canada, and involves participation from approximately 80 scientists in Canada, the United States and the United Kingdom. Through the use of heavy water, SNO will be able to detect neutrinos from a number of reactions, including one specifically sensitive to solar electron neutrinos and another which is sensitive to all neutrino types. With these two interactions, the detector will be able to search for neutrino flavour change without the requirement of electron neutrino flux normalization by solar model calculations. It will also provide unusual sensitivity for other measurements of solar neutrino properties, atmospheric neutrinos and supernova neutrinos.

## INTRODUCTION

The Sudbury Neutrino Observatory (1) has been constructed to study fundamental properties of neutrinos, in particular the mass and mixing parameters. At the heart of the SNO detector is 1000 tonnes of heavy water ($D_2O$), which when compared to light water ($H_2O$) offers the advantages of a higher interaction cross section and a wider range of possible neutrino reactions. In order to measure neutrinos in both the light and heavy water regions, the SNO detector relies on four principal reactions:

1. Charge Current (CC) reaction: $\nu_e + d \rightarrow p + p + e^-$, which is sensitive exclusively to electron neutrinos.

2. Neutral Current (NC) reaction: $\nu_x + d \rightarrow \nu_x + p + \mathbf{n}$, which is sensitive to all non–sterile neutrino types.

3. Elastic Scattering (ES) reaction: $\nu_x + e^- \rightarrow \nu_x + e^-$, which is mainly sensitive to electron neutrinos since they have about six times greater cross section than other neutrino types.

4. The reaction $\bar{\nu}_e + d \rightarrow \mathbf{n} + \mathbf{n} + e^+$, which is sensitive to electron antineutrinos.

---

* The SNO Collaboration includes participants from Queen's University, University of Guelph, Laurentian University, University of British Columbia, Centre for Research in Particle Physics at Carleton University, University of Washington, University of Pennsylvania, Los Alamos National Laboratory, Brookhaven National Laboratory, Lawrence Berkeley National Laboratory, University of Oxford.

Energetic electrons and positrons in the above reactions produce Čerenkov light that is detected by a surrounding array of photomultiplier tubes. One of the most important aspects of the SNO detector is the ability to detect neutrons, and a number of methods will be discussed below. Neutrino signatures from these reactions will allow SNO to measure properties of both solar and non–solar neutrinos in a way that no other experiment which is currently running can do.

## THE DETECTOR

A schematic picture of the Sudbury Neutrino Observatory is provided in Figure 1. The detector consists of 1000 tonnes of heavy water contained within a spherical acrylic vessel that is 12 m in diameter and 5.5 cm thick. Surrounding the acrylic vessel is a geodesic support structure (PSUP) 18 m in diameter, which houses 9438 inward–looking and 91 outward–looking 20 cm diameter photomultiplier tubes (PMTs). Each of the inward–looking PMTs is equipped with a light collector, helping to increase the effective geometric coverage to 65% of $4\pi$. Between the acrylic vessel and the photomultiplier tube support structure is approximately 1700 tonnes of ultra–pure light water. The volume beyond the support structure contains 5300 tonnes of light water, and is used mainly for mechanical support and as a background shield for gamma rays and neutrons from the rock.

The SNO detector is designed to provide very low levels of radioactive background for the reactions that will be used to detect neutrinos. The entire detector is con-

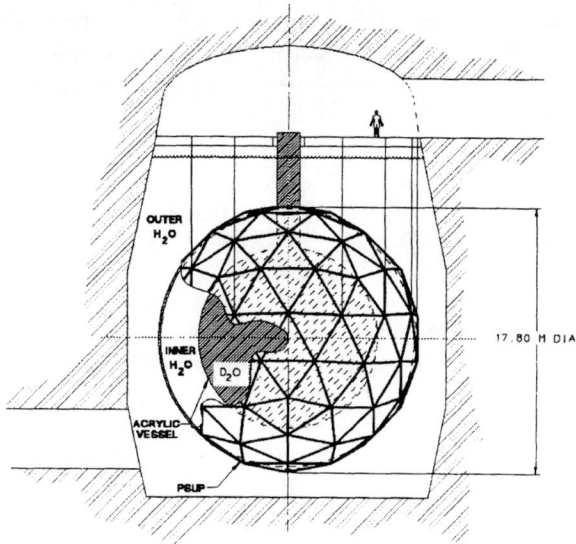

**FIGURE 1.** Cross section schematic of the SNO detector.

structed from materials which have been carefully selected for low $^{238}$U and $^{232}$Th content, since daughter particles from these decay chains can mimic the expected neutrino signals. Table 1 highlights the radioactivity levels in a number of key areas, starting in the surrounding rock and moving inward to the heavy water region. These limits have been achieved by use of very pure materials and careful manufacture. Furthermore, the detector was constructed under ultra clean conditions and continues to be operated in the same manner. The air is filtered using HEPA filters and all personnel wear clean–room clothing to maintain an air quality on the order of Class 2000. An extensive water purification system is used to reduce the radioactivity in both the heavy and light water volumes.

**Table 1.** Radioactivity levels in various regions of the SNO detector, where the most crucial areas are the heavy water and inner light water regions. In those volumes, the equivalent concentrations of $^{238}$U and $^{232}$Th are within a factor of 10 of the design objective level of $\sim 10^{-14}$. Continued work on the $N_2$ cover gas system above the heavy water is expected to help lower the radon levels.

| Region | Radioactivity Level (g/g) |
|---|---|
| Norite rock | few $\times 10^{-6}$ |
| PMT Support Structure | $10^{-8}$ |
| PMTs | $10^{-8}$ |
| Acrylic Vessel | $10^{-12}$ |
| $H_2O$ (inner) | $< 10^{-13}$ |
| $D_2O$ | few $\times 10^{-13}$ |

The PMT support structure is 99.99% impermeable to water, enabling water flow to be directed outwards from the region between the acrylic vessel and the PMTs into the outer region where it is removed, purified, stripped of radon and radium and re–injected in the inner region. This process maintains a lower level of radioactivity in the light water in the inner region than in the less critical outer region.

During the initial running stages, no enhanced neutron detection techniques will be employed, and SNO will rely on neutron capture in pure $D_2O$:

- $n + d \rightarrow t + \gamma \cdots \rightarrow e^-$ in pure $D_2O$.

Neutrons from the NC dissociation of the deuteron can subsequently capture on a another deuteron to produce an energetic 6.3 MeV gamma ray that will in turn Compton scatter resulting in a detectable electron. However, the capture efficiency for this method is relatively low at 24%.

To increase the sensitivity to neutrons resulting from the neutral current reaction, as well as the $\bar{\nu}_e$ reaction on deuterium, two separate detection methods will be used:

- $n + {}^{35}\text{Cl} \rightarrow {}^{36}\text{Cl} + \Sigma_\gamma \cdots \rightarrow e^-$ using $MgCl_2$ in $D_2O$
- $n + {}^3\text{He} \rightarrow p + t$ using proportional counters.

In the first case, $MgCl_2$ salt will be added to the heavy water up to a concentration of 0.2% to increase sensitivity to neutrons. Neutron capture in $^{35}$Cl results in a cascade of gamma rays of energy up to 8.6 MeV, which are detectable from Čerenkov light produced following Compton scattering. The capture efficiency in this case is about 83%. In order to remove the salt from the detector, a reverse osmosis filtration system has been developed. Monte Carlo simulations have indicated that it should also be possible to distinguish CC events from NC events based on the event topology. The distribution of light from CC events arising from a single energetic electron will be different from an NC event resulting from a gamma ray cascade.

A second technique that will be used for the measurement of neutrons from the NC reaction will be the introduction of an array of over 100 $^3$He gas–filled proportional counters. The counters are being constructed using ultra pure nickel tubing fabricated by a chemical vapour deposition process. Neutrons will be detected through energetic protons and tritons produced by the neutron capture on $^3$He with an efficiency of 45%. This separate neutron signal will allow for the ability to distinguish CC events from NC events on an event–by–event basis. The distinctive time evolution of these events will enable them to be distinguished from (low) alpha particle radioactivity from the walls of the ultra pure nickel tubes used as the bodies of the proportional counters.

## PHYSICS OBJECTIVES

A primary objective of the SNO detector will be the observation of solar neutrinos via the charged current and neutral current reactions. The threshold for the CC reaction is expected to be around 5 MeV in electron energy, corresponding to about 6.4 MeV in neutrino energy. The threshold for the NC reaction will be 2.2 MeV. Because of these energy restrictions, both of these reactions will be sensitive mainly to the neutrinos from $^8$B decay in the Sun[1]. A comparison of the flux measured by these two reactions should provide a determination of whether the $^8$B neutrinos are changing from electron neutrinos to another non–sterile type.

The CC reaction will provide an accurate measure of the shape of the $^8$B neutrino energy spectrum. For this reaction, the outgoing electron carries away much of the energy of the incoming neutrino, less the Q–value of 1.4 MeV. Therefore, the $^8$B spectrum can be observed with an energy resolution of better than about 20%. This is an advantage compared to the ES reaction where the incident neutrino energy is shared between the outgoing electron and the recoiling neutrino with comparable energy. Since the CC reaction has a relatively large cross section, the SNO detector will observe about 4450 counts per year for 50% of the flux calculated by typical standard solar models (2) for the CC reaction and about 550 counts per year for the ES reaction.

The NC reaction will provide a measure of the total flux resulting from all neutrino types. In this case, the full standard solar model flux would yield about 930 counts per year using neutron capture in pure $D_2O$ and 3100 counts per year using neutron capture on salt. As mentioned above, neutrons from the NC reaction will be detected by three different techniques. Each technique will have different sensitivity to systematic effects and provide somewhat independent measurements of the NC flux.

Using knowledge obtained from both the NC and CC reactions, including the $^8$B spectral shape and temporal information on the event rates, it will be possible to examine a region of oscillation parameters for electron neutrinos ranging from:

- $\Delta m^2 = 10^{-4}$ eV$^2$ to $10^{-11}$ eV$^2$
- $\sin^2(2\theta) = 1$ to $10^{-4}$

With this set of information, it will be possible to determine whether electron neutrinos are making transitions to sterile or non–sterile neutrinos for each of the possible solutions allowed by the existing set of solar neutrino data (3). For these solutions, the different regions observable by SNO are discussed below.

### *Small Angle (Non–adiabatic) MSW:*

- $\Delta m^2 \sim 5 \times 10^{-6}$ eV$^2$, $\sin^2(2\theta) \sim 5 \times 10^{-3}$

For the non–adiabatic MSW region allowed by existing experiments, we would expect the following. For electron neutrinos converting into non–sterile neutrinos there would be a distortion in the $^8$B spectrum, and the ratio of CC/NC counts would be smaller than the normalized ratio expected for a flux containing purely electron neutrinos. For $^8$B electron neutrinos making a transition to sterile neutrinos, the CC spectrum would still be distorted but the normalized CC/NC ratio would be unity.

### *Large Angle MSW and LOW MSW:*

- LMA: $\Delta m^2 \sim 20 \times 10^{-6}$ eV$^2$, $\sin^2(2\theta) \sim 0.8$
- LOW: $\Delta m^2 \sim 0.1 \times 10^{-6}$ eV$^2$, $\sin^2(2\theta) \sim 0.8$

This region of transition for electron neutrinos to sterile neutrinos may be ruled out by Big Bang nucleosynthesis. For a coupling constant this large, the effect is similar to the addition of a fourth neutrino species during the Big Bang. For electron neutrinos making a transition to non–sterile neutrinos, the signature would be a normalized CC/NC ratio smaller than 1 and no distortion of the $^8$B spectrum as observed by the CC reaction.

### *Region Sensitive to Earth Regeneration:*

- $\Delta m^2 \sim 5 \times 10^{-6}$ eV$^2$, $\sin^2(2\theta) \sim 10^{-2}$ to 1

For electron neutrino transitions to non–sterile neutrinos, regeneration effects in the earth can result in day–night effects, which vary according to season and can show some variation in magnitude according to the time of night. This distinctive temporal pattern would indicate neutrino flavour change within this parameter region.

### *Vacuum Oscillations:*

- $\Delta m^2 \sim 5 \times 10^{-11}$ eV$^2$, $\sin^2(2\theta) \sim 0.8$

One last solution that agrees with the existing solar neutrino measurements is that of vacuum oscillations. In this case, the distance from the Earth to the Sun is an appropriate value which causes strong supression of flux

---

[1] The SNO detector should also be sensitive to the so–called "hep" solar neutrinos, which have higher energies compared to $^8$B solar neutrinos.

for some of the existing experiments and less significant supression for the others.

The peak to peak variations in neutrino flux will be about 7% due to the change in the solid angle from the small eccentricity of the Earth's orbit. With electron neutrino flavour change to either sterile or non–sterile neutrinos, there will be significant spectral distortion observable in the CC reaction and a different seasonal variation in the integrated CC rate above threshold. For some parameter values, the spectral distortion can be very large. Observation of these seasonal effects would be an indication of such vacuum oscillations, and the NC reaction can then be used to determine whether the transitions are to sterile or non–sterile neutrinos.

## BACKGROUNDS

Background radioactivity in the detector must be minimized and the residual radioactivity must be measured accurately. The threshold for the CC reaction will be determined by the level of radiation from beta and gamma rays from the uranium and thorium chain elements in the components of the detector. The minimization of this radioactivity is the motivation for constructing all components from ultra pure materials and controlling cleanliness during construction. In addition, it is particularly important to restrict and determine the gamma ray flux in the heavy water with energies greater than 2.2 MeV, which is the threshold for photodisintegration of deuterium. These gamma rays can produce a free neutron in the $D_2O$ which looks exactly like neutrons generated by neutrinos interacting through the NC neutrino reaction. In order to address this background, extensive water purification systems have been developed for the light water, heavy water and $MgCl_2$. Sensitive techniques have been developed for the measurement of the principal sources of such gamma rays, namely decay products from the $^{238}U$ and $^{232}Th$ chains. These techniques include degassing and chemical extraction from recirculated water to determine the effective levels of these elements— $^{222}Rn$ and $^{226}Ra$ are monitored in the case of the uranium chain, while $^{224}Ra$ and $^{212}Pb$ are used to indicate the levels of thorium daughters. Information on radiation in the detector can also be gathered from observations of the Čerenkov light patterns on the PMTs. In the case of the $^3He$ neutron detectors, the materials used in construction have been carefully selected to have very low levels of uranium and thorium and their daughters. Techniques have been developed for the measurement of inherent radioactivity in these detectors which might lead to photodisintegration of deuterium.

One other form of background for neutrino experiments is high energy cosmic ray muons. However, at a depth of over 2000 m, the SNO detector is subjected to only 2–3 cosmic ray muons per hour. Consequently, there is essentially no dead time caused by gating off the detector to avoid decay of spallation particles, which are the by–products of the muon interacting with water nuclei.

## CALIBRATION

Calibration of the detector is accomplished by using various sources, including built–in electronics calibrations and dedicated optical and energy calibration sources.

The electronics are calibrated using charge pulsers on all the channels to give more than 600 000 charge offset (pedestal) values plus charge and time calibration slopes. These constants are measured weekly and are quite stable. In order to perform optical calibrations of the detector to get information about reflectivity, attenuation and relative PMT timing, a number of sources are used, including a laser light source, a number of light–emitting diodes and a sonoluminescent source. The "laserball" delivers photons with wavelengths in the range 337–700 nm at a rate of up to 45 Hz with variable intensity. The light is passed through a difusing sphere, which can be positioned almost anywhere within the $D_2O$ volume. Light–emitting diodes are mounted in fixed positions on the PMT support structure, and give photons of 480 nm wavelength with a rate of up to 1 kHz and with variable intensity. These sources can be activated at any time to monitor detector stability. For precise time calibration, a sonoluminescent source will be used to provide short pulses. Preliminary analysis shows that the PMT timing resolution is ~1.7 ns, which is very near the expected value. A variety of gamma ray sources with be used for the absolute energy calibration of the detector. To calibrate near the expected analysis threshold for electrons, $^{16}N$ is produced which subsequently decays most of the time to an excited state of $^{16}O$ giving a 6.13 MeV gamma ray. Preliminary analysis of the $^{16}N$ data shows the energy calibration to be near the Monte Carlo estimate of ~9 PMTs/MeV. The higher energy reponse of the detector will be determined using 19.8 MeV gamma rays from the reaction $^3H(p,\gamma)^4He$. Finally, a $^8Li$ source is being developed to deliver electrons which have an energy spectrum shape which is closely related to the $^8B$ decay spectrum. The $^{16}N$ and $^8Li$ sources are produced by nuclear reactions using a compact neutron generator and will be transported to the heavy water volume via capillary tubing. A manipulator system allows these sources to moved to different locations within the $D_2O$ region. In addition to the ongoing electronics calibrations, the laserball, the $^{16}N$ source, the sonoluminescent source and the light–

emitting diode sources have all been used to calibrate the detector.

## DETECTOR OPERATION

The initial operation of the detector has been relatively quiet, allowing for stable detector running at an NHIT threshold 17–18 PMTs for the 100 ns coincidence trigger (corresponding to an electron energy threshold close to 2 MeV). This is a preliminary indication that the cleanliness programme through the detector construction has been effective at limiting the total radioactivity. Individual PMT channel thresholds are continually being optimized, and currently the average threshold is < 0.5 photoelectrons. Furthermore, cooling the light water to 10 °C has helped keep the PMT noise rate low at about 500 Hz. Preliminary analysis shows that there are typically only about 2 PMTs per event due to electronic noise in the phototubes. Table 2 shows the contribution to the observed event count rate from various hardware triggers. The average overall rate in this configuration is about 10–15 Hz.

**Table 2.** Summary of the settings and event rates for various SNO hardware triggers during the initial operation of the detector. Backgrounds are monitored continuously using the pulsed trigger and at low energies using the prescaled trigger. The energy sum trigger gives information about the charge or total energy of an event.

| Trigger Type | Hardware Threshold | Trigger Rate (Hz) |
|---|---|---|
| Pulsed trigger | zero bias | 5 |
| 100 ns coincidence | 17 PMTs | 3–5 |
| 20 ns coincidence | 15 PMTs | 2–3 |
| Energy Sum | ∼200 p.e. | < 1 |
| Prescaled (1:10000) | 12 PMTs | < 1 |

A number of physics events have been observed in this early running. Most easily recognized are the through-going muons which trigger most of the phototubes in the detector. As mentioned above, there are a few such events per hour. Atmospheric neutrinos are also seen giving clearly defined ring patterns with several hundred hit PMTs. Finally, events are seen which are most probably solar neutrinos, an example of which is shown in Figure 2.

At the moment, greater than 98.5% of all PMT channels are fully operational and taking data. In order to increase the charge collection efficiency of the phototubes, magnetic field compensation coils are currently being used, which are installed in the cavity rock that sur-

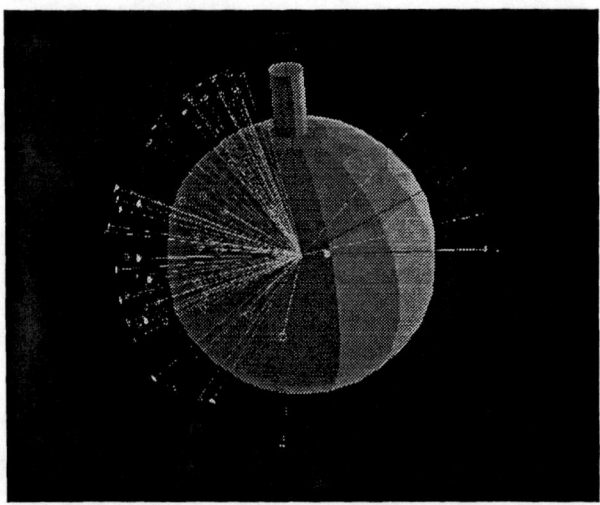

**FIGURE 2.** SNO Event display showing a solar neutrino candidate event. Rays from the centre of the detector indicate which PMTs have been hit by a photon. On average, a typical solar neutrino event will have approximately 80 detected photons.

rounds the detector volume. Initial estimates show that they have provided a 5–10% increase in PMT efficiency.

One final note is that re–gassing the light water volume with $N_2$ has effectively eliminated the connector high voltage breakdown problems that occurred during early operation of the detector.

## NON-SOLAR NEUTRINO PHYSICS

The SNO detector has sensitivity to neutrinos besides those that are produced in the Sun, including atmospheric neutrinos and supernova neutrinos.

Atmospheric neutrinos are produced by the decay of mesons created by the interaction of high energy cosmic rays with nuclei in the upper atmosphere. The study of these neutrinos by the Super–Kamiokande (SK) detector and others has produced what is possibly the first real evidence of neutrino mixing (4). While the SNO detector is relatively small compared to SK, it does have the advantage of being able to detect neutrons. This capacity can, in turn, yield an estimate of the $\nu/\bar{\nu}$ ratio in the atmospheric neutrino flux. One possible set of reactions that can be initiated by high energy atmospheric neutrinos is:

$$\nu_l + {}^{16}O \rightarrow l^- + p + {}^{15}O, \qquad \bar{\nu}_l + {}^{16}O \rightarrow l^+ + n + {}^{15}N.$$

On the whole, antineutrinos are far more likely to produce a neutron than neutrinos are. The $\nu/\bar{\nu}$ ratio has not been measured before and will provide a valuable check on flux calculations. Furthermore, the fact that Sudbury is located at a high geomagnetic latitude (and consequently low geomagnetic cut–off) means that very low energy atmospheric neutrinos can be observed (5).

The SNO detector is also in an excellent position to observe neutrinos from the next galactic supernova. For a type–II supernova just beyond the centre of our galaxy at 10 kpc, SNO expects to see on the order of 800 neutrino events (6). This translates into 600–800 counts, depending on the neutron detection method in place. Even at the far edge of the Milky Way galaxy, the number of observed events is still high at roughly 100 counts over the course of tens of seconds. All currently operating real–time neutrino detectors can measure the dominant $\bar{\nu}_e$ signal, whereas the SNO detector, with its ability to detect neutrons, can also measure the non–electron type neutrinos, which are expected to carry away most of the energy from the supernova explosion. Moreover, with a deuterium target, SNO will have good sensitivity to the expected prompt $\nu_e$ signal via the CC reaction. As well, different neutrino flavours from a supernova may arrive in bursts if they have different masses; therefore a supernova signal may help in determining the absolute masses of neutrinos, whereas information from solar neutrinos only constrains the mass differences. Finally, efforts have been made to ensure that SNO will be ready for a supernova even during calibration activities in almost all cases.

## CONCLUSION

The SNO detector as been taking data since May and is now in a stable state where high quality physics data is being collected. The detector livetime over this period has been approximately 80%, and the hope is very soon to approach 100%.

From the discussion above, it is apparent that the detection abilities of the SNO detector provide a wide ranging sensitivity to neutrinos from many sources. The full set of measurements could provide definitive answers to the remaining questions of whether neutrino flavour change takes place for solar neutrinos and with what associated parameters, including transitions to sterile or non-sterile neutrinos. The detector capabilities provide the opportunity to determine the total flux of $^8B$ neutrinos from the Sun whether or not neutrino flavour change has taken place, and so can contribute significant astrophysical information as well.

## ACKNOWLEDGMENTS

The SNO project is supported by the Natural Sciences and Engineering Research Council (Canada), INCO, Industry Canada, National research Council of Canada, Northern Ontario Heritage Fund, the Department of Energy (USA) and the Particle Physics and Astronomy Council (UK). The heavy water is on loan from AECL with the cooperation of Ontario Hydro.

## REFERENCES

1. SNO Collaboration, *nucl–ex/9910016 v2*, Nov. 1999; G. Ewan, *Nucl. Instr. Meth.* **A314**, 373 (1992); SNO Collaboration, *Physics in Canada* **48**, 112 (1992); SNO Proposal, SNO–87–12, Oct. 1987.

2. See, for example, J.N. Bahcall, S. Basu and M.H. Pinsonneault, *Phys. Lett.* **B433**, 1 (1998).

3. J.N. Bahcall, P.I. Krastev, A.Y. Smirnov, *Phys. Rev.* **D58**, 096016 (1998).

4. The Super–Kamiokande Collaboration, *Phys. Rev. Lett.* **81** 1562-1567 (1998).

5. Y. Tserkovnyak, R. Komar, C. Nally, C. Waltham, *nucl–ph/9907450*, Oct. 1999.

6. See, for example, A. Burrows, D. Klein, R. Gandhi, *Phys. Rev.* **D45**, 3361–3385 (1992); also, J.F. Beacom and P. Vogel, *Phys. Rev. D*, **58**, 093012 (1998).

# A Novel Supernova Detector

David B. Cline

*University of California Los Angeles, Dept. of Physics & Astronomy, Box 951547, Los Angeles, CA 90095-1547 USA*

**Abstract.** We discuss the prospects for detecting $\nu_{\mu,\tau}$ and $\nu_\tau$ neutrinos from Type II supernovas using the novel detector at the Supernova Burst Observatory (SNBO) or OMNIS that is being designed for an underground laboratory in the USA. This detector would collect ~2000 flavor selected events from a Galactic supernova and could probe neutrino mass down to a few eV, as well as the dynamics of the supernova process. We believe this is essential to further our understanding of the neutrino section of elementary particle physics.

## INTRODUCTION

The issue of whether or not neutrinos have masses is important for astrophysics and cosmology. Astrophysical considerations may represent the best hope for determining neutrino masses and mixings. In this paper, we examine how proposed neutral-current-based, supernova neutrino-burst detectors, in conjunction with the next generation water-Čerenkov detectors, could use a galactic supernova event to either measure or place constraints on the $\nu_{\mu,\tau}$ masses in excess of 5 eV[1,2]. Such measurements would have important implications for our understanding of particle physics, cosmology, and the solar neutrino problem and would be complementary to proposed laboratory vacuum-oscillation experiments.

A light neutrino mass between 1 eV and 100 eV would be highly significant for cosmology. In fact, if a neutrino contributes a fraction $\Omega_\nu$ of the closure density of the Universe, it must have a mass $m_\nu \approx 92\, \Omega_\nu\, h^2$ eV, where $h$ is the Hubble parameter in units of 100 km s$^{-1}$ Mpc$^{-1}$. Reasonable ranges for $\Omega_\nu$ and $h$ then give 1 eV to 30 eV as a cosmologically significant range. A neutrino with a mass in the higher end of this range (*i.e.*, $10 \leq m_\nu \leq 30$ eV) could contribute significantly to the closure density of the Universe. The cosmic background explorer (COBE) observation of anistropy in the microwave background, combined with observations at smaller scales, and the distribution of galaxy streaming velocities, have been interpreted as implying that there are two components of dark matter: hot ($\Omega_{HDM} \sim 0.3$) and cold ($\Omega_{CDM} \sim 0.6$). The hot dark matter (HDM) component could be provided by a neutrino with a mass of about 7 eV.[3-5]

## MEASURING THE NEUTRINO MASS BY TIME OF FLIGHT

Perhaps the most straightforward and obvious nature of a massive neutrino would come from the lengthening in flight time from a distant supernova. For example, the flight time difference between $\nu_\tau$ and $\nu_e$ ($\bar{\nu}_e$) in seconds is

$$\Delta t = 5.14 \times 10^{-2} R_{kpc} \left[ \left( \frac{m_{\nu_e}}{E_{\nu_e}} \right)^{-2} - \left( \frac{m_{\nu_x}}{E_{\nu_x}} \right)^{-2} \right] s ,$$

where $E_{\nu_x}$ is the neutrino energy in MeV, $m_{\nu_x}$ is in eV, and $R_{kpc}$ is the distance to the supernova in units of 10 kpc. A finite neutrino mass would alter the neutrino spectra in characteristic ways that could result in broadening and flattening of the observed signal.

Some arguments, which arose during this meeting, for detecting the neutrinos are given in Table 1. Event rates for various detectors for a galactic supernova are given in Table 2.[7] We believe the detection of these supernova neutrino signals will be essential to our understanding of the neutrino sector.

Thus, neutrino masses might be obtained by comparing the observed neutrino signal with the signal expected from supernova models. Since detectors such as Superkamiokande (SK) are relatively insensitive to $\nu_\mu$ and $\nu_\tau$, they are unlikely to measure cosmologically significant neutrino masses for these flavors. One of the neutral-current-based detectors being built at present is the Sudbury Neutrino Observatory (SNO). A general comparison of the methods of measuring neutrino mass is given in Table 3.[1,7] The rate of interaction for the world's detectors is shown in Fig. 1.

**Table 1.** Experiments for ν-Mass/Mixing

| Scheme | Tests | | | | | Nucleosynthesis | |
|---|---|---|---|---|---|---|---|
| | $\nu_\odot$ | $\nu_{atms}$ | LBL | SBL | SN ν's | BBN | SNN |
| I<br>3ν mixing<br>No LSND | Yes<br>$\nu_e \to \nu_{\mu,\tau}$ | Yes<br>$\nu_\mu \to \nu_\tau$ | Yes<br>$\nu_\mu \to \nu_\tau$ | No<br>$\nu_\tau \to \nu_e$<br>$\nu_\mu \to \nu_e$<br>τ appearance? | √ | OK<br><br>$\nu_\mu \to \nu_e$ | OK |
| II<br>4ν mixing<br>$\nu_\mu \to \nu_\tau$<br>(Doublet)<br>$\nu_\odot \to \nu_s$<br>LSND | Yes<br>$\nu_e \to \nu_s$<br>(No extra<br>N.C. signal) | Yes<br>$\nu_\mu \to \nu_\tau$ | Yes<br>$\nu_\mu \to \nu_\tau$ | No?<br>$\nu_e \to \nu_\tau$<br>τ appearance?<br>$\nu_e \to \nu_\tau$?<br>$\nu_e \to \nu_\mu$? | √<br>Hot<br>?<br>Maybe!<br>Maybe! | D/N<br>$\nu_e$-spectrum<br>? | ??<br>Good or bad<br>?? |
| III<br>4ν mixing<br>$\nu_\mu \to \nu_s$<br>Doublet<br>No LSND | Yes<br>$\nu_e \to \nu_\mu$ | Yes<br>$\nu_\mu \to \nu_s$ | Yes<br>$\nu_\mu \to \nu_s$ | No<br>$m_{\nu\tau}$<br>$\nu_\mu \to \nu_\tau$?<br>? | √ | ?<br><br>T of F | r-process<br>constraint |

**Table 2.** Requirements of a Supernova Observatory

- Life of Observatory ≥ rate (yr) for SNII on Milky Way Galaxy ≥ 20 – 40 yr
- Event Rate: ~ 5 – 10 K    $\bar{\nu}_e + P \to e^+ + n$
  ~ Few K    $\nu_x + N \to \nu_x + N$
  $\nu_x = \nu_\mu + \nu_\tau$

To: ○ Fit model of SNII process
 ○ Extract a neutrino mass or neutrino oscillation
 ○ Learn about SNII explosion process

**Table 3.** Methods for Measuring Neutrino Mass

1. Time of flight from an SNII:
$$\Delta t = \frac{1}{2}\left[\left(\frac{M\nu_1}{E_1}\right)^2 - \left(\frac{M\nu_2}{E_1}\right)^2\right]D \text{ sec}$$
Only active neutrinos can separate $M\nu_1, M\nu_2$ by varying $E_1, E_2$.

2. Neutrino oscillation: $P = \sin^2 2\theta \sin^2\left[1.27\left(\frac{L}{E}\right)(M_1^2 - M_2^2)\right]$
Can have sterile neutrinos.

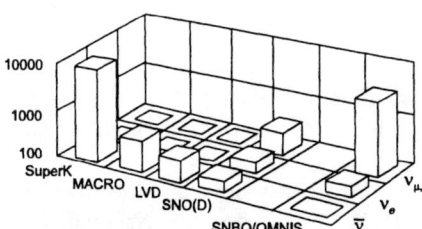

**FIGURE 1.** Comparison of world detectors (event numbers for supernovae at 8 kpc).

## DIFFERENT TYPES OF SN NEUTRINO DETECTORS AND NEUTRINO MASSES

Recently there has been real progress in SN simulations giving an explosion. These calculations give interesting predictions for the neutrino spectra. Detectors like the SK and SNBO/OMNIS may be able to detect such effects, however the SNBO/OMNIS detector may be of crucial importance for this study. Using these various detectors, it should be possible to detect a finite neutrino mass.[1] The characteristics of this detector are listed in Table 4.

In this analysis, we have assumed the existence of a very massive neutral-current detector (SNBO/OMNIS), which we discuss next. By using these different detectors it will be possible to measure the μ or τ neutrino masses, as shown in Table 3, which could determine a mass to ~ 10 eV. To go to lower mass, we need to use the possible fine structure in the burst; we have shown that it may be possible to reach ~ 3 eV with very large detectors in this case.[7] The detection of two-neutron final states, as illustrated in Fig. 2 would be useful for Pb detection.[6]

**Table 4.** Properties of the (Proposed) OMNIS/SNBO Detector

| Targets: | NaCℓ (WIPP site) |
|---|---|
| | Fe and Pb (Soudan and Boulby sites) |
| Mass of Detectors: | WIPP site ≥ 200 ton |
| | Soudan/Boulby sites ≥ 200 ton |
| Types of Detectors: | Gd in liquid scintillator |
| | $^6$Li loaded in the plastic scintillators that are read out by scintillating-fiber–PMT system |

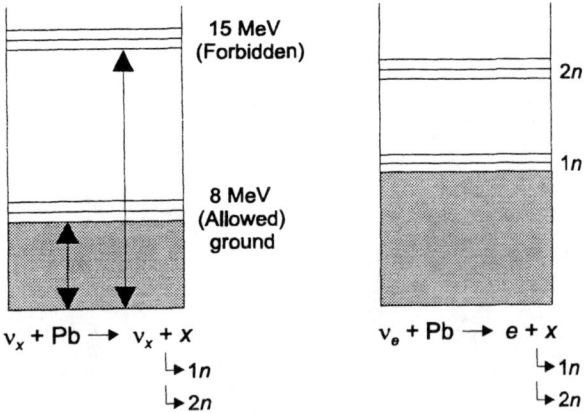

**FIGURE 2.** The two-neutron signal for $\nu_{\tau,\mu} \rightarrow \nu_e$. The $2n/1n$ signal is much larger for $\nu_e$ interactions. This is a signature for $\nu_{\tau,\mu} \rightarrow \nu_e$ in the SNII environment.

## THE PROPOSED SUPERNOVA BURST OBSERVATORY (SNBO/OMNIS)

The major problem of supernova detection is the uncertain period of time between such processes in this Galaxy. In addition, complimentary detectors should be active when the supernova goes off in order to gain the maximum amount of information possible about the explosion process and neutrino properties. In Table 2, we list some of the requirements of such an ideal supernova observatory.

Lacking an ideal observatory, a group of us have been studying a very large detector, SNBO/OMNIS.[1] Table 4 gives some of the guidelines for this detector.[7] We have located a possible site for the observatory near Carlsbad, NM, which is the WIPP site (shown in Fig. 3). We have studied the radioactive background at this site (measured by the OSU-UCLA group) and find it acceptable for a galactic supernova detector. We find less than one neutron per hour detected in a 6-ft $BF_3$ counter. This leads to the expectation that the background for a galactic supernova is much smaller than the signal at this site. A schematic of the SNBO/OMNIS detector is shown in Fig. 4.

## SEARCH FOR THE INTEGRATED FLUX OF SUPERNOVA NEUTRINOS

Another kind of relic neutrinos are the neutrinos that arise from the integrated flux from all past type-II supernovae. Figure 5 shows a schematic of these (and other) fluxes.[7] These fluxes could be modified by transmission through the SNII environment, as discussed recently.[7]

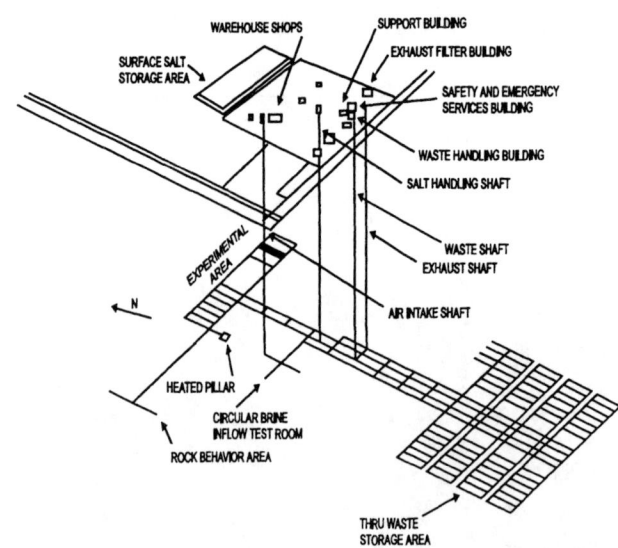

**FIGURE 3.** Illustrative arrangement of detector modules along rock tunnel; also shown is the use of supplementary iron or lead for gamma shielding or as an alternative neutrino target.

The detection of $\bar{\nu}_e$ from the relic supernovae may someday be accomplished by the SK detector. It would be as interesting to detect $\nu_e$ with an ICARUS detector, as illustrated in Table 5. High-energy $\nu_e$ would come from $\nu_{\mu,\tau} \rightarrow \nu_e$ in the supernova.[7] A window of detection occurs between the upper solar neutrino energy and the atmospheric neutrinos, as first proposed by D. Cline and reported in the first ICARUS proposal (1983-1985). The ideal detector to observe this is a large ICARUS liquid-argon detector.[7]

## ACKNOWLEDGMENTS

I wish to thank G. Fuller, D. Boyd, K. Lee, and P. F. Smith for discussions.

## REFERENCES

1. Cline, D. B., Fuller, G. M., Hong, W.P., Meyer, B., and J. Wilson, *Phys. Rev.* **D50**, 720 (1994).
2. Presentations to this workshop were made by members of the SNO, SK, AMANDA, SNBO, and KARMAN groups and are to be published in the proceedings (American Institute of Physics, 2000).
3. Smooth, G. F., *et al., Astrophys. J. Lett.* **396**, L1 (1992).
4. Schaefer, R. K., and Shafi, Q.,*Nature* **359**, 199 (1992).
5. Bond, J. R., Efstathiou, G. E., and Silk, J., *Phys. Rev. Lett.* **45**, 1980-1983 (1980).
6. G. Fuller, UCSD, private communication (1998).
7. Cline, D. B. "Search for Relic Neutrinos and Supernova Bursts," in *Proceedings, Eighth Intl. Wksp. on Neutrino Telescopes* (Venice, Feb. 23-26, 1999), ed. M. Baldo Ceolin, 1999, Vol. II, pp. 309-320.

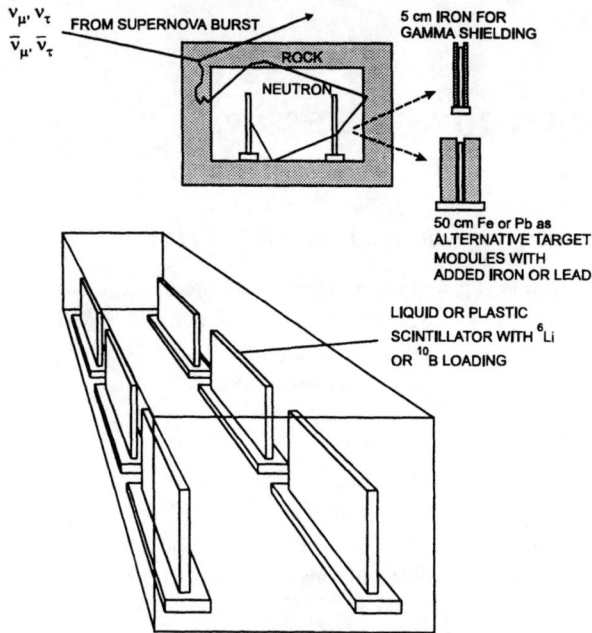
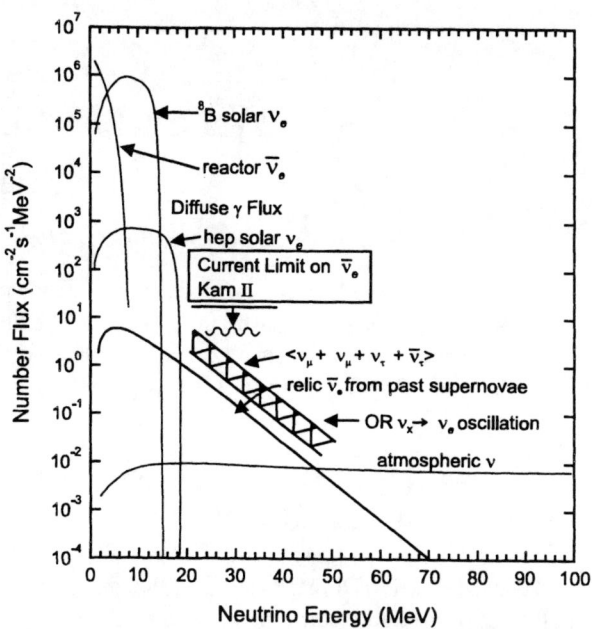

**FIGURE 4.** Isometric view of the surface and underground (looking toward the Northeast) of the WIPP site near Carlsbad, NM.

**FIGURE 5.** Relic neutrinos from past supernova. Note: $v_x \to v_e$ in the supernova can boost the energy of the $v_e$ if we find $<E v_e> >> <E\bar{v}_e>$. This will be a signal for neutrino oscillation in supernovae! and measure $\sin^2 \theta x_e$.

Table 5. Detection of $v/\bar{v}_e$ Relic Neutrino Flux from Time Integrated SNII

1. Relic $v/\bar{v}_e$ from all SNII back to $Z \sim 5$: $<E_v> \sim 1/(1+Z)<E_v>$
2. Detection would give integrated SNII rate from Universe
   - Window of detetion [D. Cline, ICARUS proposal, 1984]
3. Neutrino oscillations in SNII would give $v_x \to v_e$ with higher energy than $\bar{v}_e$
4. Detect $\bar{v}_e$ with SK or ICARUS. Attempt to detect $v_x/v_e$ detection.

# SNEWS and Future Supernova Detectors

Kate Scholberg

*Boston University Department of Physics, 590 Commonwealth Ave., Boston, MA 02215*

**Abstract.** World-wide, several detectors currently running or nearing completion are sensitive to a prompt core collapse supernova neutrino signal in the Galaxy. The SNEWS system will be able to provide early warning of a supernova's occurrence to the astronomical community using a coincidence of neutrino signals around the world. Here we describe the status of SNEWS, and point out some possibilities for using SNEWS for the next generation of detectors.

## NEUTRINOS FROM CORE COLLAPSE SUPERNOVAE

When the core of a massive star at the end of its life collapses, nearly all of the total gravitational binding energy of a neutron star is emitted in the form of neutrinos, some $E_b \sim 3 \times 10^{53}$ ergs. Less than 1% of this energy is expected to be released in the form of kinetic energy and optically visible radiation. The remainder is radiated in neutrinos, of which approximately 1% will be electron neutrinos from an initial "neutronization" burst and the remaining 99% will be neutrinos from the later cooling reactions, equally distributed among flavors. Average neutrino energies are expected to be about 12 MeV for electron neutrinos, 15 MeV for electron antineutrinos, and 18 MeV for all other flavors. The neutrinos are emitted over a total timescale of tens of seconds, with about half emitted during the first 1-2 seconds, and with the spectrum eventually softening as the proto-neutron star cools. Reference (1) summarizes the expected neutrino signal. The basic features of neutrino emission models were well confirmed in 1987A with the observation of neutrinos from SN1987A. We await the next Galactic supernova to learn more.

## NEUTRINOS FROM CORE COLLAPSE SUPERNOVAE

The neutrino burst produced by the core collapse emerges promptly from the stellar envelope. However, the the shock wave produced by the collapse takes some time to travel outwards from the core to the photosphere of the star. The time of first shock breakout of a supernova is highly dependent on the nature of the stellar envelope, and can range from minutes for bare-core stars to hours for red giants. Therefore, the detection of a neutrino burst can give an early warning that light from a supernova explosion is about to appear. For SN1987A, first light was observed about 2.5 hours after the neutrino burst; the first observable photons probably occurred about one hour earlier than that.

## SUPERNOVA NEUTRINO DETECTORS

There are several classes of detectors capable of detecting a burst of neutrinos from a gravitational collapse in our Galaxy. More details can be found via reference (2). Table 1 gives an overview of detector types. Table 2 lists some specific supernova neutrino detectors and their capabilities.

## SNEWS: THE SUPERNOVA EARLY WARNING SYSTEM

There are several benefits from a system which coordinates neutrino signals from two or more different detectors. All detectors are subject to false alarms from bursts of events due to detector pathologies or other non-Poissonian phenomena (for example, flashing phototubes or other sources of spurious light, electronic noise, correlated radioactivity events due to muon spallation of nuclei, etc.). Therefore, if an individual experiment is to issue an alarm, a human operator must first check the event burst to confirm its supernova-like nature, which can take significant time even when a fast-response human alert system is set up. Requiring a coincidence between independent detectors will add great confidence to the detection of a supernova neutrino burst, to the extent that a completely automated alert may be possible. The automation could save enough time that important early observations would not be lost.

**Table 1.** Supernova neutrino detector types.

| Detector type | Material | Energy | Timing | Pointing | Flavor sensitivity |
|---:|:---:|:---:|:---:|:---:|:---:|
| scintillator | C,H | y | y | n | $\bar{\nu}_e$ |
| water Cherenkov | $H_2O$ | y | y | y | $\bar{\nu}_e$ |
| heavy water | $D_2O$ | NC: n | y | n | all |
|  |  | CC: y | y | y | $\nu_e, \bar{\nu}_e$ |
| long string water Cherenkov | $H_2O$ | n | y | n | $\bar{\nu}_e$ |
| liquid argon | Ar | y | y | y | $\nu_e$ |
| high Z/neutron | NaCl, Pb, Fe | n | y | n | all |
| radio-chemical | $^{37}Cl, ^{127}I, ^{71}Ga$ | n | n | n | $\nu_e$ |

**Table 2.** Supernova neutrino detectors. "Online" indicates that the detector is both running and connected to SNEWS.

| Detector | Type | Mass (kton) | Location | Number of events at 8.5 kpc | Status |
|:---:|:---:|:---:|:---:|:---:|:---:|
| Super-K | $H_2O$ Ch. | 32 | Japan | 5000 | online |
| MACRO | scint. | 0.6 | Italy | 150 | online |
| SNO | $H_2O$, $D_2O$ | 1.4, 1 | Canada | 300, 450 | running |
| LVD | scint. | 0.7 | Italy | 170 | online |
| AMANDA | long string | Meff $\sim$0.1/pmt | Antarctica |  | running |
| Baksan | scint. | 0.33 | Russia | 50 | running |
| Borexino | scint. | 0.3 | Italy | 100 | 2001 |
| KamLAND | scint. | 1 | Japan | 300 | 2001 |
| OMNIS (Pb/Fe) | high Z | 5 | USA | 2000 | 2000+ |
| LAND (Pb) | high Z |  | Canada |  | 2000+ |
| Icanoe | liquid argon | 9 | Italy |  | 2000+ |

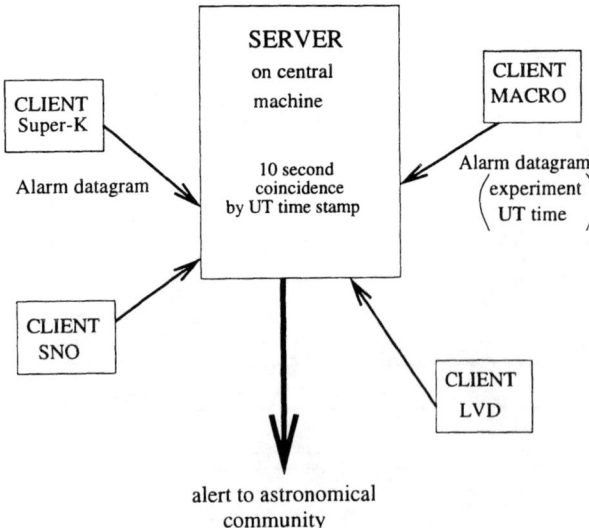

**FIGURE 1.** SNEWS setup.

Software for a prototype international supernova watch coincidence system has been designed by Alec Habig and Kate Scholberg. It is written in standard C and uses a standard UDP protocol client/server setup to make direct network connections using sockets. Dedicated phone lines could be used to increase reliability if it proves necessary. Figure 1 shows the setup.

A central machine runs a "server" program, which sits and waits for input from the outside. The individual experiments participating in the project run "client" programs. Whenever an experiment detects a candidate burst, the client program makes a connection to the server machine and sends it an alarm datagram via direct socket connection. The alarm message contains information about which experiment observed the burst, along with the time stamp information. The datagram will be expanded in the future to include information about the significance and size of the burst.

When the server receives an alarm message from any experiment, it places the alarm in a queue sorted by UT time, and searches through all alarm messages in the queue for a coincidence within a given time window (currently 10 seconds). If there are two or more different experiments in coincidence, it sends out an alarm.

## SNEWS STATUS

A test coincidence server has been set up at the Super-K site in Mozumi, Japan. Currently, MACRO, Super-K and LVD are online, sending automated alarm datagrams in test mode. SNO and AMANDA are expected to join soon. Privacy is maintained, and security precautions are taken. Additional servers can be set up at other sites. There is no automated alert to astronomers yet; we expect the automated alert to be activated after a test period.

## SNEWS AND THE NEXT GENERATION OF DETECTORS

So far, SNEWS has been intended to provide an early warning for astronomers (see, e.g. reference (5)). However, it has another potential role, which should be considered when designing the next generation of supernova-sensitive detectors.

If a core collapse supernova happens in the Galaxy, it will be an unprecedented opportunity for science, and all possible data – neutrinos, electromagnetic, gravitational waves, anything – will be extremely valuable. But many detectors which are capable of providing useful information are not necessarily capable of triggering themselves on a supernova burst and may not be continuously archiving information. They may be noisy and/or may not know what kind of signal to look for from a supernova. Some examples of detectors in this category would be:

- Some of the long string water Cherenkov detectors (AMANDA, Antares, Baikal). These are sensitive to supernova neutrinos via a coincident increase of the single photoelectron count rates of many phototubes (3). However, although some may be quiet enough to trigger themselves on a supernova, others may not be.

- Gravitational wave detectors. The gravitational wave signature of a core collapse supernova is poorly known, and it is not clear that such detectors can recognize a supernova burst by themselves. If not all data is archived, supernova-related information could be lost.

- Surface neutrino-sensitive detectors with a high rate of cosmic ray background. For instance, accelerator-based neutrino detectors use beam spill times to window their data and are not continuously sensitive.

- Radiochemical detectors, which could perform prompt extractions after a supernova occurrence (see, e.g. reference (4)). For these detectors, the earlier the extraction happens after core collapse, the better, and the SNEWS alert may be the earliest robust alert.

- Perhaps others? One could consider a wide range of detectors for various electromagnetic radiation as well as neutrinos.

- One could also consider lowering the threshold on existing supernova-sensitive neutrino detectors, and saving the information only for a SNEWS burst case.

The SNEWS neutrino coincidence will be a high confidence indication that a supernova has occurred. Noisy SN detectors could therefore arrange to use the SNEWS coincidence as an input. One could set up a buffering system, which would record data (for hours or days, depending on resources available) that would routinely be overwritten. For the case of a SNEWS coincidence, these normally uninteresting data could be saved to permanent storage, and analyzed in detail later during the time window of the supernova burst.

This use of the SNEWS coincidence could greatly enrich the world's supernova data sample.

Another thing to note is that core collapses are not necessarily followed by spectacular fireworks, or even any fireworks at all– it is not clear that a significant explosion always happens, or the supernova may be in an optically obscured region of the Galaxy. For such cases, waiting for optical confirmation could result in significant loss of information (e.g. for radiochemical detectors). So neutrinos may be not only be the earliest, but may be the only indication of a core collapse event.

## ACKNOWLEDGMENTS

The author wishes to thank all the members of the SNEWS inter-experiment working group (in particular Alec Habig). Larry Sulak pointed out the idea of using SNEWS to trigger a surface detector.

## REFERENCES

1. Burrows A. *et al.*, *Phys. Rev.* **D45**, 3362 (1992).
2. The SNEWS web page, http:/hep.bu.edu/~snnet/.
3. Halzen F. *et al.*, *Phys. Rev.* **D49**, 1758 (1994).
4. Lande, K. *et al.*, *Nucl. Phys.* **B77** (Proc. Suppl.), 13 (1999).
5. J. Bahcall (P.I.), "Observing the next nearby supernova", HST proposal 8404.

# OMNIS, The Observatory for Multiflavor Neutrinos from Supernovae

R.N. Boyd[a,b] and A.St.J. Murphy[a]

[a]Department of Physics, Ohio State University, Columbus, OH 43210 USA  [b] Department of Astronomy, Ohio State University, Columbus, Oh 43210 USA

**Abstract.** OMNIS, the Observatory for Multiflavor NeutrInos from Supernovae, is being planned for siting in the Center for Applied Repository and Underground Research, CARUS, in New Mexico. OMNIS will consist of 14 kT of lead and iron which, when radiated by neutrinos from a supernova, will produce secondary neutrons. Detection of the neutrons then will signal the arrival of the supernova neutrinos. A supernova at the center of the Galaxy, will produce about 2000 events in OMNIS, mostly from neutral current interactions. OMNIS' combination of lead and iron modules gives it particular sensitivity to neutrino oscillations of the type $\nu_\mu \rightarrow \nu_e$ or $\nu_\tau \rightarrow \nu_e$. Its intrinsic timing capability, better than 1 ms, gives it the capability to measure neutrino mass from the time-of-flight shifts in the luminosity curves of the neutrinos of different flavors to a few eV/c$^2$. OMNIS will also be able to detect differences in the luminosity cutoffs of the different flavors in the event of the fairly prompt collapse to a black hole, which might allow diagnostics on that collapse process.

## INTRODUCTION

The final state of evolution of a massive star results in a collapse which ultimately produces a few times $10^{53}$ ergs of neutrinos (1). The standard model of this process suggests that $\nu_e$'s have a mean energy of around 11 MeV, $\bar{\nu}_e$'s around 16 MeV, and all others, $\nu_\mu$'s, $\bar{\nu}_\mu$'s, $\nu_\tau$'s, and $\bar{\nu}_\tau$'s, around 25 MeV. This is a result of the fact that the $\nu_e$'s and $\bar{\nu}_e$'s interact with matter through both the charged- and neutral-current interactions, whereas all the other neutrinos, at least at the energies at which they are produced in supernovae, interact only through the neutral-current interaction. Since the $\nu_e$'s and $\bar{\nu}_e$'s therefore have more options for sharing their energy, they will emerge with a lower energy than will the neutrinos of the other flavors.

Detection of the neutrinos from a supernova can provide diagnostics of the environment from which they are produced and by which they are trapped, at least for a time span of a second or so. Indeed, the observation of the $\bar{\nu}_e$ signal from SN 1987a produced a qualitative confirmation of the theoretical description of the trapping process, as the neutrinos are thought to be produced on a time scale of milliseconds (1), but were observed over several seconds. However, much more can be learned both by observation of a much larger statistical sample of SN neutrinos and by observation of the luminosities of neutrinos other than $\bar{\nu}_e$'s.

## TECHNICAL CHARACTERISTICS OF OMNIS

With this latter point in mind, we are designing OMNIS, the Observatory for Multiflavor NeutrInos from Supernovae. OMNIS involves collaborators from Ohio State University; the Universities of California at Los Angeles, San Diego, and Irvine; the University of Texas; the University of Wisconsin; and Lawrence Livermore and Oak Ridge National Laboratories. OMNIS was originally conceived (2, 3) as utilizing the nuclei in the walls of an underground facility to convert the neutrinos into neutrons, which would then be detected in neutron detectors. However, the conversion efficiency of possible types of rock was found to be much less than that of iron or lead (4). Thus the present version of OMNIS consists of 4 kT of lead and 10 kT or iron. Slabs of the metal will be alternated with vertical racks of neutron detectors. The neutrinos will be detected when they interact with the iron or lead to emit secondary neutrons which, in turn, will be observed in the neutron detectors.

Lead is a particularly efficient converter of the neutrinos to neutrons, as its threshold for neutron emission via neutral current interactions is only 7.37 MeV. In addition, $\nu_e$'s can interact with lead through the charged current interaction to produce $^{208}$Bi, a process which has a slightly higher threshold for neutron emission of 9.77 MeV, but a considerably larger cross section (Fuller, Haxton, and McLaughlin, 1999; hereafter denoted FHM99). Indeed, it has been calculated (6) that for a supernova occuring at a distance of 10 kpc, lead will produce about 880 neutron events per kT for the standard model neutrino spectrum for all flavors, but primarily from neutral current interactions induced by $\nu_\mu$'s, $\bar\nu_\mu$'s, $\nu_\tau$'s and $\bar\nu_\tau$'s. Lead has an additional interesting feature: sufficiently high-energy neutrinos can produce events in which two neutrons are emitted. The threshold for that process is 14.98 MeV. OMNIS has the capability to identify such events with fairly high efficiency. Although the yield of events from the charged current process is expected to be a small fraction of the total, ~55 events per kT, that yield is extremely dependent on the energy of the $\nu_e$'s and $\bar\nu_e$'s, a point to which we return below.

Iron, by contrast, has a high threshold for neutron emission via neutral current interactions, 11.20 MeV, and a sufficiently high threshold for charged current processes that such production will be negligible. This is both a positive and a negative characteristic; it results in a lower efficiency, but calculation of its yield is simpler than for lead. Iron has been calculated (7, 8) to produce around 100 events per kT for the Standard Model neutrino spectrum for a supernova at 8 kpc, with virtually all events coming from $\nu_\mu$'s, $\bar\nu_\mu$'s, $\nu_\tau$'s and $\bar\nu_\tau$'s.

Thus the two types of converters will actually provide three time dependent spectra: one from the single-neutron events from lead, another from the two-neutron events from lead, and the third from the events from iron.

## SENSITIVITY TO OSCILLATIONS

The yield from the lead is particularly sensitive to some types of neutrino oscillations. Specifically, either $\nu_\mu \rightarrow \nu_e$ or $\nu_\tau \rightarrow \nu_e$ oscillations would produce much more energetic $\nu_e$'s than would be expected from the SN. This would produce dramatic effects in the lead detector; if maximal mixing occurred the yield from the lead detector would be enhanced (6) by about a factor of four, and the two-neutron event yield would be enhanced by a factor of about 40. Thus the ratio of one- to two-neutron events from the lead would produce a clear signature of oscillations of this type.

## FAST TIMING CAPABILITY OF OMNIS

Another important feature of OMNIS is its fast timing capability. Once the neutrons are produced they lose little energy until they reach the scintillator used to detect them. There they lose most of their energy in their first few scatterings from the protons in the scintillator, a time (as indicated by extensive Monte-Carlo simulations (5)) of less than 200 ns, and a time since their production that is even shorter. The subsequent thermalization of the neutrons requires about 30 $\mu$s, at which point they are captured by a Gd nucleus, a 0.1% additive to the scintillator. This capture produces four $\gamma$-rays with a total energy of almost 8 MeV. The combination of these fast-slow signals provides the signature for the neutron-induced events, but also produces a remarkably fast timing capability for OMNIS.

## Measuring Neutrino Mass

Because of OMNIS' intrinsic timing capability, the timing of the onset of any neutrino luminosity curve will actually be limited by statistics. This limit will be roughly a few ms for a SN at the galactic center (9). This level of timing could produce a measurement of neutrino mass from the effect of the time-of-flight of the neutrinos on their distributions. If one neutrino is very light, the arrival time difference between those neutrinos and more massive neutrinos is

$$\Delta t = 0.515(m/E)^2 D, \quad (1)$$

where m is the mass of the heavy neutrino in units of eV/c$^2$, E is its energy in MeV, and D is the distance to the SN in units of 10 kpc. This level of accuracy would be difficult to attain in any other way. It should be noted that a closer supernova might produce many more events (Betelgeuse would produce several million events in OMNIS!), which could determine the onset of the distributions better, but would also reduce the intrinsic time separation. A SN at the galactic center would be expected to determine the onset of the distributions to a few ms, which in turn would determine the mass of the heavier neutrino to a few eV/c$^2$ (9).

## Diagnosing Collapse to a Black Hole

Perhaps the most dramatic manifestation of such fast timing, though, would occur if collapse went fairly promptly to a black hole. If this did occur, the neutrino emission would be terminated by infalling matter near the black hole, after which both matter and neutrinos would be swallowed by the black hole. However, because the $\nu_e$'s and $\bar\nu_e$'s interact more strongly with the infalling matter (because they interact with both neutral and charged-current interactions) than do neutrinos of other flavors, they would be trapped farther from the hole than would the $\nu_\mu$'s and $\nu_\tau$'s and their respective antineutrinos. This would suggest that the $\nu_e$ and $\bar\nu_e$ luminosities would be terminated before those of the $\nu_\mu$'s, $\bar\nu_\mu$'s, $\nu_\tau$'s and $\bar{\nu}_\tau$'s. This difference, estimated (10) to be of the order of 10 ms, might be expected to depend on the details of collapse, e.g., angular momentum. Thus the timing properties of OMNIS, together with the statistics associated with 2000 events (11), would allow for diagnosis of the process of collapse to a black hole!

## ADDITIONAL ASPECTS OF OMNIS

We are planning to site OMNIS in the Center for Applied Repository and Underground Science, CARUS, in southeastern New Mexico. The lead modules will occupy about 50 linear meters of the site, while the iron modules will occupy somewhat over 150 meters. The height of the modules is restricted by the "back" (roof) of the drifts to be about 4.8 meters, while the width of the drifts restricts the module width to about 3.0 meters. The detectors will consist of alternating slabs of lead or iron and of racks of scintillators with photomultipler tubes at both ends. The ends of the modules would also be covered by doors of lead or iron, which will slide on rails. This will allow access to photomultiplier tubes or to the ends of the scintillators should maintenance be required.

At present we are testing two types of neutron detectors, one of plastic scintillator and the other of liquid scintillator. The individual detectors will either be 20 cm diameter by 2 m long cylindrical tubes of liquid scintillator or plastic scintillators of comparable size. Either type of scintillator would be loaded with a small amount of Gd, Li, or B to produce the signatures of neutron induced events.

## CONCLUSIONS

The unique physics and astrophysics that could be obtained from a statistically meaningful sample of neutrinos from the next Galactic supernova, including luminosities of all neutrino flavors, argue strongly for building an observatory to provide those data. We are planning OMNIS to fulfill that need. Furthermore, the fast timing characteristics of OMNIS give it the capability to measure neutrino masses at levels that would be difficult to achieve with any other technique, as well as possibly diagnose the process of collapse to a black hole.

## ACKNOWLEDGEMENTS

The support of the National Science Foundation through grants PHY9513893 and PHY9901241 is gratefully acknowledged.

## REFERENCES

1. Burrows, A. *Ann. Rev. Nucl. Part. Sci.* **40** 181–212 (1990)
2. Cline, D.B. et al., *Astrophys. Lett. Commun.* **27**, 403–409 (1990)
3. Cline, D.B. et al., *Phys. Rev.* **D50**, 720–729 (1994)
4. Smith, P.F., *Astropart. Phys.* **8**, 27–42 (1997)
5. Zach, J.J., Murphy, A. St.J., Marriott, D., and Boyd, R.N., private communication, 1999
6. Fuller, G.M., Haxton, W.C., and McLaughlin, G.C., *Phys. Rev.* **D59**, 085005-1 – 085005-15 (1999)
7. Woosley, S.E., Hartmann, D.H., Hoffman, R.D. and Haxton, W.C., *Astrophys. J.* **356** 272–301 (1990)
8. Kolbe, E., Langanke, K., Martinez-Pinedo, G. *Phys.Rev.* **C60** 052801 (1999)
9. Beacom, J. private communication, 1999
10. A. Mezzacappa, private communication, 1999
11. J. Beacom, R.N. Boyd, and Mezzacappa, A., private communication, 1999

# Atmospheric Neutrino Fluxes

T.K. Gaisser

*Bartol Research Institute, University of Delaware,
Newark, DE 19716, USA*

**Abstract.** The starting point for interpretation of evidence for neutrino oscillations from atmospheric neutrino data is a knowledge of the neutrino beam. In this talk I briefly review the status of calculations of the atmospheric neutrino beam. A more extensive version is given in Ref. (1), on which this talk is based.

## INTRODUCTION

The spectrum of the atmospheric $\nu_\mu$ beam decreases monotonically with energy above 100 MeV. In the few GeV range, its shape is not unlike typical neutrino beams produced with proton synchrotrons such as the AGS. Unlike accelerator beams, however, the $\nu_e$ flux is roughly half the $\nu_\mu$ flux in the sub-GeV energy range because most muons with $E_\mu < 2$ GeV decay in flight before reaching the ground. Fig. 1 compares two calculations of the atmospheric neutrino beam ($\nu + \bar{\nu}$) to the AGS charge-separated neutrino beam (4) ($\nu_\mu$ only, arbitrary normalization).

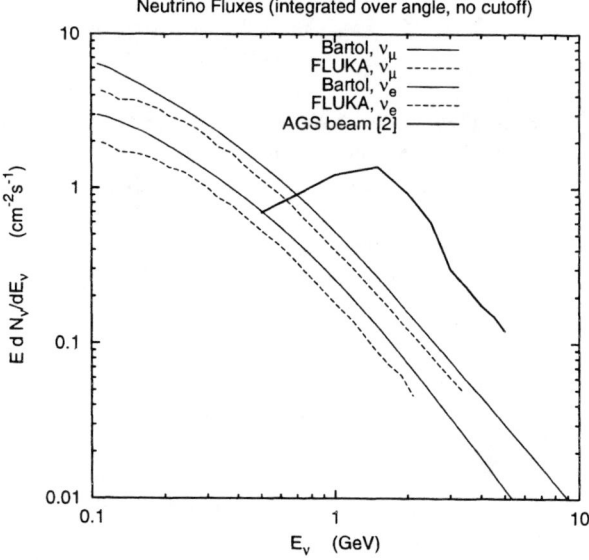

**FIGURE 1.** Fluxes of atmospheric neutrinos from two calculations: FLUKA (2), dashed; Bartol (3), solid. A charge-separated $\nu_\mu$ beam from the AGS (4) is shown for comparison (heavy line).

Pursuing the analogy with accelerator neutrino experiments, we have
Accelerated beam ⇒ Production on target (the atmosphere) ⇒ Decay path (also the atmosphere) ⇒ Beam dump (the Earth) ⇒ Neutrino beam at underground detector ⇒ Production of charged particles in detector.

Cosmic accelerators in the Galaxy provide an isotropic beam of cosmic-ray protons, α-particles and heavier ions. Roughly 80% of the nucleons in this beam are free protons and most of the rest are in helium nuclei. The geomagnetic field acts as a filter for the low energy primaries, but its influence gradually diminishes above ∼ 10 GeV. The consequent isotropy of the neutrino beam leads to the possibility of a conceptually simple neutrino oscillation experiment with a single underground detector. Neutrinos are produced around the globe at a range of altitudes peaking about 15 km. Thus the distribution of pathlengths ranges from ∼ 10 to ∼ 100 km for downward neutrinos to ∼ $10^4$ km for upward neutrinos (5). (Detailed plots of pathlength distribution as a function of zenith angle are given in Ref. (6)). Therefore the simplest and most robust evidence for oscillations is a deviation from up-down symmetry in an energy range high enough so geomagnetic effects are small.

Because of its large size, SuperKamiokande has succeeded in measuring the zenith angle dependence of multi-GeV neutrino interactions (7). These events come from primary cosmic rays of sufficiently high energy that the geomagnetic field effects are negligible. Because uncertainties in neutrino production and interaction are largely independent of zenith angle, the angular dependence of the multi-GeV $\mu$-like events constitutes the strongest, model-independent evidence (8) for oscillations involving $\nu_\mu$.

The sub-GeV events show the well-known atmospheric neutrino "anomaly" observed now for many years (9, 10, 11, 12). The ratio of muon-like to electron-like events is significantly lower than expected. In addition, a zenith-angle dependence of the sub-GeV events also shows up in SuperKamiokande (12, 8). For these events, the distortion of the primary beam, and hence also of the neutrino beam, is significant, leading to differences in the up-down rates expected at different detectors.

There is also an azimuthal geomagnetic effect, which arises from the east-west asymmetry of the primary cosmic-ray beam at low geomagnetic latitudes (13, 14, 15, 16). Since neutrinos in a given zenith angle band have the same distribution of pathlengths independent of azimuth, comparison of measured and calculated azimuthal dependence serves as a very satisfactory check on the entire measurement (17).

## UNCERTAINTIES IN THE ATMOSPHERIC NEUTRINO FLUX

An important recent development is the publication of two new calculations (2, 18). Both are 3-dimensional, thus removing the principal simplification made in previous calculations (3, 15, 19), which assumed all neutrinos in the forward hemisphere to propagate in the direction of the parent primary cosmic-ray. A conclusion of Ref. (2) is that differences between the three-dimensional and one-dimensional version of the same calculation are in practice small. This means that the simpler one-dimensional versions can be used for tracking sources of uncertainty through the calculation.

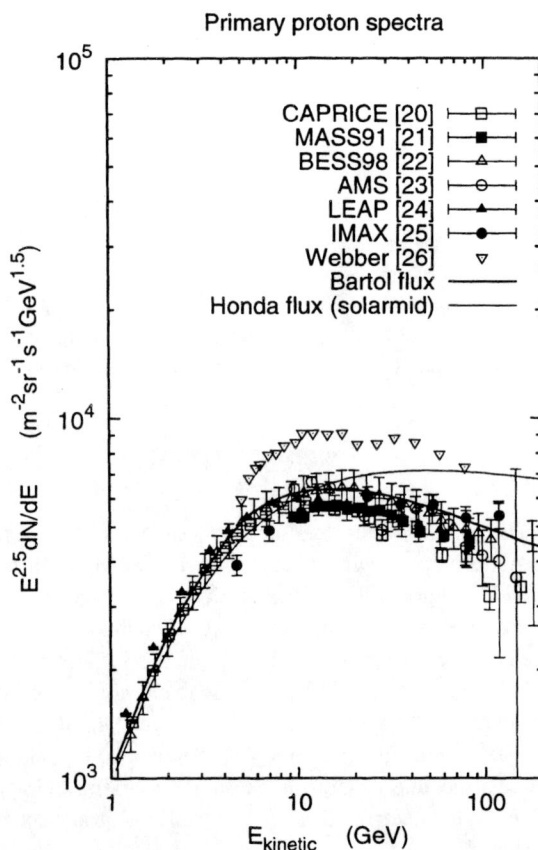

**FIGURE 2.** Data summary with curves showing the representation of the proton spectrum used in Refs. (3) (lower line) and (15) (upper line). Differential fluxes are multiplied by $E^{2.5}$.

### Primary spectrum

Another recent development is the publication in the last two years of a series of measurements of the primary spectrum, (20, 21, 22, 23). These confirm some earlier results (24, 25) showing a relatively low normalization of the proton spectrum as compared to an earlier standard reference (26). (See Fig. 2.) Thus the recent measurements reduce significantly the primary spectrum as a source of uncertainty in calculations of the flux of atmospheric neutrinos.

The Bartol calculation (3) of the neutrino flux is compared in Fig. 1 with the calculation of Battistoni et al., (2), which uses the FLUKA (27) cascade code. The FLUKA flux is approximately 15% lower than the Bartol flux. Since both calculations assume the same primary spectrum (heavy solid line in Fig. 2), the difference apparently results mainly from the representation of pion production in the two calculations.

Previouly (28) the Bartol neutrino flux has been compared with that of Honda et al. (15). These two calculated neutrino fluxes have been used extensively for comparison and interpretation of measurements of the atmospheric neutrino flux, and they agree rather closely with each other. However, the calculation of Honda et al. uses a different (higher) primary spectrum which compensates for its somewhat lower production of pions in proton interactions in the atmosphere.

### Pion production

Data on pion production comes mainly from measurements made of pion production on light nuclei, often for the purpose of designing accelerator neutrino beams. Frequently quoted examples are the measurements of Allaby et al. (29) and Eichten et al. (30), both of which measure interactions of protons with light nuclei, including beryllium and aluminum, with beams of approximately 20 GeV/c. A more recent measurement is that of Abbott et al. with a beam of 14.6 GeV/c. The range of interaction energies from 10 to 100 GeV produces most of the sub-GeV events in SuperKamiokande. None of the accelerator experiments measures all of phase space, so the event generators used for calculations of atmospheric neutrinos (which necessarily produce secondaries in all phase space) require considerable interpolation and extrapolation. The available accelerator data have been reviewed recently in the context of the atmospheric neutrino calculations in Ref. (31).

Present fits to the SuperKamiokande data leave the normalization as a free parameter. (8) The best fit to $\nu_\mu \leftrightarrow \nu_\tau$ oscillations requires an upward renormalization of the calculated nuetrino flux by 8%. If the lower pion production suggested by FLUKA or the model of Honda *et al.* is adopted, together with the lower primary spectrum, then the calculated neutrino flux will be correspondingly reduced. In this case the required upward renormalization could increase to $\sim 20\%$.

## Muon measurements

Because of the close genetic relation between muons and neutrinos, measurements of muons, particularly at high altitude where most neutrinos originate, have long been used as a check on the calculations of atmospheric neutrinos, and Perkins uses the muons as the starting point for his calculation (32). Many of the same experiments that measure the primary spectrum also measure the muon flux during balloon ascent. There have been several such measurements of muons recently in which some discrepancies with calculations have been noted. (21, 33, 34) Generally the calculated muon fluxes agree better at the top of the atmosphere (under only a few g/cm$^2$) than deeper where most neutrinos are produced. The calculations of the muons are, however, more sensitive to details than are calculations of the neutrino flux. This point is illustrated in Ref. (35). Examples of effects that affect the flux of muons at a particular altitude are bending of the muon trajectories in the geomagnetic field and angular acceptance of the spectrometer. The latter has a different effect at different altitudes because of the evolution of the angular distribution of the muons. Neutrinos, on the other hand, integrate over the whole atmosphere, so uncertainties in height of production, for example, have only a second order effect.

## CONCLUSIONS

Although there are several loose ends in the comparison between calculated and measured neutrino fluxes that remain to be understood, the evidence for neutrino oscillations is strong. This is because of the observed pathlength dependence, especially at high energy, and the fact that several sources of uncertainty cancel. In principle, measurements of the flux of atmospheric neutrinos are less sensitive to details of the calculation of the atmospheric cascade than are muons because the measurements integrate over the entire atmosphere. Already the exposure of SuperKamiokande is sufficient to begin to realize this potential, although there remains some uncertainty in the low-energy neutrino cross sections (36). The situation will be even better with a new generation of larger experiments and/or experiments with different systematic effects.

**Acknowledgements.** I am grateful to Todor Stanev and Ralph Engel for collaboration on this work. I thank Giuseppe Battistoni and Alfredo Ferrari for providing information about the calculation of Ref. (2) and I thank M. Honda for providing information about the calculation of Ref. (15). This work is supported in part by the U.S. Department of Energy under Grant DE-FG02 91ER 40626.

## REFERENCES

1. T.K. Gaisser, hep-ph/0001027, to appear in Proc. TAUP99 (Nucl. Phys. B, Proc. Suppl.).
2. G. Battistoni *et al.*,, hep-ph/9907408, Astropart. Phys. (to be published).
3. Vivek Agrawal, T.K. Gaisser, Paolo Lipari & Todor Stanev, Phys. Rev. D53 (1996) 1314.
4. L.A. Ahrens *et al.*, Phys. Rev. D34 (1986) 75.
5. D. Ayres *et al.*, Phys. Rev. D29 (1984) 902.
6. T.K. Gaisser & Todor Stanev, Phys. Rev. D57 (1998) 1977.
7. Y. Fukuda *et al.*, Phys. Lett. B436 (1998) 33.
8. Y. Fukuda *et al.*, (Super-Kamiokande Collaboration) Phys. Rev. Letters 81 (1998) 1562. See also Kate Scholberg in Proc. 8th Int. Workshop on Neutrino Telescopes (ed. Milla Baldo Ceolin, 1999) p. 183.
9. R. Becker-Szendy *et al.* (IMB Collaboration), Phys. Rev. D46 (1992) 3720 and references therein.
10. Y. Fukuda *al.* (Kamiokande Collaboration) Phys. Lett. B335 (1994) 237 and references therein.
11. W.W.M. Allison *et al.* (Soudan Collaboration), Phys. Lett. B391 (1997) 491.
12. Y. Fukuda *et al.*, Phys. Lett. B433 (1998) 9.
13. Thomas Johnson, Phys. Rev. (1933) 834.
14. L.W. Alvarez & A.H. Compton, Phys. Rev. (1933) 835.
15. M. Honda, T. Kajita, K. Kasahara & S. Midorikawa, Phys. Rev. D52 (1995) 4985.
16. Paolo Lipari, Todor Stanev & T.K. Gaisser, Phys. Rev. D58 (1998) 073003.

17. T. Futagami *et al.*, Phys. Rev. Letters 82 (1999) 5194.
18. Y. Tserkovnyak *et al.*, hep-ph/9907450.
19. E.V. Bugaev & V.A. Naumov, Phys. Lett. B232 (1989) 391.
20. M. Boezio *et al.*, Ap.J. 518 (1999) 457.
21. R. Bellotti *et al.* Phys. Rev. D60 (1999) 052002.
22. T. Sanuki *et al.* Proc. 26th Int. Cosmic Ray Conf. (Salt Lake City) vol. 3 (1999) 93.
23. AMS Collaboration, Phys. Letters (to be published).
24. E.S. Seo *et al.* (LEAP) Ap.J. 378 (1991) 763.
25. W. Menn *et al.*, Proc. 25th Int. Cosmic Ray Conf. (Durban) vol. 3 (1997) 409.
26. W.R. Webber, R.L. Golden & S.A. Stephens, Proc. 20th Int. Cosmic Ray Conf. (Moscow) vol. 1 (1987) 325.
27. A. Fassò, A. Ferrari, J. Ranft and P.R. Sala, Proc. of the 3rd Workshop on Simulating Accelerator Radiation Environment, SARE-3, KEK-Tsukuba, May 7-9, 1997, (H. Hirayama, ed.) KEK Report 97-5, p. 32 (1997).
28. T.K. Gaisser, M. Honda, K. Kasahara, H. Lee, S. Midorikawa, V. Naumov & Todor Stanev, Phys. Rev. D54 (1996) 5578.
29. J.V. Allaby *et al.*, CERN Yellow Report No. 70-12 (unpublished).
30. T. Eichten *et al.*, Nucl. Phys. B44 (1972) 333.
31. Ralph Engel, T.K. Gaisser & Todor Stanev, hep-ph/9911394, Physics Letters B (to be published).
32. D.H. Perkins, Astroparticle Physics 2 (1994) 249.
33. M. Boezio *et al.*, Phys. Rev. Letters 82 (1999) 4757.
34. S. Coutu *et al.*, Proc. 26th Int. Cosmic Ray Conf. (Salt Lake City) vol. 2 (1999) 68.
35. T. Stanev, S. Coutu, T.K. Gaisser & G. Barr, Proc. 26th Int. Cosmic Ray Conf. (Salt Lake City) vol. 2, p. 96 (1999).
36. Paolo Lipari, hep-ph/9905506.

# Results from Atmospheric Neutrinos

J. G. Learned

*Department of Physics and Astronomy, University of Hawaii*
*Honolulu, HI 96822 USA*

**Abstract.** With the announcement of new evidence for muon neutrino disappearance observed by the Super-Kamiokande experiment, the more than a decade old atmospheric neutrino anomaly moved from a possible indication for neutrino oscillations to an apparently inescapable fact. The evidence is reviewed, and new indications are presented that the oscillations are probably between muon and tau neutrinos. Implications and future directions are discussed.

## INTRODUCTION

Herein we attempt to review the present understanding of the resolution of the atmospheric neutrino anomaly, put it in context with other results, and speculate upon future directions.

This paper documents the talk given by the author at the NNN99 Workshop at SUNY Stony Brook in September 1999. The subject to be covered was generally the situation with respect to the atmospheric neutrino oscillations. The understanding of this phenomenon is now dominated by the data announced by the Super-Kamiokande Collaboration in June 1998, of which group the present author is a member. Much of this report dwells upon those results and updates to them, and so credit for this work is due to the whole Collaboration, listed in the Appendix, who have labored hard to bring this experiment to fruition and who have been ably lead by Prof. Yoji Totsuka of the University of Tokyo. That said, some of this report is the personal opinion of the author, particularly in matters of the previous history, the interpretation and future prospects for this line of research.

## The Atmospheric Neutrino Anomaly

We will not dwell upon the past history, but note that the atmospheric neutrino anomaly has been around for some time, roughly fifteen years. Indeed the first notice of something peculiar going on was in the 1960's when the seminal underground experiments in South Africa and South India first detected the natural neutrinos and observed somewhat of an absolute rate deficit, but not convincingly as the flux predictions were rough and the statistics small.

The problem became serious after the activation of the IMB experiment and by 1983 the realization that the number of events containing muon decays was lower than expected[1]. Soon this was confirmed by the Kamioka experiment, which group extended the results with good particle identification giving a redundant measure of the relative muon deficit (as also did IMB). Some members of the IMB group[2] and the Kamioka group[3] began to proclaim that oscillations were the cause of the deficit, but the claim was not widely taken seriously. This author acknowledges being one of the skeptics at that time.

The deficit in the ratio is characterized usually as an $R$ value, the ratio of muon to electron neutrinos, observed to expected. This ratio of ratios is thus independent of the 20-25% uncertain absolute flux prediction, and itself systematically uncertain by less than 10%.

With the initial evidence, the oscillations could have been from muon neutrinos to others (eg. $\nu_\tau$) or between muon and electron neutrinos, as it was the ratio that was in deficit: one could not be sure whether there was an excess of electron neutrinos, a deficit of muon neutrinos, or some of both. This lead to suggestions of other "physics" causes, such as nucleon decay favoring electron modes (since the anomaly was not detected above the nucleon mass), or an excess of extraterrestrial electron neutrinos. See Table 1 for a graphical summary of the situation. There were also suggestions of systematic problems, such as problems in muon identification, something wrong with flux calculations or neutrino interaction cross sections, entering backgrounds, or with the water Cherenkov detectors.

Over the intervening years between the emergence of this "atmospheric neutrino anomaly", as it became known, and last year's announcement, a great deal of effort went into study of the possible systematic causes of the anomaly. One troubling problem was that two European experiments, the NUSEX and the Frejus Detectors, did not observe any anomaly. Hence some people suspected a peculiarity of water as a target or with the employment of the Cherenkov radiation in vertex location. Not only were the statistics of the European detectors rather small, but as indicated by more recent work from the similar type of instrument in the US, the Soudan II detector, the presence of a surrounding veto counter is vital for the more compact type of slab detectors. As well, the MACRO experiment has elucidated the production of low energy (hundred MeV) pions by nearby cascades in rock, which particles enter cracks in non-hermetic detectors and appear to be neutrino interactions. In any case the Soudan II with now significant exposure (several kiloton-years) finds an $R$ value close to that of SuperK (and IMB and Kamioka).

## THE SUPER-KAMIOKANDE REVOLUTION

We now proceed to summarize the new evidence for oscillations which comes from the SuperK experiment. Before going on it may be worthwhile to point out what permitted the big break-through, which is not so obvious. The increase in size of detector, from near kiloton fiducial volumes for Kamioka and Soudan, and three kilotons for IMB to the twenty two kilotons of SuperK is not the whole story. As will be seen below, the progress comes from the recording of muon events with good statistics in the energy region above 1 $GeV$. This is due to detector linear dimensions as well as gross target volume: muon events with energy more than 1 $GeV$ and thus 5 $m$ range were not likely to be fully contained in the Kamioka detector (or the IMB detector). SuperK in contrast has decent muon statistics up to almost 5 $GeV$, and this is turns out to be crucial.

Most of the data to be discussed below is the "fully contained" ($FC$) event sample, consisting of those events in which both vertex and track ends remain in the fiducial volume. There are also "partially contained" ($PC$) events, in which a muon may exit the fiducial volume from a contained vertex location. Such events are useful even though the total energy is not known, the energy observed being a lower limit. Of course this is the case even with $FC$ events, though to a lessor degree, because the observed particles are not of the same energy (or direction) as the incident neutrino, which of course is what one would desire to observe.

The particle types are identified by pattern recognition software, now well tested and verified by experiment with known particle beams at the accelerator. Fortunately most of the events (roughly 2/3) are single (Cherenkov radiating) tracks, in which the identification is quite clean (at the 98% level). To be clear and cautious we usually refer to the reconstructed events as "muon-like" and "electron-like", though a safe approximation is that these represent muon and electron neutrino charged-current interactions.

The other two categories of events of which we shall report are the through-going upwards moving muons ($UM$), produced by neutrino interactions in the rock or outer detector, and which are coming from directions below the horizon (as those from above the horizon can be confused with down going muons from cosmic ray interactions in the atmosphere near overhead). Another category of event is the entering-stopping muon ($SM$). It is useful that these event categories probe approximately three different energy ranges of neutrinos: $FC \simeq 1\ GeV$; $PC$ and $SM \simeq 10\ GeV$; $UM \simeq 100 GeV$. It should be understood that as far as we know, these neutrinos are all produced in the upper atmosphere by cosmic ray interactions, and are reasonably well described by models in content, energy, and angular dependence (to a few percent)(5).

We shall not take up limited space here with the description of the SuperK detector, which is well documented elsewhere. The interested reader would do well to look at some of the theses from SuperK, which, though large files, are available on the web(6). The short summary is that the SuperK detector consists of a large stainless steel cylinder (37 $m$ high by 34 $m$ diameter inside the inner detector) with extremely high photo-tube coverage (40%), ten times more pixels than any earlier instrument, and a remarkable sensitivity of roughly eight photoelectrons per MeV of deposited (Cherenkov radiating) energy. The latter permits detection of events down to about 5 $MeV$, so for the present discussion detection efficiency versus energy is not important because the events we are discussing are all above $\simeq 100\ MeV$. The inner volume is also well protected by a 2 $m$ thick, fully-enclosing veto counter, populated by the recycled IMB photomultipliers and wavelength shifters. The inner "fiducial" volume is further taken as 2 $m$ inside the photomultiplier surface, resulting in the 22 kiloton volume used for most reported data.

The SuperK oscillations claim was first formally presented to the physics community in June 1998 at the $NEUTRINO$98 meeting in Takayama, in a talk by Professor Takaaki Kajita, leader of the on-site contained event analysis group. The data was presented in several papers to the community(7, 8, 9), building upon past data from Kamioka(3) and IMB(2), and culminating in the claim of observation of oscillations of muon neutrinos, published in Physical Review Letters in August 1998(10). We now proceed to review the evidence, which has changed little except for new indications that the $\nu_\mu$ oscillating partner is probably the $\nu_\tau$, and not a sterile neutrino.

## Up-Down Asymmetry

One way to look at the *FC* (and *PC*) data is in terms of a dimensionless up-to-down ratio, difference over sum (which has symmetrical errors in contrast to just up/down)(12). This quantity is exhibited as a function of charged particle momentum in Figure 1, for both electrons and muons, with the *PC* data shown as well (for which we know only a minimum momentum). One sees that the electron data fits satisfactorily to no asymmetry, whilst the muon data shows strong momentum dependence, starting from no asymmetry to about -1/3 ($-0.311 \pm 0.043 \pm 0.01$) above 1.3 $GeV$.

From this Figure alone, without need for Monte Carlo simulation, assuming the cause to be neutrino oscillations, one can deduce that:

1. The cause of the atmospheric neutrino anomaly is largely due to disappearing muons, not excess electrons.

2. There is little coupling of the muon neutrino to the electron neutrino in this energy/distance range.

3. The oscillations of the muon neutrinos must be nearly maximal for the asymmetry to approach one third.

4. The scale of oscillations must be of the order of 1 $GeV$/200 $km$, plus or minus a factor of several.

In fact, as seen by the dashed lines overlying the data points, the simulations do produce an excellent fit to the muon neutrino oscillation hypothesis, while the no-oscillations hypothesis is strongly rejected. The latter is so strong that statistical fluctuations are not in question, one must look for systematic problems to escape the oscillations explanation.

One concern for some people has been the fact that the asymmetry is indeed maximal, which makes it appear that we are very lucky that the earth size and cosmic ray energies are "just so" to produce this dramatic effect. This appears to this author to fall in the category of lucky coincidences, such as the angular diameter of the moon and sun being the same from earth. (There is another oscillations related peculiar coincidence that the matter oscillation scale turns out to be close to one earth diameter, and this depends upon the Fermi constant and the electron column density of the earth.) The phase space for "coincidences" is very large, and we humans are great recognizers of such patterns.

## East-West Asymmetry

The effect of the earth's magnetic field is a little complicated. For example for energies to a few GeV, it provides some shielding from straight downwards going charged cosmic rays in regions near the magnetic equator. For higher energies and incoming trajectories near the horizon, the magnetic field still prevents some arrival paths. As the SuperK detector location is not on the magnetic equator the effect is not up-down symmetric, and this spoils the symmetry otherwise expected from the neutrinos about the horizontal plane (where there is some peaking due to longer flight paths for pions in the atmosphere). However, the effects are mostly limited to neutrino energies below about 1 $GeV$, corresponding to cosmic ray primaries below about 10 $GeV$. The picture is made a bit more complicated by the earth's magnetic field not being a nice symmetrical dipole. Fortunately there are good models of the magnetic field and the people who have made flux calculations take this into account. The SuperK group has published a paper(9) showing the azimuthal variation of the SuperK data ($\pm 30$ deg about the horizon) for intermediate to higher energies (400 – 3000 $MeV$), where the calculations are reliable. (Certain simplifications such as a one dimensional cascade model have been regularly used, which surely is not a good approximation at the lowest energies). The SuperK data exhibit significant variation from uniformity yet fit the flux predictions very well, giving one confidence in the modeling(9).

## Natural Parameters for Oscillations: $L/E$

In an ideal world, one would assuredly present this data as a function of distance divided by energy, $L/E$, since that is the parameter in which one expects to see oscillatory behavior. Since we observe only the secondary charged particle's energy and direction, badly smeared at the energies available, plots in which one would wish for visible oscillations can at best show a smooth slide from the no-oscillations region to the oscillating regime. This is illustrated in Figure 2, where the ratio of numbers of events observed to those expected with no-oscillations is plotted versus "$L/E$"(13), for muon and electron (type) events.

The plot is not "normalized", and since we see somewhat of an excess of electron type events overall, the solid circle indicated electron points are a bit greater than one on average (+14%). This is a little worrisome, but acceptable since (as already noted) the absolute flux is uncertain to a larger value. In contrast to the electron data, the muon points fall with increasing $L/E$, reaching a plateau at about one half their initial value, again consistent with oscillations. Muon neutrino oscillations in the Monte Carlo simulation are indicated by dotted line, and fit the data reasonably well.

As noted, the data does not show oscillations, presumably due to convolutions washing out the oscillatory behavior. It was this smooth falloff that got us to wondering if another model might fit the data, one in which one component of the muon neutrino decays with distance. We wrote a paper(14), and a second version(15), suggesting decay to explain the atmospheric neutrino anomaly. I will not discuss details here, but note that in order to get a model that fit the available facts we had to push on all available limits, and we invoke neutrino mass and mixing in any case. Consequently such models do not pass the economy test of Occam's Razor, though most annoyingly they remain not ruled out as yet.

Considering future experiments, this is one area in which improvement may indeed be made. A hypothetical detector, such as a megaton version of the Aqua-RICH instrument studied by Ypsilantis and colleagues could have the resolution to see a multi-peaked $L/E$ plot(16).

## Fits in Energy and Angle

The SuperK Collaboration's preferred method of fitting the ensemble data is to employ a $\chi^2$ test to numbers of events binned by particle type, angle, and energy, a total of 70 bins. The bin choices may seem a bit peculiar, but they have historical precedent (they are as employed for Kamiokande) and though not optimal for the new data set, this choice allows us to avoid paying any statistical (or confidence) penalty for choosing arbitrary bins. The fit employs a set of parameters to account for potential systematic biases. Details cannot be presented here, but it has been shown that the numerical results are quite insensitive to the selection of the parameters or their supposed "errors" (except for the overall normalization)(17).

Figure 3 illustrates the data plotted for two energy intervals ($sub - GeV$ and $multi - GeV$, more or less than $1.3\ GeV$) for single track events identified as either electron-like or muon-like. The partially contained data is added to the multi-GeV muon data. The data is shown as a function of the cosine of the zenith angle, with $+1$ being down-going. One sees that the data very well fits the curves gotten from the Monte Carlo simulation, at the values gotten from the grand ensemble fit ($\delta m^2 = 0.003\ eV^2$ and $sin^2(2\theta) = 1.0$).

The results of the fits are often presented in terms of an inclusion plot, showing an acceptable region(s) in the space of mixing angle ($sin^2 2\theta$) and mass squared difference ($\delta m^2$), as presented in Figure 4. The minimum in $\delta m^2$ has moved a little upwards with accumulated statistics, though not much, (good news for long baseline experiments anyway) but remains uncertain to about a factor of two.

## Muon Decay Events

It is not often emphasized, but the original indication of the anomaly, a deficit in stopped-muon decays ($\simeq 2.2\ \mu sec$ after the initial neutrino event), remains with us, and constitutes a nice alternate, almost independent, sample with quite different systematics. It is not so clean a sample and the statistics are lower, but the complete consistency of the muon decay fraction remains a reassuring complement to the energy and angle analysis employing track identification.

## Through-going and Entering-Stopping Muons

Another cross check comes from the *UM* and *SM* samples, particularly nice as the source energies are factors of 10 and 100 higher and the detector systematics rather different (for example, the target is mostly rock not water). In going from the earlier instruments to SuperK, however, the gain is not so great (about a a factor of 2.5 times over IMB, for example), since the rate of collection of through-going muons depends upon area not volume. However, the much greater thickness of the detector and the good tagging of entering and exiting events in the veto layer yields many more stopping (*SM*) events.

The angular distribution for *UM* events from below the horizon is shown in Figure 5, where one sees that the angular distribution is nicely consistent with oscillations and not with no-oscillations. However, since much of the effect is close to the horizon, where oscillations for the energies in question are just setting in, one worries about contamination of the near horizon events with in-scattered events from the much greater numbers of down going muons. There is no room for details here, but we find no evidence for significant contamination(11).

The *SM* sample was predicted to be 35-40% of the *UM* sample, as indicated in Figure 6, yet in fact we see only about $24\% \pm 3\%$. Fitting the data to the oscillation hypothesis one can make the now usual inclusion plot, which shows that the *UM* and *SM* results are completely in accord with those from the *FC* and *PC* data. However, as the statistics are smaller and the physics leverage not as great, the muon result does not add much to the *FC* and *PC* constraints, though it does stiffen the lower bound in $\delta m^2$.

There is a lengthy tale about an $SM/UM$ analysis from the IMB experiment(2), which claimed an exclusion region very close to the now preferred solution. This result seems to have been flawed due to older flux models and Monte Carlo simulations. Work is in progress to reassess the old data with new flux calculations and an updated quark model. Thus there remains a cloud upon the horizon, but one which I expect will fade away in reanalysis.

## The Muon Neutrino's Oscillation Partner

Given that the muon neutrino is oscillating, is it oscillating with a tau neutrino or a new sterile neutrino which does not participate in either the charged (CC) or neutral current (NC) weak interaction? Fortunately we have several means to explore this with SuperK data. The NC events should show an up-down asymmetry for sterile neutrinos but not for tau neutrinos (as the NC events for all ordinary neutrinos are the same). Another avenue for discrimination is that sterile neutrinos would have an additional oscillation effect due to "matter oscillations". The consequence would be a unique signature in the angular distribution of intermediate energy muons.

Early SuperK efforts focussed upon the attempt to collect a clean sample of $\pi^o$ events. As it turns out, this has been frustrating because the rings (from the two decay $\gamma$s) cannot be separated at energies above $\simeq 1\ GeV$, and in net there are not so many reconstructed events as to permit a good discrimination. In fact the absolute rate is consistent with expectations, but the cross section is uncertain to about 20% making the hint at tau coupling not significant.

More recently, tests have been devised employing the $PC$ event sample and the $UM$ sample. The $PC$ sample can be cut on energy to yield a somewhat higher mean source energy, and the upwards going number compared to downwards number of events. For the muons a near horizontal number can be compared to number of nearly straight upcoming events. Preliminary results from SuperK give no encouragement for sterile neutrino model builders. It appears that the tau neutrino hypothesis fits the data, while the sterile neutrino hypothesis is rejected at about the 2 standard deviation level. A publication will be forthcoming from SuperK.

## Hypotheses to Explain Anomaly

We conclude with a summary Table 1 of all hypotheses put forth to explain the atmospheric neutrino anomaly. Space does not permit a full discussion here, but it is the case that with the SuperK data we now have eliminated almost all alternate hypotheses to explain the results. The only exception of which the author is aware involves the peculiar decay model, but it is one that nobody takes very seriously (including the author who is co-author of the model). The only hypothesis which fits the evidence, and it fits very well, is that muon neutrinos maximally mix with tau neutrinos with a $\delta m^2$ in the range of $2 - 5 \times 10^{-3}\ eV^2$.

## IMPLICATIONS

The ramifications of the atmospheric neutrino anomaly are great and span the known realms of fundamental physics from large to small. We have not discussed in this short paper the links to solar neutrinos(18), nor the LSND results(19). Certainly there is no conflict between the atmospheric muon neutrino results and the possible (nay likely) solar oscillations. If, however, the LSND results are correct, then we have surely some interesting physics to untangle, as it is generally admitted that no simple three neutrino model can incorporate all three neutrino anomalies, and that new degrees of freedom would be required.

## Astrophysics and Cosmology

The implications of the oscillations results have been explored in other talks at this meeting as well. First, it appears that neutrinos with summed masses of the order of $0.1\ eV$ will not make any major contribution to resolving the dark matter quandary. Nonetheless with a ratio of 2 billion to one for photons (and neutrinos) to nucleons from the Big Bang, even such a small neutrino mass may be greater in total than all the visible stars in the sky. Hence one must account for neutrino mass in further cosmological modeling, but neutrinos are not likely to constitute the bulk of the "missing matter". However if the neutrinos should be nearly degenerate in mass and all have masses in the range near $1\ eV$ (and hence we are observing only small splittings with the oscillations), then neutrino mass may dominate the universe. While neutrinos are not favored by astrophysical modelers (fitting the spatial fluctuations in the cosmic microwave background for example), large neutrino masses are not ruled out. Nearly degenerate neutrino masses would not present a consistent picture with the quark and charged lepton masses, which make large mass jumps between generations. But who knows? We do not have a viable GUT with mass predictions, so an open mind is appropriate.

The other major area of significance, perhaps of the deepest significance has to do with baryogenesis, the ori-

**Table 1.** List of hypotheses invoked to possibly explain the atmospheric neutrino anomaly. The first 3 columns are criteria available prior to SuperK, and the last 4 after the 1998 SuperK publication. The hypotheses divide into 5 systematics issues and 7 potential physics explanations. As indicated in the text, the only remaining likely hypothesis is the oscillation between muon and tau neutrinos. The "x" schematically indicates which evidence rules out the hypothesis in that row.

| Evidence | Old | | | | New | | |
|---|---|---|---|---|---|---|---|
| Hypothesis | $R$ ($E<1$ GeV) | $\mu$ decay Frac | Vol Frac | $R$ ($E>1$ GeV) | $A_e \simeq 0$ | $A_\mu < 0$ | $R(L/E) \simeq 0.5$ |
| Atm. Flux Calc. | xx | | | x | | x | x |
| Cross Sections | xx | | | x | | x | |
| Particle Ident. | | xx | xx | | | | |
| Entering Bkgrd. | | | xx | | | x | |
| Detector Asym. | | | xx | | | | |
| X-Ter. $\nu_e$ | | | | | | x | x |
| Proton Decay | | | | x | | x | x |
| $\nu_\mu$ Decay | | | | | | | $\simeq$x |
| $\nu_\mu$ Abs. | | | | | | | x |
| $\nu_\mu$ - $\nu_e$ | | | | | x | | |
| $\nu_\mu$ - $\nu_s$ | | | | | | x | |
| $\nu_\mu$ - $\nu_\tau$ | | | | | | | |

gin of the predominance of matter by one part per billion over anti-matter at Big Bang time. There are claims that the old idea of accumulation of net baryon number will not survive the early stages of universe expansion(20, 21). If that is indeed the case, it may be that neutrinos provide the avenue for net baryon asymmetry generation, relatively late in the game(22).

Neutrino masses and possible sterile neutrinos have also been invoked to help resolve problems in understanding heavy element synthesis in supernovae.

## Theoretical Situation: Why So Important

There have been a number of talks at this Workshop about the particle theory situation, so I can add little. In Figure 7, I show the masses of the fundamental fermions in three generations, on a logarithmic scale in mass. Dramatically, one sees that if the neutrino masses are near the lower bounds (that is at the presumed mass differences from present atmospheric and solar results), they lie 10-15 orders of magnitude below the other fundamental fermions (charged quarks and leptons). Graphically one notes the spacing between the neutrino masses and the charged fermion masses is just about the same as the distance (on the log scale) to the unification scale. This is a pictorial representation of the see-saw prediction, as we noted more than ten years ago(4). This points up the task for grand unification, and highlights the deep link between neutrino masses and nucleon decay.

## Future

During the last year the physics community seems to have largely accepted the inevitability of neutrino mass and oscillations(23). Of course the game has hardly begun and many a sublety may await our exploration. But if the LSND claims will go quietly away, the mass and mixing may settle into the simple hierarchical pattern explored in the bi-maximal mixing scenario (or similar versions). To my taste this highlights the importance of experiments to follow up on the LSND results as one of the first agenda items in the neutrino business.

Given present indications, it would seem that the K2K and MINOS experiments should confirm the SuperK results and make the oscillation parameters more precise. Of course, one would really like to see tau appearance, not just muon disappearance to be sure we are not being misled. There are many arguments in the community as to what constitutes appearance. Because of the complexity of tau final state identification, this author would prefer to see a real tau track recorded. In any case plans are in progress in the US, Japan and Europe for the obvious follow up experiments to nail things down.

More interesting for the long range physics is filling in the MNS matrix (lepton equivalent of quark CKM matrix) for neutrinos. This is not an easy business. The atmospheric neutrino measurements really are only defining, at best, three of the nine elements! Solar neutrinos get us another, perhaps a constraint on two. Measuring the tau related components directly seems pretty hopeless. Of course if we can assume the matrix to be unitary and real we are, or soon will be, in good shape as there are then only three independent parameters (plus the masses). But we do not know this, and if there exist CP violations we then have a total of three angles and two phases (but only one measurable). If there are more (heavy or sterile) neutrinos, then things could be much more complicated (as the 3 by 3 sub-matrix will not be unitary). By analogy with the quarks (where the 3 by 3 CKM with small mixing angles and one CP violating phase seems to do the job), perhaps we should not worry too much, except for lack of any guidance whatsoever from theory. CP violation is only very weakly constrained experimentally in the neutrino sector at present, so we could be in for big surprises, and given the neutrino connection with cosmology and baryogenesis, one should indeed be suspicious, I believe. As a whole, the particle physics community is just beginning to explore this avenue, but it looks as though muon colliders may provide our best route for exploring this new realm. Measuring absolute mass remains a frustrating problem, which will not be resolved in the near future it seems.

It seems to me that a next generation (megaton) scale nucleon decay instrument would do wonders for advancing this line of investigation. To my view, simply building a larger version of SuperK will not suffice because we need greater resolution as well as size. The only candidate I see at the moment is something like the AQUA-Rich style of imaging water Cherenkov detector(16). In any case we can expect a long and interesting exploration into neutrino mass and mixing now that the door has been opened.

## ACKNOWLEDGMENTS

The author wishes to thank the organizers for an excellent and fruitful workshop. As noted earlier, the Super-Kamiokande Collaboration deserves the credit for the work reported herein, and any errors in interpretation are those of the author. Thanks to Sandip Pakvasa, Chris Wiebusch and Brett Viren for various discussions and help.

# REFERENCES

1. "Calculation of Atmospheric Neutrino-Induced Backgrounds in a Nucleon-Decay Search", IMB Collaboration (with T. J. Haines et al.), *Phys. Rev. Lett.* **57**, 1986 (1986).

2. R. Clark, *et al.*, (IMB Collaboration), Phys. Rev. Lett., **79**, 345 (1997); R. Becker-Szendy *et al.*, Phys. Rev. **D46**(1992) 3720; D. Casper *et al.*,, Phys. Rev. Lett. **66**(1991) 2561.

3. Y. Oyama, *et al.*, (Kamiokande Collaboration), hep-ex/9706008 (1997); K.S. Hirata *et al.*, Phys. Lett. **B205**(1988) 416; K.S. Hirata *et al.*, Phys. Lett. **B280**(1992) 146; Y. Fukuda *et al.*, Phys. Lett. **B335**(1994) 237.

4. "Neutrino Mass and Mixing Implied by Underground Deficit of Low-Energy Muon-Neutrino Events", John G. Learned, Sandip Pakvasa, and Thomas J. Weiler, *Phys. Lett.* B **207**, 79 (1988). Also see V. Barger and K. Whisnant, Phys. Lett. B **209**,365 (1988); K. Hidaka, M. Honda and S. Midorikawa, Phys.Rev.Lett., **61**, 1537 (1988).

5. see "Geomagnetic Effects on Atmospheric Neutrinos", Paolo Lipari, T.K. Gaisser, Todor Stanev *Phys. Rev.* D **58**, 73003 (1998), preprint astro-ph/9803093, and references therein. See also report of Atmospheric Neutrino Working Group, D. Casper, *et al.* in these Proceedings.

6. http://www-sk.icrr.u-tokyo.ac.jp/doc/sk/pub/

7. "Measurement of a Small Atmospheric $\nu_\mu/\nu_e$ Ratio", The Super-Kamiokande Collaboration: Y.Fukuda *et al.*, *Phys. Lett.* B **433**, 9 (1998), preprint hep-ex/9803006.

8. "Study of the atmospheric neutrino flux in the multi-GeV energy range", The Super-Kamiokande Collaboration, Y. Fukuda, *et al.*,, *Phys. Lett.* B **436**, 33 (1998), hep-ex/9805006.

9. "Observation of the East-West Anisotropy of the Atmospheric Neutrino Flux", The Super-Kamiokande Collaboration: T.Futagami, *et al.*, PRL **82**, 5194 (1999), preprint astro-ph/9901139.

10. "Evidence for Oscillation of Atmospheric Neutrinos", The Super-Kamiokande Collaboration, Y. Fukuda, *et al.*, *Phys. Rev. Lett.* **81**, 1562 (1998), preprint hep-ex/9807003.

11. "Measurement of the flux and zenith-angle distribution of upward through-going muons by Super-Kamiokande", The Super-Kamiokande Collaboration: Y. Fukuda, *et al.*, *Phys. Rev. Lett.* **82**, 2644 (1999), preprint hep-ex/9812014.

12. "Up-Down Asymmetry: A Diagnostic for Neutrino Oscillations" John W. Flanagan, John G. Learned, and Sandip Pakvasa, *Phys. Rev.* D **57**, 2649 (1998), preprint hep-ph/9709438. Also see J.W. Flanagan dissertation, UH 1997, available at the SK web page.

13. See PhD dissertation of M. Messier, BU 1999 for derivation of the correction used for translating observed energy into $L/E$. This author has used a different form and definition in the past. The figure presented is the SuperK official plot.

14. "Neutrino Decay as an Explanation of Atmospheric Neutrino Observations", V.Barger, J.G. Learned, S. Pakvasa, T.J. Weiler, *Phys. Rev. Lett.* **82**, 2640 (1999), preprint astro-ph/9810121.

15. "Neutrino Decay and Atmospheric Neutrinos", V. Barger, J.G. Learned, P. Lipari, M. Lusignoli, S. Pakvasa, T.J. Weiler, *Phys. Lett.* B **462**, 109 (1999), preprint hep-ph/9907421

16. "The AQUA-RICH atmospheric neutrino experiment", P. Antonioli, it et al., to be published in NIM in the *Proceedings of the RICH98 Workshop in Israel*, preprint CERN-LAA/99-03, 5/5/99, and talk of T. Ypsilantis at this Workshop.

17. Nonetheless, this author (personal opinion, not collaboration) suspects that the process pulls the minima slightly towards lower values of $\delta m^2$. The reasoning is that the introduced parameters un-weight the effect of $R$ on the fitting, which pulls upwards, while the shape pulls downwards. The author's bet remains that $\delta m^2$ settles at around $5 \times 10^{-3} \ eV^2$, whereas the official fits give 2.5 to 3.5.

18. See the paper by J. Bahcall in these Proceedings.

19. See the papers by G. Grexlin and R. Tayloe in these Proceedings.

20. A.D. Sakharov, JETP Lett., **5**, 24 (1967).

21. V.A. Kuzmin, JETP Lett. **12**, 228 (1970).

22. "Baryogenesis via neutrino oscillations", E.Kh. Akhmedov, V.A. Rubakov, A.Yu. Smirnov, *Phys. Rev. Lett.* **81**, 1359 (1998). There are many papers dealing with this topic, as a search on "baryogenesis" in the preprint server will reveal.

23. See the papers of P. Ramond, J. Pati and P. Langacker in these Proceedings.

# APPENDIX

## Super-Kamiokande Collaboration, 6/99

- **Institute for Cosmic Ray Research, University of Tokyo** Y. Fukuda, K. Ishihara, Y. Itow, T. Kajita, J. Kameda, S. Kasuga, K. Kobayashi, Y. Kobayashi, Y. Koshio, M. Miura, M. Nakahata, S. Nakayama, Y. Obayashi, A. Okada, K. Okumura, N. Sakurai, M. Shiozawa, Y. Suzuki, H. Takeuchi, Y. Takeuchi, Y. Totsuka (spokesman), S. Yamada

- **Gifu University** S. Tasaka

- **National Laboratory for High Energy Physics (KEK)** T. Ishii, H. Ishino, T. Kobayashi, K. Nakamura, Y. Oyama, A. Sakai, M. Sakuda, O. Sasaki

- **Department of Physics, Kobe University** S. Echigo, M. Kohama, A.T. Suzuki

- **Department of Physics, Kyoto University** T. Inagaki, K. Nishikawa

- **Niigata University** W. Doki, M. Kirisawa, S. Inaba, K. Miyano, H. Okazawa, C. Saji, M. Takahashi, M. Takahata

- **Department of Physics, Osaka University** K. Higuchi, Y. Nagashima, M. Takita, T. Yamaguchi, M. Yoshida

- **Bubble Chamber Physics Laboratory, Tohoku University** M. Etoh, A. Hasegawa, T. Hasegawa, S. Hatakeyama, K. Inoue, T. Iwamoto, M. Koga, T. Maruyama, H. Ogawa, J. Shirai, A. Suzuki, F. Tsushima

- **The University of Tokyo** M. Koshiba

- **Tokai University** Y. Hatakeyama, M. Koike, M. Nemoto, K. Nishijima

- **Department of Physics, Tokyo Institute of Technology** H. Fujiyasu, T. Futagami, Y. Hayato, Y. Kanaya, K. Kaneyuki, Y. Watanabe

- **Boston University** M. Earl, A. Habig, E. Kearns, M.D. Messier, K. Scholberg, J.L. Stone, L.R. Sulak, C.W. Walter

- **Brookhaven National Laboratory** M. Goldhaber

- **University of California, Irvine** T. Barszczak, D. Casper, W. Gajewski, W.R. Kropp, L.R. Price, M. Smy, H.W. Sobel, M.R. Vagins

- **California State University, Dominguez Hills** K.S. Ganezer, W.E. Keig

- **George Mason University** R.W. Ellsworth

- **University of Hawaii** A. Kibayashi, J.G. Learned, S. Matsuno, V.J. Stenger, D. Takemori

- **Los Alamos National Laboratory** T.J. Haines

- **Louisiana State University** E. Blaufuss, B.K. Kim, R. Sanford, R. Svoboda

- **University of Maryland** M.L. Chen, J.A. Goodman, G.W. Sullivan

- **State University of New York, Stony Brook** J. Hill, C.K. Jung, K. Martens, C. Mauger, C. McGrew, E. Sharkey B. Viren, C. Yanagisawa

- **University of Warsaw** D. Kielczewska

- **University of Washington** J.S. George, A.L. Stachyra, R.J. Wilkes, K.K. Young

- **Department of Physics, Seoul National University** S.B. Kim

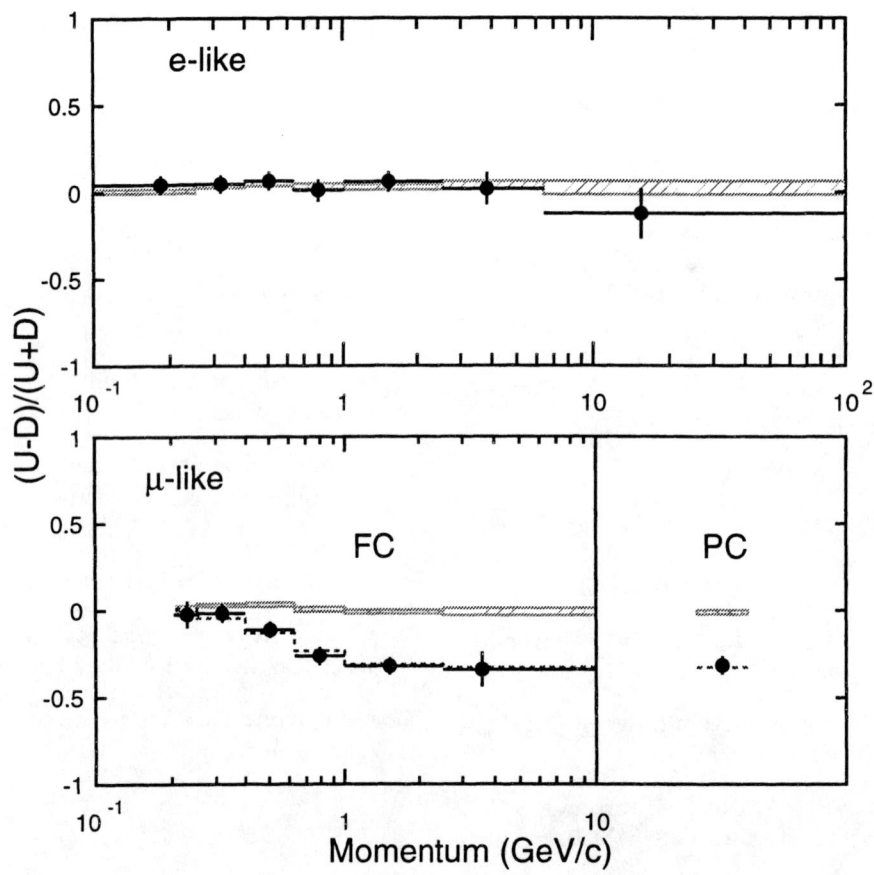

**FIGURE 1.** The up-down asymmetry for muon and electron type events in SuperK, from 848 days of live time (analyzed by 6/99), as a function of observed charged particle momentum. The muon data includes a point for the partially contained data (*PC*), which is more than about 1 *GeV*.

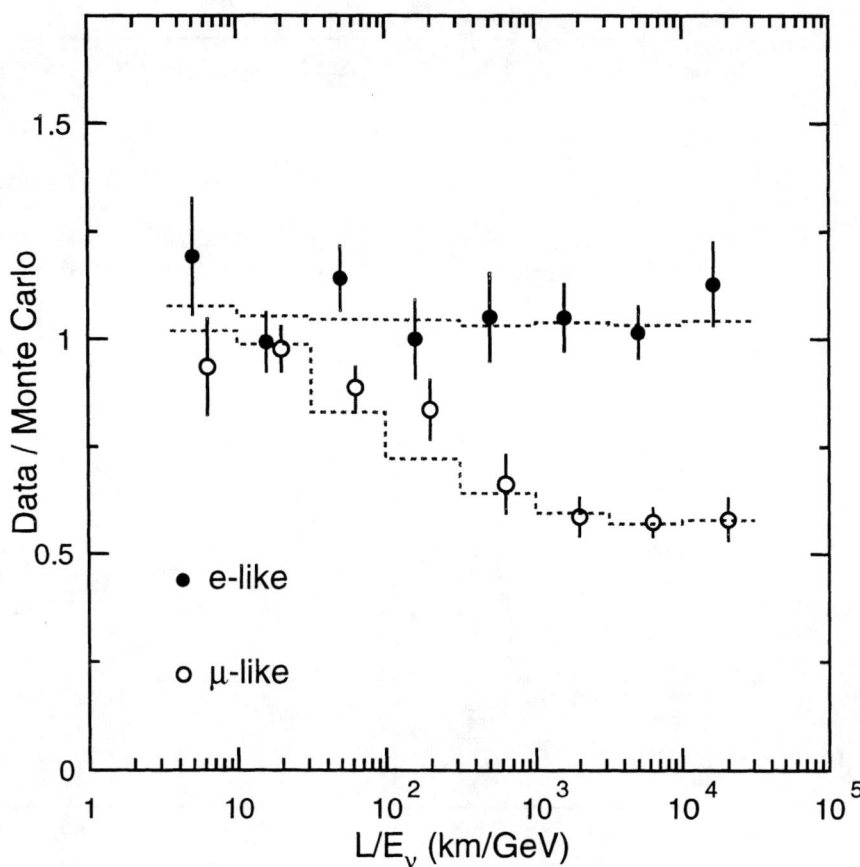

**FIGURE 2.** The ratio of numbers of events observed compared to predicted as a function of the natural oscillations parameter, distance divided by energy. The results are not normalized and overall there is a slight excess (not significant) compared to expectations. Electrons show no evidence for oscillations, while muons exhibit a strong drop with $L/E$. This is consistent with oscillations, as indicated by the dashed lines from simulations.

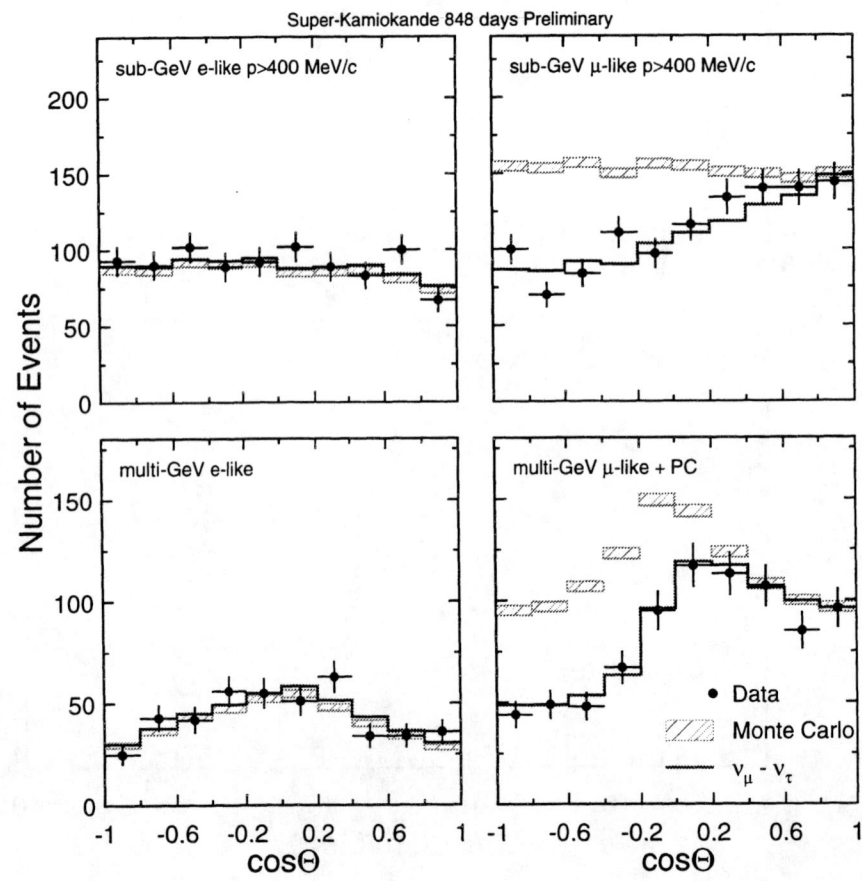

**FIGURE 3.** Cosine of zenith angle distributions of the contained event data for two different energy ranges, electron and muon events. Shaded region shows no-oscillations simulation, and heavy line for oscillations between muon and tau neutrinos.

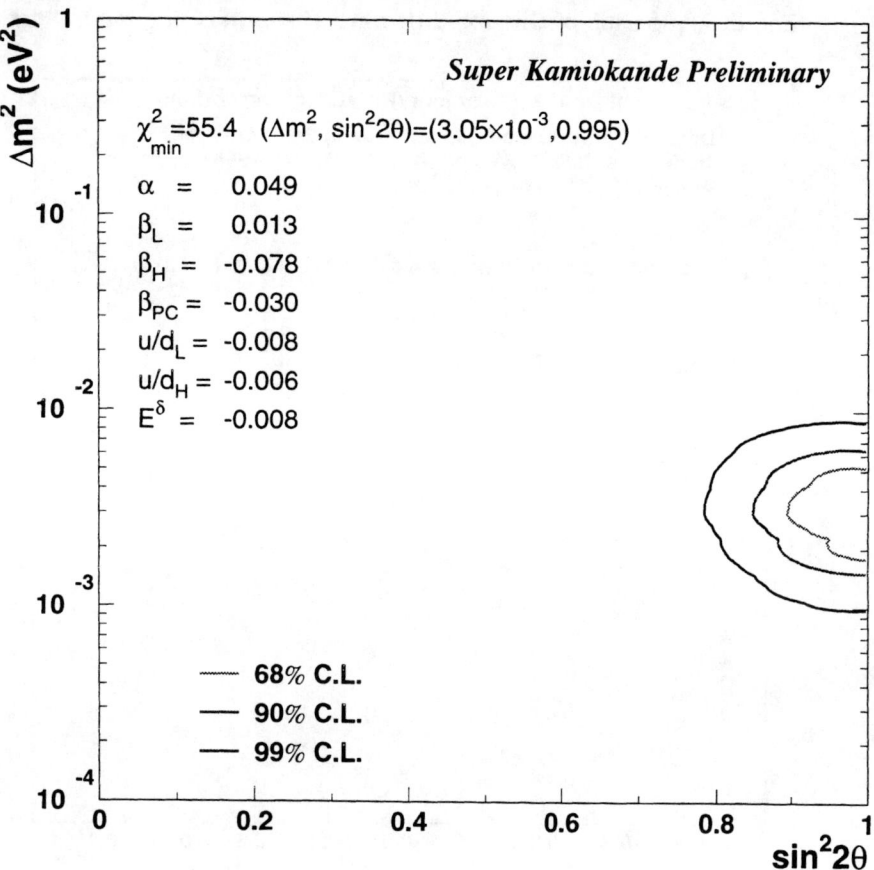

**FIGURE 4.** Inclusion plot, showing the regions for various degrees of statistical acceptability in the plane of mixing angle and mass squared difference. Contained event data analysis of June 1999.

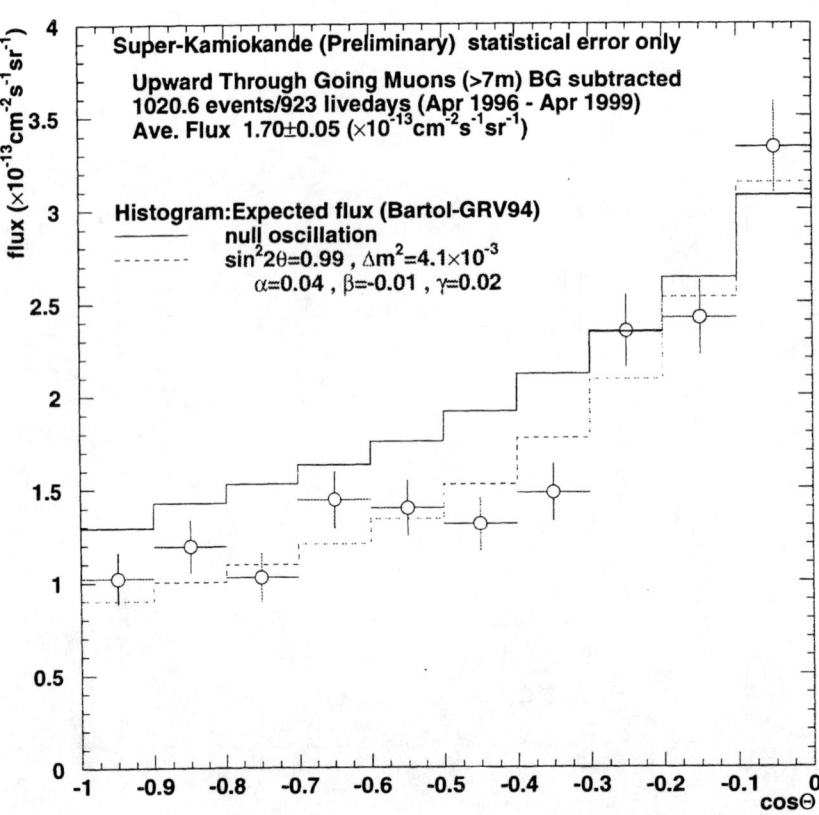

**FIGURE 5.** The angular distribution of upcoming muons from 902 days of SuperK data. Expectations for no-oscillations and best fit from contained events are indicated.

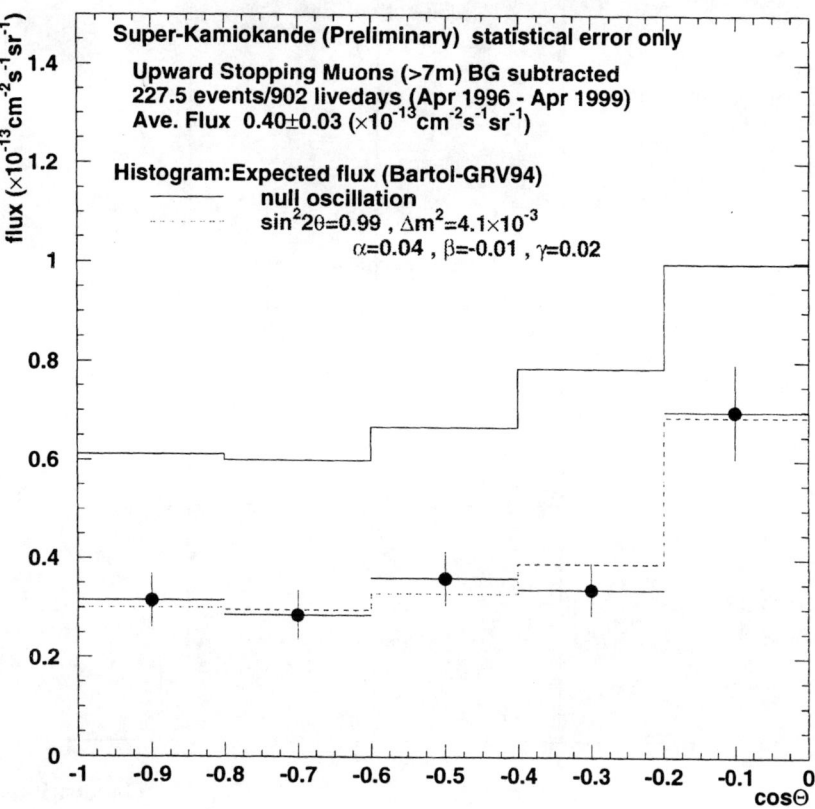

**FIGURE 6.** The rate of stopping muons versus zenith angle in 902 days of SuperK data. The large deficit is less significant than it appears because of a 20% uncertainty in absolute flux.

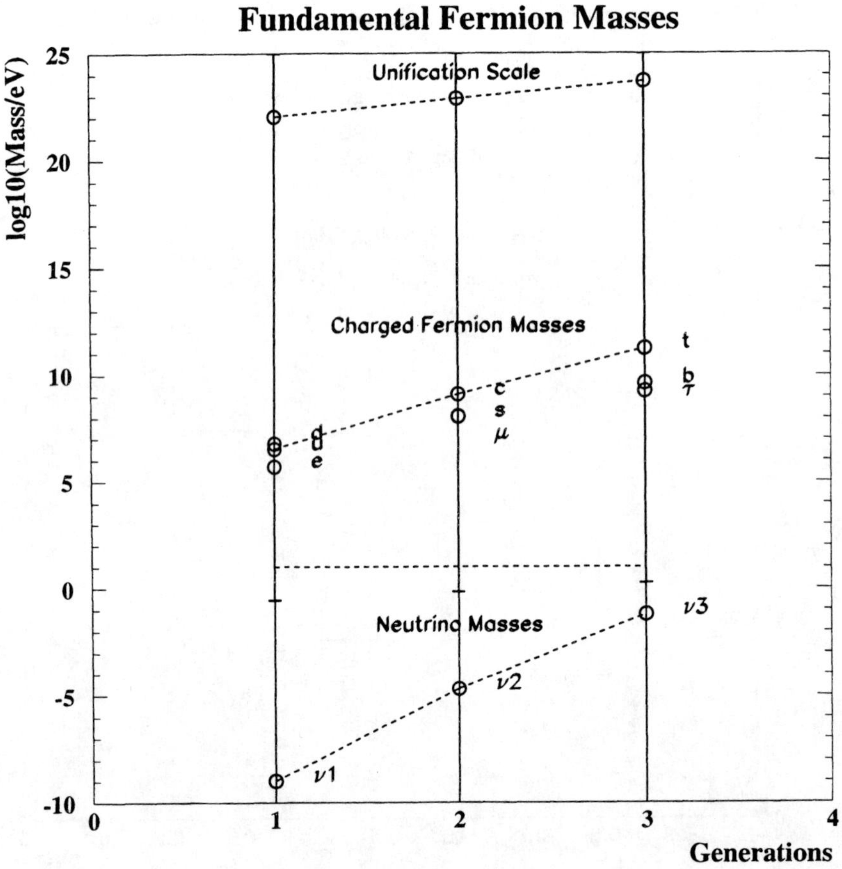

**FIGURE 7.** The masses of the fundamental fermions.

# Neutral current $\pi^0$ production and the $\nu_\mu \to \nu_\tau$ - $\nu_\mu \to \nu_s$ debate.

S. B. Boyd

*Department of Physics, University of Washington, Seattle*

**Abstract.** Discrimination between the $\nu_\mu \to \nu_\tau$ and $\nu_\mu \to \nu_s$ oscillation modes, in the context of the Super Kamiokande experiment, is discussed. It is shown that the up-down production asymmetry of neutral current $\pi^0$ events cannot currently be used for this purpose. The dominant source of error in the measurement of the neutral current $\pi^0$ production rate is shown to be the cross section iuncertainty and means of measuring this to a precision of 10% are discussed, with the conclusion that MiniBooNE may be able to achieve this goal. The role of the K2K long baseline experiment in discriminating between the two oscillation hypothese is considered.

## INTRODUCTION

In 1998 the Super-Kamiokande (Super-K) experiment published results claiming the observation of neutrino flavour oscillations(1) in the neutrino flux from interactions of cosmic ray hadrons in the atmosphere. This evidence took the form of a suppression of the observed number of muon neutrino events with respect to expectation coupled with no significant suppression of the electron neutrino event numbers. As striking as these results are, the question of whether the muon neutrinos are oscillating to a tau neutrino or an, as yet, undetected sterile neutrino still remains. One way to discriminate between these two hypotheses is to study the rate of neutral current event production. The oscillation of muon neutrinos to tau neutrinos would not affect the neutral current event rate, since the neutral current interaction is flavour blind. However, if the muon neutrinos were oscillating to sterile neutrinos, a suppression of the neutral current event rate would be observed. In the context of the Super Kamiokande experiment, the best way of observing neutral current interactions is to search for events with a single $\pi^0$ and no other particles.

This paper discusses methods of determining the oscillation mode evinced in the Super-K data. The up-down asymmetry of neutral current $\pi^0$ production is dicussed as a means of discriminating between then $\nu_\tau$ and $\nu_s$ oscillation modes. Following this the issue of the measurement of the $\pi^0$ production rate and its main systematic errors is explored and the role that the long baseline experiment, K2K, will play in determining the oscillation mode is investigated.

**Table 1.** Measurement of the neutral current asymmetry for different oscillation modes. $U$ signifies the number of neutral current events induced by upward going neutrinos and $D$ signifies the number of neutral currente events induced by downward going neutrinos.

| Mode | Effect |
|---|---|
| $\nu_\mu \to \nu_\tau$ | $\frac{U-D}{U+D} = 0$ |
| $\nu_\mu \to \nu_s$ | $\frac{U-D}{U+D} < 0$ |

## THE $\pi^0$ PRODUCTION ASYMMETRY

The cleanest way to distinguish oscillations to $\nu_\tau$ from oscillations to $\nu_s$ in water Cerenkov atmospheric neutrino detectors is to measure the neutral current interaction up-down production asymmetry. For a given neutrino energy, the probability for an upward going $\nu_\mu$, which was produced in the atmosphere on the side of the earth opposite that of the detector and consequently has a baseline of approximately 12,000 km, to oscillate is less than for a downward going $\nu_\mu$ which was produced in the atmosphere directly over the detector and which has a baseline of about 10 km. As the neutral current interaction is flavour blind and since, by definition, sterile neutrinos do not interact with normal matter, the up-down asymmetry in the neutral current event production rate can be used to distinguish the oscillation mode, as displayed in Table 1.

As discussed in the introduction, the easiest way for Super-K to distinguish neutral current events is through

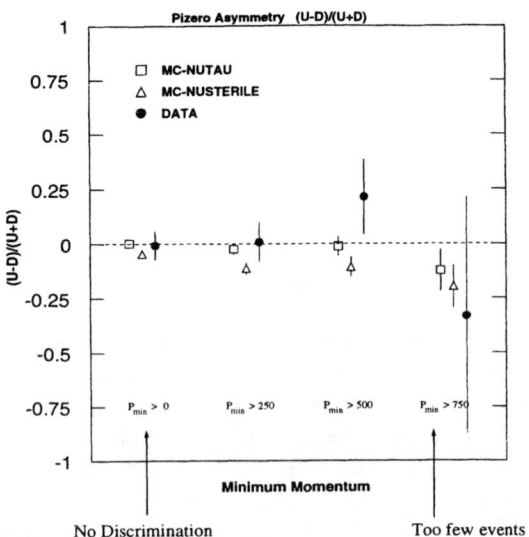

**FIGURE 1.** The Up-down $\pi^0$ production asymmetry for different $\pi^0$ momentum cuts. Only the very highest bin ($P_{min} >$ 750 MeV) has any discriminating power and it lacks statistics. The error bars are statistical.

the production of single $\pi^0$, and so the problem devolves to the measurement of the single $\pi^0$ up-down production asymmetry.

Conceptually simple, this method is insensitive to uncertainties in the neutral current $\pi^0$ production cross section, to the model of the atmospheric neutrino flux and to reconstruction effects. However, it suffers from two inadequacies that substantially limit its use:

- The effect of intra-nuclear pion scattering on the direction of the $\pi^0$ is not known precisely

- The $\pi^0$ direction is, in any case, a poor estimator of the direction of the initial neutrino. The $\pi^0$ arises as a decay product of a $\Delta$ baryon which is generated in the hadronic system of a neutrino interaction. Unless the boost of the $\Delta$ in the laboratory frame is high, the $\pi^0$ will not retain a memory of the direction of the incident neutrino. In general, only very high energy $\pi^0$ events are able to correctly tag their parent neutrinos as upward- or downward-going. Unfortunately, the number of events in this high energy sample are too few to distinguish between the two oscillation modes, as shown in Figure 1.

Although it is not possible, with current data, to distinguish between oscillation of $\nu_\mu$ to $\nu_\tau$ or to $\nu_s$ using the $\pi^0$ up-down production asymmetry, some discrimination may be had by studying the absolute neutral current $\pi^0$ production rate.

# THE ABSOLUTE $\pi^0$ PRODUCTION RATE

In order to be as insensitive as possible to uncertainties in the shape of the atmospheric neutrino flux when determining the $\pi^0$ production rate is desirable to form the "double ratio"

$$R = \frac{(\frac{\pi^0}{e-like})_{Data}}{(\frac{\pi^0}{e-like})_{MC}} \quad (1)$$

The numerator and denominator are normalised by the electron-like event sample. This minimises systematics due to reconstruction effects, since both the e-like events and the neutral current $\pi^0$ events initiate electromagnetic showers in the detector.

In the presence of the $\nu_\mu \to \nu_\tau$ oscillation mode, $R \approx 1.0$, whereas for $\nu_\mu \to \nu_s$, $R$ is substantially reduced. For the best fit oscillation parameters reported by Super-K, $R \approx 0.75$. There appears to be significant difference between the $\nu_\mu \to \nu_\tau$ and the $\nu_\mu \to \nu_s$ oscillation hypotheses. However, the number measured by Super-K is

$$R = 1.03 \pm 0.06 \pm 0.24 \quad (2)$$

where the first uncertainty is statistical and the second is the systematic uncertainty. A table of the systematic uncertainties is presented in Table 2. The dominant source of systematic error is the uncertainty on the neutral current $\pi^0$ cross section.

In order that the difference between $R$ when calculated assuming $\nu_\mu \to \nu_\tau$ and when using the $\nu_\mu \to \nu_s$ hypothesis be better than 3 $\sigma$ the cross section must be known to better than 11%.

## An independent measurement of the neutral current $\pi^0$ cross section

There are two ways to measure a cross section. One is to estimate the total neutrino beam flux delivered over the life time of the experiment and compute the cross section from the observed number of events of the process of interest; the other is to compare the observed number of events of the process of interest with a reference sample of events from some other well known process for which the cross section is known. Each method presents a number of issues for consideration. For the first

- can the selection efficiency for the process of interest be calculated accurately?

- how well is the beam flux known?

**Table 2.** Source of systematic uncertainty in the $\pi^0$ double ratio.

| Source of Uncertianty | | Source of Undertainty | |
|---|---|---|---|
| Data statistics | ±0.06 | MC statistics | ±0.02 |
| Flux estimate | ±0.03 | Reconstruction | ±0.07 |
| Nuclear reinteractions | ±0.07 | NC $\pi^0$ cross section | ±0.20 |

Since the normalisation of the beam flux is usually known only to about 20% this method will not be discussed furthur. For the second method

- can the selection efficiency for the process of interest be calculated accurately?

- does the reference process have a well-known differential cross section?

- does the reference process provide a large statistics sample in the energy range under study?

The major source of uncertainty in the calculation of the selection efficiency for neutral current $\pi^0$ events is the effect of pion absorption and scattering in the nuclear medium. Theoretical models for pion absorption and scattering for water predict a decrease in the observed pion yield from 30% to 50% (2) and this would certainly limit the study of neutral current $\pi^0$ events in water Cerenkov experiments. However, for lighter materials the effect of the nuclear medium on pions has been extensively studied (3) and it seems that the selection efficiency should be known to better than 10%.

The precision to which the reference process is understood is also a source of systematic uncertainty. Most experiments use the quasielastic process, $\nu + n \rightarrow \mu^- + p$, where $n$ is a neutron and $p$ is a proton. This process has the advantage of a large cross section at energies of interest to atmospheric neutrino experiments. However, the quasielastic cross section is only known to about 10% and the nuclear effects at low momentum transfer are not well understood. This implies that the neutral current $\pi^0$ cross section cannot be determined to the required precision if quasielastic events are used to normalise the event sample.

An intriguing possibility would be to use the *elastic* process, $\nu + e \rightarrow \nu + e$, as the reference sample. This has a precisely known cross section (with an uncertainty of much less than 1%) and is robust to nuclear effects. However, the elastic production rate is very small (approximately a factor of 200 less than the production rate of quasielastic events) and there is significant background from $\nu_e$ quasielastic events and from neutral current $\pi^0$ events. In order to use this process a experiment must have excellent $e/\pi^0$ discrimination and a high flux beam. Of all the currently proposed neutrino experiments, only MiniBooNE (and later BooNE) has the capability for this measurement. The MiniBooNE proposal (4) claims a probability of misidentifying a $\pi^0$ as an electron of less than 0.005%. Combined with the high flux, low $\nu_e$ component (less than 0.2% with respect to the $\nu_\mu$ component) beam from the Fermilab booster, MiniBooNE may observe approximately 130 $\nu_\mu$ elastic events and 65,000 neutral current $\pi^0$ events. This may allow the neutral current $\pi^0$ cross section to be determined to the required 11%, the error being dominated by the low statistics in the elastic event sample.

## THE ROLE OF THE K2K EXPERIMENT

The K2K(6) experiment is the worlds first operating long baseline neutrino oscillation experiment. Based at KEK in Japan, it directs a high purity muon neutrino beam, generated by the 12 GeV KEK proton synchrotron, through a near detector system on the KEK site to the 250 km distant Super-K detector. The near detector includes a 1 kt water Cerenkov detector using the same operating principles as Super-K and a fine grained tracking detector for beam studies.

The most precise measurement that could be made with the K2K experiment is to measure the double ratio of neutral current $\pi^0$ events normalised to the quasielastic events at both the near and far detectors :

$$R = \frac{\left(\frac{NC\pi^0}{CCQE}\right)_{far}}{\left(\frac{NC\pi^0}{CCQE}\right)_{near}} \quad (3)$$

Assuming the Super-K oscillation parameters, this statistic is approximately 1 for oscillations of $\nu_\mu$ to $\nu_\tau$ and is about 2 for oscillations of $\nu_\mu$ to $\nu_s$. It is robust to nuclear effects, cross section uncertainties and uncertainties in the absolute flux. The major problem with this measurement is the low event rate at the far detector. At the most optimistic, the number of neutral current $\pi^0$ events recorded in Super-K over the entire running period of the experiment is less than 40, not enough to make a significant choice between the two oscillation scenarios.

The measurement of neutral current $\pi^0$ to quasielastic rate in the near detector of K2K may, however, be used to calculate a prediction for the number of neutral current $\pi^0$ events observed in the atmospheric neutrino flux at Super-K. The shapes of the energy spectra are different but are both known to about 5%. By normalising the $\pi^0$ samples observed in Super-K and in the K2K front detector to the number of observed $\nu_\mu$ quasielastic events, systematic effects from nuclear reinteractions and from incomplete knowledge of the absolute neutrino flux are minimised. It is, perhaps, more desirable to normalise to the $\nu_e$ induced quasielastic events as the reconstruction properties of electron and $\pi^0$ showers contain similar systematic effects. However, the number of $\nu_e$ events is very small (less than 1% of the $\nu_\mu$ component) and the statistical uncertainty on the number of $\nu_e$ quasielastic events would dominate the total error in the predicted number of neutral current $\pi^0$ events at Super-K. Using $\nu_\mu$ quasielastics it seems possible to obtain a prediction of the number of neutral current $\pi^0$ events which should be observed in Super-K which has an accuracy of better than 10%.

## CONCLUSION

Discrimination between the hypotheses of $\nu_\mu \to \nu_\tau$ oscillations and $\nu_\mu \to \nu_s$ oscillations may be found in the rate of neutral current interactions and (for atmospheric neutrino detectors) in the up-down asymmetry of the neutral current event production rate. In the context of the existing Super Kamiokande data, the best method of distinguishing a neutral current event is through the production of a single $\pi^0$. The uncertainty on the absolute neutral current $\pi^0$ rate in Super-K is dominated by the 20% uncertainty in the single $\pi^0$ cross section, which much be reduced to 5% to obtain results significant at the 99% confidence level. This may be achieved in two ways : by measuring the cross section using an independent experiment or by determining the ratio of neutral current $\pi^0$ events to quasielastic events in another water Cerenkov detector with similar systematic effects and applying this to the Super Kamiokande data. It does not seem possible that any independent experiment currently proposed can measure the cross section to better than 10%, with the possible exception of MiniBooNE which may be able to use the $\nu_e$ elastic events to normalise the neutral current $\pi^0$ event sample. The long baseline neutrino experiment, K2K, which uses Super Kamiokande as the far detector and a smaller water Cerenkov detector in the near detector should be able estimate the number of neutral current $\pi^0$ events observed at Super-K in the atmospheric neutrino flux to an accuracy of approximately 10%.

It is worthwhile emphasising that it is the ratio of neutral current to charged current events that is the statistic of interest in discriminating between the $\nu_\tau$ and $\nu_s$ oscillation modes. Experiments such as Monolith(5) and MiniBooNE(4) will have greater sensitivity to this statistic than Super Kamiokande due to their capability of studying more neutral current interaction modes. However, none of these experiments are scheduled to start data taking before 2004 and K2K is already operational. The first results must come from K2K.

## REFERENCES

1. The Super–Kamiokande Collaboration, "Evidence for oscillation of atmospheric neutrinos", *Phys. Rev. Lett.* **82**, 1562 (1998)

2. S.K. Singh, "Nuclear Physics Issues in Neutrino Oscillation Experiments", Valencia 1997 Beyond the Standard Model, 311 (1997); S.K. Singh *et al*, *Phys. Lett.* **B416**, 23 (1998)

3. D. Ashery *et al*, *Phys. Rev.* **C23**, 2173 (1981)

4. The BooNE Collaboration "A proposal for an experiment to measure $\nu_\mu \to \nu_e$ oscillations and $\nu_\mu$ disappearance at the Fermilab Booster." (1997).

5. The Monolith Collaboration "Atmospheric Neutrino Oscillations with a Massive Magnetised iron Detector" (1999)

6. The K2K Collaboration, "Proposal for Participation in the Long-Baseline Neutrino Oscillation Experiment E362 at KEK" (1996)

# The ANTARES Project

S. Navas-Concha

*(On behalf of the ANTARES Collaboration)*
*Centre de Physique des Particules de Marseille, 163. Av de Luminy – Case 907, 13288 Marseille, France*

**Abstract.** ANTARES is a project to build a deep-sea neutrino telescope. The scientific program includes neutrino astronomy, indirect detection of non barionic dark matter in the form of neutralinos, and the search for neutrino oscillations. The site, the procedures and the performances of the proposed detector have been evaluated. The deployment of detectors and underwater electrical connections have been tested successfully at a depth of 2400 m in the French Mediterranean sea, near Toulon, and the optical characteristics of the proposed site have been measured. A detector with an effective area of about 0.1 km$^2$ will be deployed in 2002-2003.

## INTRODUCTION

The ANTARES collaboration (1) was formed in 1996, based on a close collaboration between particle physicists, astrophysicists, and experts in marine technology. Since then, most of the technical aspects relating to the construction and operation of this device have been addressed, and its sensitivity has been evaluated for the principal physics goals.

The ANTARES scientific program includes astrophysics (neutrino astronomy), cosmology (searches for dark matter in the form of neutralinos), and particle physics (neutrino oscillations). The main motivation comes from *neutrino astronomy*, since the detection of the cosmic neutrinos would lead to a better understanding of several astrophysical environments, from the supposed accelerators of the ultra high-energy cosmic rays, to high density regions, opaques to photons and hadrons. The existence of astrophysical sources of high-energy neutrinos can be inferred from the existence of high-energy cosmic-ray protons. The neutrinos would result from the charged pions produced by the interaction of accelerated protons with matter ($pp$) or radiation ($p\gamma$). Candidate sources of high-energy neutrinos include X-ray binaries and supernova remnants, active galactic nuclei (2) (AGN) and gamma-ray bursters (GRB).

*Dark matter* could be detected in high-energy neutrinos if super-symmetric neutralinos make up part of the missing mass. Relic neutralinos from the early Universe would accumulate at the core of heavy bodies such as the Earth, the Sun or the centre of our Galaxy (3). Their annihilation would lead to a constant flux of neutrinos, with an angular distribution depending on the neutralino mass (4). ANTARES would be sensitive to this effect because of its good angular resolution and its low energy threshold.

Atmospheric neutrinos are an irreducible background for the study of high energy cosmic neutrinos. However, at lower energy (typically 10–100 GeV), they can be a very interesting signal for the *neutrino oscillation study*. For this study, ANTARES benefits from an oscillation baseline length of the order of the Earth diameter. For the Super-Kamiokande oscillation parameters ($\Delta m^2 = 3.5 \cdot 10^{-3}$ eV$^2$, $\sin^2 2\vartheta = 1$) (5), the first dip in the survival probability of the neutrinos crossing the Earth occurs at 350 km/GeV, at the maximum of the ANTARES sensitivity.

A first phase of the R&D program finished in 1999, when the construction of a detector with an effective area of about 0.1 km$^2$ was approved.

## DETECTOR DESIGN

A schematic of the future ANTARES detector is shown in figure 1. It consists of an array of $\sim$1000 photo-multiplier tubes (housed in pressure resistant glass spheres) in 13 vertical strings, spread over an area of about 0.1 km$^2$ and with an active height of about 350 m. The strings, arranged in a randomised spiral with a minimum horizontal spacing of about 60 m, are composed of standard clusters consisting of three optical modules (oriented down at 45$^0$), a local control module (LCM) and an electro-optical cable. At the base of each string there is a string control module (SCM) linked to a common junction box by electro-optical cables. The vertical spacing of the clusters is 12 m.

The design ensures a good efficiency for up-going tracks and a marginal acceptance for the down-going ones. In addition, the overlap in the active solid angle for

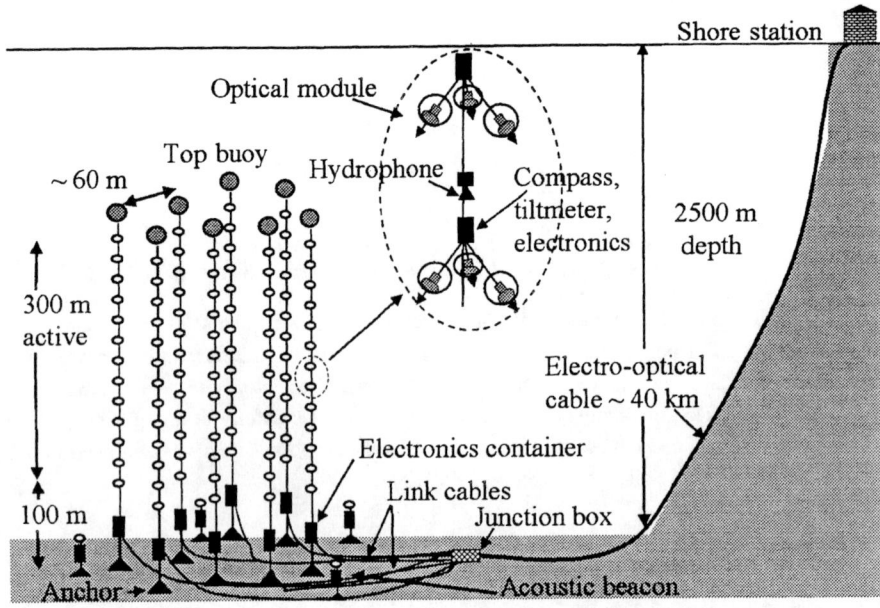

**FIGURE 1.** Schematic diagram of the first phase of the ANTARES detector.

the photo-multipliers in a cluster allows an event trigger based on local coincidences in a storey. The mechanical structure is ensured by an electro-mechanical cable.

Positioning and time calibration are of special importance for a neutrino telescope. For neutrino source pointing, an absolute orientation with respect to the sky of about $0.2^0$ is necessary. It will be obtained by triangulation of acoustical beacons (one at the bottom of each string) with the surface boat, coupled with Dinamic Global Positioning System (DGPS). Supplementary ones will be distributed around the detector in order to have a precise positioning of the outer strings. The relative positions and orientations of all optical modules in the detector are given in real time by a positioning system based in a compass and a tilt-meter housed in each LCM. A maximum error of $\sim 1$ m on the reconstructed shape is estimated. The master clock on shore, linked to Universal Time (UT) through the GPS network, will allow ANTARES events to be matched to transient astronomical phenomena such as gamma-ray bursts. Special optical beacons containing high intensity pulsed light sources will illuminate several strings simultaneously, providing the time calibration with an error better than 1 ns.

The off-shore trigger logic will be as simple and flexible as possible. The first level trigger requires a coincidence of two of the three optical modules serviced by a single local control module (rate of $\sim 150$ kHz). The second-level trigger is based on combinations of the first-level triggers (rate of a few kHz). If a second-level trigger occurs, the full detector will be read out. The third-level trigger will be made in a farm of processors on shore and will select about 100 Hz of events to be recorded for off-line analysis.

The off-shore electronics is based on front-end digitalisation and digital data transmission. An ASIC called the Analogue Ring Sampler (ARS) is under development. For single-photo-electron pulses (99% of the photo-multiplier pulses), timing information will be stored in a pipeline memory until the second-level trigger arrives.

## PROTOTYPE STRING

More than twenty different deployment/recovery cycles have been completed in order to test the technique, instruments and slow-control. The recovery of the string is activated by an acoustical signal from the surface which disconnects the string from its anchoring weight. The undersea electrical connection of the string-control modules to the junction box has been tested successfully by the IFREMER crew of the "Nautile" submarine during one of these operations.

A prototype string, 350 m long with 16 pairs of optical modules frames (eight of them fully equipped with photo-multipliers and positioning systems) will be deployed during autumn'99. The slow-control network will control the power distribution, the red-out electronics, the acoustic system, and the readout of sensor data. The experimental data will be sent to the shore station, 40 km away, with an electro-optical cable.

# SITE EVALUATION STUDY

A suitable site for a neutrino telescope requires deep, transparent water and proximity to shore facilities. The chosen site for the deployment of a 0.1 km$^2$ undersea detector is located 20 nautical miles off the French Mediterranean coast, near Toulon, at a depth of ~2400 m. Optical background, biological fouling of optical surfaces, undersea currents, and meteorological conditions have been studied at this site, using special test strings.

### Optical background

Due to the $^{40}$K decay and to the life activity in sea water (the bioluminescence), the optical background has important constraints on the trigger logics and the electronics as well as the mechanical layout of the photo-multipliers. A low level background rate varying from 20 to 47 kHz was measured with an 8-inch photo-multiplier tube housed in a 17-inch pressure-resistant Benthos glass sphere. Bursts of luminescence lasting about one second and giving peak counting rates up to several MHz were observed, correlated with the speed of the undersea currents. When integrated over the measured distribution of counting rates, the dead time induced on the electronics for the whole detector is less than 5% randomly distributed over the detector due to the local character of the bioluminescence bursts.

### Water transparency

Different systems have been used to measure the water transparency. Initially, the light attenuation as a function of the distance was measured with a 33 m long rigid structure holding a collimated LED source located at a variable distance from a glass sphere containing an 8-inch photo-multiplier tube. The effective attenuation length for blue light (466 nm wavelength) was measured to be $41 \pm 1 \pm 1$ m (December 97). Since this magnitude results from a combination of absorption and scattering, a second test was designed to disentangle the contributions from these two components.

Light scattering was measured using a pulsed isotropic light source, consisting of a set of six blue LEDs (466 nm), located at a distance of either 24 m or 44 m away from a fast 1-inch photo-multiplier tube. The recorded time distributions (shown in figure 2) exhibit a clear peak coming from direct photons, and a tail extending to larger delays due to scattered photons. At 24 m (44 m), the ratio of scattered to direct photons is less than 5% (10%). An effective attenuation length could be determined from the ratio of the integrated spectra measured at the two distances, yielding:

$$\lambda_{att,eff} = \begin{vmatrix} 60.0 \pm 0.4 \text{(stat) m (July 98)} \\ 52.2 \pm 0.7 \text{(stat) m (March 99)} \end{vmatrix}$$

A systematic uncertainty of a few meters might affect these estimates due to the fact that the LED luminosity is not monitored and yet assumed to be the same for the time distributions collected at the two distances. The different measurements seem to indicate a time variation of the water transparency at the ANTARES site.

The data were also analysed in order to extract the scattering length and the mean photon scattering angle. Because of the $\sigma = 4.5$ ns time resolution of the electronics, only the water properties of for "large" scattering angles were measured (which are, however, the photons expected to affect the performances of a neutrino telescope). The data were nicely reproduced with an absorption length in the range (55-65) m, a scattering length at large angle greater than 200 m and a correspondingly roughly isotropic scattering angle distribution.

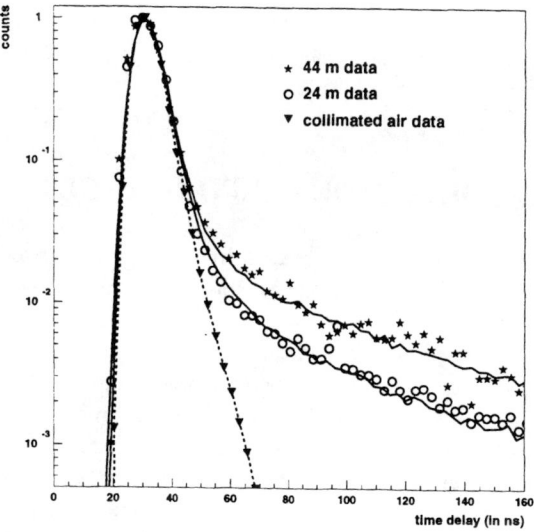

**FIGURE 2.** Distribution of photons arrival times for 24 m, 44 m and a calibration in air, with Monte-Carlo curves superimposed.

### Biofouling

Fouling occurs on the modules, due to the sedimentation and to the growth of bacteria on the surfaces. The measurement of the fouling rate is crucial for a long-term project planing to leave optical modules immersed for several years. A long-term measurement has been performed in order to study the loss of transmission on a Benthos sphere as a function of the polar angle. At the equator, the fouling produces a loss of transmission of 1.5% after eight months of immersion, which is an upper limit of what the fouling is expected to be on the actual detector whose optical module axes will be oriented at a polar angle of 135 degrees.

Salinity and temperature (13.2$^0$) are extremely stable. The highest recorded value for the current speed is 17 cm/s (the average is less than 3 cm/s).

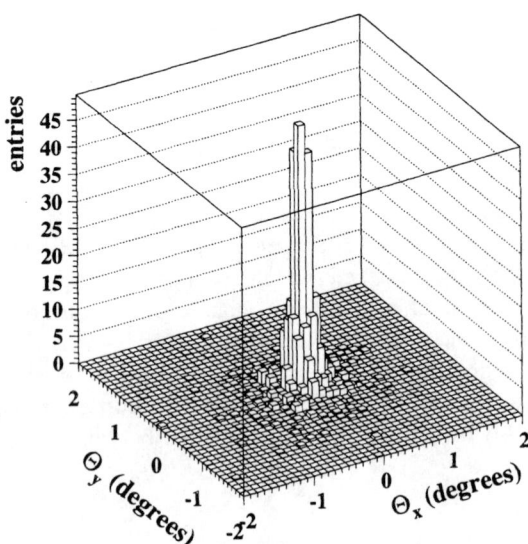

**FIGURE 3.** Angular resolution for a $1/E_\nu^2$ spectrum, for $\sigma_t = 1.3$ ns.

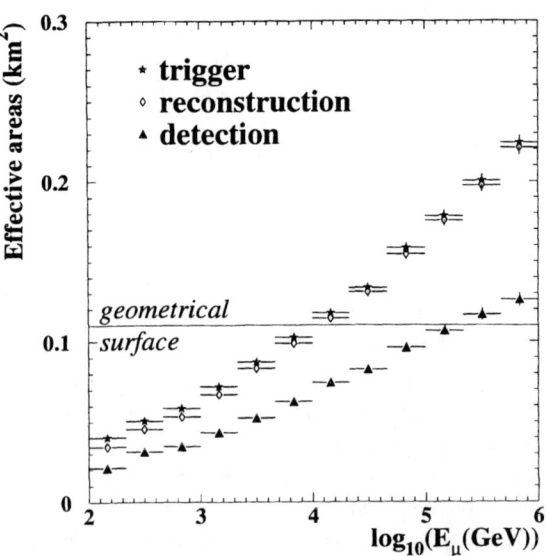

**FIGURE 4.** Effective area for triggering, reconstructed and selected events, averaged for muons coming from the lower hemisphere.

## DETECTOR PERFORMANCES

Special software packages for physics and detector simulation were developed in order to estimate the detector performances and to optimise its geometry for the interesting physics channels.

Low energy neutrinos (below a few hundred GeV for oscillation studies and from the annihilation of neutralinos) are contained, whereas high energy neutrinos (above 1 TeV from astrophysical sources) are completely dominated by the non-contained events, when the muon is generated in a neutrino charged current interaction far from the detector volume, but it has enough energy to be detected (and, eventually, to leave the detector).

All simulated events include random hits from $^{40}$K in the sea water. Special filters have been developed to remove them. The string deformation due to the current has also been simulated. The event reconstruction is made by using the position and the recorded time for each hit photo-multiplier.

The angular resolution for high energy neutrinos with a $1/E_\nu^2$ spectrum and for $\sigma_t = 1.3$ ns is shown in figure 3: supposing a point source of neutrinos, half of the reconstructed and selected events would point back at less than 0.2 degrees from the real neutrino source. Above 1 TeV the muon energy can be estimated by an algorithm which combines the reconstructed trajectory and the sum of the recorded pulse heights. The error on the energy is a factor of 3 (factor of 2) for energies below 10 TeV (above 10 TeV), which allows the reconstruction of the energy spectra. Below 100 GeV the muon energy is estimated by measuring its range.

### Astrophysics

Figure 4 shows the effective area for the events which trigger the detector, for those which are reconstructed and those which satisfy the selection criteria. The surfaces are averaged over the whole lower hemisphere and the error bars are only statistical.

Different models predict diffuse neutrino fluxes from active galactic nuclei (AGN) which could dominate the atmospheric neutrino flux at energies above 10-100 TeV. The event rates expected for these neutrino fluxes have been estimated for events which satisfy our selection criteria and are given in table 1 for 10 TeV and 100 TeV thresholds on the reconstructed energy.

The detection of point-like sources of neutrinos does not require a high energy threshold; the good angular resolution reduces the atmospheric background to a very low level, as soon as a small angular region of the sky is selected. The search for neutrinos from sources of gamma-ray bursts uses the narrow space and time window from the gamma ray information. With the flux given on Ref. (12) and the effective detector area shown in figure 4, the expected rate is $\sim 10$ signal events per year with a background rate of less than 0.001 events per year.

### Neutrino oscillations

The principal source of neutrinos for neutrino oscillation studies is the decay of charged pions produced by cosmic rays interacting in the Earth's atmosphere.

**Table 1.** Number of accepted muons reconstructed as upward-going above the estimated energy $E_\mu^{rec}$ per year.

| Model | $E_\mu^{rec} \geq$ 10 TeV | 100 TeV |
|---|---|---|
| Atmospheric: | | |
| ATM (8) | 68±13 | 0.8±0.1 |
| Generic AGN models: | | |
| SDSS (2) | 251±12 | 134±10 |
| NMB (9) | 217±9 | 64±4 |
| Blazars: | | |
| PRO (10) | 34±2 | 21±2 |
| MRLB (11) | 7.8±0.4 | 2.6±0.3 |

Charged-current interactions of $\nu_\mu$ producing upward-going muons with energies below 100 GeV are selected so that the background from muons produced in the atmosphere is absorbed by the Earth. The analysis carried out to estimate the sensitivity of the proposed experiment to the neutrino oscillation parameters $\Delta m^2$ and $\sin^2 2\vartheta$ is based in partially contained events. The number of signal events was calculated using the presently known values for the flux of atmospheric neutrinos an the expected experimental acceptance. The neutrino fight distance $L_\nu$ is related to the zenith angle of the track, $\theta$, by $L_\nu = L_0 \cos\theta$, where $L_0$ is the diameter of the Earth, 12740 km.

The histogram in figure 5 shows the distribution of accepted events as a function of $E_\mu/(L_\mu/L_0)$, where $E_\mu$ is the energy and $L_\mu$ the oscillation length corresponding to the reconstructed muon tracks, in absence of oscillations. The statistics corresponds to three years of data taking. The "data" points in the same figure represent the same simulated events, but in an oscillation scenario where $\Delta m^2 = 3.5 \cdot 10^{-3}$ eV$^2$ and $\sin^2 2\vartheta = 1$, i.e. weighting each event with the survival probability $P = 1 - \sin^2 2\vartheta \cdot \sin^2(1.27 \Delta m^2 L_\nu / E_\nu)$ (with the true neutrino values $E_\nu$ and $L_\nu$).

To explore the $\Delta m^2$ and $\sin^2 2\vartheta$ parameter space and to simulate different data-taking experiments, a series of three-year simulations were carried out for different points in the parameter phase-space, each one producing a distribution of "data" points like the ones in figure 5. Each of these distributions were compared to a high statistics simulation without oscillations. Figure 6 shows the region of parameters where neutrino oscillations could be excluded at 90% confidence level in 80% of simulated experiments (solid curve), as well as the region where both $\Delta m^2$ and $\sin^2 2\vartheta$ could be measured with an error less than 33% in more than 80% of the experiments (dashed curve).

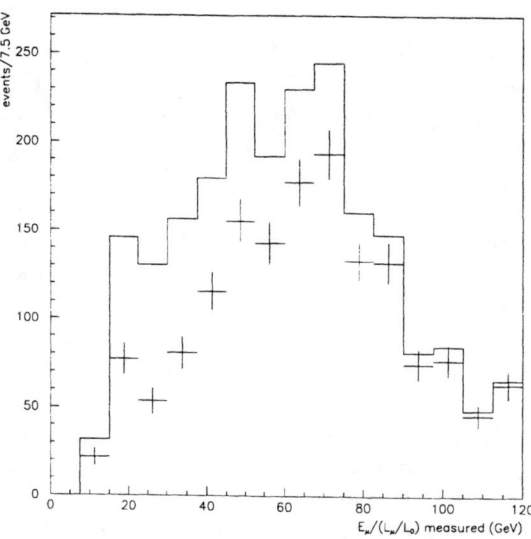

**FIGURE 5.** Simulated number of events in three years with no oscillations (histogram) and with oscillations for $\Delta m^2 = 3.5 \cdot 10^{-3}$ eV$^2$ and $\sin^2 2\vartheta = 1$ (points).

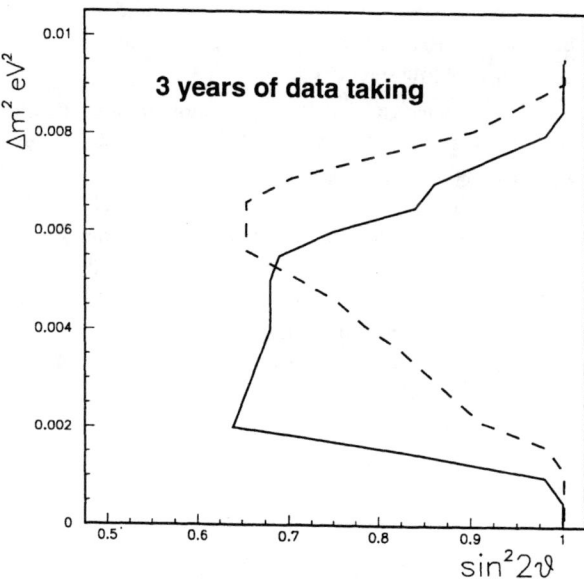

**FIGURE 6.** For three years of data taking, neutrino oscillations could be excluded at 90% confidence level in 80% of simulated experiments in the region of the right of the solid curve. With the same statistics, the oscillation parameters $\Delta m^2$ and $\sin^2 2\vartheta$ can be measured with fractional error less than 33% in more than 80% of experiments in the region to the right of the dashed curve.

The measurement region is shifted up in $\Delta m^2$ compared to the exclusion region because of the low-energy cut-off in the acceptance. A range of $E/L$ values around the minimum of the survival probability is needed to find the precise position of the minimum and make an accurate determination of $\Delta m^2$, whereas a measurement of a number of events at the calculated position of the minimum is sufficient for exclusion.

It is important to notice that the present analysis is exploratory and can be refined in many ways. The $\nu_\mu$ quasi-elastic events (estimated to about 15% of the charged current $\nu_\mu$ events) were not included in these simulations. Inclusion of these events (which have a better correlation between the produced muon and the parent neutrino energy than the deep-inelastic events used here) will improve the experimental sensitivity to low $\Delta m^2$. The backgrounds have so far not been taken into account in the analysis. However, making a known background subtraction with the magnitude of backgrounds presently estimated would have little impact on the results. The backgrounds from $\nu_e$ and from non-contained $\nu_\mu$ are related to the atmospheric flux and are well simulated, so a background subtraction can be made. The present simulations indicate the backgrounds from muons originating in the atmosphere are small. These simulations must be refined by including light diffusion in the water, and extended to higher statistics. The detector will have a significant acceptance for events above the horizontal, and these events can be used to control the muon background simulations for events below the horizontal.

The knowledge of the flux of atmospheric neutrinos in the energy range from 10 GeV to 1 TeV is of crucial importance for the study of neutrino oscillations, because observed deviations from the expected flux could be interpreted as a possible oscillation signal. The comparison of the total flux shows variations of the order of 20% between the different calculations (8) (13) (14). This is likely to be the limiting systematic on neutrino oscillation measurements by ANTARES, and studies of theoretical calculations and archive data are under way to reduce it as much as possible.

## CONCLUSIONS

Since its beginning in Spring 1996, considerable efforts have been done by the ANTARES collaboration to design a deep underwater neutrino detector.

The scientific programme includes neutrino astronomy, indirect detection of non barionic dark matter and the search for neutrino oscillations. Extensive computer simulations have been pursued to optimise the detector geometry and to estimate its performances. Due to the very good angular resolution, the atmospheric neutrino background can be reduced to very low values allowing the detection of signals from AGNs and from sources of gamma ray bursts. On the other hand, the analysis of low energy events will allow the study of the neutrino oscillations in a region of $\Delta m^2$, $\sin^2 2\vartheta$ which fits with the recent values reported by the Super-Kamiokande experiment.

During the R&D phase of the project, the program of the site evaluation has shown that the quality of the environmental parameters of the chosen site is quite satisfactory to consider the construction of the next stage detector consisting of $\sim 1000$ optical modules covering a detector area of $\sim 0.1$ km$^2$.

A prototype string, with 8 photo-multipliers and connected to the shore, has been conceived to study the problems of deployment and operating of such a structure.

The very first strings would be deployed at the end of 2000 and the whole 0.1 km$^2$ detector completed by the end of 2003.

## REFERENCES

1. ANTARES proposal, ASTRO-PH–9907432
2. F.W.Stecker et al., *Phys. Rev. Lett.* **69** (1992) 2738
3. G.Jungman, M.Kamionkowski and K.Griest, *Phys. Rep.* **267** (1996) 195
4. J.Edsjö, *PhD Thesis*, "Aspects of neutrino detection of neutralino dark matter", University of Upsala (1997)
5. F.Fukuda et al., *Phys. Rev. Lett.* **81** (1998) 1562
6. N.Palanque-Delabrouille, proceedings ICRC 99, HE 6.3.20
7. F.Hubaut, *PhD Thesis*, "Optimisation et caracterisation des performances d'un télescope sous-marin à neutrinos pour le projet ANTARES", Université de la Méditerranée, Aix-Marseille II, CPPM-T-1999-01
8. L.V.Volkova, *Sov. J. Nucl. Phys.* **31** (1980) 1510
9. L.Nellen, K.Mannheim and P.L.Biermann, *Phys. Rev* **D47** (1993) 5270
10. R.J.Protheroe, ADP-AT-96-7 and ASTRO-PH–9607165
11. K.Mannheim, *Astrop. Phys.* **3** (1995) 295
12. E.Waxman and J.Bahcall, *Phys. Rev.* **D59** (1999) 023002
13. A.V.Butkevich, L.G.Dedenko and I.M.Zheleznykh, *Sov. J. Nucl. Phys* **50** (1989) 90
14. V.Agrawal, T.K.Gaisser, P.Lipari and T.Stanev, *Phys. Rev.* **D53** (1996) 1314

# Baseline Concept for a Precise Measurement of Atmospheric Neutrino Oscillation

M. Aglietta[a], M. Ambrosio[b], E. Aprile[c], G. Bologna[a,d],
M. Bonesini[e], G. Bencivenni[d], M. Calvi[e], A. Castellina[f],
A. Curioni[c], W. Fulgione[f], P.L. Ghia[f], C. Gustavino[g],
R.P. Kokoulin[h], G. Mannocchi[d,f], F. Murtas[d], G.P. Murtas[d],
P. Negri[e], M. Paganoni[e], L. Periale[f], A.A. Petrukhin[h],
P. Picchi[d,f,a], A. Pullia[e], S. Ragazzi[e], N. Redaelli[e],
L. Satta[d], T. Tabarelli de Fatis[e], F. Terranova[e],
A. Tonazzo[e], G. Trinchero[f], P. Vallania[f], B. Villone[f]

[a] *Dipartimento di Fisica, Università di Torino, Torino, Italy*
[b] *INFN, Sezione di Napoli, Napoli, Italy*
[c] *Physics Department and Columbia Astrophysics Laboratory, Columbia University, New York, NY 10027, USA*
[d] *Laboratori Nazionali di Frascati, INFN, Frascati, Italy*
[e] *Dipartimento di Fisica, Università di Milano Bicocca and INFN, Milano, Italy*
[f] *Istituto di Cosmogeofisica, CNR, Torino, Italy*
[g] *Laboratori Nazionali del Gran Sasso, INFN, Assergi, Italy*
[h] *Moscow Engineering Physics Institute, Moscow, Russia*

**Abstract.**
A high-density calorimeter, consisting of magnetized iron planes interleaved by Resistive Plates Chambers (RPCs, Ref. (1)), as tracking and timing devices, is a good candidate for a new experiment on atmospheric neutrinos. With 34 kt of mass and in four years of data taking, this experiment will be sensitive to $\nu_\mu \to \nu_x$ oscillation with $\Delta m^2 > 6 \times 10^{-5}$ and large mixing, covering the region suggested by the SuperKamiokande results. Moreover, the experimental method will enable to measure the oscillation parameters from the modulation of the $L/E$ spectrum ($\nu_\mu$ disappearance). For $\Delta m^2 > 3 \times 10^{-3}$ eV$^2$, this experiment can also establish whether the oscillation occurs into a tau or a sterile neutrino, by looking for an excess of muon-less events at high energies produced by upward-going tau neutrinos ($\nu_\tau$ appearance).

## INTRODUCTION

SuperKamiokande data (2) exhibit a zenith angle dependent deficit of muon neutrinos inconsistent with expectations based on calculation of the atmospheric neutrino fluxes and well explained in terms of a two-flavour $\nu_\mu \to \nu_x$ oscillation with maximal mixing and mass-square difference $5 \times 10^{-4} \text{eV}^2 < \Delta m^2 < 6 \times 10^{-3} \text{eV}^2$. The absence of a corresponding deficit in the electron neutrino fluxes and data from reactor experiments (3) suggest that muon neutrinos either oscillate into a tau neutrino or in a new sterile neutrino[1]. This interpretation, given its relevance, needs to be tested by an independent experiment with sensitivity to the same region of oscillation parameters and with enough redundancy to be able to prove, or disprove, that an observed anomaly in atmospheric neutrino fluxes be due to neutrino oscillations.

These requirements are fulfilled by an experiment on atmospheric neutrinos based on a large mass and high-density tracking calorimeter (4, 5, 6), which has the capability to reconstruct in each event the $L/E$ ratio of the neutrino path length to its energy. As formerly suggested in Ref. (7), an unambiguous signature of neutrino oscillations would be provided by detection of an oscillation pattern in the $L/E$ spectrum of atmospheric $\nu_{mu}$. This method ($\nu_\mu$ disappearance) has sensitivity to $\nu_\mu$ oscillations with $\Delta m^2 > 6 \times 10^{-5}$ eV$^2$ and large mixing. Moreover, the appearance of $\nu_\tau$ interactions at high energies can be searched for with the same detector to establish

---

[1] At the same time, the detection of significant anomalies in the atmospheric $\nu_e$ flux probably requires impracticable improvements in sensitivity after SuperKamiokande.

whether the $\nu_\mu$ oscillation occurs into a tau or into a sterile neutrino ($\nu_\tau$ appearance).

In this paper, the experimental method, the basic detector parameters and characteristics and its capabilities to detect oscillations of atmospheric neutrinos are reviewed[2]. With respect to previous studies, a higher sensitivity in the region of $\Delta m^2 > 3 \times 10^{-3}$ eV$^2$ is obtained by means of a magnetized iron detector[3].

## EXPERIMENTAL METHOD

For neutrino energies above 1.5 GeV, atmospheric neutrinos fluxes are to a good approximation up/down symmetric (9). At these energies and for $\Delta m^2 < 10^{-2}$ eV$^2$, neutrino oscillations would result in a modulation of the $L/E$ distribution of upward-going neutrinos, while downward-going neutrinos are almost unaffected by oscillations. Downward neutrinos can therefore provide a *near* reference source to which compare the $L/E$ distribution of upward-going neutrinos (*far* source), detected in the same apparatus. For upward neutrinos the path length $L$ is determined by their zenith angle as $L(\theta) = R_{Earth}(\sqrt{1 - k^2 \sin^2\theta} - k|\cos\theta|)$, where $k = R_{Earth}/(R_{Earth} + \Delta R) \simeq 0.995$ and $\Delta R$ represents an average neutrino production height in the atmosphere, while the reference distribution is obtained replacing the actual path length of downward neutrinos with the mirror-distance $L'(\theta) = L(\pi - \theta)$. The ratio $N_{up}(L/E)/N_{down}(L'/E)$ will correspond to the survival probability given by

$$P(L/E) = 1 - \sin^2(2\Theta)\sin^2(1.27\Delta m^2 L/E) \quad (1)$$

with $L$ in km, $E$ in GeV, $\Delta m^2$ in eV$^2$. A smearing of the modulation is introduced by the finite $L/E$ resolution of the detection method.

Three remarks:

i. the results obtained by this method are to a large extent insensitive to systematics arising from calculations of atmospheric fluxes, neutrino cross sections and detector inefficiencies [4].

ii. the method does not work with neutrinos at angles near to the horizontal ($|\cos(\theta)| < 0.07$), since the path lengths corresponding to a direction and its mirror-direction are of the same order (mixing the *reference* and the *oscillated samples*).

iii. in case of $\nu_\mu \to \nu_{sterile}$ the oscillation pattern described by eq. 1 is distorted. Such a distortion can be taken into account and offers valuable information in separating tau and sterile scenarios. Strictly speaking, results presented here as "disappearance of muon neutrinos" are valid only in the tau scenario.

If evidence of neutrino oscillation from the study of $\nu_\mu$ disappearance is obtained, a method based on $\tau$ appearance can be used to discriminate between oscillations $\nu_\mu \to \nu_\tau$ and $\nu_\mu \to \nu_{sterile}$. This method consists in measuring the upward/downward ratio of muon-less events as a function of the neutrino energy. Oscillations of $\nu_\mu$ into $\nu_\tau$ would in fact result in an excess of muon-less events produced by upward neutrinos with respect to muon-less downward. Due to threshold effects on $\tau$ production this excess would be important at high energy. Oscillations into a sterile neutrino would instead result in a depletion of upward muon-less events. Discrimination between $\nu_\mu \to \nu_\tau$ and $\nu_\mu \to \nu_{sterile}$ is thus obtained from a study of the ratio of upward to downward muon-less events as a function of the energy. Because this method works with the high energy component of atmospheric neutrinos, it becomes effective for $\Delta m^2 > 3 \times 10^{-3}$ eV$^2$.

## THE DETECTOR

The outlined experimental method requires that the energy $E$ and direction $\theta$ of the incoming neutrino be measured in each event. The latter, in the simplest experimental approach, can be estimated from the direction of the muon produced in the $\nu_\mu$ charged-current interaction. The estimate of the neutrino energy $E$ requires the measurement of the energy of the muon and of the hadrons produced in the interaction. In order to make the oscillation pattern detectable, the experimental requirement is that $L/E$ be measured with a FWHM error smaller than half of the modulation period. This translates into requirements on the energy and angular resolutions of the detector. As a general feature the resolution on $L/E$ improves at high energies, mostly because the muon direction gives an improved estimate of the neutrino direction. Hence, the ability to measure high momentum muons (in the multi-GeV range), which is rather limited in the ongoing atmospheric neutrino experiments, would be particularly rewarding.

These arguments led to consider in previous papers (5, 6) a large mass and high density tracking calorime-

---

[2] A much more comprehensive exposition and further developments of these ideas can be found in Ref.(8).

[3] The feature of a magnetized detector of measuring the muon charge offers additional information on the oscillation mechanism, from studies of matter effects related to the propagation of upward-going neutrinos through the Earth. The potentiality of this latter method will not be further addressed in this paper.

[4] In the hypothesis of an *up/down symmetric* detector: requiring up/down symmetry instead of isotropy simpifies the structure of the detector.

ter as a suitable detector. A large mass is necessary to provide enough neutrino interaction rate at high energy, while the high density provided muon energy measurement by range. Here we consider a detector of the same structure as in Ref. (6), but with the addition of a magnetic field which improves muon acceptance at high momenta, and correspondingly efficiency at small $L/E$.

Thus, in the experiment simulation presented hereafter, a detector consisting of a stack of 120 horizontal iron planes 8 cm thick and $15 \times 30$ m$^2$ surface, interleaved by planes of sensitive elements has been considered. The sensitive elements (tracking devices) are housed in a 2 cm gap between the iron planes and provide two coordinates with a pitch of 3 cm. The detector has a total height of 12 m and a total mass exceeding 34 kt. The total surface of sensitive planes is 54,000 m$^2$; the number of read-out channels is 180,000. A magnetic induction of toroidal shape exceeds 1 T over most of the iron volume.

The elements of the sensitive planes should also enable to identify the flight direction of the incoming neutrino. In fact in the $\nu_\mu$ disappearance method, if the interaction vertex is not identified, the identification of the muon flight direction with high efficiency and high purity is required. This can be obtained by means of RPCs, given their time resolution of about 2 ns (1).

## DETECTION OF ATMOSPHERIC NEUTRINO OSCILLATIONS

As outlined in section 2, detection of oscillation of atmospheric neutrinos and measurement of their parameters will rely on two main techniques:

- disappearance of events with a high-energy muon pointing upward;
- comparison of rates upward and downward muonless events of high energy.

The first technique ($\nu_\mu$ disappearance) will test the hypothesis of $\nu_\mu$ oscillations and measure $\Delta m^2$; the second one ($\nu_\tau$ appearance) will be used to discriminate between oscillations into a sterile or a tau neutrino.

A full simulation of the experimental apparatus has been implemented. Neutrino interactions, according to the differential flux distribution predicted at the Gran Sasso, have been kindly provided by G. Battistoni and P. Lipari. Each interaction has been tracked in the detector using the GEANT 3.21 package.

The muon direction is obtained by a best fit procedure to the muon track, which accounts for effects of detector resolution, scattering and magnetic field. The muon energy is mainly determined by range for stopping muons, by track curvature for outgoing muons. The hadronic energy is estimated from the hit multiplicity in the calorimeter. The detector has a coarse hadronic energy resolution and essentially no capability of reconstructing the hadronic energy flow.

## Disappearence of muon neutrinos

Oscillation parameters are not known *a priori*, therefore a unique set of event selections and a unique analysis method have been defined in order to make the oscillation pattern detectable for every possible experimental outcome.

In order to select a pure $\nu_\mu$ charged current sample, only events with a reconstructed track corresponding to a muon of at least 1.5 GeV were retained in the analysis. This energy cut also insure a good up/down symmetry in absence of oscillations. In order to reject – in a real experiment – the background due to incoming muons, a further selection required the events to be either fully contained in a fiducial volume corresponding to about 85% of the detector, or to have a single outgoing track (muon) with a reconstructed range greater than 4 metres; in both samples the muon was required to hit at least seven layers. Further selections, based on the quality of muon track fit, and on event kinematics, were then applied in order to guarantee that the final sample had the required $L/E$ resolution (better than 50% FWHM) over the whole $L/E$ range (see Fig.1).

Altogether, these selections reduce the charged-current interaction rate of "unoscillated" downward muon neutrinos to about 7 kt$^{-1} \cdot$y$^{-1}$ (20% of the total rate of muon neutrinos above 1 GeV). The presence of a magnetic field, which allows to include in the sample events with an outgoing muon, increases by a factor 2 the acceptance for $L/E$ less than 300 km/GeV.

Three $L/E$ distributions obtained with the outlined selections are shown in Fig. 2.

The figures also show the discovery potential (allowed regions of the oscillation parameter space) of the experiment after four years of exposure, as derived from a fit to the $L/E$ spectra of a predictive curve folded with the detector resolution.

We also notice that if $\Delta m^2$ were larger than a few $10^{-2}$ eV$^2$, upward neutrinos – at large $L/E$ – would be in complete oscillation, while the oscillation pattern would become detectable in the downward sample. In this limit, a mirror distance $L'(\theta) = L(\pi - \theta)$ can be assigned to upward neutrinos, which can be used as a reference $L/E$ distribution for downward neutrinos. In this case, due to the uncertain estimate of the neutrino pathlength for downgoing neutrinos related to our ignorance

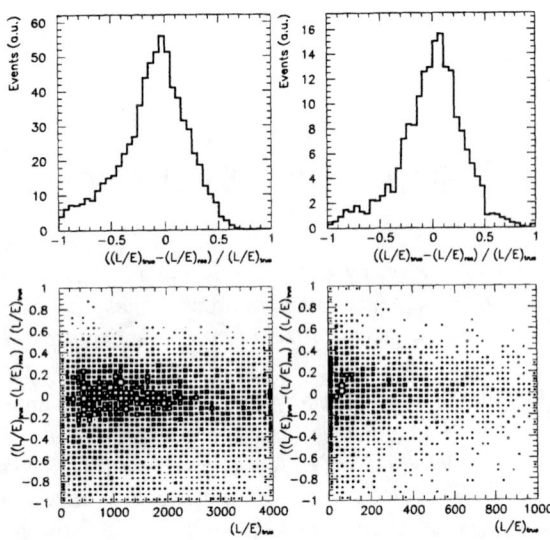

**FIGURE 1.** $L/E$ resolution for contained (left) and partially contained (right) events after the selections discussed in the text.

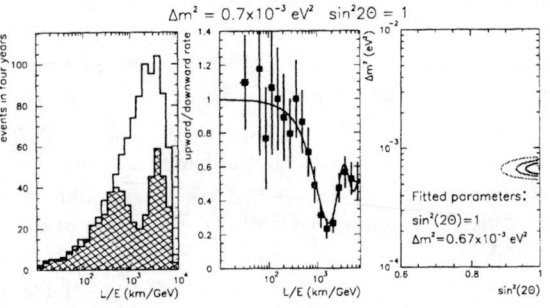

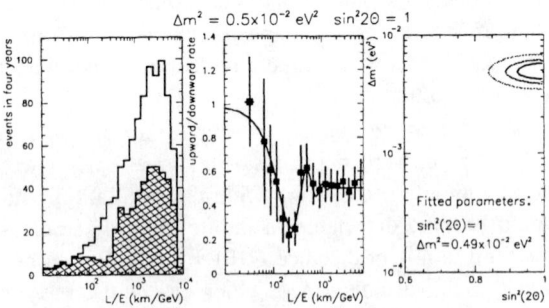

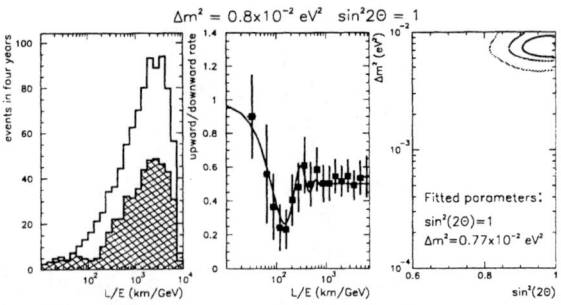

**FIGURE 2.** Results of the $L/E$ analysis on a simulated sample in presence of $\nu_\mu \to \nu_x$ oscillations, with parameters $\Delta m^2 = 7 \times 10^{-4}\ eV^2$, $\Delta m^2 = 5 \times 10^{-3}\ eV^2$ and $\Delta m^2 = 8 \times 10^{-3}\ eV^2$ and $\sin^2(2\Theta) = 1.0$. The figures show from left to right: $L/E$ spectra for upward muon events (hatched area) and downward ones (open area); their ratio with the best-fit superimposed (the first point is integrated over the first six bins) and the result of the fit with the corresponding allowed regions for oscillation parameters at 68%, 90% and 99% C.L.. Simulated statistics correspond to 25 years of data taking, rate normalisation, error bars and errors entering in the best fit procedure correspond to 4 years

of their production height in the atmosphere, there would be some model dependence in the determination of oscillation paramenters. Nonetheless, the observation of an oscillation pattern would still firmly test the oscillation hypothesis.

In absence of neutrino oscillations, these arguments can be used to exclude a region of oscillation parameters. The exclusion limits at 90% and 99% C.L. that this experiment will be able to set after an exposure of three years are shown in Fig. 3. A 2% systematic uncertainty in the knowledge of the up/down ratio of atmospheric neutrino fluxes has been assumed.

## Appearance of tau neutrinos

This method consists in measuring the upward/downward ratio of muon-less events, as a function of the visible energy. Charged-current $\nu_\tau$ interactions would in fact result in an excess of muon-less events in the upward sample at high energies, due to the large tau branching ratio into muon-less channels ($BR \simeq 0.8$). Moreover, because of threshold effect on tau production, events of large visible energy must be selected, in order to enhance the relative $\nu_\tau$ contribution to the muon-less event sample.

The selection of muon-less events is based on the ratio of the visible energy to the event length and on the recognition of a penetrating track, defined as muon candidate

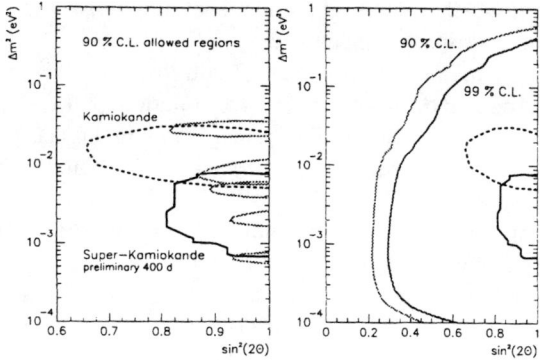

**FIGURE 3.** *Left*: Expected allowed regions of oscillation parameters after four years of exposure, as obtained by the analysis described in the text: the results of the simulation for $\Delta m^2 = 0.7, 2, 5, 8, 30 \times 10^{-3}$ eV$^2$ and maximal mixing are shown. *Right*: Exclusion curves at 90% and 99% C.L. after three years of data taking assuming no oscillations. The full (dashed) black line shows the results of the Super-Kamiokande (Kamiokande) experiment.

of energy greater than 1.5 GeV (taken from simulation truth). The sample of muon-less candidates selected by this criteria has a slightly larger contamination of $\nu_\mu$-CC events than the muon-less sample of Ref. (6), based on hand scanning selection[5]. In addition, events with less than 5 layers fired and visible energy less than 4 GeV have been rejected.

These selections have an efficiency on $\nu_\tau$ interactions followed by muon-less tau decays of about 60%, while the purity of the upgoing muon-less sample, assuming perfect up/down discrimination, is about 25%. The $\nu_\mu$-CC and $\nu_e$-CC background accounts for about 10% and 25% of the sample, while the remaining events are genuine NC interactions, which also carry useful information for the $\nu_\tau/\nu_{sterile}$ discrimination.

The up/down discrimination algorithm is not yet as effective as the result of hand scanning and optimisation is in progress. After rejection of events near the horizontal with ambiguous determination of the neutrino direction, the efficiency to muon-less $\nu_\tau$ interactions is reduced to around 40%. On average 85% of the $\nu_\tau$-CC events have their sense of direction correctly assigned.

Fig. 4 shows the result of the outlined analysis. The differential up/down asymmetry of muon-less events as a

---

[5] The comparison of visible energy and event length does not require pattern recognition, while hand scanning shows that muons of 1.5 GeV are easily visible and in the reach of the progresses in pattern recognition algorithms.

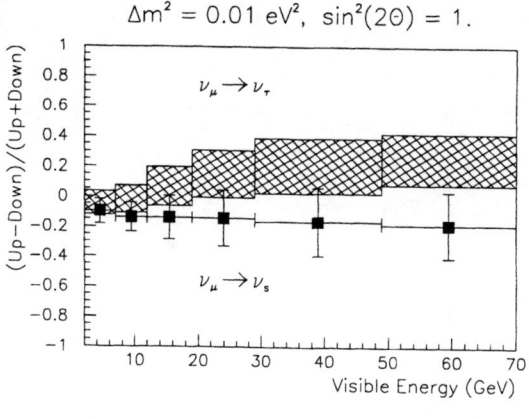

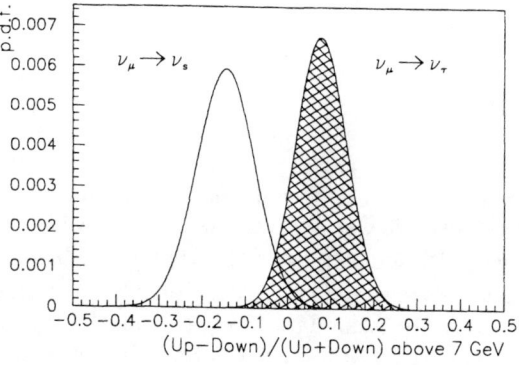

**FIGURE 4.** *Top*: Up/down asymmetry of muon-less events as a function of the visible energy, for maximal mixing and $\Delta m^2 = 10^{-2}$ eV$^2$. The expectations for $\nu_\mu \to \nu_\tau$ (hatched area) and for $\nu_\mu \to \nu_{sterile}$ oscillations (dots) are compared. Events have been generated with high statistics, error bars correspond to four years of data taking. The rightmost bin also integrates the contribution of events with reconstructed energy larger than 70 GeV. *Bottom*: Probability density function for the experimental outcome of the up/down asymmetry of events above 7 GeV after four years of data taking. Results are given for four years of data taking: the two alternative hypotheses are discriminated at 90% (95%) C.L. with a rejection power of about 1% (3%).

function of the visible energy, defined as

$$A(E) = \frac{U(E) - D(E)}{U(E) + D(E)}, \quad (2)$$

with clear meaning of symbols, is shown for maximal mixing and $\Delta m^2 = 10^{-2}$ eV$^2$ for the two alternative oscillation hypotheses (top). In the $\nu_\mu \to \nu_\tau$ case there is an excess of muon-less events with high visible energy from the bottom hemisphere due to the tau decay into muon-less channels that produce neutral current like events and, hence, a positive asymmetry; in the $\nu_\mu \to \nu_{sterile}$ case there is a lack of neutral currents from the bottom hemi-

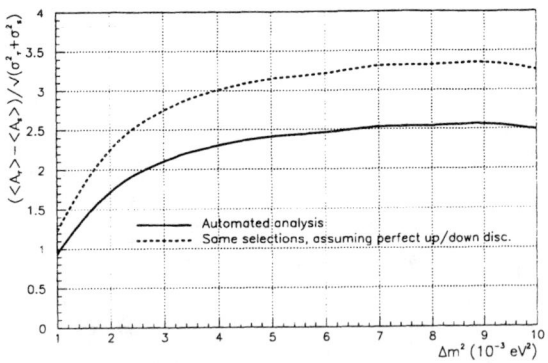

**FIGURE 5.** Significance of the $\nu_\tau/\nu_s$ separation in four years as a function of $\Delta m^2$.

sphere at all visible energies, for the sterile neutrino does not interact. At low energies a negative asymmetry is observed also in the $\nu_\mu \to \nu_\tau$ case. This is due to the residual background of $\nu_\mu$-CC events at small energies which gets depleted in the upgoing sample because of oscillations.

Figure 5 shows the significance (number of sigmas) of the $\nu_\tau/\nu_s$ separation in four years, defined as the difference between the expectation values of the asymmetry for the two hypotheses normalised to the expected error. Below $3 \times 10^{-3}$ eV$^2$ this method loses sensitivity, since it relies on high energy neutrinos which does not oscillate enough. If a perfect up/down discrimination is assumed, a separation of more than $3\sigma$ on the same sample of muonless events is obtained [6].

## CONCLUSIONS

A high density calorimeter of 34 kt with rough sampling and good tracking capability is a good candidate for a next generation experiment on atmospheric neutrinos. The detector has an estimated cost of about 20 MEuro and can be built in a short time.

In three years of data taking, this experiment will be sensitive to $\nu_\mu \to \nu_x$ oscillation with $\Delta m^2 > 6 \times 10^{-5}$ and mixing near to maximal and fully cover the region of oscillation parameters suggested by the SuperKamiokande results. Moreover, the experimental method will enable to measure the oscillation parameters from the modulation of the $L/E$ spectrum ($\nu_\mu$ disappearance) and to establish whether the oscillation occurs into a tau or into a sterile neutrino ($\nu_\tau$ appearance).

The major improvement with respect to SuperKamiokande relies on the exploitation of the high energy component of the atmospheric muon neutrino spectrum, which reflects in a better $L/E$ resolution in the range around $10^3$ km/GeV.

This experiment, completely devoted to the atmospheric neutrino study, offers a physics yield comparable to experiments designed for long-baseline neutrino detection with artificial neutrino beams. In fact, in the energy range considered here, atmospheric neutrinos are still an unexploited source of potential discovery, since they cover a wider $L/E$ range than long-baseline beams, and offer a unique possibility of precise determination of the oscillation parameters.

## REFERENCES

1. G. Bencivenni et al., *Nucl. Instrum. Methods* A **345**, 456 (1994)
2. SuperKamiokande Collaboration, Y. Fukuda et al., *Phys. Rev. Lett.* **81**, 1562-1567 (1998)
   SuperKamiokande Collaboration, Y. Fukuda et al., *Phys. Lett.* B **436**, 33 (1998)
   SuperKamiokande Collaboration, Y. Fukuda et al., *Phys. Lett.* B **433**, 9 (1998)
3. M. Apollonio et al., *Phys. Lett.* B **420**, 397-404 (1998)
4. G. Mannocchi et al., CERN/OPEN-98-004
5. A. Curioni et al., hep-ph/9805249
6. M. Aglietta et al., CERN/SPSC 98-28 SPSC?M615, Oct. 1998
7. P. Picchi and F. Pietropaolo, "Atmospheric Neutrino Oscillations Experiments",ICGF RAP. INT. 344/1997, Torino 1997, (CERN preprint SCAN–9710037).
8. MONOLITH collaboration,LNGS-LOI 20/99, CERN/SPSC 99-24, 26/08/96
9. P. Lipari, T. K. Gaisser and T. Stanev, astro-ph/9803093
10. OPERA collaboration, CERN-SPSC-99-20; SPSC-M-635; LNGS-LOI-19-99, ICARUS and NOE collaborations, CERN-SPSC-99-25; SPSC-P-314; INFN-AE-99-17

---

[6] It is fair to conclude this paragraph saying that an eventual signal of tau appearance obtained by the proposed experimental searches with a long-baseline neutrino beam (10) would be considered a much more significant evidence of $\nu_\mu \to \nu_\tau$

# Radio detection of ultra high energy neutrinos ($E_\nu > 10^{18}$ eV).

D. Seckel

*Bartol Research Institute, University of Delaware, Newark DE 19716, USA*
*E-mail: seckel@bartol.udel.edu*

**Abstract.**
A summary is given of the particle and astrophysics motivations to search for ultra high energy cosmic neutrinos. The general requirements for detector mass and geometry are discussed, along with an overview of possible techniques. The RICE (Radio Ice Cherenkov Experiment) technique is discussed in greater detail, including current efforts taking place at the South Pole.

## INTRODUCTION.

It is widely anticipated that high energy neutrinos are produced in nature(1), either as a result of the same processes that make high energy cosmic-rays, or as a by product of the interactions of those cosmic rays with intervening target matter. It follows that their detection would be of wide interest to both astrophysicists and particle physicists.

The prime astrophysical motivation for measuring the cosmic ultra high energy (UHE) neutrino flux is the potential to shed light on the source of UHE cosmic rays (UHECR) with energies above the GZK cutoff. There are two broad classes of source models for UHECRs. In astrophysical 'bottom up' models, ordinary matter is accelerated to high energies in a source, which may be compact (AGN, GRB's) or diffuse (termination shocks of AGN jets in the intergalactic medium). In compact models, cosmic rays are accelerated as protons, but escape as neutrons after a charge changing collision. Such interactions produce at least one charged pion. One may roughly estimate the flux of neutrinos as one per escaped nucleon, but at an average energy about a factor of 10 smaller than that of the nucleon, e.g. for each proton with energy above 10 EeV there will be roughly one neutrino with energy above 1 EeV.

In diffuse models, escape from the acceleration region occurs by diffusion, and neutrino production may be inefficient.

The other type of models are the top down models, in which UHECR's result from energy released in the decay or annihilation of supermassive particles, or via the evaporation of topological defects left over from the early universe. In these models neutrinos may be produced directly in the energy release process; for example as direct decay products of X bosons of a grand unified theory. Quark and gluon jets may arise at the same level, but in this case nucleon production would be only a few percent of meson production. Thus, in such models for each nucleon observed, one expects a larger number of neutrinos and they may be higher in energy as well.

Whatever the source of UHECR's, once produced they will propagate through the intergalactic medium. Specifically, they pass through the cosmic microwave background, and for energies much greater than 10 EeV this implies interactions with CMBR photons via the Griessen, Zatsepin, Kuzmin (GZK) process. This causes energy loss for the nucleons in a way that necessarily produces $\pi$'s and neutrinos of energy approximately one tenth that of the GZK cutoff energy.

These ideas can be summarized by forming the ratio of $\nu$ flux to nucleon flux,

$$R(E) = \frac{\int_E \phi_\nu(E')dE'}{\int_{10E} \phi_p(E')dE'}. \quad (1)$$

In general, one may separate neutrinos produced in the source from those produced in propagation and write $R = R_p + R_s$. For $E > 1$ EeV, we may expect $R_p = 1$, corresponding to GZK neutrino production. Whereas the behavior of $R_p$ is insensitve to the source model, $R_s$ may be used as a diagnostic. For compact astrophysical sources where escape is dominated by charge changing collisions, $R_s \simeq 1$. In diffuse astrophysical models, $R_s \ll 1$. In top-down models, $R_s \gg 1$. Thus, by observing the flux of UHE neutrinos and comparing to the UHECR flux, one may distinguish between the different source models.

As given, $R$ is determined at the source, but if one desires the ratio at Earth one must account for propagation effects. Above the GZK cutoff, one can only see UHECR's from sources that are closer than $\sim 10$ Mpc, approximately 1% of the column depth to the horizon.

Thus, for these super high energies, the diffuse flux of UHECRs is suppressed by about a factor of 100.

It is straight forward to list some of the particle physics applications of UHE neutrino detection. Cosmic neutrinos would provide a very long baseline for neutrino oscillations. If neutrinos could be associated with a gamma ray burst, the timing information would provide new tests of relativity. Measurement of $\nu$ absorption by the Earth, as a function of nadir angle, would test models of $\nu N$ cross-sections at high energy. If top down models turn out to be correct, neutrino astronomy would provide a direct test of the ideas of grand unified models.

## DETECTING GZK NEUTRINOS

Since $R \gtrsim 1$, it is useful to plan UHEN detectors on the basis of observing GZK neutrinos which may be estimated in a fairly reliable way by comparing directly to the UHECR data. Required detector sizes are large, but since particle energies are also high, instrumentation of the detector may be fairly sparse and based only on the ability to detect the hadronic and electromagnetic cascades caused by UHEN interactions.

### Detector mass

Detecting neutrinos is a challenge due to the small cross-sections with matter. At high energies though, the cross-sections increase. Above 1 EeV the interaction path length for neutrinos is between 100 and 1000 km of water, i.e. the efficiency of a 1 km thick water detector is somewhat less than a percent. This is comparable to the 1% column depth suppression of the super GZK cosmic rays. It follows that in order to measure the ratio $R$, one needs comparable area detectors for both the UHECR's and for the UHEN's. Indeed, taking a conservative cosmological model for source evolution from Yoshida and Teshima(2), and utilizing the cross-sections of Ghandi, et al.(3), one estimates an event rate of $\sim 10^{-2}$ km$^{-3}$ yr$^{-1}$. To go beyond discovery level science, one may desire 100's of events per year, suggesting a detector area of approximately $10^4$ km$^2$, i.e. a detector comparable in size to the Pierre Auger cosmic ray observatories.

### Detector geometry and technology

At high energies (> 100 TeV) the Earth becomes opaque to neutrinos, and so the overall detector design for UHEN's should be one optimized to detect neutrinos incident from above, not below. The obvious possiblity is to take advantage of the optical properties of ice or deep ocean water. Indeed, to detect neutrinos in the PeV range this requires a detector with size of order 1 km$^3$. Efforts are underway at South Pole (AMANDA/ICECUBE) and in the Mediterranean (ANTARES, NESTOR). However, for EeV neutrino astronomy, instrumenting $10^4$ km$^3$ with phototubes is a dubious proposition. An alternative is to look for cherenkov radiation at radio frequencies, 100 MHz - 1 GHz, and this is the basis for the RICE (Radio Ice Cherenkov Experiment) projects described below. Another proposal is to look for 'horizontal' air showers, i.e. air showers initiated deep in the atmosphere but with slant depths that allow development before hitting the ground. Large areas can be monitored efficiently by atmospheric flourescence techniques or by large area ground arrays. The difficulty is that the column depth of the atmosphere is only equivalent to 10 m of water. Thus, instead of a $10^4$ km$^2$ array one needs a detector 1000 km on a side. To gain larger area, the OWL/AIRWATCH team proposes to utilize the air-flourescent technique from space. Lastly, an effort has been made to utilize the radio cherenkov technique by searching for pulses of radio emission from the limb of the moon(4).

In addition to total target mass, it is also important to have sufficiently dense instrumentation to allow energy and angular resolution on an event by event basis. Of the above technologies, this may favor RICE and air shower techniques.

### Interaction and flavor identification

UHE neutrino detection will occur through deep inelastic scattering from nucleons. Such processes produce an outgoing lepton of reduced energy, and a spray of hadronic particles in the recoil jet. At EeV energies it is plausible to separate neutral current (NC) events from charged current (CC) events. Specifically, visible energy in NC events comes solely from a hadronic cascade, whereas CC events are tagged by the charged lepton. Since the different flavor leptons produce distinctive event topologies, the neutrino flavor may be identified.

If the lepton is an electron, an electromagnetic cascade is produced. Above a few PeV (in water/ice) the radiation length increases as $\sqrt{E}$ by the Landau-Pomeranchuk-Migdal (LPM) effect. At EeV energies the EM cascade in $\nu_e$ charged current events is an order of magnitude longer than the hadronic cascade. For $\nu_\mu$'s, $dE/dx$ losses to pair production, bremsstrahlung and photoproduction over $\sim$ km would tag charged current interactions. For $\nu_\tau$, $dE/dx \sim 0.1$ that for a muon(5). One may also observe the $\tau$ decay, but at an EeV $\tau$'s have a decay length of some 50 km. Charged current $\nu_\tau$ events may be difficult to separate from neutral current events, but a fraction will have clean $\tau$ decay signatures.

## RICE

It has been proposed for many years that neutrino induced electromagnetic cascades in dense media may be detected via radio cherenkov emission. The basic mechanism relies on development of a charge asymmetry in the evolution of an electromagnetic cascade in matter. This occurs via upscattering of electrons from the medium into the cascade and annihilation of positrons on electrons in the medium, resulting in a net negative charge flowing relativistically with the cascade. The charge asymmetry is of order 20%(6). At shower max there is roughly one particle per 80 MeV of energy, so the net charge of the cascade is $Q \sim .25E/\text{GeV}$, where E is the energy in the cascade. For wavelengths large compared to its transverse dimensions, the cascade will radiate as a point particle moving relativistically through the medium. If the index of refraction is greater than 1, cherenkov radiation occurs with a very large source strength. These conditions are satisfied for radio frequencies in ice.

Peak power occurs for frequencies of a few GHz, but other considerations may lead to an experiment operating in the range of several hundred MHz. For the technique to be effective, the medium must be transparent. Salt water doesn't work, nor will rock with significant water content. Cold fresh water ice, however, is promising. The absorption lengths depend on frequency and temperature. Generally the colder the better. The best ice is the top 1500 m. Below that, proximity to the geothermal heat source warms the ice and radio transparency is poor. Above a GHz, opacity increases as well, but below a GHz the absorption length is greater than a km.

Since the radiating cherenkov path is finite in length, and the individual particles are not exactly parallel, the radiation pattern has some width around the cherenkov angle. These effects have been modeled extensively by Zas, Halzen and Stanev(6) (ZHS) for electromagnetic cascades with energy of 1 PeV. Those calculations were used by Frichter, Ralston and Mckay(7) to show that an extensive array of radio antennas could be used effectively to detect high energy astrophysical neutrinos. More recently, Zas and Muniz-Alvarez(8) have extended the calculations of ZHS to higher energy including LPM effects and also modeling hadronic cascades. Seckel and Frichter(9) have used these improvements to study the array parameters appropriate to detection of EeV neutrinos using the radio technique.

The result of these studies is that three versions of RICE are being considered. Basic RICE is an exploratory effort underway at South Pole, taking advantage of the infrastructure developed for AMANDA. RICE[3] would extend the RICE effort to the km$^3$ scale, possibly in conjunction with ICECUBE. X-RICE would deploy antennas on a spacing of 1 per km over $10^4$ km$^2$ to carry out an EeV neutrino astronomy program.

## RICE

The RICE collaboration has been active since 1995. The goal has been to deploy an array of antennas with a data acquisition system that allows one to distinguish neutrino induced radio pulses from other noise and background events. As of 1999, the array consists of 12 in ice receivers, 3 surface horns and 4 in ice transmitters that can be used for calibration and ice transparency studies. The DAQ requires at least four amplitude discriminated hits from 12 channels in order to form a trigger. Time delays are calculated for each trigger, and with four hits an event vertex can be generated. By doing a vertex reconstruction cut a real time veto can eliminate electronic noise pulses generated at the surface. The number of hits also helps lower the signal to noise threshold required on each channel. These systems are now in place and tested(10). The array sensitive volume is roughly .01 km$^3$ at 10 PeV. It is not likely that an array this small will detect sources of astrophysical neutrinos: a more pragmatic goal is technology demonstration and upper limits on some of the more optimistic models for PeV neutrinos.

## RICE$^3$

The second level of development for the radio technique would be an array occupying roughly 1 km$^3$. This is still not large enough to do EeV astronomy, but a strawman design including 330 antennas achieves PeV sensitivty. By co-locating with ICECUBE one expects a reasonable number of events seen in both detectors, providing confirmation of the events and validating both techniques. The dual observations would also allow for improved flavor identification, and source location. An important result would be an *in situ* calibration of the radio technique, an essential task before commiting the resources necessary for an EeV neutrino detector.

## X-RICE

The largest experiment, X-RICE, would consist of roughly 1 antenna per km$^2$ with a sensitive volume of $10^4$ km$^3$ for $E_\nu$ above an EeV. Compared to RICE[3], the higher threshold allows a much sparser array. As a test of the idea, Seckel and Frichter[9] simulated arrays with varying separation of the ice holes in which antennas would be deployed and the number of antennas per hole. An array spacing of one antenna per hole at 1/2 km spacing or four antennas per hole at one km spacing both gave roughly 1 km effective depth for the array, i.e. an array of $10^4$ km$^2$ would have an effective volume of $10^4$ km$^3$.

It appears that the design concept is valid in terms of sensitivity. More difficult issues concern logistics, data acquisition over such an extended area, trigger formation, operation in the polar winter, etc. Such an experiment would draw on the polar experiences of AMANDA/ICECUBE and large scale logistics and data acquisition techniques pioneered by AUGER.

## ACKNOWLEDGMENTS

This work was partially supported by DOE grant DE-FG02-91ER40626.

## REFERENCES

1. Gaisser, T.K., F. Halzen, and T. Stanev, *Phys. Rep.* **258**, 173 (1995).
2. Yoshida, S. and M. Teshiima 1993, Prog. Theor. Phys. 89, 833.
3. Gandhi, R., C. Quigg, H. Reno, I. Sarkevic 1998, Phys. Rev. D58 (093009).
4. Gorham, P.W., K.M. Liewer and C.J. Naudet 1999, Proceedings 26th Int. Cos. Ray Conf., Salt Lake City (astro-ph/9906504).
5. Reno, M.H., I. Sarkevic, and D. Seckel 1999, In preparation.
6. Zas, E., F. Halzen, and T. Stanev 1992, Phys. Rev. D 45, 362.
7. Frichter, G., J. Ralston and D. McKay 1996, Phys. Rev. D 53, 1684.
8. Alarez-Muniz, J. and E. Zas 1997, Phys. Lett. B 411, 218; Alarez-Muniz, J. and E. Zas 1998, Phys. Lett. B 434, 396.
9. Seckel, D. and G. Frichter 1999, Proc. 26th Int. Cos. Ray Conf., Salt Lake City.
10. Frichter, G. et al. 1999, Proceedings 26th Int. Cos. Ray Conf., Salt Lake City.

# Studies and Site Characterisation for a $km^3$ scale Underwater Neutrino Telescope in the Mediterranean Sea

G. Riccobene *

*Department of Physics, University of Catania and Laboratori Nazionali del Sud INFN, Italy*
*riccobene@lns.infn.it*

**Abstract.**
NEMO (NEutrino Mediterranean Observatory) is an INFN Collaboration aiming at a feasibility study of a $km^3$ scale underwater Neutrino Detector to be located in the Mediterranean Sea. The Project concerns, at the present stage, an R&D program for marine sites characterisation, low power data acquisition electronics and development of Montecarlo codes to study the best performing layout of the detector in collaboration with a team of engineering consultants to study the detector mechanical structure. The major effort of the collaboration was, up to now, the study of deep sea properties, in several sites close to the Italian Coasts. Measurements were performed together with oceanogragraphic Institutions (Consiglio Nazionale delle Ricerche, Osservatorio Geofisico Sperimentale) and the Italian Military Navy. Within the end of year 2000 NEMO is going to deploy part of a deep sea test station at 2000m depth.

## INTRODUCTION

The interest in the field of Neutrino Physics is continuously growing since many years. One of the most promising field for the Physics of the next decades seems to be Neutrino Astrophysics. Several theoretical models suggest that Extra-galactic accelerators, such as AGNs and GRBs, emit high-energy neutrinos through the production and decay of charged pions within their core (1) (2). The emitted neutrinos interact weakly with intergalactic matter and are not deflected during their journey towards the Earth. Therefore neutrinos could be used as probes in identifying astrophysical sources tracking back their trajectory. Unfortunately high energy neutrino fluxes reaching the Earth from cosmic sources are expected to be so faint ($< 10^{-15} cm^{-2} sec^{-1} sr^{-1}$) and the $\nu$ interaction cross section is so low ($< 10^{-35} cm^2$ at 1 $TeV$) that the detection of a sufficient number of events needs a large mass target. Calculations show that an apparatus suitable for neutrino astronomy should reach a detection area of about 1 $km^2$. Such a huge detector is usually called "$km^3$ Neutrino Telescope". An underwater(-ice) array of phototubes should efficiently detect Cherenkov light propagating from UHE muons produced by neutrino CC interactions in water or in within sea-bottom rocks. The track characteristics, energy and direction, can be reconstructed if good detector resolution and low background conditions are satisfied. The study of deep sea is, in this framework, as important as the detector characterization in a typical HEP experiment. Careful studies of candidate deployment sites must be carried out in order to identify the most suitable one. The Mediterranean Sea offers optimal conditions for locating an underwater neutrino telescope. Since 1998 NEMO Collaboration started a research program to characterize four sites close to the Italian Coasts that could be appropriate for the construction of a deep-sea high-energy neutrino detector. We have performed several measurements in the sites of interest to investigate deep sea water optical properties (light absorption and diffusion) and oceanographic properties: water temperature, salinity, biological activity, water current, and sedimentation rate.

## THE QUEST FOR A DEEP SEA SITE

The choice of a suitable place for the Telescope needs the knowledge of many parameters (3). The main characteristics required are: depth, proximity to the coast, good optical transmission, low sea-currents intensity, low biological activity and low optical background. The site has to be deep enough to filter out the background due to down-going cosmic muons. Some sites of the central Mediterranean Sea, have a depth of about 3500m: there the muon flux is reduced by 5 orders of magnitude relatively to Earth's surface. The site sea bottom has to be flat over a large area and far enough from canyons and

---
* for the NEMO Collaboration

mountains. These conditions make the deployment and positioning easier and reduce the impact of destructive phenomena such as seismic events and turbid currents. The water transparency is one of the most important parameters for the site selection since the detection efficiency is a function of light propagation in water. Sedimentation and biological fouling deposited upon optical modules affect light collection and reduce the global detector efficiency. The water currents have to be limited in intensity and stable in direction to avoid special requirements of the mechanical structure and to improve the easiness of detector deployment and positioning. Optical noise must be taken into account. Increasing the photon detection thresholds causes a reduction of the detector effective area. The previous requirements have to be matched with the strong constraint represented by the distance of the site from the shore. Proximity to the coast is fundamental for data and power transmission feasibility, deployment and maintenance operations.

Having in mind these fundamental requirements, the NEMO Collaboration has identified four areas, close to the Italian Coasts (see figure 1):

- $35°50'$ N, $16°10'$ E in the Jonian Sea, South-East of "Capo Passero",

- $39°05'$N, $13°20'$ E in the Tyrrhenian Sea, North-East of "Ustica" island,

- $39°05'$N, $14°20'$ E in the Tyrrhenian Sea, North of "Alicudi" island,

- $40°40'$N, $12°45'$ E in the Tyrrhenian Sea, South-West of "Ponza" island

Moreover, an interesting region for the deployment of a test site station has been identified $20 km$ East of the port of Catania.

## Light Propagation

In order to measure deep sea water optical properties, a compact size device is commercially available: the AC9 trasmissometer manufactured by Wetlabs (4). The AC9 performs absorption and attenuation (i.e. the combined effect of absorption and scattering) measurements independently, using two different light paths, in the range $412 - 715$ nm, spanning over nine different wavelengths ($412, 440, 488, 510, 532, 555, 650, 676, 715 nm$). During several cruises onboard the Research Oceanographic Vessel "*Urania*" we collected data in the four above mentioned sites and in the vicinity of Matapan Abyss (Latitude: $36°30'$ N, Longitude: $21°12'$ E, depth= $5005 m$). In

**Table 1.** Measured attenuation and absorption coefficients values at $\lambda = 440 nm$. At this wavelength the ligth trasmission in water reaches its maximum value.

| Site | attenuation $[m^{-1}]$ | absorption $[m^{-1}]$ |
|---|---|---|
| Ponza | $0.0351 \pm 0.0027$ | $0.0143 \pm 0.0023$ |
| Capo Passero | $0.0246 \pm 0.0022$ | $0.0136 \pm 0.0015$ |
| Matapan * | $0.0341 \pm 0.0019$ | |

* In Matapan site only attenuation measurements were carried out.

the deep sea measurement set-up the AC9 is connected to an oceanographic CTD (Conductivity, Temperature Depth) probe Ocean MK-317 manufactured by IDRONAUT (5). Using this set-up the sea water optical measurements can be carried out as a function of depth. In table 1 we quote the value of attenuation and absorption at 440nm (blue light) measured in Ponza, Capo Passero and Matapan in the depth interval between $3000 m$ and $3300 m$. This could be the useful interval for $km^3$ deployment. New measurement in Capo Passero and the values measured in Alicudi and Ustica are still under analysis. Quoted errors are calculated combining in quadrature statistical and systematic errors. Measurements show very good optical characteristics: absorption seems to be comparable (within errors) to quoted pure water values. In the same part of the spectrum the attenuation length is larger than 30 m reaching, in Capo Passero (Latitude: $36°09'$ N, Longitude: $16°20'$ E, depth= $3350 m$), the excellent value of $42 m$. This implies that blue light travels for more than $40 m$ in deep sea water with a probability equal to $\frac{1}{e}$ to be absorbed or scattered even for very small angles (the angular acceptance of the device in the attenuation channel is $0.7\ deg$). Using AC9 data, the scattering coefficient can be calculated by subtracting the absorption value from the attenuation value at each given wavelength. It is worth to mention that this information must be coupled to the knowledge of the scattered light angular distribution in order to estimate the effect of diffusion. A scattering angular distribution characterized by a forward peak slightly affects photons trajectory even if the scattering coefficient is large, on the contrary a wide angular distribution could seriously affect light transmission. The collaboration is developing an apparatus devoted to blue light diffusion measurements "in situ". We plan to deploy the device in the region of Capo Passero during year 2000.

## Deep Sea Currents

Deep sea current measurements are performed by means of chains of current meters moored to the sea

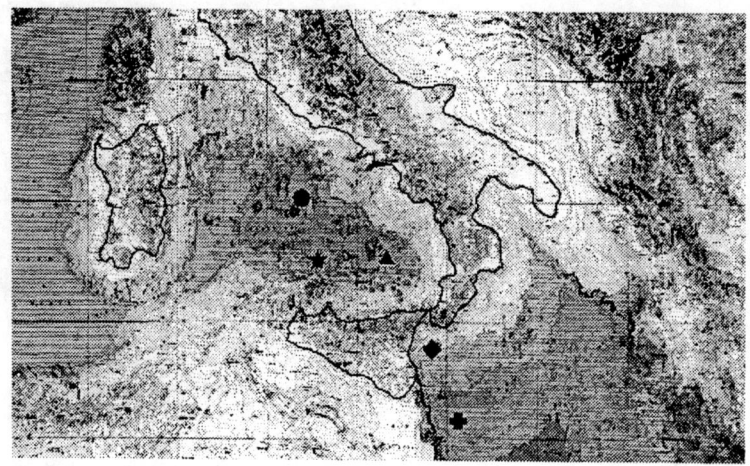

**FIGURE 1.** Location of the four sites selected by the NEMO Collaboration as candidates for the $km^3$ deployment: Alicudi (triangle), Ustica (star), Capo Passero (cross), Ponza (circle). The Catania "test site" point is also shown (diamond).

bottom and hanged vertically by the pulling of a buoy. The used instrument is the RMC8 manufactured by Aanderaa(6). Deep sea current intensity and direction have been measured in Capo Passero region during the period July 1998 - December 1999. From July 98 to August 99 current meters were positioned at about $3200m$ depth. Data show an average value about 3.6 $cm/s$ and RMS of 2.5 $cm/s$. The mean current flows from **SE** to **NW** at average angle of $8^o$ **NW** (see figures 2 and 3). The maximum value of the current intensity is never larger than about 10 $cm/sec$ and a seasonal dependence is observed. In August 1999 we moored another chain, $20km$ **SE** far from the previous one. Preliminary analysis shows that in this station, at same depth, currents are even lower. The measured low intensity values allow us, under the suggestions of expert engineers, to believe in the deployment feasibility and mechanical stability of an apparatus moored in that region.

## Biofouling and Sedimentation

The coverage of optical surfaces, by sediments, bacteria and other biological creatures living in deep sea environment can strongly reduce the light collection efficiency of the optical modules, as shown by BAIKAL (7) and ANTARES Collaborations (8). A commercial sediment trap is collecting data in Capo Passero ($3000m$ depth) since August 1999. Moreover in order to study bio-fouling effects the NEMO Collaboration constructed a deep Sea autonomous measurement station and deployed it, at $3200m$ depth in Capo Passero, in December 1999 (figure 4). The bio-fouling measurement system is composed by an array of 14 HAMAMATSU photodi-

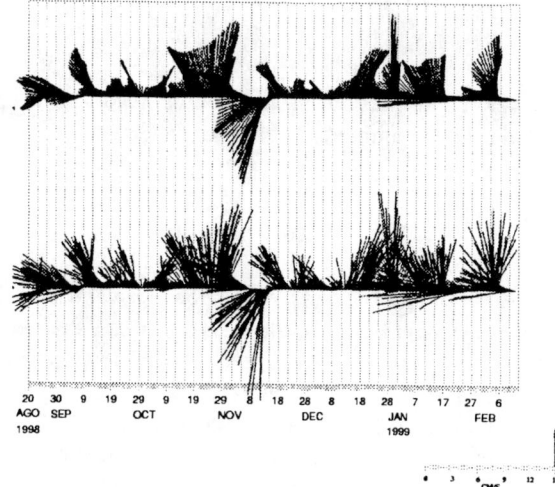

**FIGURE 2.** Stick diagrams, representing deep sea current velocity vector as a function of time, measured in Capo Passero from August 1998 to February 1999. Raw data are plotted in the lower part. In the upper line the tidal effect is filtered out. North direction and velocity scale are also shown.

odes coupled with custom pre-amplifier cards. The array is placed inside a $43cm$ diameter Benthos sphere, facing its inner surface. The PDs, positioned at different angles (ranging from upward vertical to downward vertical), detect the light emitted by an external blue LED. The expected reduction, day-by-day, of the PDs signals, with respect to a reference external light detector, measures the variation of the benthos surface transparency. Supplying instrumentations (current meter, CTD) monitor the deep-

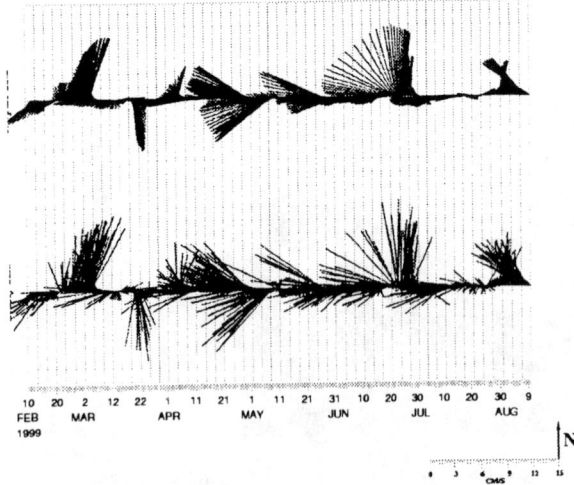

**FIGURE 3.** Stick diagrams of deep sea current (same as in figure 2) measured in Capo Passero from February 1998 to August 1999. Plots from OGS.

sea water conditions during acquisition. Data will be uploaded using a commercial acoustic modem.

## Sea Bottom Profiles

The bottom profile of a large region of several squared kilometres $50km$ SE in "Capo Passero" area, studied in collaboration with CNR, shows a smooth downhill profile descending from North-West (Lat: $36°00'$ N, Long: $15°50'$ E, depth= $3262m$) to South-East (Lat: $36°20'$ N, Long $16°10'$ E, depth= $3304m$) with a gradient of about 1%. A $10-20m$ deep dusty layer at ground overcomes the rocky bottom. This result may indicate a low deep sea activity and characterizes the region as a calm area. The thick layer of sediments can be also helpful for the mooring of mechanical structure.

## THE TEST SITE

The NEMO Collaboration is working to build, within two years, a deep sea test-site in proximity of the LNS (one of the four italian INFN Laboratories). The station, moored at $2000m$ depth, will be used to test optical modules, electronics, data transmission, electrical and optical connections. The Collaboration has already acquired the 20km long electro-optical submarine cable that will connect the shore to deep sea. A submarine splitter box will separate the cable into two branches. One branch will be devoted to NEMO and the other one to a seismo-

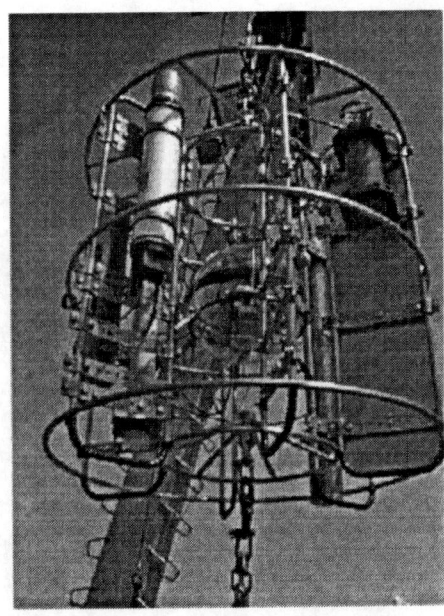

**FIGURE 4.** The deep sea station for biofouling measurements hanging from the crane of the R/V *"Urania"*. From left to right: the CTD probe, the LED houing, the Current Meter and the battery pack. The sphere is placed in the center of the structure.

logic monitoring station (GEOSTAR) operated in collaboration with ING (Istituto Nazionale di Geofisica, Italy). GEOSTAR will be connected to a network of sensors for the environmental and seismologic monitoring of the Eastern Sicily. Sea bottom profiles have been carried out in a large area in order to study the best access from the shore to the site, for the cable. Oceanographic and optical properties have also been measured in the same area.

## REFERENCES

1. Gaisser T. K., Halzen F., Stanev T., *Physics Reports* **258**, (1995) 173.
2. Halzen, F., *astro-ph* 9703004 (1997).
3. Capone, A., XXVI International Cosmic Ray Conference (1999), *e-proc* H.E. 6.3.20.
4. WETlabs inc., www.Wetlabs.com.
5. Idronaut s.r.l., www.Idronaut.it.
6. Aanderaa Instruments, www.aanderaa.com.
7. Wiebush, C., Ph.D. thesis, RWTH Aachen, Dec. 1995.
8. The ANTARES Collaboration, *A Deep Sea telescope for High Energy Neutrinos*, Proposal, May 1999.

# REACTOR AND ACCELERATOR NEUTRINOS

# Muon Storage Rings - Neutrino Factories

Z. Parsa*

*Brookhaven National Laboratory, Physics Department 510A, Upton, NY 11973-5000*
*E-mail: parsa@bnl.gov*

**Abstract.** The concept of a muon storage ring based Neutrino Source (Neutrino Factory) has sparked considerable interest in the High Energy Physics community. Besides providing a first phase of a muon collider facility, it would generate more intense and well collimated neutrino beams than currently available. The BNL-AGS or some other proton driver would provide an intense proton beam that hits a target, produces pions that decay into muons. The muons must be cooled, accelerated and injected into a storage ring with a long straight section where they decay. The decays occurring in the straight sections of the ring would generate neutrino beams that could be directed to detectors located thousands of kilometers away, allowing studies of neutrino oscillations with precisions not currently accessible. For example, with the neutrino source at BNL, detectors at Soudan, Minnesota (1715 km), and Gran Sasso, Italy (6527 km) become very interesting possibilities. The feasibility of constructing and operating such a muon-storage-ring based Neutrino-Factory, including geotechnical questions related to building non-planar storage rings (e.g. at 8° angle for BNL-Soudan, and 31° angle for BNL-Gran Sasso) along with the design of the muon capture, cooling, acceleration, and storage ring for such a facility is being explored by our growing Neutrino Factory and Muon Collider Collaboration (NFMCC). We present overview of Neutrino Factory concept based on a muon storage ring, its components, physics opportunities, Possible upgrade to a full muon collider, latest simulations of front-end, and a new bowtie - muon storage ring design.

## INTRODUCTION

Although many of the recent, exciting results in neutrino physics have been obtained by non-accelerator experiments, the neutrino mass and mixing parameters appear to require a new generation of accelerator based experiments. For this, an intense source of well-collimated neutrinos is needed.

Excitement is high in the accelerator physics community because atmospheric-neutrino results suggest that the long-baseline accelerator experiments such as MINOS [3], K2K [2], and NGS [4] should also find neutrino oscillations. Further, the LSND experiment that was conducted at a short-baseline accelerator facility, can be confirmed by future accelerator experiments such as MiniBooNE [5], ORLanD [6], and CERN P311 [7]. Moreover, physics associated with some interpretations of the solar-neutrino deficit may be accessible to studies in accelerator-based experiments, if neutrino-beam fluxes can be improved by 1-2 orders of magnitude.

To obtain a factor of 100 improvement in neutrino flux, the best prospect appears to be neutrino-beams derived from a muon-storage-ring, rather than from direct pion decays. However, such an approach requires considerable development before it can be realized in the laboratory. The idea of muon storage rings has been discussed since at least 1960 [8]. However, storage rings with enough circulating muons to provide higher intensity neutrinos than from conventional horn beams have only been considered more recently, in the context of muon collider technology [9].

The neutrino fluxes from the proposed muon-based beams would be higher than ever previously achieved with a much better-understood flavor composition. In addition, since the neutrino beams from these sources would be secondary beams from high energy muon decays, they would be extremely well collimated. Distances between production and detection could, therefore span the globe. Using the precisely known flavor composition of the beam, one could envision an extensive program to measure the neutrino oscillation mixing matrix, including possible CP violating effects.

A schematic concept of a Neutrino Factory Facility based on a muon storage ring, its components and physics opportunities are briefly discussed in section 2. A possible upgrade to a full muon collider is discussed in section 3. The examples described are based on some of the scenarios being explored by our Neutrino Factory and Muon Collider Collaboration (NFMCC),[10].

---

* Supported by US Department of Energy contract DE-AC02-98CH10886

# NEUTRINO FACTORY

A neutrino factory based on a muon storage ring is a challenging extension of present accelerator technology. Conventionally, neutrino beams employ a proton beam on a target to generate pions, which are focused and allowed to decay into neutrinos and, muons [3]. The muons are stopped in the shielding, while the muon-neutrinos are directed toward the detector. In a neutrino factory, pions are made the same way and allowed to decay, but it is the decay muons that are captured and used. The initial neutrinos from pion decay are discarded, or used in a parasitic low-energy neutrino experiment. But the muons are accelerated and allowed to decay in a storage ring with long straight sections. It is the neutrinos from the decaying muons (both muon-neutrinos and anti-electron-neutrinos) that are directed to a detector.

## Components

In a Neutrino Factory, a proton driver of moderate energy ($< 50$ GeV) and high average power,(e.g., 1-4 MW, similar to that required for a muon collider, but with a less stringent requirements on the charge per bunch and power is needed. This is followed by a target and a pion-muons capture system. A longitudinal phase rotation is performed to reduce the muon energy spread at the expense of spreading it out over a longer time interval. The phase rotation system may be designed to correlate the muon polarization with time, allowing control of the relative intensity of muon and anti-electron neutrinos. Some cooling may be needed, to reduce phase space, about a factor of 50 in six dimensions. This is much smaller than the factor of $10^6$ needed for a muon collider. Production is followed by fast muon acceleration to 50 GeV (for example), in a system of linac and two recirculating linear accelerators (RLA's), which may be identical to that for a first stage of muon collider such as a Higgs Factory. A muon-storage ring with long straight sections could point to one or more distant neutrino detectors for oscillation studies, and to one or more near detectors for high intensity scattering studies.

Figure 1 illustrates components of a Neutrino Factory based on a racetrack - shaped muon storage lattice [10]. Alternately a planar bowtie - shaped ring can be designed and oriented to send neutrino beams to any two detector sites. Since,there is no net bending, the polarization may be preserved. (A disadvantage of the Bowtie - shaped ring is that it may need extra bending. Since there is geometry constrains on the ratio of short to long straight sections, the ring circumference may increase.) With the ring in a tilted plane, both long straight sections would point down into the earth, such that neutrinos can be directed

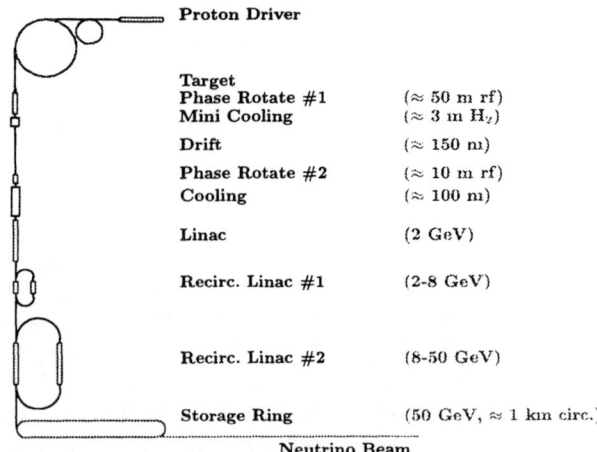

**FIGURE 1.** Overview of a Neutrino Factory Concept, with a Racetrack Muon - Storage Ring

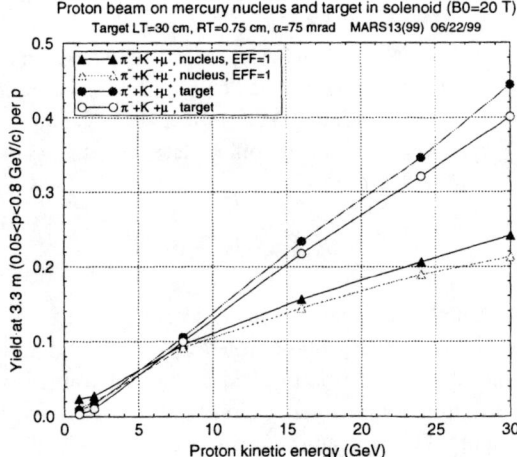

**FIGURE 2.** The number of pions produced per proton incident on a mercury target vs. proton energy. The yield at the target is shown by the circles, and the yield 3 m downstream of the target in a solenoid capture system is shown by the triangles.

into two very distant detectors. Triangular-shaped storage rings also have this advantage. In the following sections, a description of the targets, a simulation of target through cooling-channel and a new example of a bowtie-shaped muon storage lattice will be discussed.

## Driver

The number of pions per proton produced with an optimized system varies linearly with the proton energy, as shown in Fig. 2. Thus, the number of pions, and the number of muons into which they decay, is essentially proportional to the proton beam power. The total six-

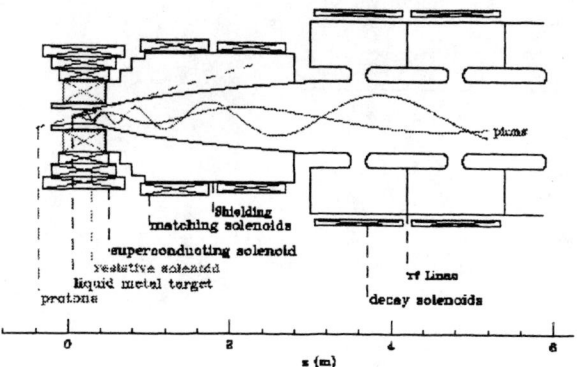

**FIGURE 3.** A Schematic of Targetry, Pion Capture, and beginning of Phase Rotation.

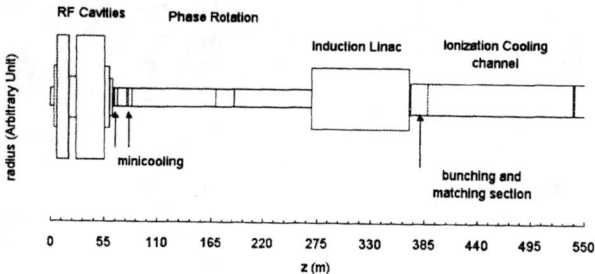

**FIGURE 4.** Schematics of the Muon Source from Target to Linac.

dimensional emittance of the produced muons depends on, e.g., the pion bunch length, and thus on the rms proton bunch length $\sigma_p$ if that length is longer than the characteristic decay process length $c\,\tau^{decay}$:

$$\tau^{decay} = \frac{1}{\gamma_\pi^2} \frac{(m_\pi - m_\mu)}{m_\pi} \tau_\pi, \qquad (1)$$

where $\tau_\pi$ is the pion lifetime and $\gamma_\pi m_\pi$ is the pion energy. The pion yield peaks at $E_\pi \approx 300$ MeV, with $\tau^{decay} \approx 1$ nsec. If the proton energy is low, this may imply a large tune shift:

$$\Delta v \propto \frac{n_p\,C}{\gamma_p^2\,\sigma_t\,\varepsilon_{transverse}} \propto \frac{n_p}{\gamma_p \langle B \rangle \sigma_t\,\varepsilon_{transverse}} \qquad (2)$$

in the proton ring before extraction, where $C$ is the circumference of the proton driver, $\langle B \rangle$ is the average bending field, and $\varepsilon_{transverse}$ is the transverse emittance of the protons. The above dependency favors a higher proton energy. The total six-dimensional emittance of the produced pions depends also on the number of proton bunches employed to fill the storage ring. This favors a smaller number of large proton bunches in the driver, and thus a larger tune shift. Table 1 presents possible parameters for proton drivers at BNL and FNAL. The target requirements are very similar to those for the muon collider, except the instantaneous shock heating is somewhat less because protons are distributed in a larger number of bunches. In the scheme presented here, it is assumed that the liquid mercury jet solution is used. The capture solenoid is likely to be the same as described in the muon collider status report [9]. Figure 3, shows the pion production target, solenoidal capture, decay channel and beginning of phase rotation. At the end of this first phase rotation stage, the bunch length increases by about a factor of 6 and the energy spread decreases by the same amount. Whether this first stage of phase rotation can be eliminated is being investigated.

## Target - Cooling Section

In this section a new integrated design for the Neutrino Factory front-end subsystem is described. Other designs and simulations are being explored by NFMCC. In the latest muon cooling simulation all the available subsystem simulations such as the target, pion capture and decay, phase rotation, matching, bunching and cooling were integrated together in such a way that the same particles generated at the target travels all the way to the end of the cooling channel. In this example [15], 16 GeV protons hits a carbon target, generates pions which then decay to muons. Figs. 4, Figs. 5 – Figs. 13, respectively, show schematics of Target to Linac Muon channel. (RF cavities are used for the 1st phase rotation and the Induction Linac for the second phase rotation). In addition the longitudinal and transverse phase distribution plots, at z=0 (target), z=370m (just before bunching), z=388m (just after bunching), and z=605m (after cooling) are shown. Particle composition in the target-to-linac channel is shown in Fig. 14, and in table 2. The muon emittance variation in the target-to-linac channel is shown in Fig. 15.

## Cooling and Acceleration

The challenges of further acceleration and storage of the muon beam will be substantially easier if we reduce the transverse phase area of the beam by an additional factor of 10. This may not be accomplished in a single step of ionization cooling, but involves alternating ionization cooling and rf acceleration, all in a magnetic channel. The acceleration from ~ 100 MeV to e.g., ~ 50 GeV is best accomplished in recirculating linacs with superconducting rf cavities, after which muons are injected into a muon storage ring. The desire for multiply directed neutrino beams with very small angular divergence may require a more novel design for the storage ring, with a plane that is far from horizontal. The R&D needs for a muon collider are very similar, but with additional challenges in cooling and storage ring design. At least four

**Table 1.** Example of parameters for various Proton driver scenarios at BNL and FNAL.

|  | $BNL_1$ | $BNL_2$ | $FNAL_1$ | $FNAL_2$ |
|---|---|---|---|---|
| Energy [GeV] | 24 | 24 | 16 | 16 |
| Power [MW] | 1 | 4 | 1 | 4 |
| Rep. Rate [Hz] | 2.5 | 5 | 15 | 15 |
| $p$'s/fill | $10^{14}$ | $2\ 10^{14}$ | $2.5\ 10^{13}$ | $10^{14}$ |
| Bunches | 6 | 6 | 4 | 4 |
| Circumference [m] | 807 | 807 | 474 | 474 |
| Bunch spacing [m] | 135 | 135 | 118 | 118 |
| $\sigma_t$ [nsec] | 1 | 1 | 1 | 1 |

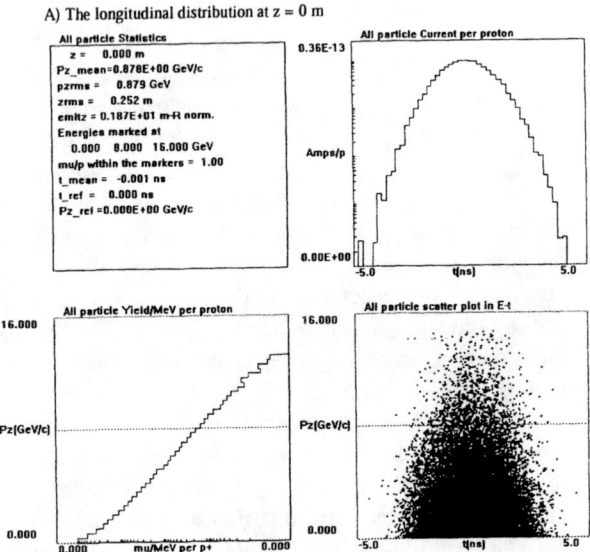

**FIGURE 5.** Longitudinal Phase distributions at z=0 (target). The scatter plot shows the distribution in $P_z$[GeV/c] vs t[ns], and the graphs above and to the left show the projection on to time and $P_z$ axis.

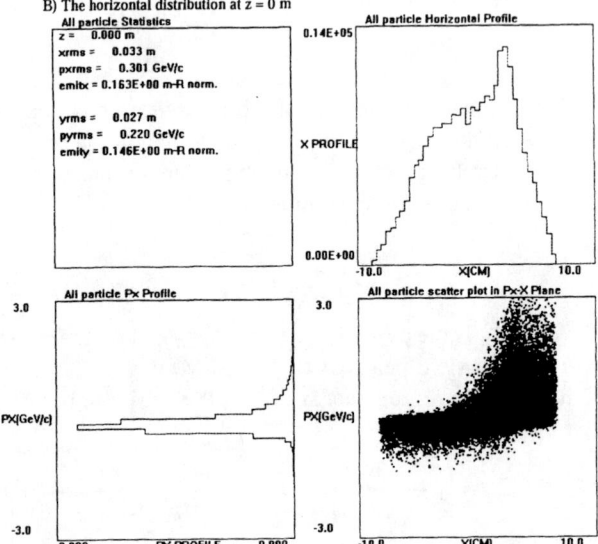

**FIGURE 6.** Horizontal Phase distributions at z=0 (target). The scatter plot shows the distribution in $P_x$[GeV/c] vs x[m] and the graphs above and to the left show the projection on to x and $P_x$ axis.

orders of magnitude more cooling (including continual exchange between transverse and longitudinal emittance) are required for a muon collider than a neutrino factory. Also, a different ring is needed to maximize collider luminosity than simply to hold the muons while they decay.

Figure 16 shows a schematic of Ionization Cooling concept. Ionization cooling that has been proposed involves passing the beam through an absorber in which the muons lose transverse- and longitudinal-momentum by ionization loss (dE/dx). The longitudinal momentum is then restored by coherent re-acceleration, leaving a net loss of transverse momentum (transverse cooling). The process is repeated many times to achieve a large cooling factor. The beam energy spread can also be reduced using

ionization cooling by introducing a transverse variation in the absorber density or thickness (e.g. a wedge) at a location where there is dispersion (the transverse position is energy dependent). Theoretical studies have shown that, assuming realistic parameters for the cooling hardware, ionization cooling can be expected to reduce the phase-space volume occupied by the initial muon beam by a factor of $10^5 - 10^6$. Ionization cooling is a new technique that has not yet been demonstrated. Special hardware needs to be developed to perform transverse and longitudinal cooling. It is recognized that understanding the feasibility of constructing an ionization cooling channel that can cool the initial muon beams by factors of $10^5 - 10^6$ is on the critical path to the overall feasibility of the

**Table 2.** Particle composition at various locations from Target to Linac, (with 16 GeV proton).

| location | Z[m] | $e^+/p^+$ | $\mu^+/p^+$ | $\pi^+/p^+$ | $k/p^+$ | $p^+$ | total/$p^+$ |
|---|---|---|---|---|---|---|---|
| Just after target | 0 | 0.000 | 0.000 | 1.000 | 0.000 | 0.000 | 1.000 |
| Just before minicooling | 62 | 0.009 | 0.407 | 0.057 | 0.000 | 0.000 | 0.472 |
| Just after minicooling | 80 | 0.003 | 0.334 | 0.031 | 0.000 | 0.000 | 0.367 |
| Just before bunching | 370 | 0.039 | 0.265 | 0.001 | 0.000 | 0.000 | 0.305 |
| Just after bunching | 388 | 0.000 | 0.222 | 0.001 | 0.000 | 0.000 | 0.224 |
| After cooling | 605 | 0.000 | 0.101 | 0.000 | 0.000 | 0.000 | 0.101 |

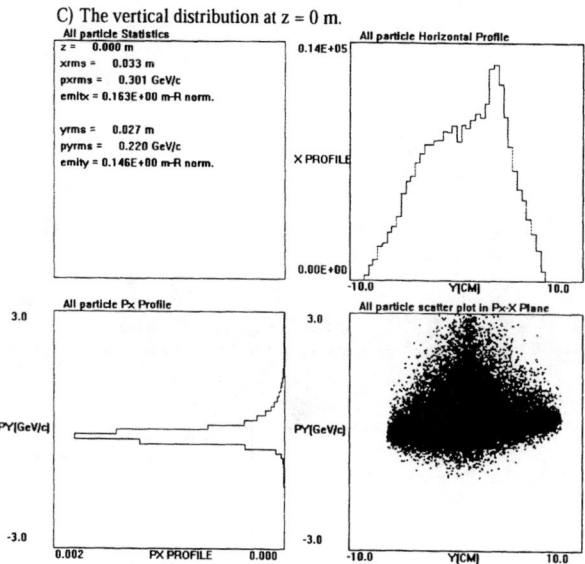

**FIGURE 7.** Vertical Phase distributions at z=0. The scatter plot shows the distribution in $P_y$[GeV/c] vs y[m] and the graphs above and to the left show the projection on to y and $P_y$ axis.

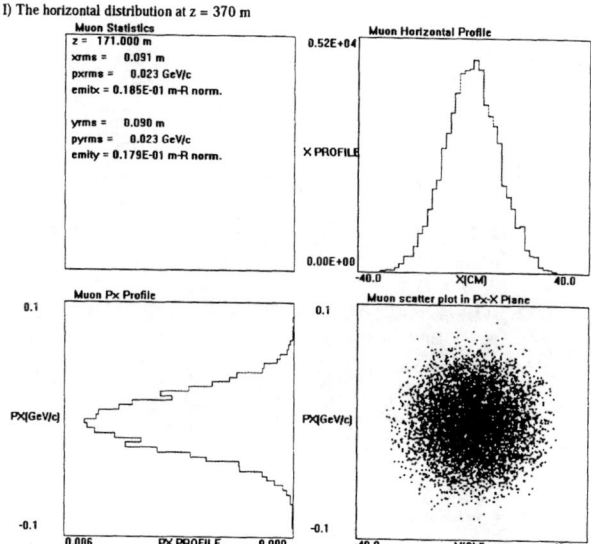

**FIGURE 9.** Horizontal Phase distributions, same as Fig. 6 but for z= 370 m (just before bunching).

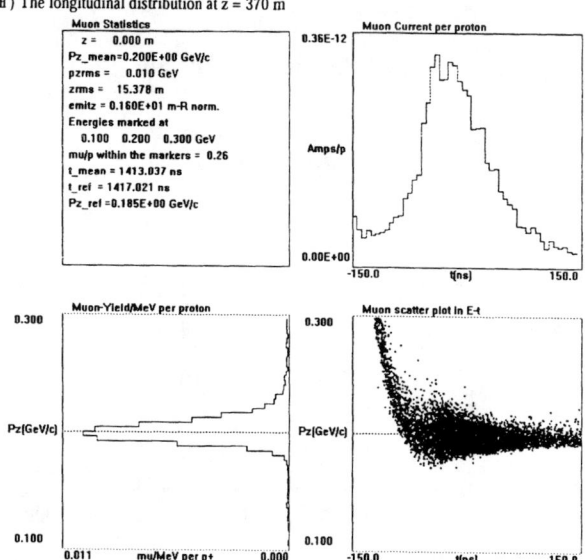

**FIGURE 8.** Longitudinal Phase distributions, same as Fig. 5 but for z= 370 m (just before bunching).

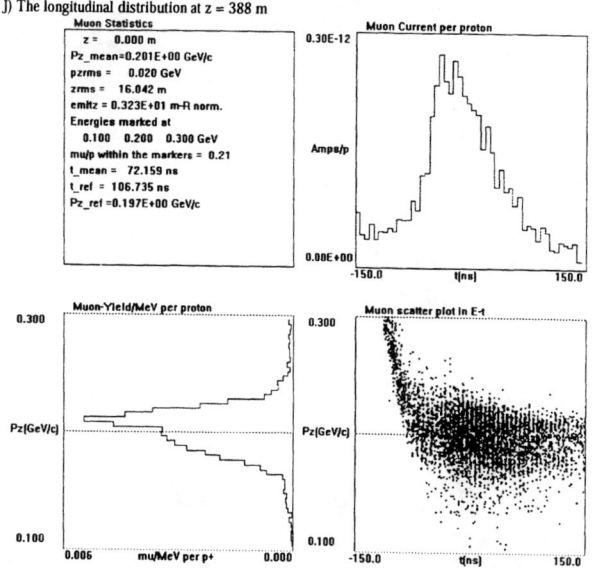

**FIGURE 10.** Longitudinal Phase distributions, same as Fig. 5, but for z= 388 m (just after bunching).

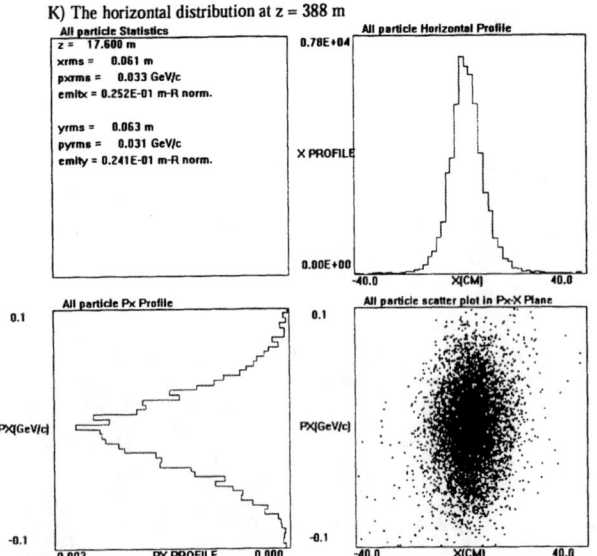

**FIGURE 11.** Horizontal Phase distributions, same as Fig. 6, but for z= 388 m (just after bunching).

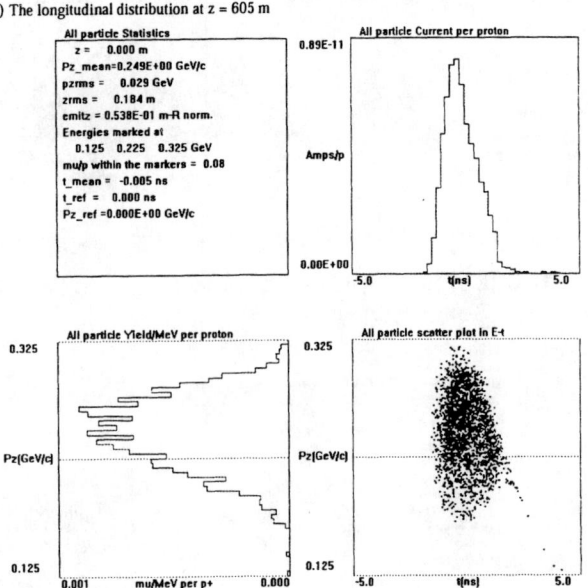

**FIGURE 12.** Longitudinal Phase distributions, same as Fig. 5, but for z= 605 m (after cooling).

muon collider concept. In Fig. 17, a schematic of the emittance exchange is shown.

## *Muon Storage Ring*

A muon collider requires as its starting point, a very intense beam of muons with a small momentum spread. Such beams would be accelerated to collider energies and be used to search for new short distance high energy

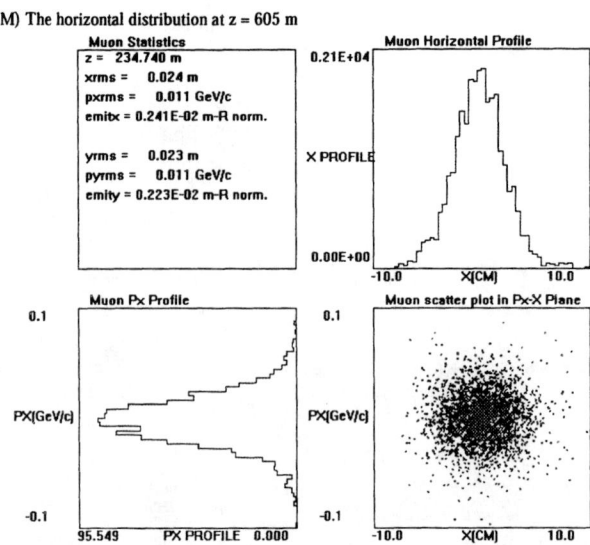

**FIGURE 13.** Horizontal Phase distributions, same as Fig. 6, but for z= 605 m (after cooling).

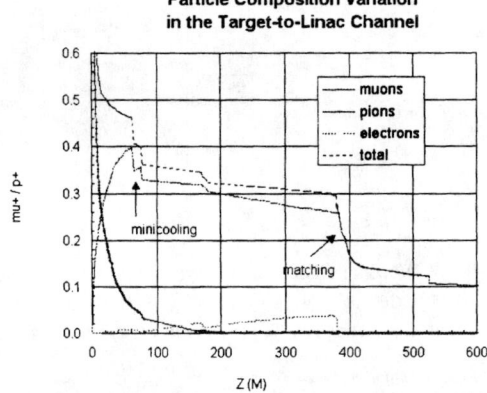

**FIGURE 14.** Particle Composition from Target to Linac.

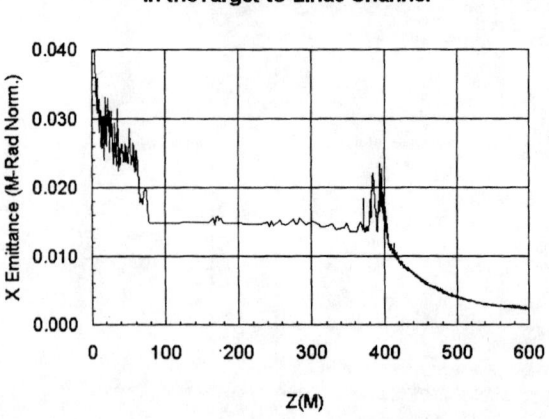

**FIGURE 15.** Muon emittance variation in Target to Linac channel.

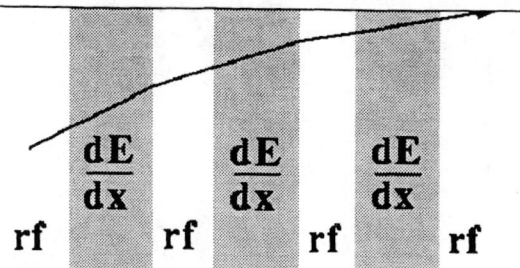

**FIGURE 16.** Schematic of Ionization Cooling concept (Ionization takes away momentum, and the RF acceleration puts momentum back along the z-axis, resulting in a Transverse Cooling).

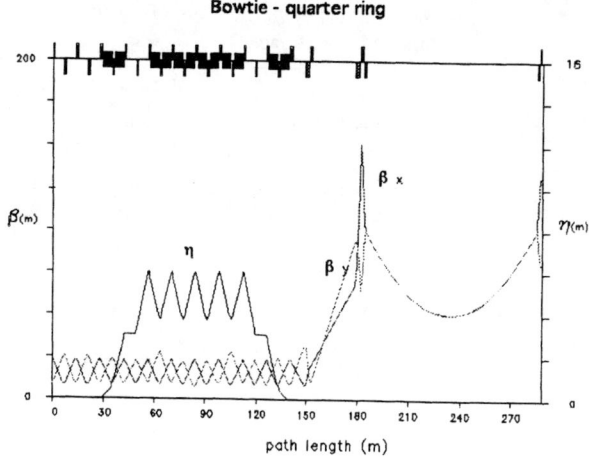

**FIGURE 18.** One Quarter of Bowtie MuonStorage Ring.

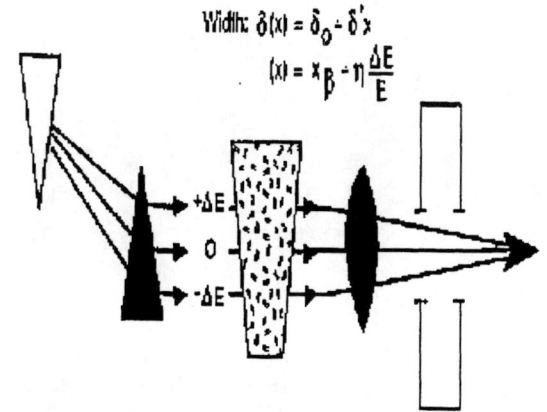

**FIGURE 17.** Schematic of the emittance exchange concept.

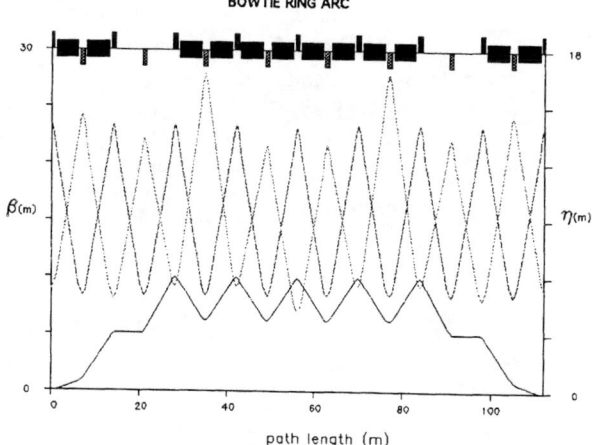

**FIGURE 19.** Bowtie-shaped Arc Lattice Functions.

phenomena. A neutrino factory based on a muon storage ring is a natural path to muon collider technology, since both facilities share essentially the same subcomponents prior to the storage ring. In previous sections we discussed some advantages and disadvantages of various shaped muon storage rings. Fig. 1 illustrated a racetrack - shaped configuration, with two long straight sections.

In this section a bowtie-shaped ring with a bypass is discussed. The planar ring can be designed and oriented to send neutrino beams to any two detector directions and with bypass(es) that could be added, to send beams to additional detector sites. In the bowtie-shaped lattice design [16], the lattice has two long-straight sections, two short-straight sections and two arcs. The description shown in Figs. 18 - 21 follows one quarter of the ring, starting at the center of the short straight section on the left side of the figure, and ends at the crossing point at the center of the bowtie. Table3 gives the parameters of the bowtie shaped ring. [Note, parameter optimizaton and additional lattice designs and simulations are being explored by NFMCC]. The short straight sections may be used for injection, and for RF. Half of the short straight section consists of two 14m arc cells without dipoles, and can be configured to provide (20m) free space for injection.

Each arc contains eight FODO cells, two without dipoles. There are 60 deg cell phase advances, and the dipole-free cells act as dispersion suppressors. Twelve 5m long dipoles each bend the beam by 10 deg, so the arc has 120 deg of bending. This amount of bending causes the long beam-lines to intersect at 60 deg (a typical angle, whose exact value depends on the selection of ring and detector sites). The long dispersion-free straight section provides a muon beam such that the decaying muons generate low divergence neutrinos. Two different configurations are shown in Fig. 20 and Fig. 21. In one, the long straight section has quadruples in the center (around the crossing point) making two beam waists, each with 50m

**Table 3.** Lattice Parameters for a Bowtie - Shaped Muon Storage Ring.

| | |
|---|---|
| Energy | 50 GeV |
| Circumference | 1150 m |
| $L_{short straight sections}$ | 20 m |
| $L_{of long straight sections}$ | 200 m |
| Dipole field | 5.82 T |
| $Gradient_{Maximum}$ | 30 T/m |
| Dipole length | 5 m |
| Arc cell length | 14 m |
| Cell phase advance | 60 deg |
| Ring tunes | 9.85, 9.23 |
| $Beta function_{Maxima}$: | |
| Arc | 28 m |
| Long straight sections | 100 m |
| Ring | 151 m |
| Dispersion: | |
| Maximum | 6 m |
| Minimum | -6 m |
| Momentum compaction | 0.025 |
| Chromaticity: | |
| Horizontal | 12.5 |
| Vertical | 11.5 |
| Beam crossing angle | 60 deg |

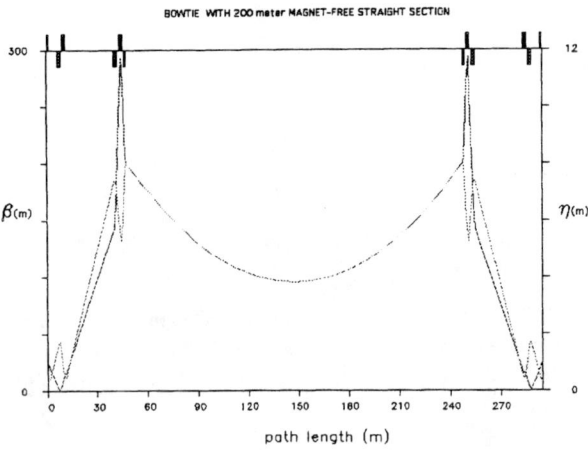

**FIGURE 21.** Lattice functions for Bowtie-shaped - Long Straight Section Alternative Configuration.

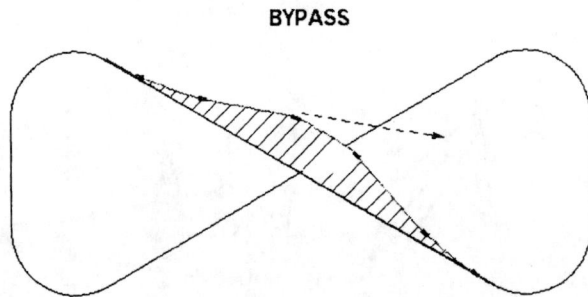

**FIGURE 22.** Lattice functions for Bowtie-shaped Ring with Bypass. The arrow illustrates direction of a neutrino beam to additional detector site(s) via the Bypass.

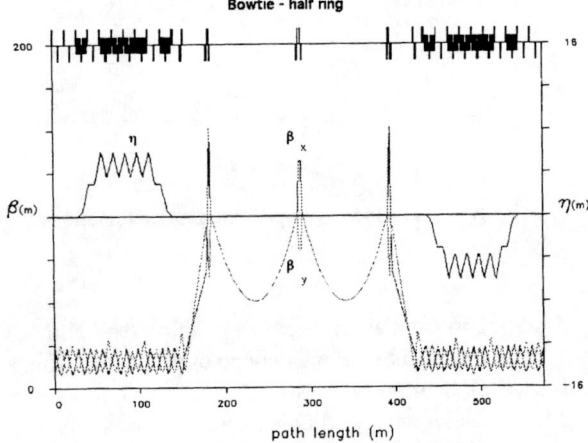

**FIGURE 20.** Lattice Functions for Bowtie-shaped Half Ring.

beta function values. In the other configuration, a 200-meter magnet-free beam-line is provided, with a beam waist at the center with 100 m beta values.

A racetrack muon storage - ring can be configured to deliver one neutrino beam to an arbitrary detector site. Bowtie - shaped, triangle shaped rings can be configured to deliver neutrino beams to two arbitrarily selected detector sites. This can be done by appropriate choice of, 1) the ring plane, 2) the orientation of the ring in that plane and 3) the angle at the crossing point between the two long straight sections. By inclusion of bypasses, additional detector sites may be accessible from a single muon storage-ring source.

A bypass would lie in a plane that includes the original long straight section (but differs from that of the ring), and begin and end on one of the long straight sections. Its magnets would be powered when one desires to send the muons along the deformed bypass path rather than along the normal straight path. In such a bypass, dipoles would produce a roughly triangular path in the bypass plane, one of whose sides would point to the desired detector. The two necessary degrees of freedom are provided by the angle between the bypass and ring planes and by the magnitude of the deflection given by the bypass dipoles. To suppress the dispersion pairs of dipoles should be placed 180 deg apart, in FODO cells.

A Schematic of Neutrino Factory concept, with a bowtie-shaped muon storage ring is illustrated in Fig. 23. The geometry of the storage ring depends on locations of the ring and detector. Table 5 shows direct distances

**Table 4.** The numbers of surviving muons after various stages in the accelerator complex.

| P driver energy | Factor | 24[GeV] $\mu/p$ | 16 [GeV] $\mu/p$ |
|---|---|---|---|
| Pions After: Match* |  | 0.66 | 0.44 |
| 1st Phase Rotation | 0.45 | 0.3 | .2 |
| 2nd Phase Rotation | 0.7 | 0.21 | .14 |
| RF Capture | 0.7 | 0.15 | .1 |
| Cooling | 0.9 | 0.13 | .09 |
| Acceleration | 0.7 | 0.092 | .061 |
| $n_\mu/(n_p E_p)$ [GeV$^{-1}$] |  | .0038 | .0038 |

* ($< 1$ GeV, forward)

from rings at BNL or FNAL to Gran Sasso, Soudan and SLAC.

## Physics Potentials - Event Rates

A neutrino factory has a strong independent physics case. It would be easier to build, less expensive than a full muon collider, and could demonstrate most of the components of a collider. For the example of Neutrino Factory Facility based on a muon storage ring (Fig. 1), the number of surviving muons, per incident proton, at various stages of the accelerator complex are summarized in Table 4, [10].

The number of neutrino interactions per unit mass of a detector at distance $L$ from a muon storage ring operating at energy $E_\mu$ scales as

$$N_{\text{events}} \propto N_\mu E_\mu^3 L^{-2}. \quad (3)$$

Table 4 illustrates the muon survival efficiencies, for the example of a proton source with 1.5 MW power, in one year ($10^7$ s) of operation, there would be about $4 \times 10^{20}$ muons per year decaying in the storage ring. Assuming the fraction of the ring pointing to a given detector to be about 0.25 (as in example of a bowtie-shaped muon storage) then the number of decays pointing to the given detector will be approximately $10^{20}$. It may be noted that the number of events with the 1.5 MW neutrino factory, in a detector at the same 730 km, is approximately 100 times that in the proposed CERN - Gran Sasso experiment (NGS [4]), and about 40 times the maximum event rate that MINOS [3] can expect. Upgrading the proton driver to 4 MW, the factors become about 300 and 100 for Gran Sasso and Soudan, respectively.

Table 5 gives charged current neutrino interaction rates (per kiloton-year) as a function of baseline length $L$ for an $E_\mu = 50$ GeV muon storage ring in which there are $1 \times 10^{20}$ unpolarized muon decays per year within a neutrino beam-forming straight section [17]. The rates are listed for oscillations:

1) $\nu_e \to \nu_\mu$: $\Delta m^2 = 3.5 \times 10^{-3}$ eV$^2$/c$^4$ & $\sin^2 2\theta = 0.1$,

2) $\nu_e \to \nu_\mu$: $\Delta m^2 = 1 \times 10^{-4}$ eV$^2$/c$^4$ & $\sin^2 2\theta = 1$,

3) $\nu_e \to \nu_\tau$: $\Delta m^2 = 3.5 \times 10^{-3}$ eV$^2$/c$^4$ & $\sin^2 2\theta = 0.1$,

4) $\nu_\mu \to \nu_\tau$: $\Delta m^2 = 3.5 \times 10^{-3}$ eV$^2$/c$^4$ & $\sin^2 2\theta = 1$.

The rates for the unoscillated neutrino interactions, the corresponding statistical significance of the disappearance signal (numbers in parenthesis), and the rates for the antineutrino interactions, are also included in Table 5

### Neutrino Oscillation

With only two massive neutrinos, with mass difference $\Delta m^2 = m_2^2 - m_1^2$, mass eigenstates $\nu_1$ and $\nu_2$ with mixing angle $\theta$, the flavor eigenstates become:

$$\begin{pmatrix} \nu_a \\ \nu_b \end{pmatrix} = \begin{pmatrix} \cos\theta & \sin\theta \\ -\sin\theta & \cos\theta \end{pmatrix} \begin{pmatrix} \nu_1 \\ \nu_2 \end{pmatrix}. \quad (4)$$

The probability that a neutrino of flavor $\nu_a$ and energy $E$ appears as flavor $\nu_b$ after traversing distance $L$ in vacuum is

$$P(\nu_a \to \nu_b) = \sin^2\left(1.27 \Delta m^2 [\text{eV}^2] \frac{L[\text{km}]}{E[\text{GeV}]}\right) \sin^2 2\theta. \quad (5)$$

Since the atmospheric neutrino data involves GeV muon neutrinos with distance scales of the Earth's diameter, this suggests $\Delta m^2$ of order $10^{-3}$ (eV)$^2$ for $\sin^2 2\theta \approx 1$. The solar neutrino data involves MeV electron neutrinos and distance scales of the radius of the Earth's orbit, suggesting $\Delta m^2$ of order $10^{-10}$ (eV)$^2$ with $\sin^2 2\theta \approx 1$ for vacuum oscillations [18]. The LSND result involves 30-MeV muon antineutrino and a distance scale of 30 m, suggesting $\Delta m^2$ of order 1 (eV)$^2$; large mixing angles are excluded by reactor data [19], thus, $\sin^2 2\theta$ can only be of order $10^{-2}$ in this case. Obviously, four different massive neutrinos are required to accommodate all three results, given their disparate scales of $\Delta m^2$. The Standard Model presently includes only three neutrinos with standard electroweak couplings and $m_\nu < m_Z/2$, so a "sterile" neutrino is required if all the data are correct [20]. Even discarding the LSND result, three massive neutrinos are required with a corresponding $3 \times 3$ mixing matrix, e.g.

$$\begin{pmatrix} \nu_e \\ \nu_\mu \\ \nu_\tau \end{pmatrix} = \begin{pmatrix} c_{12}c_{13} & s_{12}c_{13} & s_{13}e^{-i\delta} \\ -s_{12}c_{23} - c_{12}s_{13}s_{23}e^{i\delta} & c_{12}c_{23} - s_{12}s_{13}s_{23}e^{i\delta} & c_{13}s_{23} \\ s_{12}s_{23} - c_{12}s_{13}c_{23}e^{i\delta} & -c_{12}s_{23} - s_{12}s_{13}c_{23}e^{i\delta} & c_{13}c_{23} \end{pmatrix} \begin{pmatrix} \nu_1 \\ \nu_2 \\ \nu_3 \end{pmatrix}. \quad (6)$$

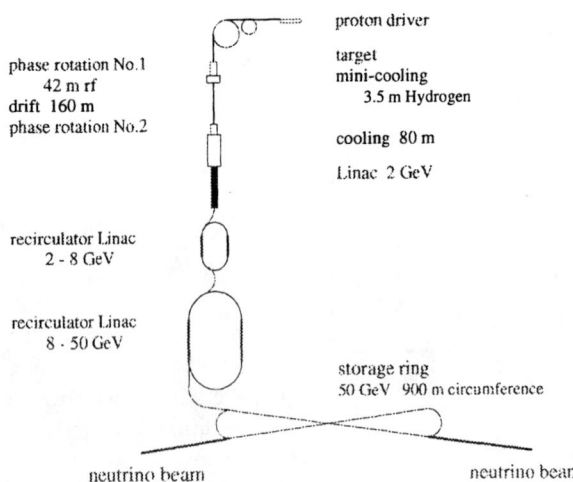

**FIGURE 23.** Neutrino Factory Concepts, with a Bowtie-Shaped Muon Storage Lattice.

a MNS matrix [21], where $c_{12} = \cos\theta_{12}$, etc.., In the three massive neutrino model, the neutrino oscillation probabilities of interest depends on six measurable parameters: three mixing angles ($\theta_{12}$, $\theta_{13}$, $\theta_{23}$); a phase $\delta$ related to CP violation as indicated in eq. (6); and two differences of the squares of the neutrino masses ($\Delta m_{12}^2$ and $\Delta m_{23}^2$ for instance). The interpretation of the solar and atmospheric neutrino data in terms of the three-neutrino oscillation hypothesis suggests $|\Delta m_{12}^2| \ll |\Delta m_{23}^2|$, with $\Delta m_{12}^2$ and $\Delta m_{23}^2$ being responsible for the transitions and/or oscillations of the solar and atmospheric neutrinos, respectively.

The description of the atmospheric neutrino data requires $\Delta m_{23}^2 \approx (2-6) \times 10^{-3}$ eV$^2$ and large mixing angle $\theta_{23}$: $\sin^2 2\theta_{23} \approx (0.9 - 1.0)$. For $|\Delta m_{12}^2| \ll |\Delta m_{23}^2|$ and with $\Delta m_{23}^2$ in the above range, the non-observation of oscillations of the reactor electron antineutrinos in the CHOOZ experiment [22] implies a limit on the angle $\theta_{13}$: $\sin^2 \theta_{13} < 0.11$. Given these constraints, the transitions/oscillations of the solar neutrinos in the three-neutrino mixing scheme under discussion depend largely on the remaining two parameters: $\Delta m_{12}^2$ and $\sin^2 2\theta_{12}$.

Further, the presence of matter may modify the oscillations of electron neutrinos because of their charged-current interaction (MSW effect [23]). In particular, the oscillations can be resonantly enhanced by the matter effects even when the oscillation probabilities are small in vacuum. This leads to additional interpretations of the solar neutrino data in which $\Delta m_{12}^2$ can be of order $10^{-5}$ (eV)$^2$ [24]. In effect at the present time, there are four viable interpretations of the solar neutrino data:

1) Vacuum oscillation (VO) solution with $\Delta m_{12}^2 \approx (0.5 - 5.0) \times 10^{-10}$ eV$^2$ and $\sin^2 2\theta_{12} \approx (0.7 - 1.0)$,

2) Low MSW solution corresponding to $\Delta m_{12}^2 \approx (0.5 - 2.0) \times 10^{-7}$ eV$^2$ and $\sin^2 2\theta_{12} \approx (0.9 - 1.0)$,

3) Small mixing angle (SMA) MSW solution with $\Delta m_{12}^2 \approx (4.0 - 9.0) \times 10^{-6}$ eV$^2$ and $\sin^2 2\theta_{12} \approx (0.001 - 0.01)$,

4) Large mixing angle (LMA) MSW solution, $\Delta m_{12}^2 \approx (0.2 - 2.0) \times 10^{-4}$ eV$^2$ and $\sin^2 \theta_{12} \approx (0.65 - 0.96)$.

With four interpretations of the solar neutrino data, and the two interpretations of the LSND data as either right or wrong, there are a total of eight scenarios for explanations of the data. The experimental challenge is to reduce these to a single scenario, and to make accurate measurements of the parameters of that scenario.

Thus, with the available experimental guidelines as to the parameters of neutrino masses and mixings, one can begin to plan for more extensive studies namely, with neutrino beams derived from the decay of muons in a storage ring. Both $\mu^-$ and $\mu^+$ can be stored in the ring, but only one sign would be used at a time. For example if $\mu^-$ are stored, their decay

$$\mu^- \to e^- \nu_\mu \bar{\nu}_e, \qquad (7)$$

leads to beams with nearly equal numbers of $\nu_\mu$ and $\bar{\nu}_e$ with spectra that are well known.

At the detectors, the neutrino and the antineutrino may or may not have changed their flavor, leading to the appearance of a different flavor or the disappearance of the initial flavor, respectively. When detected by a charged-current interaction, there are 6 classes of signatures in a three-neutrino model: 1) $\nu_\mu \to \nu_e \to e^-$ (appearance); 2) $\nu_\mu \to \nu_\mu \to \mu^-$ (disappearance); 3) $\nu_\mu \to \nu_\tau \to \tau^-$ (appearance); 4) $\bar{\nu}_e \to \bar{\nu}_e \to e^+$ (disappearance); 5) $\bar{\nu}_e \to \bar{\nu}_\mu \to \mu^+$ (appearance); 6) $\bar{\nu}_e \to \bar{\nu}_\tau \to \tau^+$ (appearance).

For operation with positive muons, a similar list of processes may be written. The 5th process where a muon

**Table 5.** Neutrino Interaction Rates at a Neutrino Factory.

| | | Source at<br>Detector at<br>L (km) | BNL<br>G. Sasso<br>6528 | BNL<br>SLAC<br>4139 | BNL<br>Soudan<br>1712 | FNAL<br>G. Sasso<br>7332 | FNAL<br>SLAC<br>2899 | FNAL<br>Soudan<br>732 |
|---|---|---|---|---|---|---|---|---|
| Case | | Mode | | | | | | |
| 1) | $\mu^+$ | $\nu_e \to \nu_\mu$ | 90 | 160 | 190 | 63 | 180 | 200 |
| | | $\nu_e \to \nu_e$ | 1400 | 3600 | 16000 | 1100 | 8000 | $1.2 \times 10^5$ |
| | | | $(2.4\sigma)$ | $(2.7\sigma)$ | $(1.5\sigma)$ | $(1.9\sigma)$ | $(2.0\sigma)$ | $(0.6\sigma)$ |
| | | $\bar{\nu}_\mu \to \bar{\nu}_\mu$ | 890 | 2200 | 9300 | 700 | 4800 | $7.0 \times 10^4$ |
| 2) | $\mu^+$ | $\nu_e \to \nu_\mu$ | $5 \times 10^{-2}$ | 0.86 | 1.5 | $3 \times 10^{-5}$ | 1.3 | 1.6 |
| | | $\nu_e \to \nu_e$ | 1500 | 3800 | 16000 | 1200 | 8200 | $1.2 \times 10^5$ |
| | | $\bar{\nu}_\mu \to \bar{\nu}_\mu$ | 890 | 2200 | 9400 | 700 | 4800 | $7.0 \times 10^4$ |
| 3) | $\mu^+$ | $\nu_e \to \nu_\tau$ | 31 | 60 | 70 | 20 | 67 | 73 |
| | | $\nu_e \to \nu_e$ | 1400 | 3700 | $1.6 \times 10^4$ | 1100 | 8000 | $1.2 \times 10^5$ |
| | | | $(2.4\sigma)$ | $(2.7\sigma)$ | $(1.5\sigma)$ | $(1.9\sigma)$ | $(2.0\sigma)$ | $(0.6\sigma)$ |
| | | $\bar{\nu}_\mu \to \bar{\nu}_\mu$ | 890 | 2200 | 9400 | 700 | 4800 | $7.0 \times 10^4$ |
| 4) | $\mu^-$ | $\nu_\mu \to \nu_\tau$ | 450 | 570 | 650 | 410 | 620 | 680 |
| | | $\nu_\mu \to \nu_\mu$ | 760 | 3100 | $1.7 \times 10^4$ | 490 | 8000 | $1.4 \times 10^5$ |
| | | | $(35\sigma)$ | $(23\sigma)$ | $(12\sigma)$ | $(40\sigma)$ | $(16\sigma)$ | $(4.6\sigma)$ |
| | | $\bar{\nu}_e \to \bar{\nu}_e$ | 770 | 1900 | 8100 | 600 | 4100 | $6.1 \times 10^4$ |

of different sign from the parent muon appears, has a very unique possibilities at a neutrino factory based on muon storage rings. Since they are the only sources of intense high energy electron (anti)neutrino beams. The $\tau$ appearance (cases 3 and 6) are practical only for neutrino beams with 10's of GeV energy.

It is anticipated that by the time a muon storage ring would be built the two angles ($\theta_{23}$ and $\theta_{12}$), and the magnitudes of two mass squared differences ($\Delta m_{23}^2$ and $\Delta m_{12}^2$) would be known, from the solar and atmospheric neutrino measurements (which would have been verified by long baseline and reactor experiments), for example, MINOS and KamLAND. The remaining pieces of the puzzle would be $\theta_{13}$, the CP-violating phase $\delta$ and the signs of the $\Delta m_{ij}^2$. Moreover, the indicated long-baseline experiments will not be sensitive to the matter effects in neutrino oscillations because the distances between the sources and detectors are not sufficiently large. Verifying the existence of matter effects in neutrino oscillations by observing directly the modification of the neutrino oscillation probabilities by these effects, would also be fundamental and interesting.

The third mixing angle $\theta_{13}$ can be measured in several channels at a neutrino factory [25]. The detector must be far to avoid background but not too far ($< 1000$ km) so that the effects of $\Delta m_{12}^2$ remain negligible and thus $\delta$ can formally be set to zero. Fig. 24 shows the achievable sensitivity to the yet-unknown value of $\theta_{13}$.

Fig. 24 illustrates sensitivity reach in the

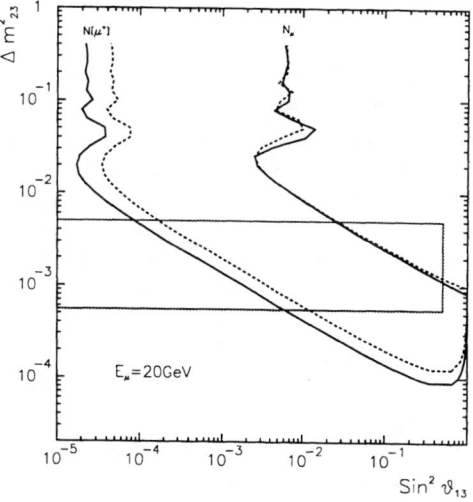

**FIGURE 24.** Sensitivity reach in the $(\sin^2\theta_{13}, \Delta m_{23}^2)$ plane.

$(\sin^2\theta_{13}, \Delta m_{23}^2)$ plane for a 10 kton detector and a neutrino beam from $2 \times 10^{20}$ decays of 20 GeV muons in a storage ring at distance 732 km. The appearance process $\bar{\nu}_e \to \bar{\nu}_\mu \to \mu^+$, shown by the lines on the left, has much greater sensitivity than the disappearance process $\nu_\mu \to \nu_\mu \to \mu^-$, shown by the lines on the right. The interior of the box is the approximate region allowed by Super-Kamiokande data [25].

## CP Violation

The three-neutrino scenario [26] can lead to CP violation in for example

$$A_{CP} = \frac{P(\nu_e \to \nu_\mu) - P(\bar{\nu}_e \to \bar{\nu}_\mu)}{P(\nu_e \to \nu_\mu) + P(\bar{\nu}_e \to \bar{\nu}_\mu)}, \quad (8)$$

or time-reversal violation

$$A_T = \frac{P(\nu_e \to \nu_\mu) - P(\nu_\mu \to \nu_e)}{P(\nu_e \to \nu_\mu) + P(\nu_\mu \to \nu_e)}. \quad (9)$$

The asymmetry (8) can be measured using wrong-sign muons and the two charges of the muon beam. However, the genuine CP violating contribution to (8) due to a non-vanishing phase $\delta$ competes with terms related to matter effects, i.e., to the different rates of evolution for $\nu_e$ and $\bar{\nu}_e$ between source and detector. The relative strength of the matter-induced asymmetry increases quadratically with distance, and dilutes the signal of CP violation in a far detector.

If the solution to solar neutrino problem involves, large mixing angles and matter enhancement (LMA MSW, $\sin^2 2\theta_{12} \approx \sin^2 2\theta_{23} \approx 1$), then there is a possibility of measuring the CP violating asymmetry (8), with expression

$$|A_{CP}| \approx \left| \frac{2\sin\delta}{\sin 2\theta_{13}} \sin\left(\frac{1.27\Delta m_{12}^2 L}{E}\right) \right|, \quad (10)$$

provided the detector is located sufficiently far and high statistics ($> 10^{21}$ muons per year) are available. For all the other solar neutrino solutions $A_{CP}$ is extremely small, being suppressed by a factor of either $\sin^2 2\theta_{12}$ or $\Delta m_{12}^2$. Figure 25 Show the CP violating asymmetry (8) divided by statistical uncertainties vs. distance $L$ for a 10 kton detector in a beam from $2 \times 10^{21}$ muon decays. A large angle MSW scenario is supposed, with $\Delta m_{12}^2 = 10^{-4}$ eV$^2$, $\Delta m_{23}^2 = 2.8 \times 10^{-3}$ eV$^2$, $\theta_{12} = 22.5°$, $\theta_{13} = 13°$, $\theta_{23} = 45°$, and $\delta = -90°$ (corresponding to large CP violation). The dashed curves ignore matter effects, while the solid curves include them; the matter effects dominate the asymmetry for distances beyond 1000 km. The lower (upper) curves are for $E_\mu = 20\,(50)$ GeV, from [hep-ph/9909254].

The asymmetry (9) is not sensitive to matter effects, but relies on distinguishing the process $\nu_\mu \to \nu_e \to e^-$ from $\bar{\nu}_e \to \bar{\nu}_e \to e^+$. In the detector, it will be very difficult to distinguish electrons from positrons but the relative $\nu_\mu$ and $\bar{\nu}_e$ fluxes can be varied by varying the polarization of the muons in the storage ring [27].

If future experiments confirm the interpretation of the LSND data that there exist more than three light neutrinos, then use of the neutrino factory flavor-rich beams would be even more crucial, because the parameter space for CP/T violating effects would be considerably enlarged

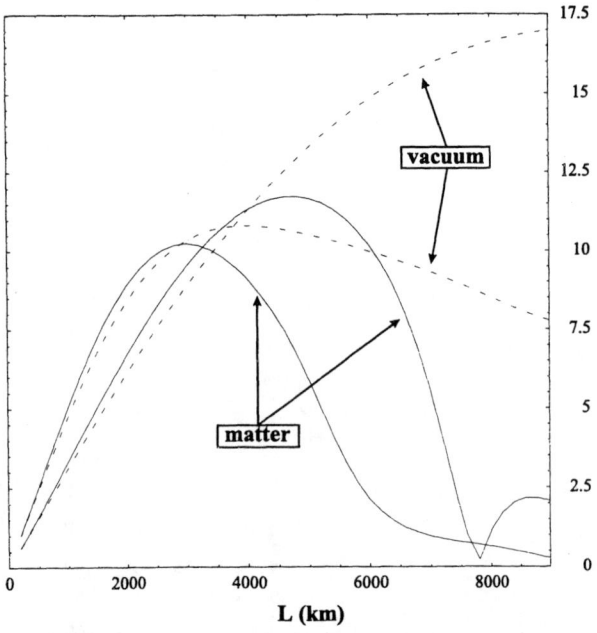

**FIGURE 25.** CP violation signal over statistical uncertainties versus distance.

and could be explored in experiments with such beams [28].

## Precision Physics

Muon storage ring based neutrino beams would bring about new neutrino oscillation measurements, and a new era for high-precision neutrino scattering experiments [29]. For example, with a detector located 30 m from a 150 m straight section of a 50-GeV, $10^{21}$-$\mu$/yr muon storage ring, the event rate is 40 million events per kilogram per year over a 10 cm radius. Oscillation-related measurements may be interpreted precision measurements of the total neutrino and antineutrino cross sections, as well as of the beam divergence. As precision probes of nuclear and nucleon structure, the neutrinos may be used to provide additional information to that obtained with charged lepton beams, in related studies. It is known that, neutrino scattering allows a clean separation of the valence and sea quark distributions, and use of a polarized target permits characterization of the spin dependence of these distributions. Thus, near detectors are the natural successor to nucleon structure measurements presently underway at HERA, HERMES, Jefferson Lab, RHIC and elsewhere. For example, scattering of the four neutrino types $\nu_\mu$, $\bar{\nu}_\mu$, $\nu_e$, and $\bar{\nu}_e$ off electrons could lead to measurements of the Weinberg angle ten times better than known at present.

Note that, a high-flux multi-GeV neutrino beam is a charm factory, in which a $\nu_\mu$ beam leads to $c$ quarks that are tagged by a final-state $\mu^-$ ($\nu_\mu d \to \mu^- c$), while $\overline{\nu}_\mu$ beam leads only to tagged $\overline{c}$ quarks. For example, for the above described beam parameters, there would be $10^7$ leptonic tagged charm decays in only 40 kg-years (not kton-years!), permitting measurements of $V_{cd}$ to fraction of a percent, and perhaps even direct observation of $D^0 - \overline{D}^0$ mixing.

## MUON COLLIDER

A muon collider with center of mass energy less than about 10 TeV can be circular and relative to NLC (a Next Linear Collider) of the same energy, it could be far smaller in size. For the same luminosity a muon collider can tolerate a far larger spot size than an electron linear collider since the muons make about 1000 crossings. Since there is little beamstrahlung, very small energy spread is easily obtainable. Fig. 26 shows a schematic of a muon collider components [9]. A high intensity proton source is bunch compressed and focused on a heavy metal target. The pions generated are captured by a high field solenoid and transferred to a solenoidal decay channel within a low frequency linac. The linac reduces, by phase rotation the momentum spread of the pions and of the muons into which they decay.

Subsequently, the muons are cooled by a sequence of ionization cooling stages. Each stage consists of energy loss, acceleration, and emittance exchange by energy absorbing wedges in the presence of dispersion. Once they are cooled the muons must be rapidly accelerated to avoid decay losses. This can be done in recirculating accelerators (as at CEBAF) or in fast pulsed synchrotrons. Muon collisions occur in a separate high field collider storage ring with a single very low beta insertion.

It is expected that the first stage, proton driver would be 16 to 30 GeV; but would be much faster pulsed, keeping the number of protons per pulse the same or smaller than the AGS, which is about $6 \times 10^{13}$ protons per pulse and with some upgrade to about $10^{14}$ protons per pulse.

Roughly one expect to get 1 muon/proton on target which would give luminosity between $10^{34}$ to $10^{35}$ the envisioned muon collider. Although the accelerating component is large, the other components can fit within it and the whole machine is compact enough to fit on existing Brookhaven or Fermilab sites.

For more information on the Muon Collider and parameters under study, see e.g. [9],[30] – [36]. Table 6 shows the parameters of potential muon colliders at 100 GeV, 500 GeV and 4 TeV center of mass energy. The 100 GeV collider would be ideal for the study of the lowest mass Higgs. The 4 TeV collider should be in the

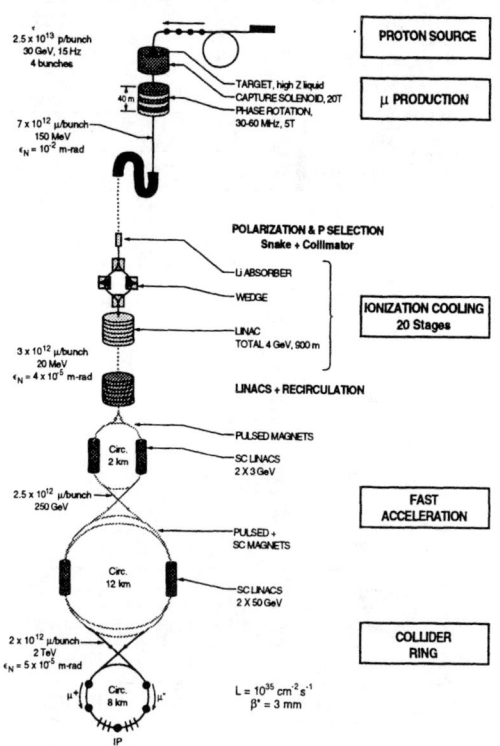

**FIGURE 26.** Schematic of a Muon Collider.

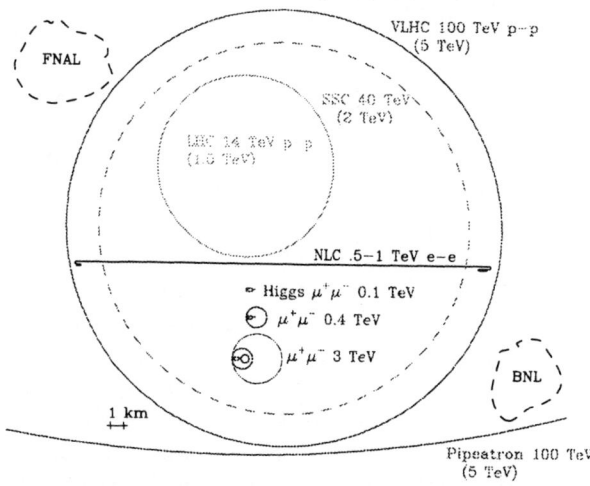

**FIGURE 27.** Comparison of relative sizes of Muon Collider, Large Hadron Collider (LHC), and Next Linear Collider (NLC), relative to the BNL and FNAL sites.

**Table 6.** Parameters of $\mu^+\mu^-$ collider Rings.

| Energy (C.M.) TeV | 4 | 0.5 | 0.1 |
|---|---|---|---|
| Beam Energy TeV | 2 | 0.25 | 0.05 |
| Beam $\gamma$ | 19,000 | 2,400 | 473 |
| Rep. rate Hz | 15 | 2.5 | 15 |
| p Energy GeV | 30 | 24 | 16 |
| p/pulse | $10^{14}$ | $10^{14}$ | $5 \times 10^{13}$ |
| $\mu$/bunch | $2 \times 10^{12}$ | $4 \times 10^{12}$ | $4 \times 10^{12}$ |
| Bunches/sign | 2 | 1 | 1 |
| Beam Power MW | 38 | 0.7 | 1.0 |
| $\varepsilon_N$ $\pi$ mm-mrad | 50 | 90 | 195 |
| Bending Field T | 9 | 9 | |
| Circumference km | 8 | 1.3 | 0.3 |
| Ave. ring field B T | 6 | 5 | 3.5 |
| Effective turns | 900 | 800 | 450 |
| $\beta^*$ mm | 3 | 8 | 9 |
| IP beam size $\mu$m | 2.8 | 17 | 187 |
| Chromaticity | 2000-4000 | 40-80 | |
| $\beta_{max}$ km | 200-400 | 10-20 | 1.5 |
| Lumin. $cm^{-2}s^{-1}$ | $10^{35}$ | $10^{33}$ | $2 \times 10^{31}$ |

energy range of most of the heavy Higgs in the minimal SUSY model (if that is the correct theory).

Although muon colliders remain a promising complement, to $e^+e^-$ and hadron colliders, much work is still needed, including demonstration of $\mu$ production and cooling, detector, and radiation.

## DISCUSSION

The muon collider concept promises to extend the high energy frontier to an unprecedented domain, with center of mass energies of 3 TeV or beyond as its goal. Considerable effort has already gone into the conceptual design of muon colliders, but much R&D remains to be carried out. Of particular importance are production targetry and cooling tests. The Muon Collider Collaboration represents a dedicated effort to address those issues and pave the way for a future muon accelerator complex.

A full high energy muon collider may take considerable time to realize. However, intermediate steps in its direction are possible and could help facilitate the process. Employing an intense muon source to carry out forefront low energy research, such as the search for muon - number non - conservation, represents one interesting possibility. For example, the MECO proposal at BNL aims for $2 \times 10^{-17}$ sensitivity in their search for coherent muon - electron conversion in the field of a nucleus. To reach that goal requires the production, capture and stopping of muon at an unprecedented $10^{11} \frac{\mu}{sec}$. If successful, such an effort would significantly advance the state of muon technology.

More ambitious ideas for utilizing high intensity muon sources are also being explored. Building a muon storage ring for the purpose of providing intense high energy neutrino beams is particularly exciting. Such neutrino factories could have their own world class research program, with neutrino oscillation studies as the primary focus. Indeed, if very high intensities, $\sim 10^{21} \frac{\nu}{year}$, are attained and nature has been kind in her neutrino mass and mixing parameters, one could envision a complete exploration of the $3 \times 3$ neutrino mixing matrix and even the detection of CP violation in the oscillation phenomena.

If a neutrino factory is successfully accomplished, it would provide a major advancement. Its ambitious goals would test essentially all aspects of the muon collider concept, muon production, collection, cooling and acceleration. Furthermore, if properly coordinated, the neutrino factory complex might be suitably expanded into the First Muon Collider, perhaps a Higgs factory with center of mass energy $\sim$ 100 GeV.

High intensity muon experiments, neutrino factories, and other intermediate steps toward the muon collider are extremely important. They will greatly expand our abilities and build confidence in the credibility of high energy muon colliders. Progress may be slower than many would prefer, but remember, Rome was not built in a day.

## ACKNOWLEDGMENTS

The author appreciates NNN99's organizers interest and invitation to write this paper. Special thanks to Dr. Diwan for extending the time, and to Dr. Paige for assisting with the latex problems; that made the completion of this report possible. This document was assembled largely from NFMCC existing sources, which have been cited among the references. Thanks to those individuals who contributed graphics, information, and assistance.

## REFERENCES

1. The Neutrino Factory and Muon Collider Collaboration (NFMCC) home page: http://www.cap.bnl.gov/mumu/
2. http://neutrino.kek.jp/ http://www.awa.tohoku.ac.jp/html/KamLAND/
3. See, for example, sec. 3.4 of the MINOS Technical Design Report, http://www.hep.anl.gov/NDK/Hypertext/minos_tdr.html
4. http://www.cern.ch/NGS/ngs99.pdf

5. The MiniBooNE project: http://www.neutrino.lanl.gov/BooNE

6. The Oak Ridge Large ν Detector http://www.orau.org/orland/

7. *Search for $\nu_\mu \to \nu_e$ Oscillations at CERN PS*, http://chorus01.cern.ch/~pzucchel/loi/

8. A.C. Melissinos unpublished note (1960).

9. C.M. Ankenbrandt *et al.*, *Status of muon collider research and development and future plans*, Phys. Rev. ST Accel. Beams **2**, 081001 (1999), and refrences therein.

10. See e.g., [11, 9, 1, 12], and references therein

11. K.T. McDonald, ed. for the Neutrino Factory and Muon Collider Collaboration, physics/9911009, 6 Nov, 1999, and references therein; ibid, Private comm.

12. http://lyoninfo.in2p3.fr/nufact99/ R.B. Palmer, C. Johnson, E Keil, BNL-66971, E. Keil, private Comm, Montauk, 99.

13. R.B. Palmer, Draft Paramets of a Neutrino Factory, MUC0046.

14. S. Kahn, private Comm; B. Autin, private Comm.

15. C. Kim, Private Comm.: received info. in Microsoft Word document format for simulation of the target to linac. These were then processed to pdf and ps format some of which is included; J. Wurtele, Private Comm.

16. A. Garren, Private Comm.: Bowtie-lattice design, simulation with code SYNCH; some of the info received is included; D.Cline, Private Comm.

17. S. Geer, *Neutrino Oscillation Rates at a Neutrino Factory*, MUC0051 (Sep. 13, 1999);

18. E.g.: S.L. Glashow and L.M. Krauss, Phys. Lett. **B190**, 199 (1987); S.P. Mikheyev and A.Y. Smirnov, Phys. Lett. **B429**, 343 (1998);V. Barger and K. Whisnant, Phys. Rev. D **59**, 093007 (1999), hep-ph/9812273; M. Maris and S.T. Petcov, Phys. Lett. **B457**, 319 (1999);
V. Berezinsky, G. Fiorentini, and M. Lissia, hep-ph/9904225; S. Goswami, D. Majumdar, and A. Raychaudhuri, hep-ph/9909453.

19. G.S. Vidaykin *et al.*, JETP Lett. **59**, 390 (1994); B. Achkar *et al.*, Nucl. Phys. **B434**, 503 (1995).

20. E.g.: A. Zee, Phys. Lett. **B93**, 389 (1980); V. Barger, P. Langacker, J. Leveille, and S. Pakvasa, Phys. Rev. Lett. **45**, 692 (1980); V. Barger, S. Pakvasa, T.J. Weiler, and K. Whisnant, Phys. Rev. D **58**, 093016 (1998); S.M. Bilenky, G. Giunti, hep-ph/9905246.

21. Z. Maki, M. Nakagawa, and S. Sakata, Prog. Theor. Phys. **28**, 970 (1962). M. Nakagawa, hep-ph/9811358.

22. http://www.hep.anl.gov/ndk/hypertext/chooz.html

23. L. Wolfenstein, Phys. Rev. D **17**, 2369 (1978); S.P. Mikheyev and A.Y. Smirnov, Sov. J. Nuc. Phys. **42**, 913 (1986).

24. N. Hata and P. Langacker, Phys. Rev. D **56**, 6107 (1997), J.N Bahcall, P.I. Krastev, and A.Y. Smirnov, Phys. Rev. D **58**, 096016 (1998), J.N. Bahcall, P. Langacker, J. Bahcall and P. Krastev, Phys. Lett. **B436**, 243 (1998), hep-ph/9807525.

25. S. Geer, Phys. Rev. D **57**, 6989 (1998), hep-ph/9712290; A. Buena, M. Campanelli, and A. Rubbia, hep-ph /9808485; and hep-ph/9811390; http://fnalpubs.fnal.gov/archive/1999 hep-ph/9905420; V. Barger, S. Geer, and K. Whisnant, hep-ph/9906487; O. Yasuda, hep-ph/9910428;
I. Mocioiu and R. Shrock, hep-ph/9910554.

26. M. Tanimoto, Prog. Theor. Phys. **97**, 9091 (1997),J. Arafune, M. Koike, and J. Sato, Phys. Rev. D **56**, 3093 (1997),S.M. Bilenky, C. Giunti, and W. Grimus, hep-ph/9705300; H. Minakata and H. Nunokawa, Phys. Lett. **B413**, 369 (1997),H. Minakata and H. Nunokawa, *CP Violating vs. Matter Effect in Long-Baseline Neutrino Oscillation* Phys. Rev. D **57**, 4403 (1998),S.M. Bilenky, C. Giunti, and W. Grimus, Phys. Rev. D **58**, 033001 (1998), M. Tanimoto, hep-ph/9906375; K.R. Schubert, hep-ph/9902215; K. Dick, M. Freund, M. Lindner, and A. Romanino, hep-ph/9903308; J. Bernabeu, hep-ph/9904474;M. Tanimoto, hep-ph/9906516; A. Donini, M.B. Gavela, P. Hernandez, and S. Rigolin, hep-ph/9909254;H. Fritzsch and Z.-Z. Xiang, hep-ph/9909304;A. Romanino, hep-ph/9909425; M. Koike and J. Sato, hep-ph/9909469; J. Sato, hep-ph/9910442. hep-ph/9910442.

27. A. Blondel, http://alephwww.cern.ch/~bdl/muon/nufacpol.ps

28. V. Barger, Y.-B. Dai, K. Whisnant and B.-L. Young, Phys. Rev. D **59**, 113010 (1999), hep-ph/9901388;A. Kalliomaki, J. Mallampi, and M. Tanimoto, hep-ph/9909301;A. Donini, M.B. Gavela, P. Hernandez, and S. Rigolin, hep-ph/9910516.

29. See, for example, B.J. King, AIP Conf. Proc. **435**, 334 (1998). D.A. Harris and K.S. McFarland, ibid. p. 505;

30. D. B. Cline (ed.), Physics Potential and Development of $\mu^+\mu^-$ Colliders AIP Conf. Proc. **352** (1996).

31. The $\mu^+\mu^-$ Collider Collaboration, $\mu^+\mu^-$ Collider Feasibility Study, BNL-52503, FERMILAB-Conf-96/092, LBNL-38946 (July 1996); http://www.cap.bnl.gov/mumu/book.html

32. Z. Parsa, ed., *Future High Energy Colliders*, AIP CP **397**, AIP-Press, Woodbury, NY (1997).

33. Z. Parsa, *Ionization cooling and Muon Dynamics*, AIP CP **441**, 289-294 (1997).

34. Z. Parsa, *New High Intensity Muon sources and Flavor Changing Neutral Currents*, World scientific Publishing, pp 147-153 (1998).

35. Z. Parsa, ed., *Beam Stability and Nonlinear Dynamics*, AIP CP **405** AIP-Press, Woodbury, NY (1997).

36. Z. Parsa, *Lasers and Future High Energy Collliders*, STS-Press, pp 823-830 (1997).

37. Kamal, B., Marciano, W., Parsa, Z., *Resonant Higgs enhancement at the first muon collider*, AIP CP **441**, pp174- (1997); ibid, AIP **435** pp567-662 (1997).

38. Z. Parsa, *Polarization and Luminosity requirements for the First Muon Collider*, Procd. of AAC98, AIP-Press, NY.(1998).

39. Z. Parsa, *Polarization Effects at a Muon Collider*, Presentation at EPAC98, Stokholm, Sweeden, June 1998.

40. Z. Parsa, *Muon Dynamics and Ionization Cooling at Muon Colliders*, Procd. of EPAC98, Stockholm, Sweden, Vol 2, pp.1055-.

41. C.N. Ankenbrandt *et al.*, *Ionization Cooling Research and Development Program for a High Luminosity Muon Collider*, FNAL-P904 (April 15, 1998), http://www.fnal.gov/projects/muon_collider/

42. Neutrino Factory Feasibility Studies at Fermilab: http://www.fnal.gov/projects/muon_collider/nu_factory/

43. N. Mokhov, *Carbon and Mercury Targets in 20-T Solenoid with Matching*, MUC0061; ibid, private comm. V. Balbekov, Private Comm.

44. Experiment E-910 at BNL-AGS; http://www.nevis.columbia.edu/heavyion/e910/

# Matter effects study in very long baseline neutrino oscillation experiments

A.Bueno, M.Campanelli,[*] A.Rubbia

*Institut für Teilchenphysik, ETHZ, CH-8093 Zürich, Switzerland*

**Abstract.** We examine the physics prospects of a neutrino oscillation experiment to be performed at a Neutrino Factory with a very long baseline in the range of 6500 km. We consider three family mixing and make the approximation that only one mass scale is relevant in our framework. We study the effect of the neutrino propagation through matter (MSW effect). Given the density of the Earth, the existence of the MSW resonance can be experimentally proven by comparing the oscillated spectra of neutrinos obtained from the decays of muons of positive and negative charges. A precision study of the oscillations of all three flavors could be performed since neutrinos are above the tau production threshold (appearance searches).

## INTRODUCTION

As well known, neutrino flavor oscillations will take place if the neutrino flavor eigenstates ($\nu_e, \nu_\mu$ and $\nu_\tau$) do not coincide with their mass eigenstates ($\nu_1, \nu_2$ and $\nu_3$). Within the two flavor oscillation framework, the transition probability in vacuum between flavor $\nu_\alpha$ and $\nu_\beta$ ($\alpha, \beta = e, \mu, \tau, \alpha \neq \beta$) is given by:

$$P(\nu_\alpha \to \nu_\beta) = \sin^2 2\theta \sin^2\left(1.27 \Delta m_{ij}^2 (eV^2) \frac{L(km)}{E_\nu(GeV)}\right) \quad (1)$$

where $\Delta m_{ij}^2 = m_i^2 - m_j^2$ ($i,j = 1,2,3, i \neq j$) is the mass squared difference between the two neutrino mass eigenstates, $\nu_i$ and $\nu_j$, $\theta$ is the mixing angle between mass and weak eigenstates, $L$ is the baseline, and $E_\nu$ the neutrino energy. When the neutrinos propagate in matter, their equation of motion must include an interaction with the medium. For a given neutrino energy which depends on the matter electron density, a "resonance" condition will be reached (the MSW-effect(1)) at which the flavor oscillation of neutrinos (antineutrinos) will then be enhanced (suppressed) compared to oscillations in vacuum.

The results on the solar and atmospheric neutrino fluxes can be naturally explained in terms of neutrino oscillations. Since the energies and distances of the solar neutrino problem are widely different from those of the atmospheric neutrino anomaly, the neutrino oscillation solutions require very different values of mass squared differences.

The solar neutrino deficit(2) can be explained via the disappearance of $\nu_e$'s due to flavor oscillation enhanced by the MSW effect. A combined fit to the experimental data implies a $\Delta m^2$ in the region $10^{-5}$ eV$^2$ and two possible angular solutions: the small (SMA) and large (LMA) mixing angle solutions. Another solution to these data would require a $\Delta m^2 \approx 10^{-10}$ eV$^2$ if the oscillation took place in vacuum.

The MSW solution of the solar neutrino problem is the more attractive one because it does not require the fine-tuning of the oscillation parameters. The conversion in the Sun is primarily a resonance phenomenon, which occurs at a specific density that corresponds to a definite neutrino energy (for a given $\Delta m^2$). Because at night solar neutrinos will cross the Earth before reaching the detector, the MSW mechanism may introduce a day-night asymmetry(3) by the phenomenon of "regeneration". However, in the relatively small density of the Earth, the resonance condition is not met. Clearly the search for the MSW effect on Earth will be of great importance.

The observation of atmospheric neutrinos by SuperKamiokande(4) and other detectors(5) (Kamiokande, IMB, Soudan-II and MACRO) has shown evidence for neutrino flavor oscillations, compatible with $\nu_\mu$ neutrinos converting to $\nu_\tau$ with maximal mixing and $10^{-3} < \Delta m^2 < 10^{-2}$ eV$^2$.

For the present generation of long baseline beams from CERN and Fermilab, the baseline of $L = 730$ km is too short to provide a strong resonance signal(6). Within this context, we explore the possibility of an experiment using neutrinos from stored muon decays at a very long baseline of 6500 km, with neutrinos of sufficient energy

---

[*] Presented by M.Campanelli

in order to produce the resonance phenomena in their passage through Earth for the set of parameters suggested by the atmospheric neutrino results. Since this phenomenology implies oscillations into tau neutrinos, the energy should also be large enough in order to observe explicitly the charged current interactions of $\nu_\tau$'s.

The synergy with future muon colliders(7) is here twofold: Muon colliders will necessarily require very intense proton sources and it will be then conceivable to obtain very intense neutrino beams(8) to compensate for the flux decrease with increasing distance from the source. In addition, the well-defined flavor composition of beams from muon decays can be exploited using a detector with charge identification capabilities to look for flavor oscillations(9).

We first consider $\nu_e \leftrightarrow \nu_\mu$ oscillations within a three flavor oscillation framework, at the $\Delta m^2_{32} = m_3^2 - m_2^2$ mass indicated by the atmospheric neutrinos. Evidence for this oscillation would imply that the (1-3) mixing between the first and the third family is non-zero. We study the effect of the neutrino propagation through the matter of the Earth (MSW effect). Given the density of the Earth, the MSW resonance occur at high energies accessible with accelerators.

With the use of neutrino beams from stored muon decays, the matter enhancement or suppression of the oscillation can be experimentally tested by comparing the event rates and energy spectra obtained from decaying stored muons of positive and negative charges.

We then consider the study of oscillations of all three flavors. Our high energy neutrino beam is well above tau threshold, therefore opening the channel of tau appearance searches. These would be optimized for the $\Delta m^2$ mass indicated by the atmospheric neutrino anomaly since the baseline is such that the relevant parameter $E_\nu/L$ is exactly in that range of $\Delta m^2$. The transition probabilities will therefore be maximized and the flavor oscillation pattern will be visible as a function of the incoming neutrino energy. This is a main difference with respect to presently planned long baseline beams where given the baseline of 730 km, the basic oscillation can be measured only in the upper part of the region indicated by the atmospheric neutrinos. Extending the baseline without a loss of rate will enlarge the explored $\Delta m^2$ region correspondingly.

# NEUTRINOS FROM DECAYS OF STORED MUONS

The neutrino beams from the decays of stored muons(8) provide an ideal configuration for the study of matter effects. First of all,

- they will contain neutrinos of the electron and muon flavors in same quantity with a well-defined helicity composition depending on the muon charge, i.e. $\mu^- \to e^- \bar{\nu}_e \nu_\mu$ or $\mu^+ \to e^+ \nu_e \bar{\nu}_\mu$.

- the charge of the muon can be easily selected, ideally within each filling of the storage ring.

These features distinguish them from traditional neutrino beams where $\nu_\mu$ dominates, where $\nu_e$ comes from kaon decays since highly suppressed in $\pi \to e \nu_e$, and in which the neutrino-antineutrino configurations are not symmetric. In addition, the fluxes of neutrino beams from muon decays can be easily predicted (since no hadronic processes involved) when the muon polarization is known and are flexible in the choice of the beam energy, since precisely determined by the muon storage ring energy.

Secondly, the well-defined flavor composition can be exploited using a detector with charge identification capabilities (see Ref. (9)). For decays of negative stored muons ($\mu^- \to e^- \bar{\nu}_e \nu_\mu$), the charged current neutrino interactions will produce leading electrons of positive charge ($\bar{\nu}_e N \to e^+ + X$) and leading muons of negative charge ($\nu_\mu N \to \mu^- + X$). In case of neutrino $\nu_e \leftrightarrow \nu_\mu$ oscillations, there will be appearance of negative leading electrons $\nu_\mu \to \nu_e N \to e^- + X$ and of positive leading muons $\bar{\nu}_e \to \bar{\nu}_\mu N \to \mu^+ + X$. For decays of positive stored muons, the charge conjugate configurations will occur.

In the case of massive detectors, it will be easier to measure the charge of leading muons than that of electrons. The golden channel to look for oscillations is therefore provided by the search for leading muons of opposite sign of the stored muons. Some information will also be available from the leading electron sample without charge discrimination but with lesser sensitivity.

# VERY LONG BASELINES AND NEUTRINO SOURCE REQUIREMENTS

Since muon colliders necessarily require very intense proton sources, it will be possible to obtain very intense neutrino beams to perform long and very long baseline oscillation experiments[1].

---

[1] In Ref. (10), we have considered a $\nu_\mu$ disappearance experiment with a baseline of 730 km. Since the neutrino beams contain both electron and

We assume here for the sake of a concrete example that the neutrino detector will be located in Europe at the Laboratori Nazionali del GranSasso (LNGS). The neutrino source could then be located in different laboratories around the world, in the American (BNL, FNAL) or Asian (KEK) continent.

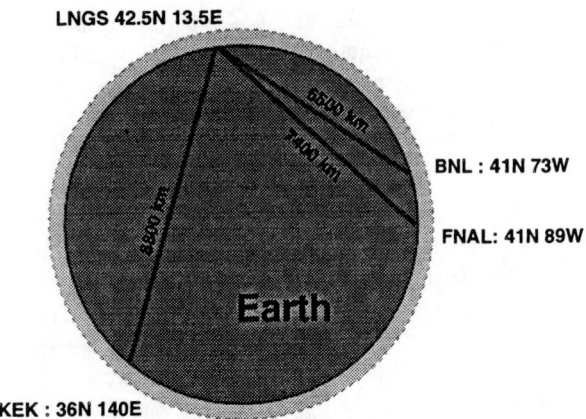

**FIGURE 1.** Possible very long baselines across the Earth (seen from above the North pole).

The BNL-LNGS has a baseline of $L = 6500$ km (see Figure 1). The beam goes to a maximum depth of 900 km and arrives at GS with an angle with respect to horizontal of about $30°$. We estimate the average density of the Earth(3) for this baseline to be $\rho = 3.6$ g/cm$^3$. Other baselines are FNAL-GranSasso ($L = 7400$ km, max. depth 1200 km, average density $\rho = 3.7$ g/cm$^3$, beam angle $36°$) or KEK-GranSasso ($L = 8800$ km, max. depth 1800 km, average density $\rho = 4.0$ g/cm$^3$, beam angle $44°$). In a neutrino beam from stored decaying muons, the neutrino event rate will grow as $E_\mu^3$ where $E_\mu$ is the energy of the muon storage ring. We are interested in a neutrino beam with energy in the range of $10 - 30$ GeV. We list in Table 1 the neutrino event rates in case of no oscillation as a function of the muon storage ring energy, for $10^{21}$ muon decays of a given charge and 10 kton target. Neutrino interactions have been divided into charged current (CC) and neutral current (NC) interactions. The CC events include inelastic scattering (DIS) and quasi-elastic (QE) interactions.

To reach $10^{21}$ muon decays, we refer to recent simulations performed for muon colliders studies(7) which consider a scheme in which the yield of accumulated muons in the storage ring per 16 GeV proton impinging on the primary target is about 10-15%.

---

muon flavors, the disappearance of muon neutrino can be performed by directly comparing electron and muon events. This method is self normalizing, i.e. without the need of a near detector to predict the original flux.

Such a storage ring would be operated for four years, alternating runs with positive and negative muons. Given the shape of the muon storage ring, about 40% of the muons would decay in the direction towards the neutrino detector (the rest decays in the bending section and in the opposite direction). To achieve an integrated intensity a total of

$$N(\mu^+) + N(\mu^-) = N_p \times 0.15 \times 0.4 \times 4 \simeq 2 \times 10^{21} \mu\text{'s}$$

requires a proton source of the order of $N_p \simeq 10^{22}$ protons/year. An upgrade of the AGS accelerator in BNL(11) could yield an integrated intensity close to this figure.

# MATTER ENHANCED NEUTRINO OSCILLATIONS

In three-family scenarios(12), the mixing between the neutrino flavors will be determined by a $3 \times 3$ unitary matrix describing the mixing between flavor and mass eigenstates. With three neutrino states, oscillations will be determined by only two independent mass square differences, say $\Delta m_{32}^2 = m_3^2 - m_2^2$ and $\Delta m_{21}^2 = m_2^2 - m_1^2$.

The current experimental results on solar and atmospheric neutrinos allow us to consider the approximation that only one mass scale is relevant, since if we assign $\Delta m_{32}^2 (\Delta m_{21}^2)$ to the atmospheric(solar) oscillations, then $\Delta m_{32}^2 \gg \Delta m_{21}^2$. This implies that the two oscillations driven by the mass differences $\Delta m_{32}^2$ and $\Delta m_{21}^2$ decouple and can be studied independently.

The three-family oscillation is then described by only three parameters: the mass difference $\Delta m_{32}^2$ and the two mixing angles $\theta$ and $\phi$. The mass eigenstate $m_1$ is defined orthogonal to the electron flavor state. Assuming CP conservation, the mixing matrix takes then the form:

$$U = \begin{pmatrix} 0 & \cos\theta & \sin\theta \\ -\cos\phi & -\sin\theta\sin\phi & \cos\theta\sin\phi \\ \sin\phi & -\sin\theta\cos\phi & \cos\theta\cos\phi \end{pmatrix} \quad (2)$$

We assume that $\nu_3$ is the heaviest state and the atmospheric neutrino data implies that the mixing is maximal between $\nu_\mu$ and $\nu_\tau$ ($\sin^2\phi = \cos^2\phi = 0.5$). The flavor oscillation probabilities for neutrinos and antineutrinos are then simply:

$$P(\nu_e \to \nu_\mu) = P(\nu_e \to \nu_\tau) = P(\bar\nu_e \to \bar\nu_\mu) = P(\bar\nu_e \to \bar\nu_\tau) = \quad (3)$$

$$\sin^2 2\theta \sin^2\left(1.27 \Delta m_{32}^2 (eV^2) \frac{L(km)}{E_\nu(GeV)}\right)$$

The negative result from CHOOZ(13) on electron disappearance and the SuperKamiokande data themselves constrain the mixing angle $\theta$. A value compatible with observation is $\sin^2\theta = 0.025(6)$.

**Table 1.** The total number of neutrinos detected in a 10 kton (fiducial) detector for a baseline $L = 6500$ km and a total number of $10^{21}$ muons decays.

| $E_\mu$ (GeV) | $10^{21}\mu^-$ decays | | | $10^{21}\mu^+$ decays | | |
|---|---|---|---|---|---|---|
| | CC($\bar{\nu}_e$) | CC($\nu_\mu$) | NC | CC($\nu_e$) | CC($\bar{\nu}_\mu$) | NC |
| 10 | 426 | 1152 | 196 | 1016 | 488 | 173 |
| 15 | 1414 | 3751 | 1129 | 3283 | 1624 | 1005 |
| 20 | 3313 | 8712 | 3542 | 7588 | 3820 | 3172 |
| 25 | 6412 | 16746 | 8221 | 14568 | 7401 | 7419 |
| 30 | 11010 | 28576 | 16008 | 24850 | 12710 | 14524 |

In matter, the modification of the flavor transition is taken into account by the mixing angle in matter $\theta_m$, which is:

$$\sin^2 2\theta_m^\mp(x) = \frac{\sin^2 2\theta}{\sin^2 2\theta + (x \mp \cos 2\theta)^2} \quad (4)$$

where the minus sign applies to $\nu$'s and the plus to $\bar{\nu}$'s and

$$x = \frac{2\sqrt{2}G_F n_e E_\nu}{\Delta m^2} \approx 0.76 \times 10^{-4} \frac{\rho(g/cm^3) E_\nu(GeV)}{\Delta m^2 (eV^2)}. \quad (5)$$

where $n_e$ is the electron density of the medium. The transition probabilities are then:

$$P(\nu_e \to \nu_\mu) = P(\nu_e \to \nu_\tau) = \sin^2(2\theta_m^-) \times \quad (6)$$

$$\sin^2\left(1.27\Delta m_{32}^2(eV^2)\frac{\lambda^-(km)}{E_\nu(GeV)}\right) \quad (7)$$

$$P(\bar{\nu}_e \to \bar{\nu}_\mu) = P(\bar{\nu}_e \to \bar{\nu}_\tau) = \sin^2(2\theta_m^+) \times \quad (8)$$

$$\sin^2\left(1.27\Delta m_{32}^2(eV^2)\frac{\lambda^+(km)}{E_\nu(GeV)}\right) \quad (9)$$

where $\lambda^\pm = L \times \sqrt{\sin^2 2\theta + (x \pm \cos 2\theta)^2}$. For neutrinos, the resonance condition will be met when $x(E_\nu, \Delta m^2, \rho) \simeq \cos 2\theta$ and the oscillation amplitude will reach a maximum. For the resonant neutrino energy, $E_\nu^{res}$, this reads

$$E_\nu^{res} \approx \frac{1.32 \times 10^4 \cos 2\theta \Delta m^2(eV^2)}{\rho(g/cm^3)} \approx 0.37 \times 10^4 \Delta m^2(eV^2) \quad (10)$$

where we have assumed a constant Earth density $\rho = 3.6$ g/cm³ and small mixing angles. For the parameters indicated by the atmospheric neutrino observation $10^{-3} < \Delta m^2 < 10^{-2}$ eV², we obtain $3.7 < E_\nu^{res} < 37$ GeV, i.e. the resonance energy lies in the region accessible to high energy accelerator neutrino beams.

The conversion from $\nu_\mu \to \nu_\tau$ flavor is independent of matter effects and is given by the probability:

$$P(\nu_\mu \to \nu_\tau) = P(\bar{\nu}_\mu \to \bar{\nu}_\tau) \approx \sin^2\left(1.27\Delta m_{32}^2(eV^2)\frac{L(km)}{E_\nu(GeV)}\right) \quad (11)$$

where we have made the approximation that $\cos^2\theta = 1$ since $\theta$ is a small angle. Given our choice of $E_\nu/L \approx \Delta m_{32}^2$, there will be large oscillation to $\nu_\tau(\bar{\nu}_\tau)$.

# DETECTION OF NEUTRINO OSCILLATIONS AND MATTER EFFECTS

We consider in more detail neutrino oscillations for $\Delta m^2 = 3 \times 10^{-3}$ eV². Our conclusion remain unchanged for values of $\Delta m^2$ in the range $10^{-2} < \Delta m^2 < 10^{-3}$ eV² (see Ref. (14)). We study the three-family oscillation of the two neutrino flavors present in the beam. In the case of a neutrino beam from negative muons, we will observe $\nu_\mu \to \nu_e$, $\nu_\mu \to \nu_\tau$, and $\bar{\nu}_e \to \bar{\nu}_\mu$, $\bar{\nu}_e \to \bar{\nu}_\tau$ (for positive muons the charge-conjugate processes).

The oscillated neutrino fluxes as a function of energy for the different flavors are shown in Figure 2 for a 30 GeV muon beam. The difference between the $\bar{\nu}_e \to \bar{\nu}_\mu$ in $\mu^-$ decays and the $\nu_e \to \nu_\mu$ in $\mu^+$ decays is clearly visible. We observe a similar effect for $\nu_e(\bar{\nu}_e)$ appearance in $\mu^-(\mu^+)$ decays. In both cases, the $\nu_\mu \to \nu_\tau$ ($\bar{\nu}_\mu \to \bar{\nu}_\tau$) conversions are maximal and correspondingly the $\nu_\mu(\bar{\nu}_\mu)$ are highly depleted. The neutrinos will be detected in their charged and neutral current interactions.

The detector should have excellent electron and muon identification and measurement capabilities. These lepton capabilities should be matched to the beams from muon decays, which provide equal amounts of electron and muon neutrinos. These features would be all met by a large detector based on liquid argon imaging technology (see Ref (15) and references therein). We also require a charge-determination by means of a muon spectrometer, to allow the identification of the leading muon charge. On the other hand, the leading electron charge cannot be identified.

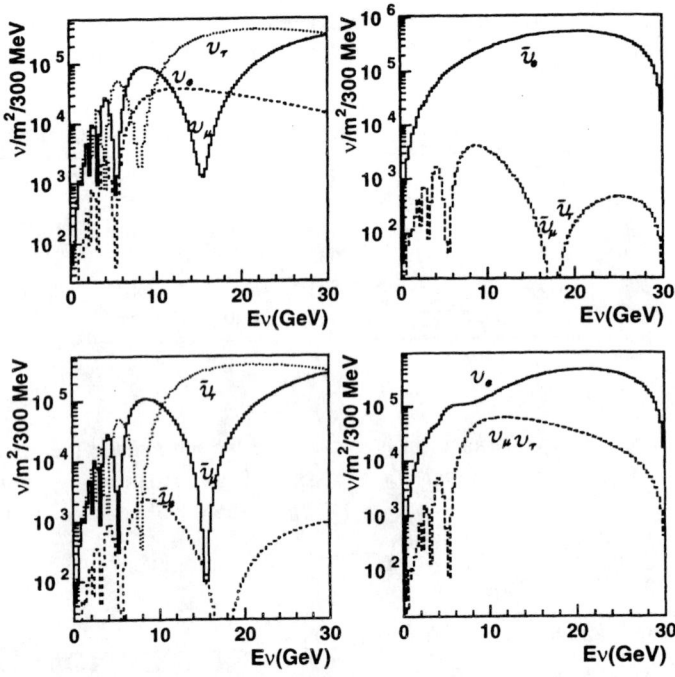

**FIGURE 2.** The oscillated neutrino fluxes reaching the detector per $m^2$ normalized to $10^{21}$ muon decays. The oscillation probability is given for three-family mixing with $\Delta m^2 = 3 \times 10^{-3} eV^2, \sin^2\phi = 0.5, \sin^2\theta = 0.025$. The two upper plots refer to decays of 30 GeV $\mu^-$, the two lower to decays of $\mu^+$. The matter enhancement(suppression) is clearly visible for the $\nu_e(\bar{\nu}_e)$ cases. The $\nu_\mu(\bar{\nu}_\mu)$ fluxes are largely suppressed with a spectacular "hole" in the spectrum due to maximal oscillation to $\nu_\tau$. The $\nu_e \to \nu_\mu$ and $\nu_e \to \nu_\tau$ ($\bar{\nu}_e \to \bar{\nu}_\mu$ and $\bar{\nu}_e \to \bar{\nu}_\tau$) contributions are also visible.

We classify the observed events in four classes:

a) events with electrons or positrons (no electron charge measured),

b) events with muons of the same sign of those circulating in the storage ring,

c) events with muons of opposite sign,

d) events without leptons.

In this study, we do not consider the possibility to directly identify appearance of $\nu_\tau$ neutrinos by means of a direct detection of the tau lepton in final states[2]. Therefore, according to the decay mode of the tau lepton, $\nu_\tau$ CC events are seen in the electron ($\tau \to e$ decays), muon ($\tau \to \mu$) or neutral current ($\tau \to h$) sample. Since for the parameters of the atmospheric neutrinos, the oscillation probabilities are large, a tau appearance signal appears as a clear excess of events, in particular in the neutral current like sample, which corresponds to the largest branching fraction of the tau.

In the following sections, we compute the number of events expected in each of the four classes for an integrated muon intensity of $10^{21}$ decays of each charge. For $\nu_\tau$ interactions, the tau lepton kinematical suppression has been taken into account and rates include quasi-elastic and deep-inelastic contributions. We will use the notation $P_{\nu_\alpha \to \nu_\beta}$ to identify the probability of neutrinos of flavor $\alpha$ to oscillate into neutrinos of flavor $\beta$.

## Events with opposite-sign muons

The muon charge is measured by a spectrometer placed behind the detector. Opposite-sign leading muons can only be produced by neutrino oscillations, since there is no component in the beam that could account for them

---

[2] These can be achieved either by "topological" searches in which a decay kink is looked for or by "kinematical" methods in which the presence of unseen neutrinos in the final states is detected by an analysis of the final state kinematics

[3]. These events are coming from oscillations of the electron component of the beam.

For the case of $\mu^-$ decays, we have appearance of opposite-sign muons

1. directly via $\bar{\nu}_e \to \bar{\nu}_\mu$ oscillations

   $P_{\bar{\nu}_e \to \bar{\nu}_\mu} \times \Phi_{\bar{\nu}_e}$

2. via $\tau$ decays after $\bar{\nu}_e \to \bar{\nu}_\tau$ oscillations.

   $P_{\bar{\nu}_e \to \bar{\nu}_\tau} \times BR(\tau \to \mu) \times \Phi_{\bar{\nu}_e}$

These contributions are listed in Table 2 from decays of positive and negative muons.

The matter effect appears clearly. For beams from decays of $\mu^-(\mu^+)$, the oscillation is suppressed(enhanced). The effect is very striking for example for $E_\mu = 30$ GeV, where we expect 24 matter-suppressed $\mu^+$ events in $\mu^-$ decays against 1703 matter-enhanced $\mu^-$ events in $\mu^+$ decays. In addition, there will be 90 $\nu_e \to \nu_\tau$ followed by $\tau^- \to \mu^-$ in $\mu^+$ and 1 $\bar{\nu}_e \to \bar{\nu}_\tau, \tau^+ \to \mu^+$ in $\mu^-$ decays.

## Events with right-sign muons

Leading muons of the same sign of those decaying in the storage ring are produced by:

1. unoscillated muon neutrinos coming from the beam

   $(1 - P_{\nu_\mu \to \nu_e} - P_{\nu_\mu \to \nu_\tau}) \times \Phi_{\nu_\mu}$

2. $\tau$ decays, where the $\tau$ is coming from $\nu_\mu \to \nu_\tau$ oscillations.

   $P_{\nu_\mu \to \nu_\tau} \times BR(\tau \to \mu) \times \Phi_{\nu_\mu}$

The right-sign muon events are strongly depleted due to the maximal oscillation to $\nu_\tau$'s. Given the proper $L/E_\nu$, the conversion to the tau flavor is maximized. The disappearance of right-sign muons will yield in the usual way the $\Delta m^2$ parameter. These oscillations are almost unaffected by matter because of the mixing angle $\sin\theta$ is much smaller than $\sin\phi$.

## Events with electrons

Events with leading electron or positron are produced by the charged-current interactions of the following neutrinos:

1. unoscillated $\bar{\nu}_e$ neutrinos from the beam

   $(1 - P_{\bar{\nu}_e \to \bar{\nu}_\mu} - P_{\bar{\nu}_e \to \bar{\nu}_\tau}) \times \Phi_{\bar{\nu}_e}$

2. muon neutrinos oscillated into electron neutrinos

   $P_{\nu_\mu \to \nu_e} \times \Phi_{\nu_\mu}$

3. tau neutrinos derived from oscillations followed by a $\tau \to e$ decay

   $P_{\nu_\mu \to \nu_\tau} \times BR(\tau^- \to e) \times \Phi_{\nu_\mu} + P_{\bar{\nu}_e \to \bar{\nu}_\tau} \times BR(\tau^+ \to e) \times \Phi_{\bar{\nu}_e}$

In case of $\mu^-$ beams, process 1) will deplete the number of leading electron events from the beam, while process 2) will increase the number of such events which come from matter-enhanced oscillations of muon neutrinos. The net effect is an increase of the number of observed events having a leading electron with respect to the case of no oscillation, as can be seen by comparing Table 3 with the non-oscillated case in Table 1. For example for 30 GeV $\mu^-$ decays, in case of no oscillations, we expect 11010 $\bar{\nu}_e$ CC events against 10967($\bar{\nu}_e$)+1366($\nu_\mu \to \nu_e$) CC in case of oscillations. For $\mu^+$ decays, we have 24850 $\nu_e$ CC events against 21457($\nu_e$)+15($\bar{\nu}_\mu \to \bar{\nu}_e$) CC in case of oscillations.

In the case of $\mu^+$ beams, process 1) involves neutrinos, so the depletion of electron events from the beam is matter-enhanced, while process 2) is suppressed. The net effect is a smaller number of observed events with leading electrons with respect to the expectations, as can be seen in Tables 3 and 1.

In conclusion, even without charge discrimination, the matter effect asymmetry is visible between leading electron events from $\mu^-$ and from $\mu^+$ decays.

## Events with no leading leptons

Events with no leading electrons or muons will be used to study the $\nu_\mu \to \nu_\tau$ oscillations. These events can be produced in

1. neutral current processes

2. hadronic $\tau$ decays

   $P_{\nu_\mu \to \nu_\tau} \times BR(\tau \to hadrons) \times \Phi_{\nu_\mu} + P_{\bar{\nu}_e \to \bar{\nu}_\tau} \times BR(\tau \to hadrons) \times \Phi_{\bar{\nu}_e}$

If there are no sterile neutrinos, the neutral current processes do not depend on the oscillations, so there is always an excess of events in this class with respect to the expectation due to the hadronic $\tau$ decays. The total number of events for these two categories is shown in Table 4, for both $\mu^-$ and $\mu^+$ beams. The large excess of neutral current like events will be a clear signal for conversion to $\nu_\tau$ flavor. A $\nu_\mu$ disappearance signal not compensated by these events would be a clear sign for a different type of conversion, like for example to a sterile neutrino.

---

[3] Opposite sign muon background comes from decays of hadrons in neutral current events and from charm decays. We note however that these muons will be soft and not isolated from the jet and can therefore be suppressed by a mild isolation and momentum cut. For charm produced in charged current events, the leading lepton should be misidentified.

**Table 2.** Contributions to leading-muon events normalized to $10^{21}\mu$ decays of each sign for a stored muon energy of 10, 20 and 30 GeV. The appearance of opposite-sign muons is enhanced(suppressed) by matter effects for decaying $\mu^+(\mu^-)$ beams.

| $E_\mu$ (GeV) $\mu^+$ decays | $\nu_e \to \nu_\mu$ $\mu^-$ | $\nu_e \to \nu_\tau$ $\tau^- \to \mu^-$ | unoscillated beam $\mu^+$ | $\bar{\nu}_\mu \to \bar{\nu}_\tau$ $\tau^+ \to \mu^+$ |
|---|---|---|---|---|
| 10 | 186 | 4 | 349 | 3 |
| 20 | 1086 | 44 | 562 | 183 |
| 30 | 1703 | 90 | 3484 | 655 |
| $\mu^-$ decays | $\bar{\nu}_e \to \bar{\nu}_\mu$ $\mu^+$ | $\bar{\nu}_e \to \bar{\nu}_\tau$ $\tau^+ \to \mu^+$ | unoscillated beam $\mu^-$ | $\nu_\mu \to \nu_\tau$ $\tau^- \to \mu^-$ |
| 10 | 4 | 0 | 580 | 6 |
| 20 | 22 | 1 | 1049 | 370 |
| 30 | 24 | 1 | 7893 | 1390 |

**Table 3.** Contributions to leading electron events normalized to $10^{21}\mu$ decays of each sign for a stored muon energy of 10, 20 and 30 GeV. The appearance of electrons muons is enhanced(suppressed) by matter effects for decaying $\mu^-(\mu^+)$ beams.

| $E_\mu$ (GeV) | unoscillated beam | $\nu_\mu \to \nu_e$ | $\tau$ decay |
|---|---|---|---|
| 10 | 418 | 252 | 6 |
| 20 | 3268 | 977 | 380 |
| 30 | 10967 | 1366 | 1423 |
| $E_\mu$ (GeV) | unoscillated beam | $\bar{\nu}_\mu \to \bar{\nu}_e$ | $\tau$ decay |
| 10 | 644 | 5 | 7 |
| 20 | 5413 | 17 | 233 |
| 30 | 21457 | 15 | 761 |

## EVENT ENERGY SPECTRA

In the previous section, we have compared the expected number of events of four classes of events and shown how they are affected by the presence of neutrino oscillations and by matter effects. More information on the nature of the oscillations can be extracted by using the total visible energy of an event, which is a measure for the incoming neutrino energy. Figure 3 shows the energy spectra of the four classes of events defined in section for decays of 30 GeV muons. The four left hand side plots correspond to $\mu^-$ decays in the storage ring, while the four right hand side plots are for $\mu^+$ decays. The dotted and full lines refer to the predicted distributions without and with oscillations with $\Delta m^2 = 3 \times 10^{-3}$ eV$^2$.

The spectacular disappearance of right-sign muon is visible in Figure 3 plots b). The enhancement(suppression) due to matter effects is clearly visible in the plots c) which contain the opposite-sign muon events. The apperance of tau can be directly observed as an excess of neutral current like events (plots d).

A fit to the observed energy distributions and rates of the right-sign muon events would allow the precise determination of the parameters $\Delta m^2$ and $\phi$ governing the oscillation.

The opposite-sign muons will directly probe the $\theta$ angle and will test the MSW effect. An indirect measurement of the $\Delta m^2$ parameter can be extracted from the energy spectrum of opposite-sign muons since the position of the peak of the distribution is directly related to the resonance energy, which is a function of $\Delta m^2$ (see section eq. 10).

## CONCLUSIONS

The observation of the MSW resonance by a terrestrial experiment for the parameters suggested by the atmospheric neutrino experiments is an integral part of the phenomenology of neutrino oscillations. Although hardly possible with the present generation of long baseline beams, it becomes possible in a synergic approach to the development of the novel muon colliders. The appearance of the MSW resonance will be truly spectacular. The enhancement(suppression) for neutrinos(antineutrinos) will be directly selected by the charge of the muons in the storage ring. The statistics in a 10 kton detector obtained from muon beams with the preliminary parameters needed for muon colliders will be sufficient to accurately measure the resonance parameters, providing a powerful tool to determine the relevant mass squared

**Table 4.** Contributions to events with no leading leptons normalized to $10^{21} \mu$ decays of each sign for a stored muon energy of 10, 20 and 30 GeV. The appearance of neutral current like events due to hadronic tau decays is clearly seen.

| $E_\mu$ (GeV) | $\mu^-$ beam | | $\mu^+$ beam | |
|---|---|---|---|---|
| | NC | $\tau \to$ hadrons | NC | $\tau \to$ hadrons |
| 10. | 196 | 21 | 173 | 25 |
| 20. | 3534 | 1383 | 3168 | 847 |
| 30. | 16061 | 5180 | 14553 | 2771 |

difference $\Delta m^2$. The large statistics at high energy will also allow the study of all neutrino flavor oscillations, including the $\nu_\tau$ apperance, and will provide an accurate determination of the $\Delta m^2$ and the mixing angles.

# REFERENCES

1. L. Wolfenstein, *Phys. Rev.* **D17**, 2369 (1978); *Phys. Rev.* **D 20**, 2634 (1979); Mikheyev and Smirnov, *Sov.J.Nucl.Phys.* **42**, 913 (1986).

2. J.N. Bahcall, P.I. Krastev and A.Y. Smirnov, *Phys. Rev.* **D58**, 096016 (1998) hep-ph/9807216 and references therein.

3. J. Bahcall and P. Krastev, *Phys. Rev.* **C 56**, 2839 (1997), hep-ph/9706239.

4. Y. Fukuda *et al.* [Super-Kamiokande Collaboration], *Phys. Rev. Lett.* **81**, 1562 (1998).

5. K.S. Hirata et al., *Phys. Lett.* B **205**, 416 (1988); *Phys. Lett.* B **280**, 146 (1992).

   Y. Fukuda et al., *Phys. Lett.* B **335**, 237 (1994).

   R. Becker-Szendy *et al.*, (IMB Collab.), *Phys. Rev.* D **46**, 3720 (1992).

   W.W. Allison *et al.* [Soudan-2 Collaboration], *Phys. Lett.* **B449**, 137 (1999) hep-ex/9901024.

   F. Ronga *et al.* [MACRO Collaboration], "Atmospheric neutrino induced muons in the MACRO detector," hep-ex/9810008.

6. P.Lipari, hep-ph/9903481.

7. C.M. Ankenbrandt *et al.*, "Status of muon collider research and development and future plans," FERMILAB-PUB-98-179.

8. S. Geer, *Phys. Rev. D* **57**, 6989 (1998).

9. A.Bueno, M.Campanelli, A.Rubbia "A medium baseline search for $\nu_\mu \to \nu_e$ oscillations at a $\nu$ beam from muon decays" hep-ph/9809252 CERN-EP/98-140 accepted by IJMP.

10. A.Bueno, M.Campanelli, A.Rubbia "Long baseline neutrino oscillation disappearance search using a $\nu$ beam from muon decays" hep-ph/9808485 ETHZ-IPP-98-05 accepted by IJMP.

11. Robert Palmer and Thomas Roser, private communications.

12. see e.g. G.L. Fogli, E. Lisi, and G. Scioscia, *Phys. Rev.* D **52**, 5334 (1995).

13. M. Apollonio *et al.* [CHOOZ Collaboration], *Phys. Lett.* **B420**, 397 (1998), hep-ex/9711002.

14. M.Campanelli, A.Bueno, A.Rubbia, "Three-family oscillations using neutrinos from muon beams at very long baseline", ICARUS-TM-99/13 hep-ph/9905240.

15. ICARUS collaboration, *"ICARUS-Like Technology for Long Baseline Neutrino Oscillations"*, CERN/SPSC/98-33 & M620, 1998 and references therein.

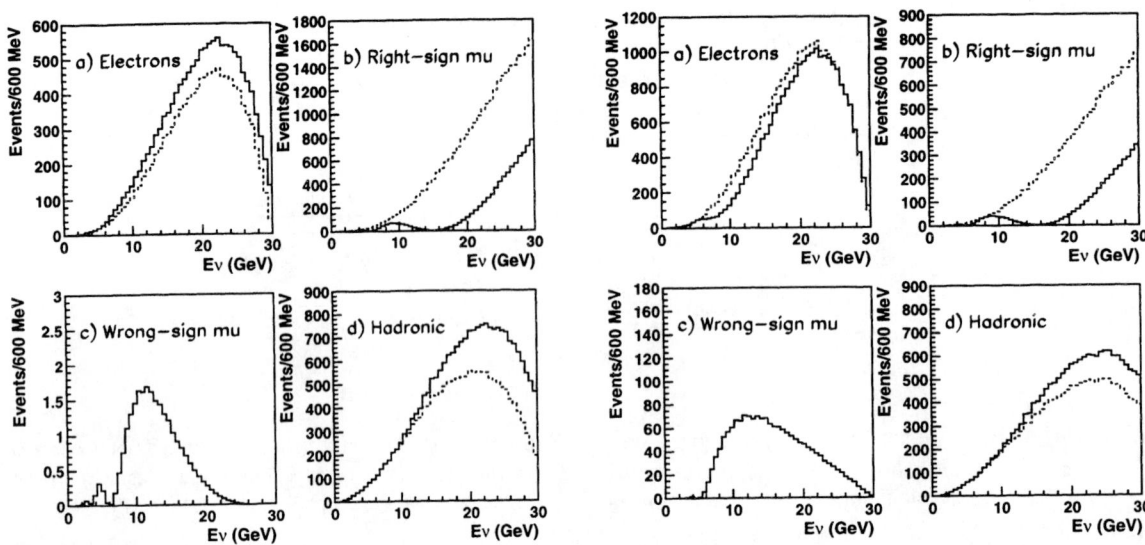

**FIGURE 3.** Predicted event energy spectrum of the four classes of events a) with leading electron or positron, b) with leading right-sign muons, c) with leading opposite sign muons d) events with no leading leptons. Full line: spectra with oscillations; dashed line: spectra without oscillations. The four upper plots are for $10^{21}\mu^-$ decays and the four lower plots are for $10^{21}\mu^+$ decays. The muon energy is 30 GeV and the target mass is 10 kton. The spectacular disappearance of right-sign muon is visible in plots b). The enhancement(suppression) due to matter effects is clearly visible in the plots c) which contain the opposite-sign muon events. The apperance of tau can be directly observed as an excess of neutral current like events (plots d).

# The Booster Neutrino Experiment: BooNE

Rex Tayloe

*Los Alamos National Laboratory*
*Los Alamos, NM 87545, USA*
*E-mail: rex@lanl.gov*
*representing the BooNE Collaboration(1)*

**Abstract.** The Booster Neutrino Experiment (BooNE) has been approved to run at Fermilab and is currently under construction. This experiment will definitively test the result from LSND that indicates neutrino oscillations. If the LSND result is due to $\nu_\mu \to \nu_e$ oscillations, the first stage of the experiment (miniBooNE) will observe approximately 1000 oscillation events in the first calendar year of operation.

## INTRODUCTION

The phenomenon of neutrino oscillations, where a neutrino of one type (e.g. $\nu_\mu$) spontaneously transforms into a neutrino of another type (e.g. $\nu_e$), has important and far-reaching consequences for particle physics and cosmology. For this phenomenon to occur, at least one neutrino must be massive and the heretofore observed lepton flavor conservation law must be violated.

There are, at present, several results that indicate neutrino oscillations and it is not clear how and if these experimental results are indicative of the situation that actually obtains in Nature. There is evidence for a deficit of electron neutrinos observed emanating from the sun. There is evidence for a deficit of electron neutrinos compared to muon neutrinos created in the upper atmosphere. And, there is an *excess* of electron antineutrinos observed in a beam of muon antineutrinos by the LSND experiment.

The Booster Neutrino Experiment (BooNE)(2) will focus on the final observation by definitively testing the LSND result.

## LSND RESULTS

The Liquid Scintillator Neutrino Detector (LSND) searches for $\bar{\nu}_\mu \to \bar{\nu}_e$ oscillations at the Los Alamos National Laboratory. Reactions initiated by a $\bar{\nu}_e$ are searched for in a virtually pure beam of $\bar{\nu}_\mu$. An excess of $\bar{\nu}_e$ events is observed by LSND indicating $\bar{\nu}_\mu \to \bar{\nu}_e$ oscillations(3). The latest results from the complete LSND data set taken 1993-1998 (1996-1998 data are still preliminary) are shown in Fig. 1. With low-background cuts ($E_{e^+} > 36$ MeV), LSND observes 33 $\bar{\nu}_\mu \to \bar{\nu}_e$ oscillation events with $9.5 \pm 0.9$ background events expected.

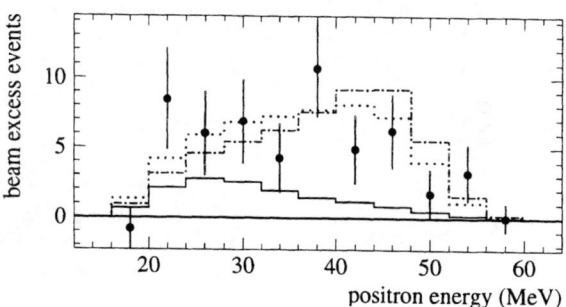

**FIGURE 1.** The $e^+$ energy distribution for $\bar{\nu}_\mu \to \bar{\nu}_e$ candidate events from the 1993-1998 LSND data (1996-1998 data are preliminary). The points show the beam excess data. Also shown are the estimated neutrino background (solid), the estimated distributions for oscillations at small $\Delta m^2$ plus background (dotted), and for large $\Delta m^2$ plus background (dot-dashed). The low-background cuts accept events with $E_{e^+} > 36$ MeV.

The LSND experiment also searches for $\nu_\mu \to \nu_e$ oscillations at higher energies where the neutrino beam is relatively pure in $\nu_\mu$. The results of this search show an excess of 40 $\nu_e$ events with an expected background of 21.9(4). This result (based on the 1993-1995 data set only), albeit less-sensitive, is consistent with the $\bar{\nu}_\mu \to \bar{\nu}_e$ result.

The results of the LSND $\bar{\nu}_\mu \to \bar{\nu}_e$ and $\nu_\mu \to \nu_e$ oscillation searches are shown together in Fig. 2 as favored regions as a function of the neutrino oscillation parameters $\Delta m^2$ and $\sin^2 2\theta$. This analysis is based on a two-generation oscillation model and uses the $\nu_\mu \to \nu_e$ oscillation probability formula,

$$P_{\nu_\mu \to \nu_e} = \sin^2 2\theta \sin^2(1.27 \Delta m^2 L/E_\nu),$$

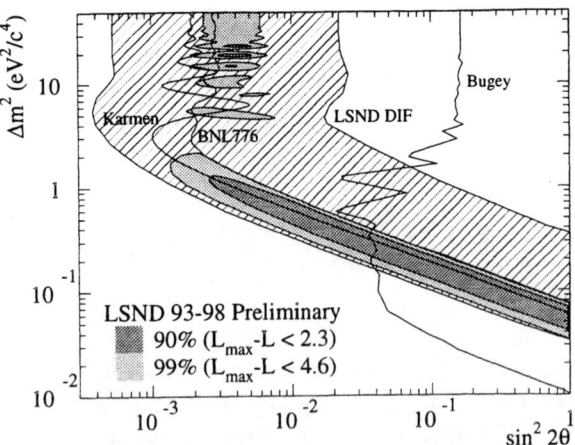

**FIGURE 2.** The LSND likelihood regions (shaded) obtained from the 1993–1998 (preliminary) $\bar{\nu}_\mu \to \bar{\nu}_e$ data set along with the 95% confidence region (lightly hatched) for 1993–1995 $\nu_\mu \to \nu_e$ data set. Also shown are the 90% confidence limits from other experiments that investigate this region of oscillation parameters.

where $\theta$ is the neutrino mixing angle, $\Delta m^2$ is the difference of the squares of the neutrino masses (in eV$^2$), $L$ is the neutrino source to detector distance (in m), and $E_\nu$ is the neutrino energy (in MeV).

These results from LSND are not statistical fluctuations of the background — the fluctuation probability of the $\bar{\nu}_\mu \to \bar{\nu}_e$ ($\nu_\mu \to \nu_e$) signal is $\sim 10^{-8}$ ($\sim 10^{-2}$). Possible backgrounds have thoroughly investigated(3, 4). Nevertheless, for a result this important, a verification experiment is required. This is the motivation for BooNE.

## BOONE

The Booster Neutrino Experiment, BooNE (FNAL-E898) will check these results from LSND with a $\nu_\mu \to \nu_e$ *appearance* oscillation search at Fermilab. This experiment will use a different neutrino production technique at a different proton energy. The detection reaction will be different from LSND as will as the signal backgrounds. Thus, BooNE will have different systematic errors and will be a stringent test of the LSND result.

The first stage of the experiment (miniBooNE) is approved and currently under construction. For this experiment to have the sensitivity to definitively confirm or refute the LSND result it will be necessary to have an intense, high-purity source of $\nu_\mu$ as well as a large, efficient detector of $\nu_e$.

## Neutrino Source

The protons for the $\nu_\mu$ source will be produced by the 8 GeV proton booster accelerator at FNAL. An average of 5 of the 15 Booster pulses per second (with $5 \times 10^{12}$ protons/pulse) will be diverted just before entering the Main Injector ring into a new extraction line to the neutrino production target.

The protons impinge on a metal target rod embedded in a electrically pulsed horn system which magnetically focuses the pions toward the detector. The pions decay ($\pi^+ \to \mu^+ \nu_\mu$) as they travel down a 50m long decay pipe, thus producing the flux of $\nu_\mu$. The length of the decay pipe is optimized to allow pion decay yet keep the background $\nu_e$ from muon decay ($\mu^+ \to e^+ \nu_e \bar{\nu}_\mu$) low. There is provision in this beam pipe to install a secondary absorber at 25m to allow a systematic study of this $\nu_e$ background.

The 8 GeV FNAL booster beam is an excellent source for this experiment. At this energy, the interaction probability of $\nu_e$ in the detector is quite large. At the same time, kaon production in the target is fairly small (keeping $K \to \nu_e$ background low) and $\pi^0$ production in the detector (a potential background) is also suppressed. Another feature of the booster beam is that the proton pulse is quite short (a few $\mu$s), so the neutrino beam arrives at the detector in a very narrow time window which makes cosmic ray backgrounds almost negligible.

The $\nu_\mu$ and $\nu_e$ fluxes as calculated in a computer simulation are shown in Fig. 3. As desired, the $\nu_\mu$ flux is intense while the $\nu_e$ flux is suppressed by more than 2 orders of magnitude.

## Neutrino Detector

The miniBooNE detector consists of a 12m spherical tank filled with 807 tons of mineral oil (CH$_2$) (see Fig. 4). A thin optical barrier divides the tank into a 445 ton inner fiducial region and an outer, 35cm-thick, veto region (see Fig. 5). Particles with $\beta > 1/n_{oil}$, $n_{oil} \sim 1.47$ will produce Čerenkov light which is viewed by 1280 8in PMTs in the fiducial region and 240 8in PMTs in the veto region. The veto region allows particles entering or exiting to be tagged.

The tank will be installed in a concrete vault below ground level which provides a temperature-stable environment as well as secondary containment for the mineral oil. A circular room will be built above the vault area to contain the detector electronics and oil plumbing equipment. The entire facility will be shielded with 3m of earth overburden to remove the hadronic component of cosmic–ray showers and to lower the muon rate in the detector to around 10KHz.

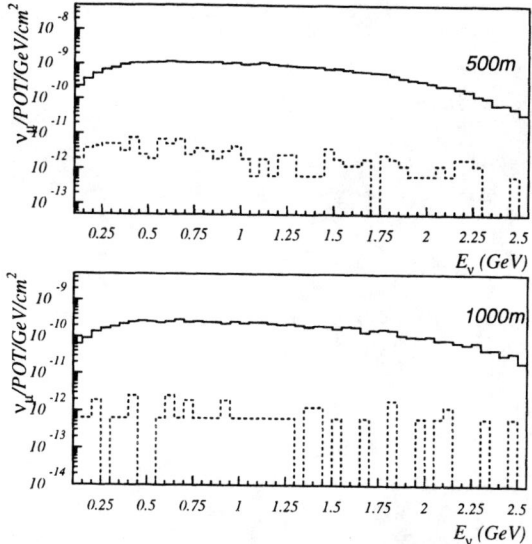

**FIGURE 3.** The calculated neutrino flux produced by the FNAL 8 GeV proton beam on the BooNE neutrino target at 500m and 1000m locations. The $\nu_\mu$ ($\nu_e$) flux is indicated by the solid (dashed) histogram.

The detector will be located ~500m from the neutrino production target on the Fermilab site. This distance was chosen so that the quantity $L/E$ for miniBooNE was the same as for LSND to optimize the sensitivity to a signal as indicated by the LSND result.

The search for appearance of $\nu_e$ will be performed by detecting the Čerenkov light from the electron produced in the $\nu_e$-induced reaction $\nu_e C \to e^- X$. Potential backgrounds of this signature are $\nu_\mu C \to \mu^- X$ and $\nu_\mu C \to \pi^0 X$ where the $\mu^-$ or $\pi^0$ is misidentified as an $e^-$. Signals from the electrons are separated from the more frequently produced $\mu^-$ and $\pi^0$ based on the pattern and timing of PMT hits induced by the distinctive pattern of Čerenkov light.

Extensive computer simulations have shown that the misidentification probability of the $\nu_\mu C \to \mu^- X$ and $\nu_\mu C \to \pi^0 X$ reactions can be kept at less than 1 part in 1000 and 1 part in 100 respectively while maintaining an efficiency for correctly identifying $\nu_e C \to e^- X$ at approximately 50%.

## Expected Results

The number of events expected in the various reaction channels after the first calendar year of running (with $2 \times 10^7$ s of accelerator time) are shown in Table 1.

First, note the prodigious number of events due to $\nu_\mu C \to \mu^- X$ and $\nu_\mu C \to \pi^0 X$. These will allow the detector to be thoroughly studied and understood (as well as producing interesting physics on their own). The number of $\nu_e C \to e^- X$ given 100% $\nu_\mu$ transmutation is provided for normalization. There will be 1200 events detected due to $\nu_\mu \to \nu_e$ oscillations if LSND is correct and the oscillation parameters are as given in the table. The last 3 rows show the backgrounds expected from $\nu_e$ due to $\mu^+ \to e^+ \nu_e \bar{\nu}_\mu$ in the decay pipe of the target, from $\mu^-$ (from $\nu_\mu C \to \mu^- X$) misidentified as $e^-$, and from $\pi^0$ (from $\nu_\mu C \to \pi^0 X$) misidentified as $e^-$. The energy distribution of this potential signal together with the three backgrounds is plotted in Fig. 6.

The backgrounds shown in Table 1 and Fig. 6 may be checked in several ways. The number of $\nu_e$ due to $\mu^+ \to e^+ \nu_e \bar{\nu}_\mu$ in the target is well-constrained by measuring the $\nu_\mu C \to \mu^- X$ reaction since the initiating $\nu_\mu$ is strongly correlated to the $\mu^-$ in the decay pipe (via $\pi^+ \to \mu^+ \nu_\mu$) which give rise to the background $\nu_e$ (via $\mu^+ \to e^+ \nu_e \bar{\nu}_\mu$). Also, as mentioned above, provision for a secondary absorber in the decay pipe at the 25m position has been added. By running with this 25m absorber inserted, the oscillation hypothesis may be checked against a background hypotheses. If a signal is truly due to $\nu_\mu \to \nu_e$ oscillations, the event rate should be halved as the decay pipe length is halved from 50m to 25m because the $\nu_\mu$ flux is reduced proportionally to decay length. If the signal is due to a error in the $\nu_e$ flux calculation, the event rate would be reduced by a factor of 4, as this signal source depends on the decay of both the pion *and* the muon. These techniques will allow the systematic error on $\nu_e$ background from the target to be held to 5%.

The backgrounds due to misidentification of $\mu^-$ and $\pi^0$ from the $\nu_\mu C \to \mu^- X$ and $\nu_\mu C \to \pi^0 X$ will be studied and understood thoroughly through the large number of correctly identified events in the detector. The extrapolation to obtain the number of events misidentified will be a straight-forward procedure. This process will keep the systematic error from these reactions below 5% as well.

The expected results after one calendar year are summarized on the oscillation parameter plot show in Fig. 7. If the LSND result is incorrect, then no signal will be seen in miniBooNE and the resulting 90% confidence limit region will be approximately that delimited by the solid line dark line shown in Fig. 7. If, on the other hand, LSND is correct and the neutrino oscillation parameters $\Delta m^2$ and $\sin^2 2\theta$ are at the values indicated by either of the dots in Fig. 7, miniBooNE will detect approximately 1000 events.

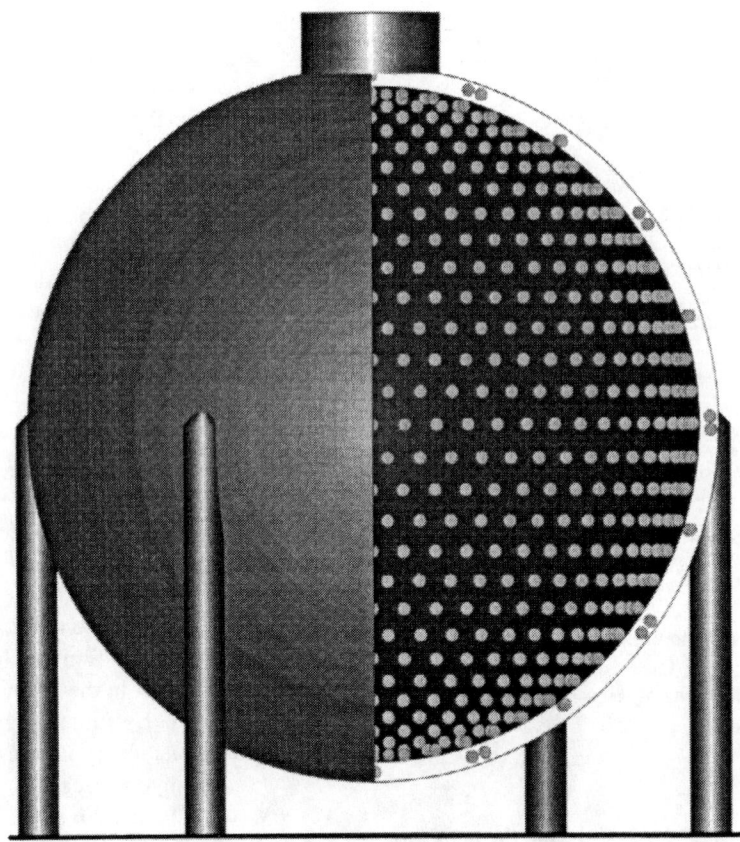

**FIGURE 4.** A schematic cut away view of the 12m miniBooNE detector tank with PMTs indicated by the array of spheres.

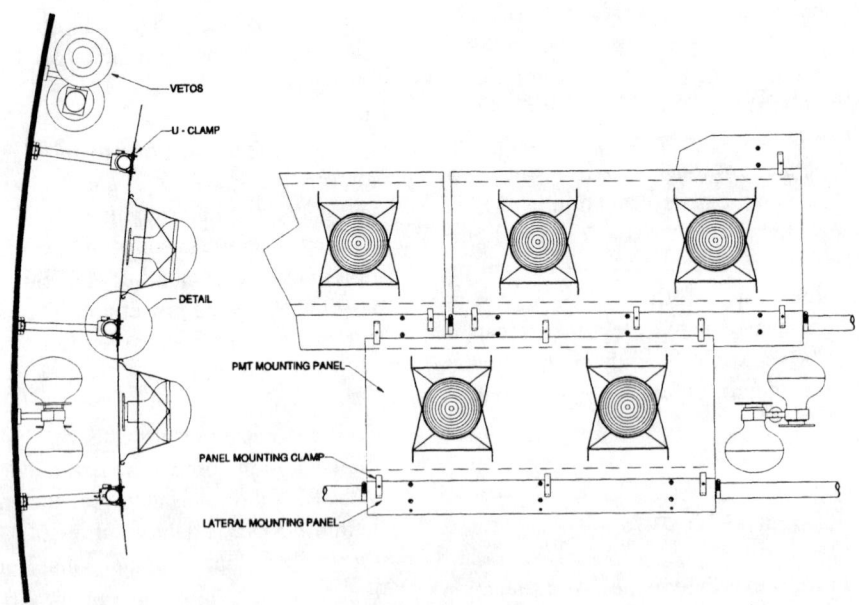

**FIGURE 5.** A section view and head-on view of the miniBooNE phototube support structure. There are 120 pairs of 8in PMTs to view the 35cm thick veto region formed by the thin optical barrier panels. There are 1280 8in PMTs to view the inner fiducial region.

**Table 1.** Number of detected events of various reaction types expected in miniBooNE after the first calendar year of operation ($2 \times 10^7$ s).

| Reaction | Number of events |
|---|---|
| $\nu_\mu C \to \mu^- X$ | 590,000 |
| $\nu_\mu C \to \pi^0 X$ | 65,000 |
| $\nu_\mu \to \nu_e, \nu_e C \to e^- X$ (given 100% $\nu_\mu$ transmutation) | 617,000 |
| $\nu_\mu \to \nu_e, \nu_e C \to e^- X$ ($\Delta m^2 = 0.4 eV^2, \sin^2 2\theta = 0.02$) | 1200 |
| background: $\nu_e$ ($\mu^+ \to e^+ \nu_e \bar{\nu}_\mu$ in target) | 1800 |
| background: $\mu^-$ mis-id ($\nu_\mu C \to \mu^- X$) | 600 |
| background: $\pi^0$ mis-id ($\nu_\mu C \to \pi^0 X$) | 600 |

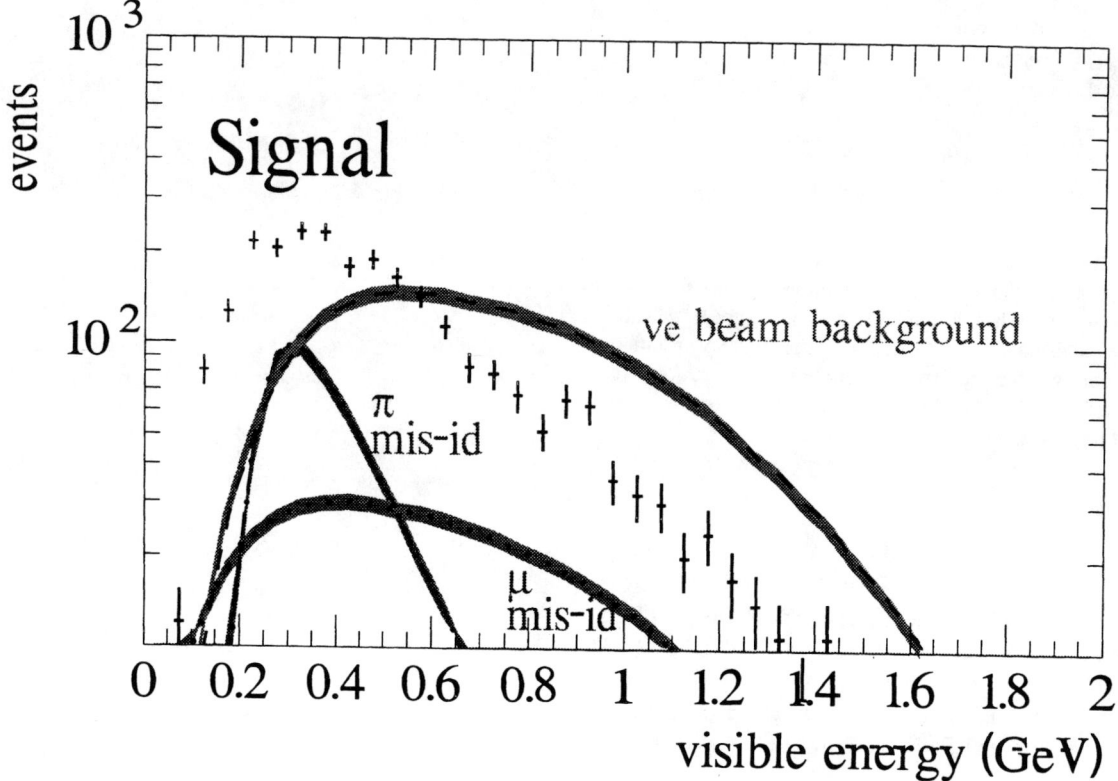

**FIGURE 6.** The expected (electron equivalent) visible energy expected in miniBooNE after the first calendar year of operation from: $\nu_\mu \to \nu_e$ oscillations with $\Delta m^2 = 0.4 eV^2, \sin^2 2\theta = 0.04$ (data points); $\nu_e$ due to $\mu^+ \to e^+ \nu_e \bar{\nu}_\mu$ in the target (dashed); $\mu^-$ misidentified as electrons (dotted); and $\pi^0$ misidentified as electrons (dot-dashed). The error bars on the data points indicate statistical error and the bands on the backgrounds indicate the 5% systematic errors.

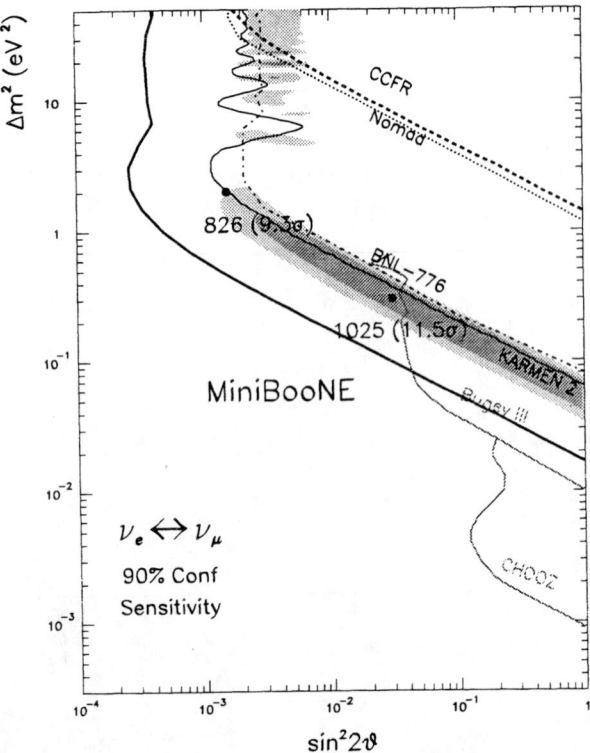

**FIGURE 7.** The miniBooNE sensitivity plot after the first calendar year of running. The 90% confidence level sensitivity is indicated by the heaviest solid line. The numbers indicate the events expected and signal significance if $\nu_\mu \to \nu_e$ oscillations are occuring at the oscillation parameters indicated. Also shown are previous LSND results (shaded regions) and relevant experiment limits (lines).

## CONCLUSIONS

The Booster Neutrino Experiment (BooNE) at Fermilab will provide a definitive test of the LSND result. The first stage of this experiment, miniBooNE, will begin operation at the end of 2001 and thus will provide this important test of the LSND result in a timely manner.

## REFERENCES

1. The BooNE Collaboration consists of the following people and institutions: S. Koutsoliotas (Bucknell University); E. Church, I. Stancu, G. J. VanDalen (University of California, Riverside); R. A. Johnson, N. Suwonjandee (University of Cincinnati); L. Bugel, J. M. Conrad, B. Fleming, J. Formaggio, M. H. Shaevitz, E. D. Zimmerman (Columbia University); D. Smith (Embry Riddle Aeronautical University); C. Bhat, B. C. Brown, R. Ford, P. Kasper, I. Kourbanis, A. Malensek, W. Marsh, P. Martin, F. Mills, C. Moore, A. Russell, P. Spentzouris, R. Stefanski (Fermi National Accelerator Laboratory); G. T. Garvey, E. Hawker, W. C. Louis, G. B. Mills, V. Sandberg, B. Sapp, R. Tayloe, D. H. White (Los Alamos National Laboratory); R. Imlay, A. Malik, W. Metcalf, M. Sung (Louisiana State University); R. Berbeco, B. P. Roe, N. Wadia, J. Yamamoto (University of Michigan); A. O. Bazarko, P. D. Meyers, F. C. Shoemaker (Princeton University).

2. E. Church et al., FNAL Proposal P898 (1997).

3. C. Athanassopoulos et al., Phys. Rev. Lett. **75**, 2650 (1995); C. Athanassopoulos et al., Phys. Rev. C **54**, 2685 (1996); C. Athanassopoulos et al., Phys. Rev. Lett. **77**, 3082 (1996).

4. C. Athanassopoulos et al., Phys. Rev. C **58**, 2489 (1998); C. Athanassopoulos et al., Phys. Rev. Lett. **81**, 1774 (1998).

# Neutrino Oscillation Experiments at Nuclear Reactors

Marco Grassi

*INFN - Pisa, Via Livornese 1291, 56010 S. Piero a Grado (Pisa), Italy*

**Abstract.** The current status of the search for neutrino oscillations at reactors is reviewed, with a particular emphasis given to the final results recently published by the CHOOZ experiment. The results of the Bugey experiments and the status of the PaloVerde experiment are also discussed.

## INTRODUCTION

Nuclear reactors are intense sources of low energy $\bar{\nu}_e$. They thus constitute a powerful tool to study the phenomenon of neutrino flavour oscillations.

In a two neutrino mixing scheme, the survival probability of a pure beam of electron neutrinos can be written, in terms of a mixing parameter $\sin^2 2\theta$ and a squared mass difference $\Delta m^2$, as follows:

$$P = 1 - \sin^2 2\theta \, \sin^2 \left( \frac{1.27 \cdot \Delta m^2 (\text{eV})^2 \cdot L(\text{m})}{E_\nu (\text{MeV})} \right) \quad (1)$$

where $E_\nu$ is the neutrino energy and $L$ is the distance between the neutrino source and the detector. The measurement of the $\bar{\nu}_e$ flux as a function of the distance from the reactor core might reveal possible $\bar{\nu}_e \to \bar{\nu}_x$ oscillations.

Past experiments at Gösgen (1), Bugey (2) and Krasnoyarsk (3) measured the $\bar{\nu}_e$ flux at various distances from the reactor cores up to $\approx 100$ m, setting therefore limits on the oscillation parameter $\Delta m^2$ down to values slightly below $10^{-2}$ eV$^2$ for full mixing.

The present generation of experiments is designed to study an oscillation parameter region completely covering the KAMIKANDE (4) atmospheric neutrino anomaly and extending the past experiment $\Delta m^2$ sensitivity over an order of magnitude. The CHOOZ experiment (5, 6) has recently published its final result (7) obtained with a detector located at $\approx 1$ Km from the two homonymous reactors. The reached sensitivity is $\Delta m^2 = 7.0 \cdot 10^{-4}$ eV$^2$ for maximum mixing. The PaloVerde experiment (8) is located at a similar distance from a group of three reactors and it is currently taking data. The analysis of the first 70 days of operation has been already presented.

The KamLAND experiment (10) will explore $\Delta m^2$ down to $10^{-5}$ eV$^2$ for maximum mixing and it will be therefore the first experiment with an artificial source of $\bar{\nu}_e$ sensitive to the large mixing angle MSW solution of the solar neutrino puzzle (11).

## REACTOR ANTINEUTRINO DETECTION

The main fuel elements of pressurized water reactors (PWR) are the two uranium isotopes $^{235}$U, $^{238}$U and the two plutonium isotopes $^{239}$Pu and $^{241}$Pu. The $\bar{\nu}_e$'s are isotropically produced by the $\beta$ decays of neutron rich fission fragments. The thermal energy emitted per fission of each of these elements (12) is reported in Table 1.

**Table 1.** Energy released per fission

| Isotope | Energy (MeV) |
|---|---|
| $^{235}$U | $201.8 \pm 0.5$ |
| $^{238}$U | $205.0 \pm 0.7$ |
| $^{239}$Pu | $210.3 \pm 0.6$ |
| $^{241}$Pu | $212.6 \pm 0.7$ |

Since the average number of $\bar{\nu}_e$'s emitted per fission is evaluated to be approximately 6, by taking 200 MeV for the average thermal energy per fission, one can easily estimate a production of approximately $2 \cdot 10^{20}$ $\bar{\nu}_e$'s per produced GW per second. The neutrino energy distribution is peaked around 4 MeV and the spectrum is rapidly decreasing being negligible above 8 MeV.

The necessity of exploring $\Delta m^2$ values below $10^{-3}$ eV$^2$ requires detectors placed at more than 1 Km from the reactors, being fixed the neutrino energy spectrum. Typical modern PWR reactors have thermal power in excess of 3 GW$_{th}$ and usually more than one reactor is located in the same nuclear power plant. Although the total number of $\nu$'s produced per second exceeds $10^{21}$, the long source-detector distance renders the measurement of $\nu$-flux at detector location a difficult experimental task. The following items have been taken into account in designing the present experiments in order to achieve the needed sensitivity:

- a large detector mass;

- a distinctive $\bar{\nu}_e$ interaction signature and
- a powerful background suppression.

The reactor neutrino energy spectrum allows only the $\bar{\nu}_e$ disappearance test and not the appearance of other $\nu$ species. The reaction commonly used for the $\bar{\nu}_e$ detection is

$$\bar{\nu}_e + p \rightarrow e^+ + n \quad (2)$$

The distinctive $\bar{\nu}_e$ reaction signature is a delayed coincidence between the prompt $e^+$ signal and the signal of the $\gamma$'s emitted after the neutron capture in the detector (neutron tag) through the process

$$n + (A, Z) \rightarrow (A+1, Z) + \gamma's. \quad (3)$$

The background rejection and a precise measurement of its residual rate are the key points for a successful neutrino oscillation experiment at reactors.

The two main sources of background are the accidental coincidence in time and in space of two distincts radioactivity events (uncorrelated component), and the moderation-absorption sequence of a fast neutron from nuclear spallation by cosmic $\mu$'s in the material surrounding the detector (correlated component). The contribution of the uncorrelated component could be reduced by shielding the detector with low activity material and by adding elements with large thermal neutron absorption cross section and large $\gamma$ de-excitation energy [1], like Lithium or Gadolinium. The detector protection from cosmic $\mu$'s with a large rock overburden directly reduces the cosmic ray $\mu$'s, which are the direct sources of fast neutrons; a further correlated background reduction is achieved by using active cosmic ray veto layers. In addition this background can be reduced by distinguishing, when possible, a positron from a $\bar{\nu}_e$ interaction from a proton recoiled during the fast neutron moderation phase.

Both the CHOOZ and the PaloVerde experiments used all these tools to increase the signal over background ratio at acceptable levels.

Although periods of time with source off would be very useful to study the background, in the case of multiple reactors, the plant optimization requires the refueling of one reactor at a time. The background is therefore measured comparing signal rates at different reactor powers. The CHOOZ experiment is a particular exception: it was operational during the slow reactor power rise for the plant commissioning, thus taking data on the full power range, including both reactors off.

---

[1] A large de-excitation energy separates the neutron tag energy from the $\gamma$ natural radioactivity and enhances the efficiency for the $\bar{\nu}_e$ interaction identification.

# THE EXPERIMENTS AT BUGEY

The main systematic uncertainty in reactor experiments is due to the knowledge of the $\bar{\nu}_e$ flux. The high statistic results of past reactor experiments have been used to accurately test the neutrino flux and energy spectrum predicted by several models.

The Bugey 3 experiment (2, 13) measured the spectrum of positrons produced in Reaction 2 at 15, 40 and 95 meters from the reactor cores. The detector consisted of three identical modules, each of 600 liters filled with a $^6$Li loaded liquid scintillator for neutron detection. The sum of the spectra at 15 and 40 meters were compared with several calculated positron spectra.

The best agreement is obtained with the phenomenological model here described. The cross section of the $\bar{\nu}_e$ interaction process 2 can be accurately calculated (14) in the standard V-A theory by making use of the neutron lifetime measurements. Neutrino spectra for each fuel element are derived from the measured $\beta$ spectra (15) for $^{235}$U, $^{239}$Pu and $^{241}$Pu and directly calculated (16) for $^{238}$U. The energy spectrum of unoscillated $\bar{\nu}_e$'s results from the average of the individual neutrino spectra over the fuel elements concentration. The expected positron spectrum is obtained by convolution of that spectrum with the cross section and the detector response function. The ratio of the data to the prediction, shown in Figure 1, confirms the validity of the uncertainties quoted for the

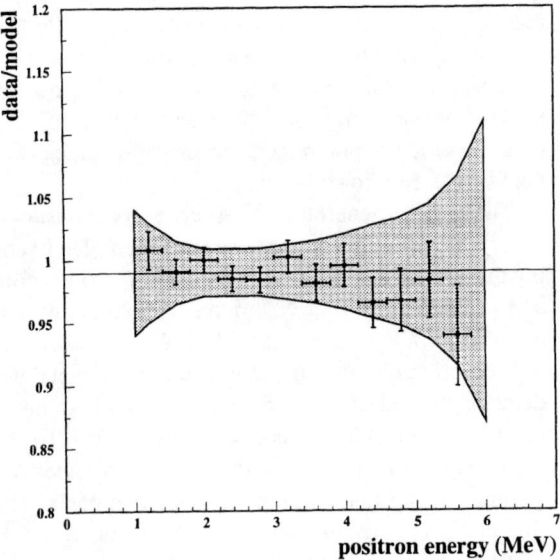

**FIGURE 1.** Ratio of the positron spectra measured at 15 and 40 m by the Bugey 3 experiment to the model described in the text

model: 2.7% for the overall normalization of the neutrino flux and a shape uncertainty ranging from 1.34% at 3 MeV to 9.2% at 8 MeV.

Another experiment (17), still at the Bugey site, used an integral type detector previously utilized at the Rovno (18) nuclear power plant, to measure just the total number of neutrons produced in Reaction 2. The detector consisted of a stainless steel tank of $130 \times 130 \times 120\,\text{cm}^3$ filled with distilled water in which 252 $^3$He counters were used to detect the neutrons. The total number of neutrino events detected $N_\nu$ can be expressed as a function of reactor thermal power ($W_{th}$); of average energy released per fission ($\langle E_f \rangle$); of distance from the reactor (R); of neutrino detection efficiency ($\varepsilon$); of the total number of protons in the target ($N_p$); of the exposure time ($T$) and of the parameter $\sigma_f$, according to the following formula:

$$N_\nu = \frac{W_{th}}{E_f} \frac{1}{4\pi R^2} \varepsilon \, N_p \, \sigma_f \, T \qquad (4)$$

The reason for using this formula is that the systematic error on the flux knowledge is separated from all the other sources of errors and is contained in $\sigma_f$. The measured quantity $\sigma_f$, named "cross section per fission", can be compared with the cross section for Reaction 2 averaged over the unoscillated neutrino spectra and integrated over the neutrino energy ($\sigma_{V-A}$).

The value of the ratio was found to be:

$$\frac{\sigma_f}{\sigma_{V-A}} = 0.987 \pm 1.4\% \pm 2.7\% \qquad (5)$$

where the first error is experimental and the second is due to the prediction. This result confirms again the validity of the model used and suggests the possibility to use the measured value and its experimental error as a normalization for the neutrino flux.

## THE CHOOZ EXPERIMENT

### The detector

A detailed description of the CHOOZ detector can be found in reference (5). The detector is located in an underground laboratory at a distance of about 1 km from two PWR's with a total thermal power of 8.5 GW$_{th}$. Considering an average $\bar{\nu}_e$ energy of $\sim$ 3 MeV the $L/E$ ratio results $\sim$ 300, of the same order of magnitude of the accelerator "long-baseline" neutrino oscillation experiments. In this sense the Chooz experiment is the first "long-baseline" experiment using an artificial neutrino source.

The underground location of the detector is an important feature of the Chooz site. The $\sim$ 300 MWE rock overburden reduces the external cosmic ray muon flux significantly decreasing the fast neutron correlated background. The detector target is a single central volume, delimited by a Plexiglas container, and filled with 5-ton of 0.09% Gd–loaded liquid scintillator. The scintillation light is collected by 192 eight-inch PMT's, and the entire detector is protected by an active cosmic-ray muon veto shield.

As usual $\bar{\nu}_e$'s are detected via Reaction 2 and the neutrons are captured on Gd, accordingly to process 3. The measured capture time, averaged on the entire target, is $\tau = 30.5 \pm 1\,\mu s$ and the total energy released is $\sim$ 8 MeV, well above the natural radioactivity.

During the data taking period, the detector slowly varied its response due to the decrease of the optical transparency of the Gd-loaded scintillator. The detector response was daily checked by $^{60}$Co, $^{252}$Cf and Am/Be sources, which provide $\gamma$-signals, neutron signals, and time correlated $\gamma - n$ signals. The reconstruction of these event samples, the study of their time evolution, and the comparison with Montecarlo method predictions, permitted a thorough understanding of the detector behaviour and a precise evaluation of the small efficiency variations on neutrino-induced and background events.

Figure 2 shows calibration data using the $^{252}$Cf source

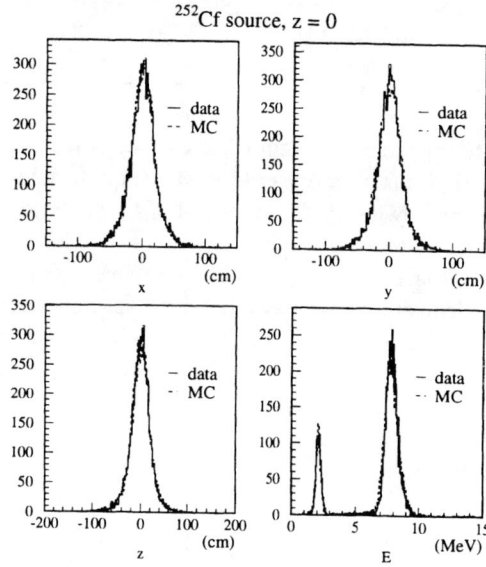

**FIGURE 2.** Visible energy and position distributions of $^{252}$Cf source calibration data (at the detector center): comparison between data and Monte Carlo simulation.

at the detector center; the n–capture lines (2.2 MeV on hydrogen and 8 MeV on gadolinium) are compared with Montecarlo predictions. Position and energy resolutions are $\sigma_x = 17.5$ cm and $\sigma_E/E = 5.6\%$ for n-captures on gadolinium. Calibrations at other locations always produced detector response in good agreement with Montecarlo predictions. The detailed knowledge of the detector parameters (light attenuation length, PMT gains, etc) allowed the collaboration to obtain an excellent stability in both energy and position reconstruction; as a matter of fact, the measured energy resolution for spallation neutrons captures averaged on the entire target volume and on the entire duration of the experiment is similar ($\sigma_E/E = 6.1\%$) to the indicated resolution at detector centre by using the Cf calibration source.

## Experimental data

The CHOOZ experiment collected data from April 1997 to July 1998, during the commissioning phase of the power plant. A summary of the data taking is presented in Table 2.

**Table 2.** Summary of the Chooz data acquisition cycle.

|  | Time (h) | $\int W\,dt$ (GW h) |
|---|---|---|
| Run | 8761.7 | |
| Live time | 8209.3 | |
| Dead time | 552.4 | |
| Reactor 1 only ON | 2058.0 | 8295 |
| Reactor 2 only ON | 1187.8 | 4136 |
| Reactors 1 & 2 ON | 1543.1 | 8841 |
| Reactors 1 & 2 OFF | 3420.4 | |

During the experiment $1.2 \times 10^7$ events were recorded on disk; weak selection criteria, based on the total charge measured by the PMT's, reduced this number to $7 \times 10^5$ events; these events were fully reconstructed in energy and in space. A space and time correlated couple of energy deposit in the detector defines a $\bar{\nu}_e$–candidate event, the first energy deposit being the $e^+$–like signal and the second energy deposit being the n–like signal. Explicitly the selection criteria applied are the following:

- energy cuts on the n–event (6 – 12 MeV) and on the $e^+$–event (from the hardware energy threshold $\sim 1.3$ MeV to 8 MeV),
- a time window on the delay between the $e^+$ and the neutron events ($2 - 100\,\mu s$),
- spatial selections on the $e^+$ and the neutron (distance from the PMT wall $> 30$ cm and relative $n - e^+$ distance $< 100$ cm)
- only one pulse satisfying the n–event criteria.

After these selections the number of candidates is reduced to 2991, including 287 candidates from reactor-OFF periods.

A good understanding of the nature of the neutrino candidates can be obtained by viewing the events on a two-dimensional plot of n-like versus $e^+$-like energy deposits as shown in Figure 3. One can observe four re-

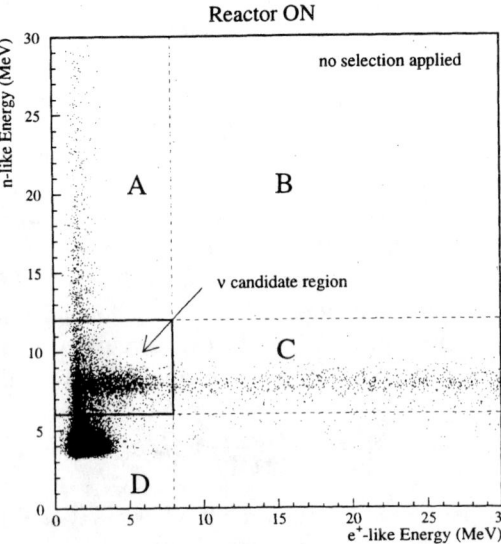

**FIGURE 3.** n-like versus $rme^+$–like energy deposits during the acquisition period with at least one reactor ON; no selection applied.

gions labelled A,B,C,D, and the $\bar{\nu}_e$–candidate region. The four regions are populated by the event categories here described:

- A accidental coincidences of a low energy natural radioactivity signal and a high energy recoil proton due to a fast neutron scattering;
- B correlated couples of a stopping muon, entering the detector through dead spaces in the anti-coincidence shield, and the muon decay products;
- C correlated events due to fast neutrons from muon spallations, with the $e^+$ signal mimicked by the proton recoil;
- D accidental coincidences of two low energy natural radioactivity signals.

The $\bar{\nu}_e$–candidate selection criteria strongly reduce the background as shown in Figure 4. The $\bar{\nu}_e$ events are clearly visible in the candidate region. The remaining background contributions are mainly caused by the tails of the event distributions in region C and D.

The effect of the selection criteria used to define the neutrino interactions was extensively studied by Montecarlo simulation and by means of $\gamma$ and $n$-source calibrations. The collaboration also investigated the small edge effect associated with the acrylic vessel containing the Gd-loaded scintillator. In Table 3 the efficiencies associated with the selection criteria and their errors are presented.

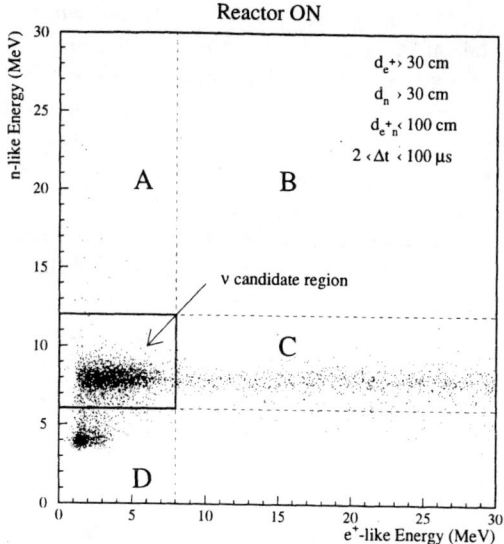

**FIGURE 4.** n-like energy versus e+–like energy deposit during the acquisition period with at least one reactor ON; selections other than energy applied.

**Table 3.** Summary of the neutrino detection efficiencies.

| selection | efficiency (%) | error (%) |
|---|---|---|
| positron energy | 97.8 | 0.8 |
| positron-geode distance | 99.85 | 0.1 |
| neutron capture | 84.6 | 0.85 |
| capture energy containment | 94.6 | 0.4 |
| neutron-geode distance | 99.5 | 0.1 |
| neutron delay | 93.7 | 0.4 |
| positron-neutron distance | 98.4 | 0.3 |
| secondary multiplicity | 97.4 | 0.5 |
| combined | 69.8 | 1.1 |

## The e+ spectrum and the $\overline{\nu}_e$ interaction yield

The measured positron spectrum for all reactor-ON data, after background subtraction as measured during the reactor-OFF periods, is shown in the upper part of Figure 5. The expected, non-oscillated positron spectrum was computed by using the model selected by the Bugey experiment (2). The measured over the expected spectra ratio is presented in the lower part of Figure 5. The average over the energy of this ratio is

$$R = 1.01 \pm 2.8\,\%(\text{stat}) \pm 2.7\,\%(\text{syst}). \quad (6)$$

The experiment was operational during the commissioning period of the new Chooz power plant. This peculiar operational period gave to the collaboration the

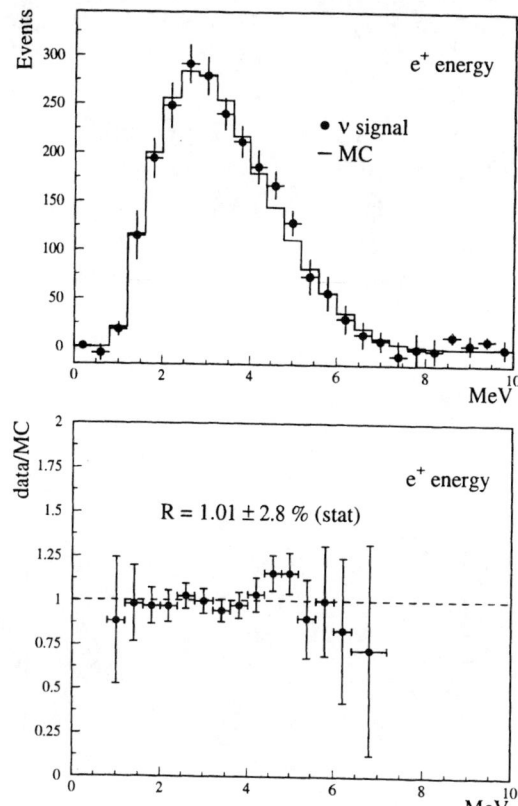

**FIGURE 5.** (above) The measured positron spectrum, obtained from the subtraction of reactor-ON and reactor-OFF spectra, compared with the expected positron spectrum for the case of no oscillations; (below) measured over expected ratio. The errors shown are statistical.

chance to determine the background rate, together with the neutrino signal rate, in a short period of time and with high precision.

For each run and for a given positron energy bin the measured $\overline{\nu}_e$-candidate rate is the sum of two signal terms linearly dependent on the reactor powers, and a common background term assumed to be constant and independent of the power. Thus the expected number of events ($\overline{N}$) is

$$\overline{N}_i(E_j) = (B(E_j) + W_{1i}X_1(E_j) + W_{2i}X_2(E_j))\Delta t_i \quad (7)$$

where $E_j$ is a positron energy bin ($j = 1, \ldots, 7$), the index $i$ labels the run number, $\Delta t_i$ is the corresponding live time, $B$ is the background rate and $X_1$ and $X_2$ are the positron yields per GW induced by each reactor. The proportionality coefficients $W_{1i}$ and $W_{2i}$) are the the reactor thermal powers corrected for the different solid angles of the reactors [2]. A likelihood function $\mathcal{L}$ can be built by the joint Poissonian probability of detecting $N_i(E_j)$ candidates when $\overline{N}_i(E_j)$ are expected

$$\mathcal{L} = \prod_{i=1}^{n_{run}} \prod_{j=1}^{n_{bin}} P(N_i(E_j); \overline{N}_i(E_j)) \qquad (8)$$

A maximum likelihood search determines simultaneously 14 positron yields $X_1(E_j)$ and $X_2(E_j)$ and the background B, with the two yields slightly correlated within the same energy bin. The average positron yields for the two reactors, summed over all energy bins, can be drawn as a function of the total reactor power, and the result is shown in Figure 6. The fitted values are $B = 1.9 \pm 0.11$ counts

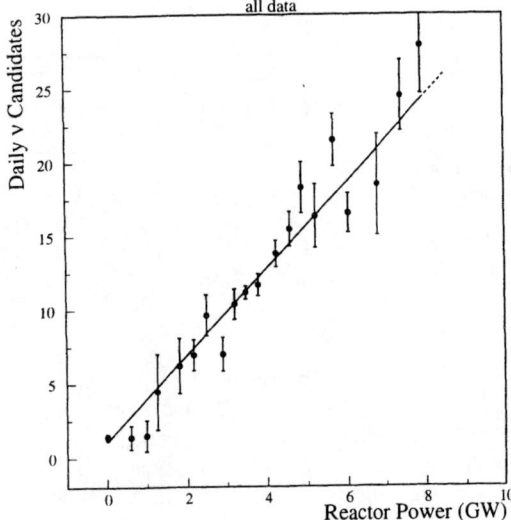

**FIGURE 6.** Number of $\overline{\nu}_e$–candidates per day as a function of the reactor power. The superimposed line corresponds to the fitted signal and background rates. The experimental points correspond to groups of runs of similar reactor power.

per day and $\langle X \rangle = 24.7 \pm 0.7$ daily neutrino candidates at full reactor power.

In conclusion the measured quantities are two positron spectra $X_1$ and $X_2$ from the two different distance reactors: Reactor 1 at 1114.6 m and Reactor 2 at 997.9 m.

## Neutrino oscillation tests

The experimental information consists in the shapes and the absolute normalizations of the two positron spec-

---

[2] Other minor corrections like the known differences in the fuel element composition and the effects of the hardware thresholds are included.

tra. The collaboration decided to use the whole experimental information, or a part of it, to explicitly distinguish the contribution of the statistical and systematic errors to the excluded regions. Three analysis are published by the collaboration (7).

The expected positron spectra are computed by using the model experimentally verified by the Bugey experiments. The associated systematic uncertainties have been already discussed. In particular, the Chooz collaboration assumes as uncertainty on the "cross section per fission" the Bugey experimental error of 1.4% added to the error on the extrapolation from the Bugey reactor working condition to the Chooz reactor working condition. A summary of the systematic and normalization uncertainties is given in Table 4. The confidence regions in the

**Table 4.** Contributions to the overall systematic uncertainty on the absolute normalization factor.

| parameter | relative error (%) |
|---|---|
| reaction cross section | 1.9% |
| number of protons | 0.8% |
| detection efficiency | 1.5% |
| reactor power | 0.7% |
| energy absorbed per fission | 0.6% |
| combined | 2.7% |

$(\Delta m^2, \sin^2 2\theta)$ plane are evaluated by following the Feldman & Cousins prescriptions (19).

### Analysis A

The analysis "A" uses all the experimental information available; the systematic uncertainties directly depend on the correct determination of the integrated neutrino flux, on the number of target protons, on the detection efficiencies and on the $\overline{\nu}_e$ cross section. The 7 experimental yields for the two reactors are used to build a $\chi^2$ here defined:

$$\chi^2(\theta, \delta m^2, \alpha, g) = \sum_{i,j=1}^{14} (X_i - \alpha \overline{X}_i) V_{ij}^{-1} (X_j - \alpha \overline{X}_j) \\ + \left(\frac{\alpha-1}{\sigma_\alpha}\right)^2 + \left(\frac{g-1}{\sigma_g}\right)^2, \qquad (9)$$

where the 14 expected positron yields $\overline{X}_i$ are functions of the oscillation parameters, of the energy bin and of the detector–reactor distances ($\overline{X} \equiv \overline{X}(gE_i, L_i, \theta, \delta m^2)$). The 14 × 14 covariance matrix $V_{ij}$ combines the statistical variances on $X_i$ and the systematic uncertainties on the shape of the $\overline{\nu}_e$ energy spectrum. The parameter $\alpha$ is the absolute normalization constant whose uncertainty is reported in Table 4; the parameter $g$ is an energy-scale calibration factor, whose uncertainty is evaluated to be 1.1% by using calibration sources. Function 9 is a $\chi^2$ with 12 degrees of freedom (16 $\chi^2$ terms and 4 parameters). The explicit contributions to the $\chi^2$ of the parameters $\alpha$ and $g$ constrain their variation ranges.

The excluded values of the oscillation parameters at the 90% C.L. are shown in Figure 7. The region al-

## Analysis B

The analysis "B" uses the ratio of the measured positron yields from the two, different distance, reactors. This analysis is almost completely independent of the absolute normalization factor (see Table 4) as well as of the expected positron spectrum shape. The statistical error dominates. The ratio $R(E_i) \equiv X_1(E_i)/X_2(E_i)$ of the measured positron spectra is used to form the following $\chi^2$ function:

$$\chi^2 = \sum_{i=1}^{7} \left( \frac{R(E_i) - \overline{R}(E_i, \theta, \delta m^2)}{\delta R(E_i)} \right)^2 \quad (10)$$

where $\delta R(E_i)$ is the statistical uncertainty on the measured ratio, and $\overline{R}$ the expectation. The resulting exclusion plot at 90% C.L. is shown in Figure 7. Although less powerful than analysis "A", the region excluded by this oscillation test nevertheless almost completely covers the one allowed by Kamiokande.

## Analysis C

The analysis "C" uses the measured positron spectra from each reactor, but assumes the absolute normalization as a free parameter; thus only the positron spectrum shape is relevant.

The $\chi^2$ function for this analysis is mathematically similar to Equation 9 for analysis "A", the only difference being the omission of the absolute normalization term. The exclusion plot, obtained according to the Feldman-Cousins prescriptions, is shown in Figure 7 and it is compared to the results of analyses "A" and "B".

# THE PALO VERDE EXPERIMENT

The detector of the Palo Verde experiment (8, 9) is installed at a distance of about 800 m from the Palo Verde nuclear plant in Arizona, where three reactors produce about 10.9 GW of thermal power. The detector is in an underground laboratory at a depth of 32 mwe, sufficiently deep to eliminate the hadronic components of cosmic rays. The detector consists of 66 acrylic cells (9m × .25m × .25m), filled with a 0.1 % Gd-loaded liquid scintillator and viewed by 5-inch PMT's for a fiducial volume of about 12 tons. The cells are surrounded by a 1 m water buffer and by an active liquid scintillator veto which will help in suppressing the fast-neutron correlated-background.

The Reaction 2 is used to detect $\overline{\nu}_e$'s and the delayed coincidence signature is obtained by absorbing the neutrons on Gd. The shallowness of the site influences the

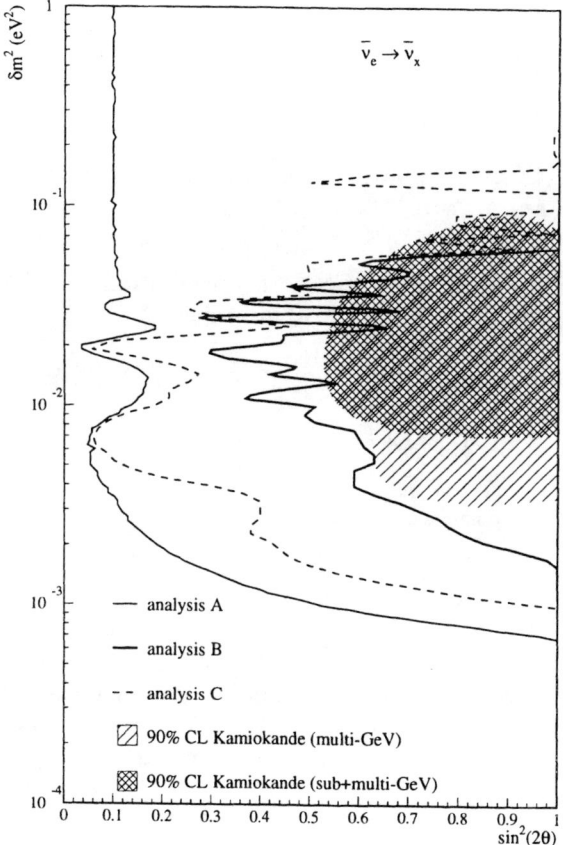

**FIGURE 7.** Exclusion plot contours at 90% C.L. obtained from the three analyses presented by the Chooz collaboration.

lowed by Kamiokande (4) for the $\nu_\mu \to \nu_e$ oscillations is also shown for comparison. The $\delta m^2$ limit at full mixing is $7 \cdot 10^{-4} \text{eV}^2$, to be compared with $9.5 \cdot 10^{-4} \text{eV}^2$ previously published (6). The limit for the mixing angle in the asymptotic range of large mass differences is $\sin^2 2\theta = 0.10$, which is better by a factor of two than the previously published value as recalculated according to (19).

detector since a large high energy neutron flux associated to cosmic rays is expected. A segmented liquid scintillation detector has been therefore designed in order to distinguish the $e^+$ signals from proton recoil signals induced by fast neutrons. Positrons are identified by requiring a fast threefold coincidence of the $e^+$ energy release with the associated 511 KeV annihilation gammas in neighboring back to back detector elements. This fast triple coincidence is furthermore required to be time correlated with the neutron capture gamma rays in Gd.

The PaloVerde experiment started the data taking (20) in 1998. The results corresponding to the first 37.9 days with full power and 32.7 days with one reactor off for refueling have been presented (20).

The $\overline{\nu}_e$-candidates are selected by the following criteria:

- a time window on the delay between the $e^+$ and the neutron events (5 – 200 μs);
- the delay since the last veto hit greather than 600 μs;
- spatial selections on the $e^+$ and the neutron events (distance along the counter axis < 100 cm and within an array of 7 counters);
- at least one of the two events ($e^+$ or n) should have more than 3 MeV;
- always 3 counters fired for each event: the most energetic counter should have an energy deposit in the range .5 – 8. MeV; the third energy release should be > 30 KeV;
- $e^+$-event identification: the second and the third energy deposit should be < 600 KeV, and the total energy not measured by the most energetic counter should be < 1.2 MeV;

These selections reduce the number of candidates to 39.06 ± 1.00(stat) events per day for reactors at full power and 32.62 ± 1.02(stat) events per day when one reactor is off. The difference is 6.44 ± 1.43(stat) events per day. The energy spectrum of positrons calculated by subtracting the spectrum with one reactor off from the the one at full power is shown in Figure 8.

The efficiencies of the selection criteria were evaluated by means of detailed Montecarlo simulations. The use of LED and movable optical fibers as well as radioactive sources like $^{22}$Na and Am-Be, were extensively used to measure the absolute positron and neutron detection efficiencies. The total systematic uncertainty due to the selection criteria results to be 12.7%. The final $\overline{\nu}_e$ detection efficiency is reported in Table 5. Taking into account efficiency corrections, the measured yields for one of the three reactors results 77 ± 17 (stat.) ± 11 (syst.) events per day. The expected candidate rate for no oscillation is 59.6 events per day.

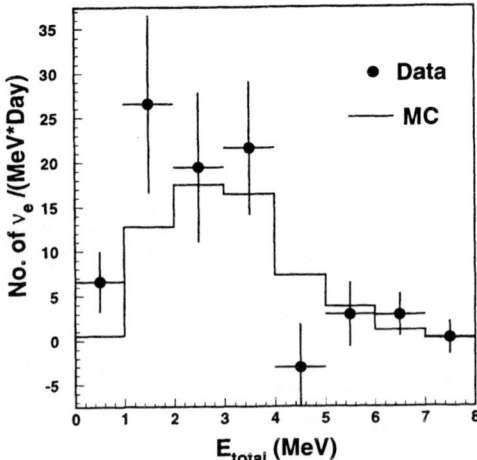

**FIGURE 8.** The measured positron spectrum of the PaloVerde experiment after 70 days of data taking.

**Table 5.** $\overline{\nu}_e$ detection efficiency and cumulative systematic errors for the PaloVerde experiment.

|  | efficiency | relative error (%) |
|---|---|---|
| selection criteria | 0.159 | 12.7% |
| DAQ efficiency | 0.820 | 5.6% |
| μ-crossing veto | 0.640 | 2.5% |
| combined | 0.083 | 14.% |

The ratio of the measured over the expected event rates is used to exclude a region of the oscillation parameters at 90% C.L. as shown in Figure 9. A large fraction of the KAMIKANDE atmospheric neutrino anomaly is excluded by this initial result of the PaloVerde experiment.

## CONCLUSIONS

The production of neutrinos at reactors is now known with a precision of the order of the percent. Neutrino oscillation experiments located at large and very large distances from the reactors are therefore being made using only one neutrino detecting station. The Chooz experiment was the first of such experiments to be performed. Its final result put a severe constraint on the possibility of explaining the atmospheric neutrino anomaly by a $\nu_\mu \rightarrow \nu_e$ oscillation. The PaloVerde experiment is currently taking data, and a final result is expected soon.

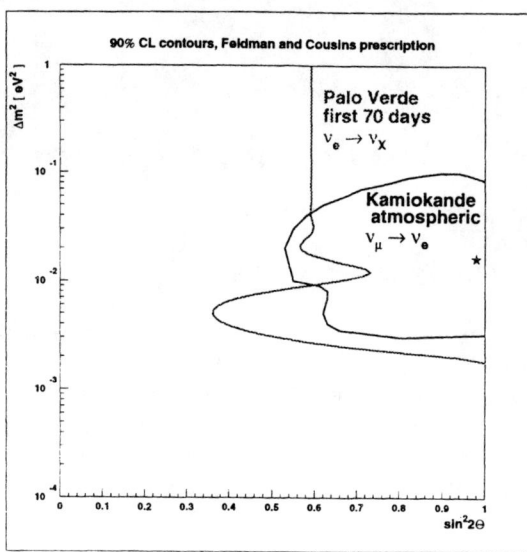

**FIGURE 9.** Exclusion plot contours at 90% C.L. obtained by the PaloVerde experiment after 70 days.

Future experiments of this kind will be sensitive to part of the oscillation parameters space indicated as a possibility for explaining the results of solar neutrino experiments.

## ACKNOWLEDGEMENTS

I would like to express my gratitude to my collaborators in the CHOOZ experiment for lots of interesting discussions on neutrino physics. I would like to tank also G. Gratta who provided me with the material about the PaloVerde experiment.

## REFERENCES

1. G. Zacek *et Al.*, Phys. Rev. **D34** (1986) 2621.
2. B. Achakar *et Al.*, Nucl. Phys. B **434** (1995) 503.
3. G.S. Vidyakin *et Al.*, JETP Lett. **59** (1994) 364.
4. Y. Fukuda *et Al.*, Phys. Lett. B**335** (1994) 237.
5. H. de Kerret *et Al.*, "Chooz Proposal", 1993, Unpublished.
6. M. Apollonio *et Al.*, Phys. Lett. B**420** (1998) 397.
7. M. Apollonio *et Al.*, Phys. Lett. B**466** (1999) 415.
8. F. Boehm *et Al.*, "Proposal for the San Onofre Neutrino Oscillation Experiment", 1994, Unpublished; F. Bohem *et Al.*, "Addendum to the San Onofre Proposal", 1996, Unpublished.
9. A. Piepke, Neutrino'98 Conference, Takayama, 1998.
10. A. Suzuki, Neutrino'98 Conference, Takayama, 1998.
11. J. N. Bahcall *et Al.*, Phys. Lett. B**433** (1998) 1.
12. V.I. Kopekin, Preprint Kurchatov Institute of Atomic Energy, Moscow IAE-4305/2 (1988).
13. B. Achakar *et Al.*, Phys. Lett. B **374** (1996) 243.
14. P. Vogel, Phys. Rev. D29 (1984) 1918.
15. K. Schreckenbach *et Al.*, Phys. Lett. B **160** 325 (1985). A.A. Hahn *et Al.*, Phys. Lett. B **218** 365 (1989).
16. H.V. Klapdor and J. Metzinger, Phys. Lett. B **112** (1982) 22.
    H.V. Klapdor and J. Metzinger, Phys. Rev. Lett. **48** (1982) 127.
17. B. Achakar *et Al.*, Phys. Lett. B **338** (1994) 383.
18. A.A. Kuvshinnikov *et Al.*, Sov. J. Nucl. Phys. **52** (1990) 300.
    A.A Kuvshinnikov *et Al.*, JETP Lett **54** (1991) 255.
19. G.J. Feldman & R.D. Cousins, Phys. Rev. **D57** (1998) 3873.
20. Y.F. Wang, XXXIV Rencontres de Moriond, Les Arcs (1999).

# ICANOE
# Imaging and Calorimetric Neutrino Oscillation Experiment

André Rubbia[1]*

[1] *Institut für Teilchenphysik, ETHZ, CH-8093 Zürich, Switzerland*

**Abstract.** The main scientific goal of the *ICANOE* detector(1) is *the one of elucidating in a comprehensive way the pattern of neutrino masses and mixings*, following the SuperKamiokande results and the observed solar neutrinos deficit. To achieve these goals, the experimental method is based upon the *complementary and simultaneous detection of CERN beam (CNGS) and cosmic ray (CR) events*. For the currently allowed values of the SuperKamiokande results, both CNGS and cosmic ray data will give independent measurements and provide a precise determination of the oscillation parameters. Since one will observe and unambiguously identify $\nu_e$, $\nu_\mu$ and $\nu_\tau$ components, the full (3 x 3) mixing matrix will be explored.

## INTRODUCTION

The reference mass for underground detectors is now set by the operating SuperKamiokande(2) detector, which is of the order of 30 ktons. However the rather coarse nature of the Cherenkov ring detection is capable to reconstruct only part of the features of the events.

New generation underground experiments are now facing new challenges, for which novel and more powerful technologies are required, with respect to the existing detectors:

1. The *long baseline accelerator neutrino oscillation* experiments require, with respect to existing short baseline detectors (like NOMAD(3) and CHORUS(4)), a large increase of the detector fiducial mass (about 3 ton in NOMAD), in order to cope with the flux attenuation due to the distance. In order to perform a comprehensive program on neutrino oscillations the fiducial mass of the detector must be increased to several ktons. Moreover, the detector must be able to tag efficiently the interaction of $\nu_e$'s and $\nu_\tau$'s out of the bulk of $\nu_\mu$ events. This requires a detailed event reconstruction that can be achieved only by means of a high granularity detector.

2. Likewise, the comprehensive investigation of *atmospheric neutrinos* events, in order to reach the level of at least one thousand events/year, also requires a fiducial mass of several ktons. The capability to observe all separate processes, electron, muon and tau neutrino charged currents (CC) and all neutral currents (NC) without detector biases and down to kinematical threshold is highly desirable.

3. *Nucleon decay*: because of the already very high limit on the nucleon lifetime ($\geq 10^{32}$ years in most of the decay channels), a modern proton decay detector should have an adequately large sensitive mass.

Event imaging should be provided by a modern bubble chamber-like technology since (1) it has to be able to provide high resolution, unbiased, three dimensional images of ionising events; (2) it has to provide an accurate measurement of the basic kinematical properties of the particles of the event, including particle identification. (3) it has to accomplish simultaneously the two basic functions of target and detector.

In the *ICANOE* design a fully sensitive, bubble chamber-like detector will permit discovery limits at the few events level and a much more powerful background rejection. A detector of this kind, already at the level of a few ktons of mass will be fully competitive with the potentialities of SuperKamiokande and in several domains will permit to extend much further the investigations.

The *ICANOE* detector fruitfully merges the superior imaging quality of the ICARUS technology(5) with the high resolution full calorimetric containment of NOE(6), suitably upgraded to provide also magnetic analysis of muons. It has a modular structure of independent supermodules and is expandable by the addition of such supermodules, each consisting of a low density 1.6 kton liquid target and of a high density 0.7 kton active solid target.

---

* On behalf of the ICARUS and NOE Collaborations.

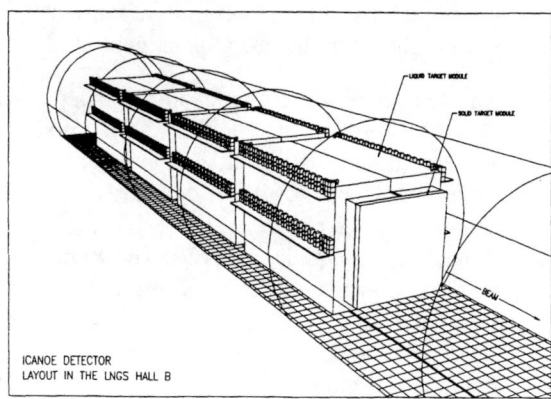

**FIGURE 1.** Perspective view of the baseline detector with 4 supermodules.

The superior quality of the event vertex inspection and reconstruction of the liquid Argon volumes is ideally complemented by the addition of an external module capable of magnetic analysis of the muons escaping the chamber. Bubble chambers have in fact often been very similarly complemented in the past by external identifiers. An iron muon tracking spectrometer would fulfill this job, but it would also introduce in between adjacent liquid argon volumes a blind region incapable of giving information on the energy and on the nature of the escaping particles. A sensitive magnetized calorimeter appears therefore as an ideal containment module to be interleaved between adjacent liquid Argon volumes.

## OUTLINE OF THE *ICANOE* DETECTOR

The *ICANOE* layout (Figure 1) is similar to that of a "classical" neutrino detector, segmented into almost independent **supermodules**. The layout of the apparatus can be summarized as follows:

- the **liquid target**, with extremely high resolution, dedicated to tracking, $dE/dx$ measurements, full e.m. calorimetry and hadronic calorimetry, where electrons and photons are identified and measured with extremely good precision and $\pi/\mu$, $K$ and $p$ separation is possible at low momenta;

- the **solid target**, with good e.m. and hadronic resolution, dedicated to calorimetry of the jet and a magnetic field for measurement of the muon features (sign and momentum);

- The **supermodule**, obtained joining a liquid and solid module, which constitutes the basic module of an *expandable* apparatus. A supermodule behaves as a complete building block, capable of identifying and measuring electrons, photons, muons and hadrons produced in the events. The solid, high density sector reduces the transverse and longitudinal size of hadron shower, confining the event (apart from the muon) within the supermodule.

The *ICANOE SuperModule* is, according to the present design, composed by a liquid Argon module, with $18.0 \times (11.3)^2$ m$^3$ of external dimensions and 1.4 kton (1.9 kton) of active (total) mass, and a magnetized calorimeter module, with $2.6 \times 9^2$ m$^3$ of external dimensions and 0.8 kton of mass. The depth consists of 2 m for the calorimetric unit, corresponding to 7.4 $\lambda_{int}$ and 59 $X_0$ and 0.6 m for the tracking units (the interleaved planes of tracking chambers).

At this stage, *four* SuperModules with a total length of the experiment of 82.5 m and a total active mass of 9.3 kton fully instrumented are being considered for the baseline option.

## PHYSICS GOALS

*ICANOE* is an underground detector capable to achieve the full reconstruction of neutrino (and antineutrino) events of *any* flavor, and with an energy ranging from the tens of MeV to the tens of GeV, for the relevant physics analyses. No other combinations can provide such a rich spectrum of physical observations, including the systematic, on-line monitoring of the CNGS $\nu$-beam at the LNGS site. The unique lepton capabilities of *ICANOE* are really fundamental in tagging the neutrino flavor. In general, the oscillation pattern of the neutrinos may be complicated and involve a combination of $\nu_\mu \to \nu_\tau$, $\nu_\mu \to \nu_e$ and $\nu_\mu \to \nu_{sterile}$ transitions. In order to fully sort out the mixing matrix, unambiguous neutrino flavor identification is mandatory to distinguish $\tau$'s from $\nu_\tau$'s and electrons from $\nu_e$'s interactions. In other words, we stress the importance of constraining the oscillation scenarios by coupling appearance in several different channels and disappearance signatures.

The sensitivity in the classic ($\sin^2 2\theta, \Delta m^2$) plot is evidenced in Figure 2, for a data taking time of 4 years, with $4.5 \times 10^{19}$ pot at each year. We remark:

1. The recent results on atmospheric neutrinos ("A" and "B" of Figure 2) can be thoroughly explored by appearance and disappearance experiments. For the current central value, both CNGS and cosmic ray data will give independent and complementary measurements and they will provide a precise ($\sin^2 2\theta, \Delta m^2$) determination.

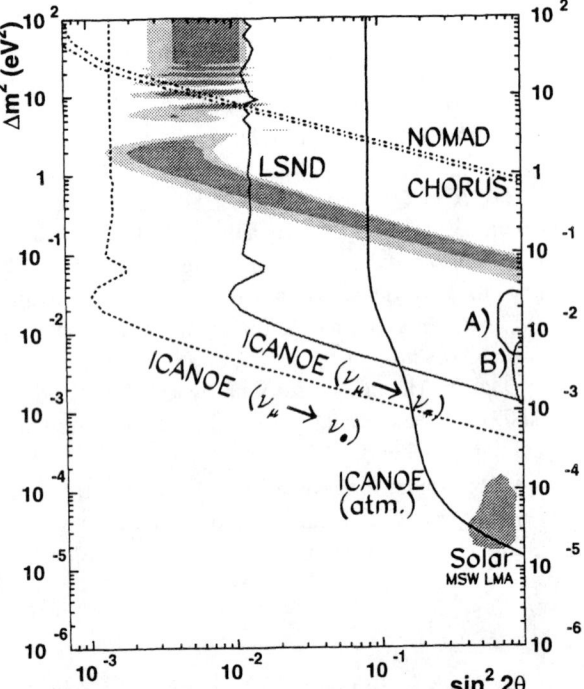

**FIGURE 2.** Overview of the status of the neutrino oscillations searches, displayed assuming two neutrino mixing schemes in the ($\sin^2 2\theta, \Delta m^2$) plane. The 90%C.L. allowed regions obtained from the Kamiokande (resp. Superkamiokande) FC and PC samples are shown as A) (resp. B)). The 90% (resp. 99%C.L.) regions consistent with the LSND excess are shown as dark (resp. light) shaded areas in the upper region of the plane. The shaded area in the region $\Delta m^2 \approx 10^{-5}$ eV$^2$ represents the large angle MSW solution of the solar neutrino deficit. CHORUS and NOMAD 90%C.L. limits on $\nu_\mu \to \nu_\tau$ oscillations are visible in the upper $\Delta m^2$ region. The *ICANOE* sensitivities at 90%C.L. are indicated by three curves: the limit by direct observation of the atmospheric neutrinos ("*ICANOE* atm"); the direct tau appearance search at the CNGS ("*ICANOE* $\nu_\mu \to \nu_\tau$"); the direct electron appearance search at the CNGS ("*ICANOE* $\nu_\mu \to \nu_e$").

2. In the mass range of LSND, the sensitivity is sufficient in order to solve definitely the puzzle.

3. At high masses of cosmological relevance for $\Delta m^2 <\approx 10$ eV$^2$, the sensitivity to $\nu_\mu \to \nu_\tau$ oscillations is better or equal to the one of CHORUS and NOMAD.

4. in the atmospheric neutrino events, one can reach a level of sensitivity sufficient to detect also the effect until now observed in solar neutrinos. This purely terrestrial detection of the LMA solar neutrino solution is performed using neutrino in the GeV range, much higher than the one of solar neutrinos.

Since we can observe and unambiguously identify both $\nu_e$ and $\nu_\tau$ components, the full (3 x 3) mixing matrix can be explored. By itself, this is one of the main justifications for the choice of the detector's mass.

In the cosmic ray channel, all specific modes (electron, muon, NC) are equally well observed without detector biases and down to kinematical threshold. The CR-spectrum being rather poorly known, a confirmation of the SuperKamiokande result requires detecting both (1) the modulation in the muon channel and (2) the lack of effect of the electron channel. The consistency of the simultaneous observation of the $L/E$ phenomenon in as many modes as they are available is a powerful tool in separating genuine flavour oscillations from exotic scenarios.

In some favourable conditions, the direct appearance of the oscillated tau neutrino may be directly identified in the upgoing events, since even a few events will be highly significant.

While in atmospheric neutrinos, the knowledge of the sign of the muon is of little relevance, in the case of the CNGS is a powerful tool to verify the neutrino nature after oscillation path, excluding for instance oscillation channels into anti-neutrinos.

For a discussion on the nucleon decay searches, see Ref. (7).

## PHYSICS AT THE CNGS

The design and performance of the CERN neutrino beam to Gran Sasso - the CNGS facility - are described in a conceptual technical design report (8). The CNGS beam performance for the new reference beam are summarized in Table 1. The rms radius of the $\nu_\mu$ CC event distribution is about 1.37 km at Gran Sasso. The expected numbers of detectable $\nu_\tau$ for $\sin^2 2\theta = 1$ and a few typical values of $\Delta m^2$ are shown in Table 2.

Events will occur in the whole *ICANOE* detector. As reference, we assume an exposure of 20 kton × year for the liquid argon. This corresponds to four years running of the CNGS beam in shared mode. For the events occuring in the solid detector, given the smaller mass, the reference exposure is 10 kton × year. The last three meters of the liquid target are defined as a transition region, since beam events occuring in this region are most likely to deposit energy in both targets. Table 3 shows the computed total event rates for each neutrino species present in the beam for the liquid, solid and in the transition region. Table 3 also shows, for three different values of $\Delta m^2$, the $\nu_\tau$ CC rates expected in case oscillations take place.

**Table 1.** Predicted performance of the new CNGS reference beam for an isoscalar target. The statistical accuracy of the Monte-Carlo simulations is 1% for the $\nu_\mu$ component of the beam, somewhat larger for the other neutrino species.

| Energy region $E_{\nu_\mu}$ [GeV] | 1 - 30 | 1 - 100 |
|---|---|---|
| $\nu_\mu$ [m$^{-2}$/pot] | $7.1 \times 10^{-9}$ | $7.45 \times 10^{-9}$ |
| $\nu_\mu$ CC events/pot/kt | $4.70 \times 10^{-17}$ | $5.44 \times 10^{-17}$ |
| $\langle E \rangle_{\nu_\mu\,fluence}$ [GeV] | | 17 |
| fraction of other events: | | |
| $\nu_e/\nu_\mu$ | | 0.8 % |
| $\bar{\nu}_\mu/\nu_\mu$ | | 2.0 % |
| $\bar{\nu}_e/\nu_\mu$ | | 0.05 % |

**Table 2.** Expected number of $\nu_\tau$ CC events at Gran Sasso per kton per year for an isoscalar target. Results of simulations for different values of $\Delta m^2$ and for $sin^2(2\theta) = 1$ are given for $4.5 \times 10^{19}$ pot/year. These event numbers do not take detector efficiencies into account.

| Energy region $E_{\nu_\tau}$ [GeV] | 1 - 30 | 1 - 100 |
|---|---|---|
| $\Delta m^2 = 1 \times 10^{-3}$ eV$^2$ | 2.34 | 2.48 |
| $\Delta m^2 = 3 \times 10^{-3}$ eV$^2$ | 20.7 | 21.4 |
| $\Delta m^2 = 5 \times 10^{-3}$ eV$^2$ | 55.9 | 57.7 |
| $\Delta m^2 = 1 \times 10^{-2}$ eV$^2$ | 195 | 202 |

**Table 3.** Expected event rates for an exposure of 20 kton × year for the liquid target and 10 kton × year for the solid target. All the rates include nuclear corrections and are computed for the proper target composition. For standard processes, no oscillations is assumed. For $\nu_\tau$ CC, we take two neutrino $\nu_\mu \to \nu_\tau$ with $sin^2 2\theta = 1$.

| Process | liquid target | transition | solid |
|---|---|---|---|
| $\nu_\mu$ CC | 54300 | 10200 | 27150 |
| $\bar{\nu}_\mu$ CC | 1090 | 200 | 545 |
| $\nu_e$ CC | 437 | 80 | 219 |
| $\bar{\nu}_e$ CC | 29 | 5 | 15 |
| $\nu$ NC | 17750 | 3330 | 8875 |
| $\bar{\nu}$ NC | 410 | 77 | 205 |
| $\nu_\tau$ CC, $\Delta m^2$ (eV$^2$) | | | |
| $1 \times 10^{-3}$ | 52 | 10 | 26 |
| $2 \times 10^{-3}$ | 208 | 40 | 104 |
| $3.5 \times 10^{-3}$ | 620 | 115 | 310 |
| $5 \times 10^{-3}$ | 1250 | 235 | 625 |
| $7.5 \times 10^{-3}$ | 2850 | 535 | 1425 |
| $1 \times 10^{-2}$ | 4330 | 810 | 2165 |

## Event kinematics and tau identification

Kinematical identification of the $\tau$ decay, which follows the $\nu_\tau$ CC interaction requires excellent detector performance: good calorimetric features together with tracking and event topology reconstruction capabilities. The background from standard processes are, depending on the decay mode of the tau lepton considered, the $\nu_e$ CC events and/or the $\nu_\mu$ CC and $\nu$ NC events.

In order to separate separate $\nu_\tau$ events from the background, two basic criteria, already adopted by the short baseline NOMAD experiment, can be used:

- an unbalanced total transverse momentum due to neutrinos produced in the $\tau$ decay,

- a kinematical isolation of hadronic prongs and missing momentum in the transverse plane.

In addition, given the baseline $L$ between CERN and GranSasso, for the lower $\Delta m^2$ values of the allowed region indicated by the atmospheric neutrino results, we expect most of the oscillation to occur at low energy. In this case, a criteria on the visible energy is also very important to suppress backgrounds.

In order to apply the most efficient kinematic selection, it is mandatory to reconstruct with the best possible resolution the energy and the angle of the hadronic jet, with a particular attention to the tails of the distributions. Therefore, the energy flow algorithm should be designed with care taking into account the needs of the tau search analyses.

A specially developed energy flow algorithm has been tested on a sample of fully simulated $\nu_e$ CC events, in order to estimate the resolution of the kinematical reconstruction on realistic events. It yields an average missing $P_T$ of 450 MeV/c. This value improves to an average of 410 MeV/c when the primary vertex is required to lie within a fiducial volume of transverse dimensions $7.8 \times 7.8 m^2$.

We used the neutrino data collected in the NOMAD detector to probe the reliability of the physics simulation. $\nu_\mu$ CC events have been fully simulated and reconstructed using NOMAD official packages. We found that the kinematics in the transverse plane are well reproduced by the Monte-Carlo model. This is clearly not the case when nuclear corrections are neglected.

**Table 4.** Rejection of the $\nu_e$ CC background in the $\tau \to e$ analysis. Figures are normalized to an exposure of 20 kton × year.

| Cuts | $\nu_\tau$ Eff. (%) | $\nu_e$ CC | $\bar{\nu}_e$ CC | $\nu_\tau$ CC $\Delta m^2 = 10^{-3}$ eV$^2$ | $\nu_\tau$ CC $\Delta m^2 = 3.5 \times 10^{-3}$ eV$^2$ | $\nu_\tau$ CC $\Delta m^2 = 10^{-2}$ eV$^2$ |
|---|---|---|---|---|---|---|
| Initial | 100 | 437 | 29 | 9.3 | 111 | 779 |
| Fiducial volume | 88 | 383 | 25 | 8.2 | 97 | 686 |
| One candidate with momentum > 1 GeV | 72 | 365 | 25 | 6.7 | 80 | 561 |
| $E_{vis} < 18$ GeV | 67 | 64 | 5 | 6.2 | 75 | 522 |
| $P_T^e < 0.9$ GeV | 54 | 31 | 3 | 5.0 | 60 | 421 |
| $P_T^{lep} > 0.3$ GeV | 51 | 29 | 2 | 4.7 | 56 | 397 |
| $P_T^{miss} > 0.6$ GeV | 33 | 4 | 0.4 | 3.1 | 37 | 257 |

## $\nu_\mu \to \nu_\tau$ appearance searches

The channel of tau decaying into an electron plus two neutrinos provides the best sample for $\nu_\tau$ appearance studies due to the low background level. The intrinsic $\nu_e$, $\bar{\nu}_e$ contaminations of the beam amount to $\approx 470$ events for an exposure of 20 kton × year.

The comparison of this figure with the expected number of $\nu_\tau$ CC events decaying into electrons shows that the search of $\tau \to e$ at the CNGS will have to be optimized *a posteriori*. Indeed the $\nu_\tau$ rate has a strong dependence on the exact value of the $\Delta m^2$ in the parameter region suggested by the Super-Kamiokande data, and the $\Delta m^2$ value is not well constrained by the atmospheric neutrino experiments.

For "large" values of $\Delta m^2$, i.e. $\Delta m^2 > 5 \times 10^{-3}$, the rate of tau is spectacular and exceeds the number of intrinsic beam $\nu_e$, $\bar{\nu}_e$ CC events, i.e. $S/B > 1$ even prior to any kinematical cuts. So the kinematical cuts can be very mild. An excess will be striking.

For our "best" value taken from atmospheric neutrino results, i.e. $\Delta m^2 = 3.5 \times 10^{-3}$ eV$^2$, the number of $\nu_\tau$ CC with $\tau \to e$ is about 110, or about a signal over background ratio of $110/470 \simeq 1/4$. Here with modest kinematical cuts, we can extract statistically significant signals, as shown in the following sections.

The most difficult region lies below $\Delta m^2 \approx 1.5 \times 10^{-3}$ eV$^2$, for which, kinematical cuts are tuned to suppress backgrounds by a factor more than 200 while keeping about half of the signal events.

In the following paragraphs, we discuss background sources and their suppression.

$\nu_e$ **CC rejection:** The main background from genuine leading electrons comes from the CC interactions of the $\nu_e$ and $\bar{\nu}_e$ components of the beam. In Table 4 we summarize the list of sequential cuts applied to reduce the $\nu_e$ and $\bar{\nu}_e$ CC backgrounds and the expected number of signal events for three different $\Delta m^2$ values. The most sensitive analysis predicts, for a 20 kton × year exposure, a total background of 4.4 events for a total $\tau$ selection efficiency of 33%.

$\nu$ **NC rejection:** Neutral current events contribute to the background from four sources: (1) electrons from Dalitz decays, (2) early photon conversions, (3) interacting charged pions and (4) $\pi^\pm/\pi^0$ overlap. Table 5 summarizes the rejection power of kinematics criteria for the four sources that contribute to $\nu$ NC background. The requirement on the electron candidate energy $E_e > 1$ GeV suppresses about one third of the Dalitz, pion overlap and $\pi^0$ conversions induced backgrounds, since electrons in the jet are soft.

**Table 5.** $\nu_\mu$ NC background to the $\tau \to e$ analysis. Results are normalized to an exposure of 20 kton × year. We illustrate background reduction by means of kinematical criteria only. Imaging and $dE/dx$ measurements reduce the NC background to a negligible level.

| Cuts | $\nu_\mu$ NC | | | |
|---|---|---|---|---|
| Initial | 17750 | | | |
| Fiducial volume | 15550 | | | |

| | Dalitz | $\gamma$ conv. | $\pi \to e$ | $\pi^\pm/\pi^0$ |
|---|---|---|---|---|
| One candidate | 275 | 4262 | 6.5 | 25 |
| $P_e > 1$ GeV | 79 | 1361 | 6.3 | 16 |
| $E_{vis} < 18$ GeV | 49 | 835 | 3.2 | 11 |
| $P_T^e < 0.9$ GeV | 46 | 794 | 1.8 | 9 |
| $P_T^{lep} > 0.3$ GeV | 24 | 429 | 1.7 | 8 |
| $P_T^{miss} > 0.6$ GeV | 19 | 350 | 1.3 | 7 |
| Imaging and $dE/dx$ | <1 | <1 | <1 | <1 |

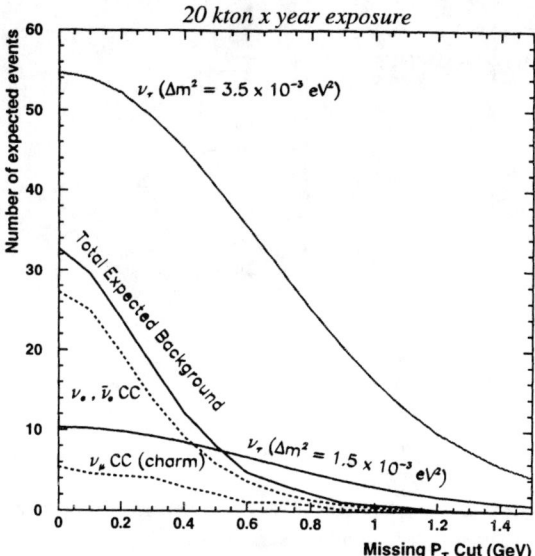

**FIGURE 3.** Number of $\nu_e$, $\bar{\nu}_e$ background and tau events expected for a 20 kton × year exposure as a function of the missing transverse momentum.

**Table 6.** $\tau \to e$ analysis summary. For a total exposure of 20 × year we show the expected number of $\tau$ events for different $\Delta m^2$ values. The last three columns show the expected background.

| $\Delta m^2$ (eV$^2$) | $\nu_\tau$ CC | $\nu_e, \bar{\nu}_e$ CC | $\nu_\mu, \bar{\nu}_\mu$ CC | $\nu_\mu$ NC |
|---|---|---|---|---|
| $1 \times 10^{-3}$ | 3 | | | |
| $2 \times 10^{-3}$ | 12 | | | |
| $3 \times 10^{-3}$ | 26 | | | |
| $3.5 \times 10^{-3}$ | 35 | 4.1 | 1.0 | < 1 |
| $5 \times 10^{-3}$ | 71 | | | |
| $7 \times 10^{-3}$ | 121 | | | |
| $1 \times 10^{-2}$ | 248 | | | |

The ultimate discrimination of these backgrounds relies primarily on the imaging capabilities and on $dE/dx$ measurements. The combination of $dE/dx$ information together with kinematics criteria is sufficient to reduce $\nu$ NC background to a negligible level.

$\nu_\mu$ **CC rejection:** Charged current events can contribute to the background in a similar way as the neutral current events described above when the leading muon escapes detection. In case the muon is not identified, the event will appear in first instance as a neutral current event. The source of electrons which can induce backgrounds are then similar to those discussed previously and are reduced to a negligible level for reasons already discussed. A more important source of background specific to charged current interactions comes from the decays of charmed mesons. At the CNGS energies $\sigma(\nu_\mu N \to \mu c X)/\sigma(\nu_\mu N \to \mu X) \approx 4\%$, therefore for a total exposure of 20 kton × year we expect to collect about 200 events where a charmed meson decays into a positron and a neutrino. These events resemble kinematically the real $\nu_\tau$ events, since they have a neutrino in the final state and possess a softer energy spectrum and a genuine sizeable missing transverse momentum. After all cuts, the expected number of charm induced background events $n^b_{CC\,(charm)}$ for a total exposure of 20 kton × year is at the level of 1 event.

## Combined $\nu_\mu \to \nu_\tau$ sensitivity

Table 6 summarizes the expectations for the $\tau \to e$ analysis once kinematics criteria and muon vetoes have been applied to every potential background source. In conclusion, we obtain for a 20 kton × year exposure, that the overall electron selection efficiency is 32% for an expected number of about five background events. The expected number of fully identified tau events at the central $\Delta m^2$ value of $3.5 \times 10^{-3}$ eV$^2$ is 35.

We show in Figure 3 as a function of the cut on the missing transverse momentum, the number of expected background and tau events for two different $\Delta m^2$ values. We see that even for a value as low as $1.5 \times 10^{-3}$ eV$^2$, a $P_T^{miss}$ cut above 0.6 GeV gives a $S/B$ ratio in excess of 1.

Finally it is crucial to study the exposures needed to obtain a statistically significant $\tau$ appearance for different neutrino oscillation scenarios. We see in Figure 4 that for $\Delta m^2 = 3.5$ eV$^2$, few months of data taking will suffice to claim a $3\sigma$ effect. However for $\Delta m^2$ values of about $10^{-3}$ eV$^2$, exposures above 20 kton × year are needed to obtain an effect in excess of $2\sigma$. We conclude that after four years of running of CNGS in shared mode or after one year of running in dedicated mode, the *ICANOE* detector will observe a statistically significant $\nu_\mu \to \nu_\tau$ oscillation signal for most of the $\Delta m^2$ values presently favored by atmospheric neutrino data.

## $\nu_\mu \to \nu_e$ oscillation search

The unambiguous detection and identification of $\nu_e$ CC events endows *ICANOE* with the ability of performing also a $\nu_\mu \to \nu_e$ oscillation search. In this case, the oscillation reveals itself as an excess on the number of expected events having a leading identified electron. Given the expected rates for a 20 kton × year exposure, the sta-

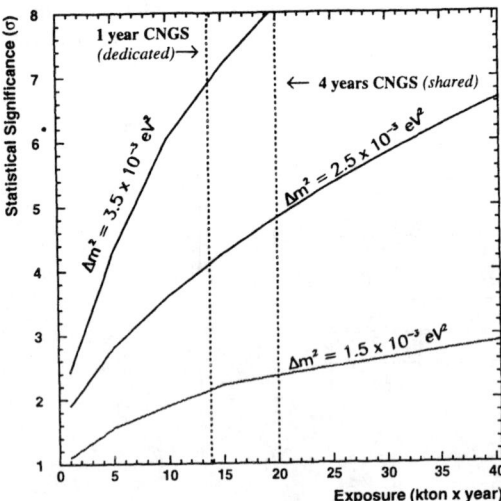

**FIGURE 4.** τ appearance statistical significance for different $\Delta m^2$ values as a function of the exposure. The arrows indicate the statistical significance achieved for one year of dedicated run of the CNGS and four years of CNGS running in shared mode.

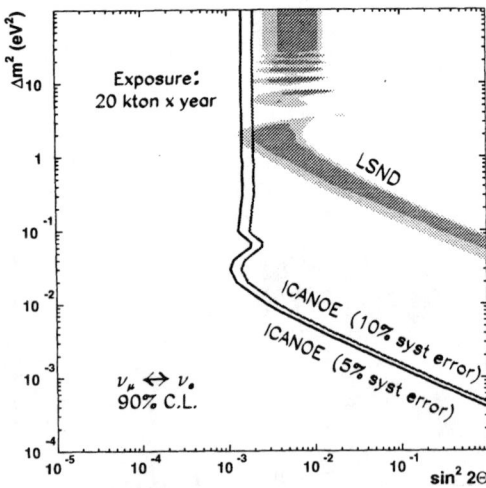

**FIGURE 5.** *ICANOE* 90% C.L. exclusion region in case no $\nu_\mu \to \nu_e$ are experimentally observed.

tistical error is about 4%. Therefore the sensitivity to $\nu_\mu \to \nu_e$ oscillations is dominated by the systematic error on the beam knowledge.

Figure 5 shows the 90% C.L. contours in case no oscillations are observed assuming overall systematic errors of 5% and 10%. We observe that nearly the whole region favored by the LSND claim is comfortably covered. The excluded values are $\sin^2 2\theta > 2 \times 10^{-3}$ at high $\Delta m^2$ and $\Delta m^2 > 4 \times 10^{-4}$ eV$^2$ for maximal mixing.

## ATMOSPHERIC NEUTRINOS

The physics goals of the new atmospheric neutrino measurements are to firmly establish the evidence of neutrino oscillations with a different experimental technique, possibly free of systematic biases, measure the oscillation parameters and clarify the nature of the oscillation mechanism. *ICANOE* will provide, in addition to comfortable statistics, an observation of atmospheric neutrinos of a very high quality. Unlike measurements obtained up to now in Water Cherenkov detectors, which are in practice limited to the analysis of "single-ring" events, complicated final states with multi-pion products, occurring mostly at energies higher than a few GeV, will be completely analyzed and reconstructed in *ICANOE*. This will be a significant improvement with respect to previous observations.

We have considered the following three methods:

- $\nu_\mu$ **disappearance**: detection of the oscillation pattern in the $L/E$ distribution, where $L$ is the neutrino pathlength and $E$ its energy;

- $\nu_\tau$ **appearance**: comparison of the NC/CC with expectation;

- **direct $\nu_\tau$ appearance**: comparison of upward and downward rates of "tau-like" events.

together with the well established ones:

- **the double ratio**, $(\nu_\mu/\nu_e)_{obs}/(\nu_\mu/\nu_e)_{MC}$;

- **up/down asymmetry**;

The tau appearance measurements can shed light on the nature of the oscillation mechanism, by discriminating between the hypothesis of oscillations into a sterile or a tau neutrino. The $\nu_\tau$ appearance method is based on $\nu_\tau$ CC interactions with $\nu_\tau$ decaying into hadrons, hence to "neutral-current-like" events of high energy. An excess of "NC-like" events from the bottom will indicate the presence of oscillation to the $\nu_\tau$ flavour. A kinematical analysis of the final state particles in the event can be used to further improve the statistical significance of the excess. Such a feature can only be obtained in a detector with the resolution of the *ICANOE* liquid target, in which all final state particles can be identified and precisely measured. The kinematical method would allow the evidence for "tau-like" events in the atmospheric neutrino beam.

Both the $\nu_\mu$ disappearance and the direct $\nu_\tau$ appearance methods are weakly depending on the predictions of

**Table 7.** Expected atmospheric neutrino rates per kton×year on Argon in case of no oscillations. The figures have been computed with FLUKA 3D atmospheric neutrino flux with geomagnetic cutoffs and include all nuclear effects. All NC events are included, even though only a fraction of quasi-elastic ones will end up in an observable final state.

| Process | elastic | single-$\pi$ | inelastic | Total | $<E>$ (GeV) |
|---|---|---|---|---|---|
| $\nu_\mu$ CC | 66.7 | 15.9 | 24.4 | 107.0 | 2.36 |
| $\bar{\nu}_\mu$ CC | 12.2 | 5.3 | 9.8 | 27.2 | 3.34 |
| $\nu_e$ CC | 39.4 | 8.4 | 12.1 | 59.9 | 1.60 |
| $\bar{\nu}_e$ CC | 5.4 | 2.1 | 4.2 | 11.7 | 2.36 |
| $\nu$ NC | 42.9 | 8.6 | 13.2 | 64.8 | 1.94 |
| $\bar{\nu}$ NC | 21.1 | 3.5 | 5.0 | 29.6 | 2.00 |

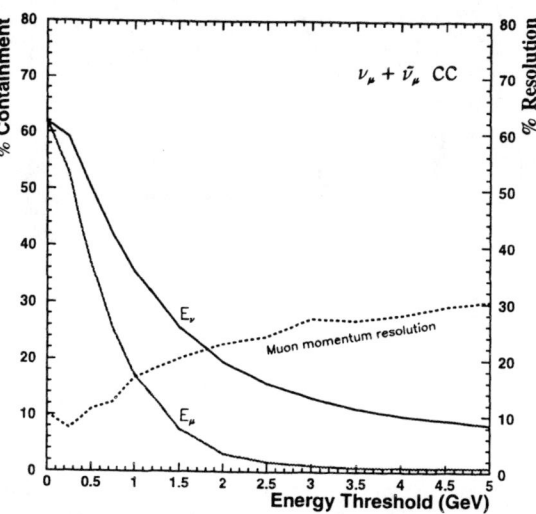

**FIGURE 6.** Integral distributions showing the containment for $\nu_\mu$ CC as a function of the neutrino energy and the leading muon momentum (solid lines). Differential distribution showing the muon momentum resolution as a function of muon momentum (dashed line), including both contained and partially contained events.

neutrino event rates, since they rely on the comparison of rates induced by a downward going and upward going neutrinos.

The *NC/CC* method, already investigated by SuperKamiokande, can be significantly improved compared to this latter measurement. In *ICANOE*, imaging in the liquid target provides a clean bias free identification of neutral-current, independent on the hadronic final state, since the identification is based on the absence of an electron or a muon in the final state.

In the following sections, we will study our results for three different exposures: 5 kton×year corresponding to 1 year of operation, 20 kton×year for 4 years and an ultimate exposure of 50 kton×year or 10 years of operation.

The computed rates for the different neutrino processes (in events/kton/year) and their mean energies are quoted in Table 7, using the FLUKA-3D atmospheric neutrino fluxes(9).

## Event containment and muon measurement

The muon measurement is crucial to most atmospheric neutrino analyses. In *ICANOE*, we achieve the required performances using the multiple scattering measurement rather than resorting to a high–density, coarser resolution detector. Keeping a low density detector, high granularity detector imaging allows in addition the identification and measurement of electrons and individual hadrons in the event.

"Fully contained events" are those for which the visible products of the neutrino interaction are completely contained within the detector volume. "Partially contained events" are $\nu_\mu$ CC events for which the muon exits the detector volume (only muons are penetrating enough).

Figure 6 shows containment of charged current events for different incoming neutrino energies or muon momentum thresholds. Clearly, because of the average energy loss of the muon in argon (about 210 MeV/m for a m.i.p.), muons produced in neutrino events are often energetic enough to escape from the subdetector volume.

It should first be noted that for contained events the muon energy resolution is 4% from $dE/dx$ measurements. For escaping muons, the high granularity of the imaging allows to collect a very precise determination of the track trajectory. Therefore the multiple scattering method can be effectively used to estimate the momentum of the escaping muons. This method requires in practice tracks in excess of 1 meter and works extremely well in the relevant energy range of atmospheric neutrino events (typically below 10 GeV).

The average muon momentum resolution as a function of the energy threshold is shown in Figure 6. This resolution has been computed using the range measurement for contained muons and multiple scattering method for the escaping ones. For energies below 1 GeV, the average muon momentum resolution is about 10%. It increases slowly as a function of the muon momentum and reaches about 30% at 5 GeV.

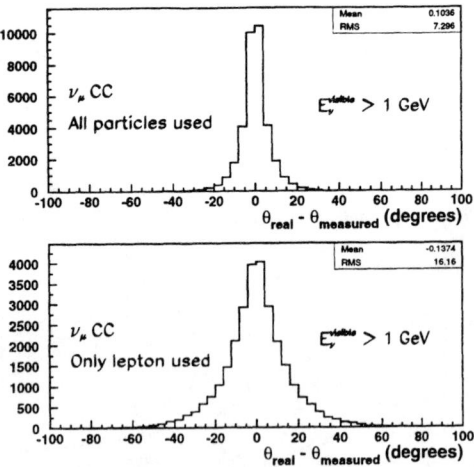

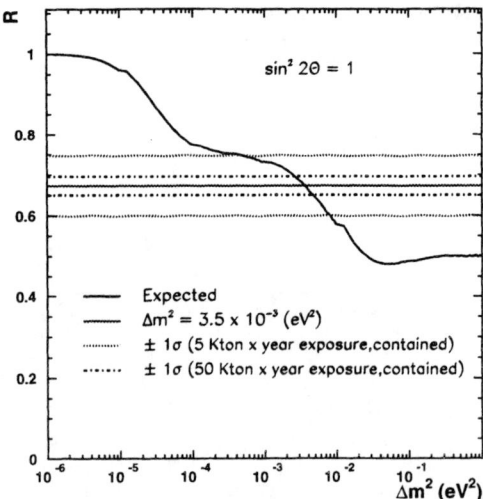

**FIGURE 7.** Zenith angle resolution. The top plot shows the resolution obtained by reconstructing the incoming neutrino direction using all particles momenta, the bottom plot shows the resolution obtained using only leading lepton momentum.

**FIGURE 8.** Expected $R$ value as a function of $\Delta m^2$. Assuming a true $\Delta m^2$ value of $3.5 \times 10^{-3}$ eV$^2$, the 68% confidence intervals are given for 5 and 50 kton×year.

## Incoming neutrino angular resolution

The reconstruction of the zenith angle of the incoming $\nu$ is of great importance in the search of oscillations in atmospheric neutrinos. *ICANOE* allows for a good reconstruction of the incoming neutrino variables (i.e. incidence angle, energy) by using the information coming from all particles produced in the final state. Figure 7 shows the distribution of the difference between the real and reconstructed neutrino angle for events with $E_\nu > 1$ GeV. The improvement on the angular resolution is visible. The RMS of the distribution improves from $\sim 16$ to $\sim 8$ degrees after the inclusion of the hadronic jet in the reconstruction.

## The flavor ratio and up/down asymmetry

Given the clean event reconstruction of *ICANOE*, the ratio $R$ of "muon-like" to "electron-like" events can be determined free of experimental systematic errors. In fact, the expected purity of the samples is above 99%. In particular, the contamination from $\pi^0$ in the "electron-like" sample is expected to be completely negligible. The measurement accuracy will be dominated by the statistical uncertainty and by the theoretical systematic error on the double ratio.

In order to estimate statistical sensitivities, we show in Table 8 the values and statistical errors of $R$ for different exposures, assuming an oscillation $\nu_\mu \to \nu_{x \neq \mu}$, with parameters $\sin^2 2\theta = 1$ and $\Delta m^2 = 3.5 \times 10^{-3}$ eV$^2$. The table lists the expected results when using all events or only fully contained events which turn out to be quite similar.

Clearly, after an exposure corresponding to about four years of running of *ICANOE*, the statistical error reaches a level below 5%. We expect that further theoretical improvements should reduce the systematic error to a level matched to statistical precision achievable in *ICANOE*.

**Table 8.** $R$ as a function of the exposure assuming maximal mixing and $\Delta m^2 = 3.5 \times 10^{-3}$ eV$^2$. In left column: results obtained by using all events; right column: results obtained by using only fully contained events. Quoted errors are of statistical nature.

| Exposure (kton×year) | $R$ all events | contained |
|---|---|---|
| 5 | $0.696 \pm 0.067$ | $0.674 \pm 0.074$ |
| 20 | $0.696 \pm 0.033$ | $0.674 \pm 0.037$ |
| 50 | $0.696 \pm 0.021$ | $0.674 \pm 0.023$ |

From the value of $R$ and from its zenith angle dependence we can obtain the allowed parameter regions of neutrino oscillations. Figure 8 shows how precisely we can determine $\Delta m^2$ in the oscillation case. The $1\sigma$ regions corresponding to 5 and 50 kton×year have been computed using contained events only.

The expected up/down asymmetries are shown in Figure 9 for three different mixing angles as a function of

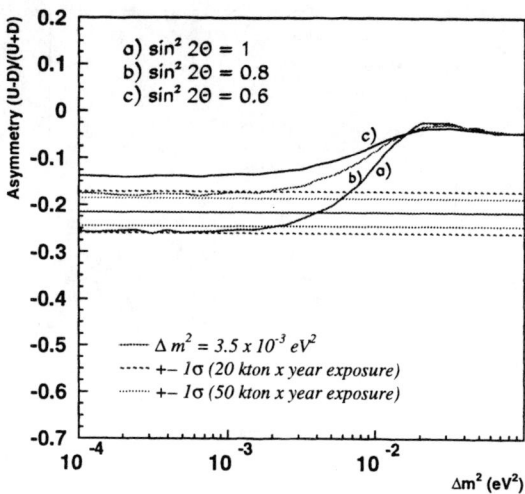
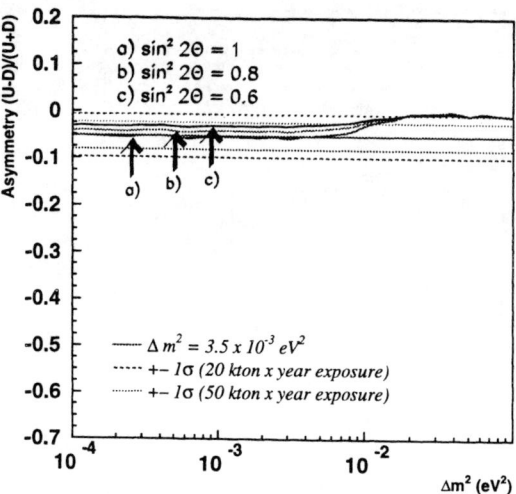

**FIGURE 9.** Expected Up-Down asymmetry for $\sin^2(2\theta) = 1, 0.8, 0.6$ as a function of $\Delta m^2$ for low momentum ($P \leq 0.4$ GeV) $\nu_\mu$ events. (left) when all particles are used to reconstruct the incoming neutrino direction (right) only the lepton is used. The error bands show the 1 $\sigma$ uncertainty for 20 kton×year and 50 kton×year exposures and a 5% systematic error assuming $\sin^2 2\theta = 1$ and $\Delta m^2 = 3.5 \times 10^{-3}$ eV$^2$.

$\Delta m^2$. Note in Figure 9 the much better asymmetry resolution of *ICANOE* for low energy muons when compared to a measurement including the lepton only.

## $\nu_\mu$ disappearance – $L/E$ studies

In order to verify that atmospheric neutrino disappearance is really due to neutrino oscillations, an effective method consists in observing the modulation given by the characteristic oscillation probability:

$$P\left(\frac{L}{E}\right) = 1 - \sin^2(2\theta)\sin^2\left(1.27\Delta m^2 \frac{L}{E}\right) \quad (1)$$

with $L$ in km, $E$ in GeV, $\Delta m^2$ in eV$^2$. This modulation will be characteristic of a given $\Delta m^2$, when the event rate is plotted as a function of the reconstructed $L/E$ of the events when compared to theoretical predictions. The ratio of the observed and predicted spectra has the advantage of being quite insensitive to the precise knowledge of the atmospheric neutrino flux, since the oscillation pattern is found by dips in the $L/E$ distribution while the neutrino interaction spectrum is known to be a slowly varying function of $L/E$. Such a method is in principle capable of measuring $\Delta m^2$ exploiting atmospheric neutrino events.

A smearing of the modulation is introduced by the finite $L/E$ resolution of the detection method. Precise measurements of energy and direction of both the muon and hadrons are therefore needed in order to reconstruct precisely the neutrino $L/E$. This is quite well achieved in *ICANOE*. The contained muons can be measured with a resolution of 4%, while the non-contained muons are measured by multiple scattering method.

The RMS reconstructed $L/E$ resolution is about 30% for events with $E_{visible} > 1$ GeV.

The $\nu_\mu$ survival probability as a function of $L/E$ let us determine the value of $\Delta m^2$ in case of oscillation is confirmed. In figure 10 we can see the survival probabilities of $\nu_\mu$ for neutrino oscillation hypothesis and four different values of $\Delta m^2$. The first minimum on the survival probability happens at highest $L/E$ values for the lowest $\Delta m^2$ values, and allows us to discriminate between them for an exposure of 50 kton×year.

The most favored solution of the atmospheric neutrino anomaly is through $\nu_\mu \to \nu_\tau$ oscillations. However, alternative explanations, like neutrino decay, cannot yet be excluded (10). For example, in a model in which one of the mass-eigenstates of neutrinos with $\nu_\mu$ flavour content decays, the disappearance probability can be described by the expression:

$$P(\nu_\mu \to \nu_{x \neq \mu}) = (\sin^2\theta + \cos^2\theta e^{-\alpha L/2E})^2. \quad (2)$$

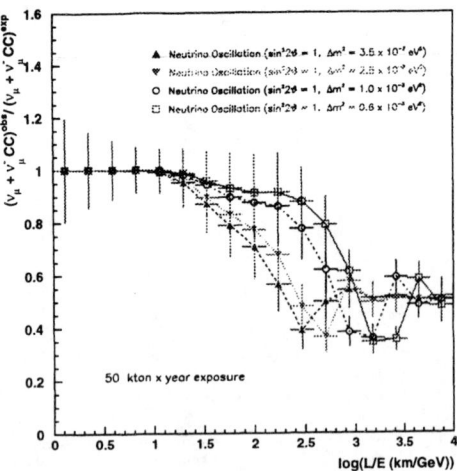

**FIGURE 10.** Survival probability as a function of the $L/E$ ratio assuming neutrino oscillation hypothesis and for various $\Delta m^2$ values and for 50 kton×year. Only statistical error has been considered.

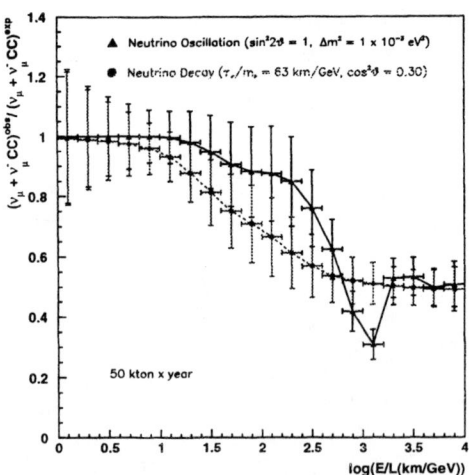

**FIGURE 11.** Survival probability as a function of the $L/E$ ratio for oscillation (triangles) and decay hypothesis (circles). Only statistical error has been considered.

Such a model gives an equally good fit for the choice of parameters $\alpha = 1/63$ GeV/km, $\cos^2\theta = 0.30$ (10).

The capability of distinguishing between the two hypothesis depends on the resolution in measuring the $L/E$ ratio, which depends on the angular and momentum resolution.

Figure 11 shows the survival probabilities as a function of $L/E$ for the neutrino decay hypothesis with $\alpha = m_\nu/\tau_\nu = 1/63$ GeV/km and $\cos^2\theta = 0.30$, and the oscillation hypothesis with $\sin^2 2\theta = 1$ and $\Delta m^2 = 1.0 \times 10^{-3}$, for an exposure of 50 kton×year. Both hypothesis are distinguishable from each other at around 2000 km/GeV within the statistical errors.

## (Direct) appearance of tau neutrinos

To discriminate between $\nu_\mu \to \nu_\tau$ and $\nu_\mu \to \nu_s$ oscillations, we measure the ratio $R_{NC/e} = \frac{NC^{obs}/\nu_e CC^{obs}}{NC^{exp}/\nu_e CC^{exp}}$. An oscillation to an active neutrino leads to $R_{NC/e} = 1$, while $R_{NC/e} \sim 0.7$ is expected for an oscillation to sterile neutrino.

Table 9 shows values and errors of $R_{NC/e}$ in case of oscillation to a sterile neutrino, for all events and fully contained events respectively.

For $\Delta m^2 \leq 10^{-2}$ eV$^2$, oscillations of $\nu_\mu$ into $\nu_\tau$ would in fact result in an excess of "neutral-current-like" events produced by upward neutrinos with respect to downward, since charged-current $\nu_\tau$ interactions would contribute to the "neutral-current-like" event sample, due to the large $\tau$

**Table 9.** $R_{NC/e}$ as a function of the exposure assuming oscillation to a sterile neutrino. Quoted errors are of statistical nature.

| Exposure (kton×year) | $R_{NC/e}$ all events | contained |
|---|---|---|
| 5 | 0.674 ± 0.086 | 0.670 ± 0.087 |
| 20 | 0.674 ± 0.043 | 0.670 ± 0.044 |
| 50 | 0.674 ± 0.027 | 0.670 ± 0.028 |

branching ratio into hadronic channels. Moreover, due to threshold effect on $\tau$ production, this excess would be important at high energy. Oscillations into a sterile neutrino would instead result in a depletion of upward muon-less events. Discrimination between $\nu_\mu \to \nu_\tau$ and $\nu_\mu \to \nu_s$ is thus obtained from a study of the asymmetry of upward to downward muon-less events. Because this method works with the high energy component of atmospheric neutrinos, it becomes effective for relatively large values of $\Delta m^2$ ($\geq 3 \times 10^{-3}$ eV$^2$).

Charged current $\nu_\tau$ rates for five $\Delta m^2$ hypothesis: $5 \times 10^{-4}$, $1 \times 10^{-3}$, $3.5 \times 10^{-3} eV^2$, $5 \times 10^{-3} eV^2$ and $1 \times 10^{-2} eV^2$ are listed in Table 10. We see that the rates saturate at about one event per kton×year for the larger $\Delta m^2$ values. Such small rates pose a major experimental challenge in the detection of $\nu_\tau$ in the cosmic ray induced neutrino flux.

The total visible energy ($E_{visible}$) is a suitable discriminant variable to enhance the $S/B$ ratio. After cuts, surviving events are classified as: $n_b$ (number of expected

**Table 10.** Expected $\nu_\tau$, $\bar{\nu}_\tau$ absolute rates for five different $\Delta m^2$ with FLUKA-3D fluxes and relative to FLUKA-1D and Bartol fluxes.

| $\Delta m^2$ (eV$^2$) | $\nu_\tau + \bar{\nu}_\tau$ CC (NUX, Fluka 3D flux) Rate (kton× year) | | | Rel. to Fluka 1D | Rel. to Bartol |
|---|---|---|---|---|---|
| | DIS | QE | Sum | | |
| $5 \times 10^{-4}$ | 0.11 | 0.11 | 0.22 | 0.96 | 0.81 |
| $1 \times 10^{-3}$ | 0.28 | 0.18 | 0.46 | 1.02 | 0.84 |
| $3.5 \times 10^{-3}$ | 0.59 | 0.21 | 0.80 | 1.00 | 0.81 |
| $5 \times 10^{-3}$ | 0.64 | 0.24 | 0.88 | 1.01 | 0.80 |
| $1 \times 10^{-2}$ | 0.70 | 0.20 | 0.90 | 0.99 | 0.78 |

**Table 11.** Number of NC and tau events as a function of the visible energy cut. The statistical sample used corresponds to an exposure of 50 kton×year.

| 50 kton×year exposure | | | | |
|---|---|---|---|---|
| $E_{Visible}$cut | $\nu$ NC top | $\tau$ bottom | $P_\alpha$ (%) | $P_\beta$ (%) |
| > 1 GeV | 327 | 22 | 55.0 | 10.8 |
| > 2 GeV | 150 | 22 | 38.6 | 3.54 |
| > 3 GeV | 95 | 21 | 30.6 | 1.6 |
| > 4 GeV | 67 | 20 | 25.3 | 0.8 |
| > 5 GeV | 51 | 17 | 27.3 | 0.9 |
| > 6 GeV | 40 | 16 | 24.6 | 0.6 |
| > 7 GeV | 33 | 14 | 26.6 | 0.8 |
| > 8 GeV | 28 | 13 | 26.7 | 0.8 |
| > 9 GeV | 23 | 12 | 26.2 | 0.7 |
| > 10 GeV | 21 | 11 | 28.3 | 0.9 |

downward going background) and $n_t = n_b + n_s$ (number of expected upward going events, where $n_s$ is the number of taus). The statistical significance of the expected $n_s$ excess is evaluated following two procedures:

- The $f_b$ and $f_t$ pdf's are integrated over the whole spectrum of possible measured $r$ values and the overlap between the two is computed: $P_\alpha \equiv \int_0^\infty min(f_b(r), f_t(r))dr$, where $f_b$ and $f_t$ are the Poisson p.d.f.'s for means $\mu = n_b$ and $\mu = n_t$ respectively. The smaller the overlap integrated probability ($P_\alpha$) the larger the significance of the expected excess.

- computing the probability $P_\beta \equiv \int_{n_t}^\infty \frac{e^{-n_b} n_b^r}{r!} dr$ that, due to a statistical fluctuation of the unoscillated data, we measure $n_t$ events or more when $n_b$ are expected.

For a 50 kton×year exposure, the results of a search based on $E_{visible}$ are shown in Table 11. We see that a cut on visible energy between 6 and 7 GeV results in: (1) an overlap integrated probability between the two distributions amounting to 25 – 26%. (2) a Poisson probability that the measured excess ("$\tau$ bottom") corresponds to a statistical fluctuation is 0.6 – 0.8%.

The search for $\nu_\tau$ appearance can be improved taking advantage of the special characteristics of $\nu_\tau$ CC and the subsequent decay of the produced $\tau$ lepton when compared to CC and NC interactions of $\nu_\mu$ and $\nu_e$, i.e. by making use of $\vec{P}_{lepton}$ and $\vec{P}_{hadron}$.

The information related to the directionality of the incoming neutrino (i.e. the beam direction!) is missing. As a result, we have three kinematical independent variables in order to separate signal from background. After a careful evaluation of the performance of different combinations of variables, we decided to use: $E_{visible}$, $y_{bj}$ (the ratio between the total hadronic energy and $E_{visible}$), and $Q_T$ (the transverse momentum of the $\tau$ candidate with respect to the total measured momentum) which contains the information on the isolation of the tau candidate from the recoiling jet.

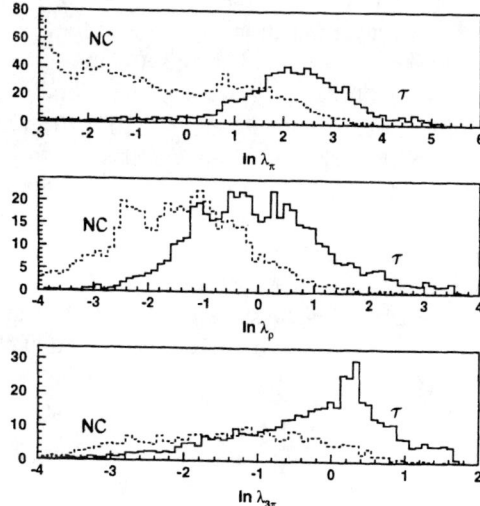

**FIGURE 12.** Likelihood ratio distributions, arbitrarily normalized, for signal and background events. The distributions are computed separately for each of the considered hadronic channels: $\pi$, $\rho$, 3 prongs.

The chosen variables are not independent one from another but show correlations between them. These correlations can be exploited to reduce the background. In order to maximize the separation between signal and background, we use three dimensional likelihood functions $\mathcal{L}(Q_T, E_{visible}, y_{bj})$ where correlations are taken into account.

Table 12 illustrates the statistical significance achieved

**Table 12.** Expected background and signal events for different combinations of the π, ρ and 3π analyses. The considered statistical sample corresponds to an exposure of 50 kton×year.

| π Cut | ρ Cut | 3π Cut | Top Evts | Bot. Evts | $P_\alpha$ (%) | $P_\beta$ (%) |
|---|---|---|---|---|---|---|
| 0   | 0.5 | 0  | 112 | 134 | 32.1 | 1.9 (2.3σ) |
| 1.5 | 1.5 | 0  | 46  | 63  | 24.8 | 0.7 (2.7σ) |
| 3   | −1  | 0  | 43  | 59  | 26.1 | 0.8 (2.6σ) |
| 3   | 0.5 | 0  | 12  | 23  | 18.3 | 0.14 (3.3σ) |
| 3   | 1.5 | 0  | 10  | 20  | 18.8 | 0.16 (3.2σ) |
| 3   | 0.5 | −1 | 30  | 45  | 21.9 | 0.4 (2.9σ) |
| 3   | 0.5 | 1  | 9   | 17  | 25.9 | 0.5 (2.8σ) |

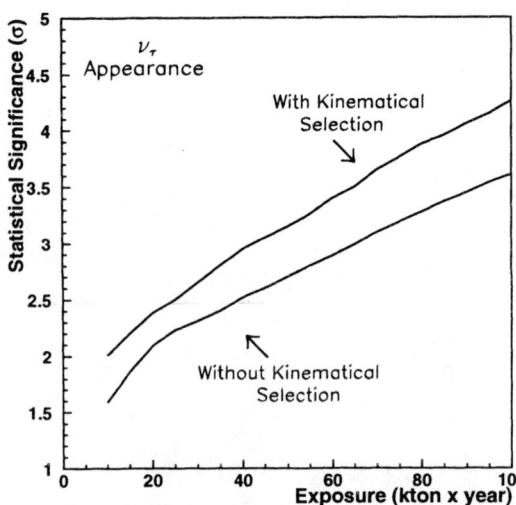

**FIGURE 13.** Probability for the measured excess of upward going events to be due to a statistical fluctuation of the data as a function of the exposure.

by several selected combinations of the likelihood ratios for an exposure equivalent to 50 kton×year. We take as the best combination the one with the lowest $P_\alpha$. This is achieved for the following set of cuts: $\ln\lambda_\pi > 3$, $\ln\lambda_\rho > 0.5$ and $\ln\lambda_{3\pi} > 0$. The expected number of NC background events amounts to 12 (top) while 12+11 = 23 (bottom) are expected. This corresponds to a $P_\alpha$ of 18.3%. In the case we consider $E_{visible}$ as the unique discriminating variable, a similar number of background events is obtained demanding $E_{visible} > 14$ GeV. With this cut, the expected number of τ events is 7 and the $P_\alpha$ is 37%. Therefore, for the same level of background, the approach using the ratio of three dimensional likelihood functions enhances the number of expected signal events by approximately 50%.

Finally, in figure 13 we present the Poisson probability $P_\beta$ for the measured excess of upward going events to be due to a statistical fluctuation as a function of the exposure. The bottom curve corresponds to the case where no kinematical selection has been applied and only a cut on $E_{visible} > 6$ GeV is used. We see that for exposures around 30 kton×year, in case we use the kinematical selection algorithm, the observed excess corresponds to a 2.6σ effect. This effect is larger than 3σ for an exposure of 50 kton×year.

## ACKNOWLEDGMENTS

I thank the organizers for preparing such a great workshop, in particular, C.K. Jung. The help of A. Bueno, M. Campanelli, A. Ferrari and J. Rico is greatly acknowledged.

## REFERENCES

1. F. Arneodo *et al.* [ICARUS and NOE Collab.], "ICANOE: Imaging and calorimetric neutrino oscillation experiment," LNGS-P21/99, INFN/AE-99-17, CERN/SPSC 99-25, SPSC/P314. Updated information can be found at *http://pcnometh4.cern.ch*.
2. see e.g. C. Walter [Superkamiokande Collab.], "Results from Super-Kamiokande and the status of K2K", to appear in the Proceedings of the EPS99 conference, Tampere, Finland, 1999.
3. J. Altegoer *et al.* [NOMAD Collaboration], Phys. Lett. **B431**, 219 (1998).
4. CHORUS Collab., E. Eskut *et al.*, Phys. Lett. B **434**, 205 (1998). CHORUS Collab., E. Eskut *et al.*, Phys. Lett. B **424**, 202 (1998).
5. ICARUS Web page: *http://www.aquila.infn.it/icarus/*
6. NOE Web page: *http://www.na.it/NOE/*
7. A. Bueno, M. Campanelli, A. Ferrari, A. Rubbia, *"Nucleon Decay studies in a large Liquid Argon detector"*, these proceedings.
8. G. Acquistapace et al., CERN 98-02 and INFN/AE-98/05; R. Bailey et al., CERN-SL/99-034(DI) and INFN/AE-99/05
9. G. Battistoni, A. Ferrari, P. Lipari, T. Montaruli, P.R. Sala and T. Rancati, *"A 3–Dimensional Calculation of Atmospheric Neutrino Flux"* hep-ph/9907408, submitted to Astroparticle Physics.
10. V. Barger et al., hep-ph/9907421; V. Barger et al., Phys. Rev. Lett. **82** (1999) 2640.

# OPERA: a long baseline $\nu_\tau$ appearance experiment in the CNGS beam from CERN to Gran Sasso

Pasquale Migliozzi

*INFN Napoli, Italy*

**Abstract.** In this paper we present the OPERA experiment(?) designed for the appearance search of $\nu_\mu \leftrightarrow \nu_\tau$ oscillations in the region indicated by Super-Kamiokande, as explaination of the zenith dependence of the atmospheric neutrino deficit.

## INTRODUCTION

The OPERA experiment is designed for a long baseline oscillation search in the proposed CNGS beam (?) from the CERN SPS to the Gran Sasso Laboratory. It aims at a high sensitivity search for $\nu_\mu \leftrightarrow \nu_\tau$ oscillations. The experiment exploits nuclear emulsion as high resolution tracking devices for the direct detection of the $\tau$ lepton, produced in the charged current interaction of the $\nu_\tau$ with the target. Thanks to its capability of identifying electrons and to the small contamination of $\nu_e$ in the beam, the experiment will also accomplish a $\nu_\mu \leftrightarrow \nu_e$ oscillation search. The technique of nuclear emulsion has found a large scale application in the target of the CHORUS experiment (?), where the automatic scanning of a large samples of events has first been applied. In OPERA emulsions are used exclusively as high precision trackers, unlike in CHORUS where they constitute the active target itself. The feasibility of OPERA requires, on the one hand, the production of emulsions at industrial scale and, on the other hand, it is linked to the impressive progress in the field of computer controlled microscopes read out by CCD cameras, with automatic pattern recognition and track reconstruction (?). After its pioneering work, the Nagoya group of the CHORUS and OPERA Collaborations has delivered a third generation automatic system about 1000 times faster than previous semi-automatic systems (?). Improvements are expected from the intense R&D programmes underway in Europe and Japan, in particular within several groups of the OPERA Collaboration.

## THE DETECTOR CONCEPT

OPERA is designed starting from the ECC concept, which combines the high precision tracking capabilities of nuclear emulsion and the large mass achievable by employing metal plates as a target. The Emulsion Cloud Chamber (ECC) is a massive emulsion detector made of a sandwich of passive material plates interspaced with nuclear emulsion layers used as high precision tracking devices. The present experience with the ECC, combined with the availability of automatic emulsion scanning devices allows us to conceive a $\sim 1000\,ton$ fine-grained vertex detector for the identification of $\nu_\tau$ appearance in the CNGS beam.

By piling-up a series of cells in a sandwich-like structure one obtains the so called *brick*, which constitutes a detector element appropriate for the assembly of more massive structures (*walls*).

One can envisage two basic cell arrangements to be used for the identification of the $\tau$ decay kink through its direct detection, *i.e.* not only by an impact parameter measurement. The first one, called *compact cell*, is shown in Fig. 1. The cell can be made of a $1\,mm$ thick metal plate (lead, in particular) followed by a thin emulsion sheet (ES). The ES is made up of a pair of emulsion layers $50\,\mu m$ thick on either side of a $200\,\mu m$ plastic base. Charged particles give two track segments in each ES. The number of grain hits in $50\,\mu m$ (15-20) is adequate for reconstruction of track segments by means of automatic scanning devices and ensures redundancy in the measurement of particle trajectories.

If a $\tau$ is produced by a $\nu_\tau$ interacting in a lead plate, it will decay either in the same plate (*short decays*) or in the following plate (*long decays*). For short decays, only an impact parameter measurement is possible. For long decays, instead, the $\tau$ is detected by measuring the angle between the charged decay daughter and the $\tau$ direction. This kink angle is caused by the invisible neutrino(s) produced in the decay. For its measurement, the directions of the tracks before and after the kink are reconstructed (in space) by means of a pair of ESs, sandwiching the lead plate where the primary vertex occurred (Fig. 1). The τ can also decay in one of the ESs downstream of the vertex plate (*e.g.* in its plastic base). Even in this case the kink angle can be reconstructed though with a lower angular resolution.

A source of background to the hadronic τ decays in the lead plates is given by hadron re-interactions. One of the primary hadrons may re-interact in the lead plate giving products not seen by the emulsion, hence simulating the charged 1-prong decay of the τ. Unlike the hadronic channels, both the electron and muon decay modes are characterised by low background, as will be shown in the following. The electron channel benefits from the dense lead/emulsion sandwich structure, which allows the electron identification through its showering in the downstream cells.

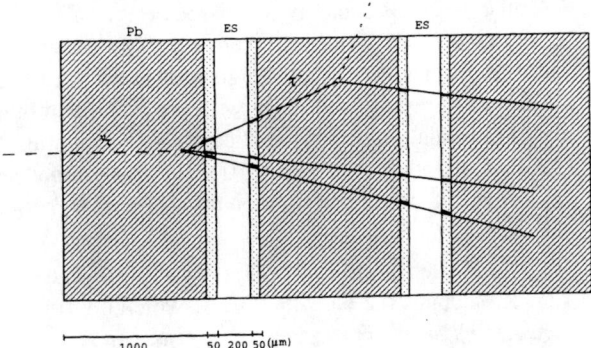

**FIGURE 1.** Schematic structure of a compact ECC cell employing thin emulsion sheets. The τ decay kink is directly reconstructed in space by using four track segments in the emulsion sheets (ES).

Another possible ECC cell structure (called *spacer cell*) foresees the insertion of a low density spacer between consecutive emulsion sheets (Fig. 2). This gap between the ESs allows direct detection of the τ decay kink, as with the compact cell of Fig. 1. In this case, however, the low density of the spacer makes the hadron re-interaction probability negligible. This results in a substantial background reduction for the τ hadronic decay channels, which have the largest branching ratio. The τ decays in the spacer are classified as *gold plated* events, with practically no background. A possible design foresees a cell composed of a 1 *mm* thick lead plate followed by an ES, a spacer of 3 *mm* and another ES, as shown in Fig. 2. The spacer consists of very low density material (Rohacell, polystyrene foam, etc.).

The design of the OPERA target can then be based on a lead-emulsion target subdivided into walls made of bricks. Two types of bricks can be used consisting of sequences of *compact* or *spacer* cells. In the following, the features of either approach will be described.

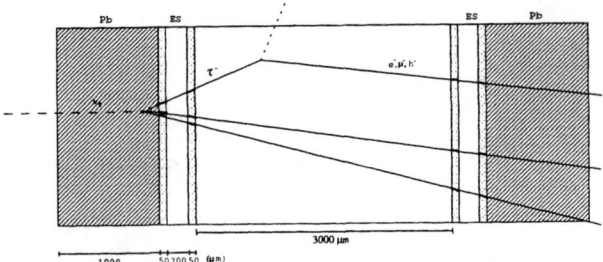

**FIGURE 2.** Schematic structure of a spacer cell.

Electronic detectors, placed downstream of each emulsion brick wall, are needed to identify the brick where the neutrino interaction took place and to guide the scanning. OPERA at the Gran Sasso is naturally shielded against non penetrating particles and it is exposed to a neutrino beam without halo muons. This leads to a relatively low density of background tracks stored in the emulsion and removes the need for a precise location of individual tracks in the ESs by electronic trackers. Therefore, a moderate space resolution can be tolerated, also imposed by practical considerations on the cost of the electronic detectors.

The main purpose of the electronic detectors is to determine the event shower-axis, hence locating the brick whose emulsion must be scanned.

Once the brick has been identified, it can be removed from the detector and its ESs developed. Scanning could then begin following analysis procedures derived from those developed in the last years for neutrino emulsion experiments.

An important requirement for the experiment is the determination of the decay transverse momentum of the daughter particle. This is a tool to reject background events. For most of the muons this measurement can be accomplished by means of the muon spectrometers. For electrons the energy is estimated from the shower development. For charged hadrons, as well as for muons, the momentum is measured by multiple scattering in the ECC sandwich structure.

## GENERAL FEATURES OF THE EXPERIMENT

The main distinctive feature of OPERA is its capability of performing a $\nu_\mu \leftrightarrow \nu_\tau$ oscillation search with low background. A few signal events are therefore meaningful. This extends the discovery potential down to the values of the $\Delta m^2$ parameter allowed by the atmospheric neutrino anomaly. Nuclear emulsions are a key ingredient to achieve this result due to the unambiguous detection of the τ decay kink. The solution offered by the ECC technique is appropriate, since it exploits the tracking capabilities of nuclear emulsions, though for a mass much smaller than the total target mass.

The τ decay channel into a muon, despite the relatively low BR and the overwhelming (reducible) background from $\nu_\mu$ CC interactions, has special features which make it attractive, in particular for a preliminary search. The presence of the penetrating (often isolated) muon track allows an easier event vertex finding. The electronic decay mode can be identified by the characteristic energy loss of the electron inside the dense ECC material. Hadronic modes have the largest BR. Their analysis completes the oscillation search.

Using ECC compact bricks, one would directly profit from the experience already gathered on the ECC technique in e.g. the DONUT experiment. A compact design has merits as far as the direct (long decays) kink detection of the muonic and electronic decay channels of the τ is concerned. Moreover, there is an optimal use of the emulsion, since only one ES is needed for each cell. ECC spacer bricks are expected to offer the appealing feature of the kink detection in a low density material. In particular, hadronic decays inside the spacer have, as shown later, negligible background from primary-hadron re-interactions and decays. However, this is a novel technique which requires further tests and R&D.

The detector design is based on a modular structure. In the present configuration, the total thickness of one spacer(compact) brick is about 14 cm (7.5 cm) along the beam direction. A matrix of adjacent bricks arranged in a plane structure forms a target wall of $\sim 6.5$ m transverse dimensions. A wall and its related electronic detector planes constitutes a *module*. A *supermodule* is defined by a target section made of several consecutive modules followed by a muon detection system. The modularity of the target is used to reach the total mass needed to meet the physics goal of the experiment. The structure of the target permits the removal of the *fired* bricks and to analyse their emulsion sheets during the run.

Supermodules can be built with either types of bricks. Each supermodule contains a given target mass and thus delivers a given number of signal (and background) events. The types and the number of supermodules will be indicated by the physics objectives, but also by the results of the test programmes underway and by the progress in the understanding of the basic properties of the ECC technique as well as by considerations of cost effectiveness.

The dimensions of the bricks are determined by conflicting requirements. A small brick size favours the optimal use of the emulsions when considering brick removals after interactions. In fact, about 25000 events are expected in the total running period and one brick has to be removed for most of the events. The integrated mass to be taken out should be small compared to the total target mass. In addition, bricks have to be small and light enough to allow easy handling by automatic machines, *i.e.* quick installation and removal. On the other hand, losses due to edge effects are reduced with larger brick cross section.

Another issue is the brick thickness in units of radiation lengths. It determines the particle identification capabilities when following tracks in consecutive cells and also the momentum measurement by multiple scattering. For instance, the actual thickness of the compact bricks is motivated by the need of electron identification via its electromagnetic showering.

In the present configuration each brick has dimensions of $15 \times 15$ $cm^2$ orthogonal to the beam direction. Compact bricks can consist of 56 cells with a total thickness of about 7 cm (10 $X_0$) and a weight of about 14.5 kg. Bricks with spacers can contain 30 *cells* with 14 cm thickness (5 $X_0$) and about 8 kg weight.

Tracks from τ candidate events will have also to be followed in consecutive (and in some cases adjacent) bricks to perform a complete kinematical analysis by measuring and identifying the largest number of particles. This requirement imposes mechanical constraints on the way brick walls are assembled and installed and on their support structures. The distance between consecutive walls has to be minimised ($\sim 5$ cm, just to permit the insertion of electronic detectors) as well as the transverse distance between bricks in the wall matrix ($\sim 5mm$). The need of connecting events in consecutive bricks is the motivation for further optimisation of the brick dimensions, once experimental information is available on the specific scanning procedure.

The wall support structure has to be of sufficiently low mass in order to reduce the amount of neutrino interactions in the uninstrumented material, thereby avoiding the possible removal and scanning of *empty* bricks.

According to (?), the target structure can be as follows. The octagonal section is assumed as a reference. A supermodule with spacer bricks consists of 18 modules. Each module is made of a plane of electronic detectors and a wall of 1340 bricks: 14 cm thick and $\sim 6.3$ m side to side. The number of bricks per wall comes from a practical estimate of the dead space between bricks (a few mm), which to some extent depends on the support structure. The space between consecutive brick walls is 5 cm, available for the electronic trackers. This makes the total length of the spacer supermodule about $\sim 3.5$ m. It comprises 24120 bricks for a total mass of 185 *ton*. A compact brick supermodule consists of 24 modules. The thickness of the bricks is in this case 7.7 cm and the supermodule length is $\sim 3$ m. Each supermodule includes 32160 bricks for a mass of 455 *ton*. Table 1 lists the features of the target supermodules. A schematic super-module layout is shown in Fig. 3.

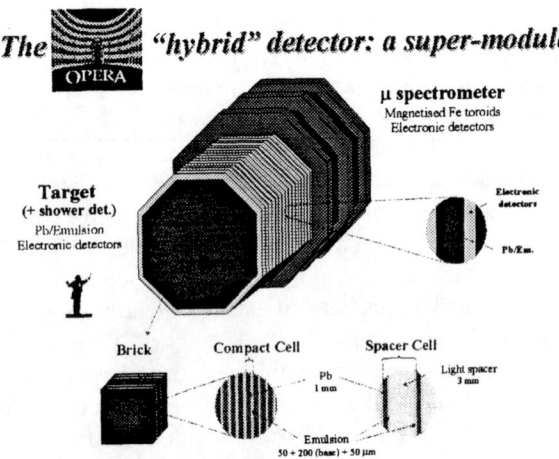

**FIGURE 3.** Schematic view of a OPERA super-module.

The electronic detectors of the target modules are arranged in planes between brick walls. They measure the transverse coordinates of tracks, locate the interaction vertex brick and contribute to the hadronic energy measurement.

The electronic detectors have to provide adequate space resolution, high efficiency and long term reliability. The large total area (about 4000 $m^2$ for each co-ordinate measurement) restricts the choice to cost effective and mature technologies previously used in applications of a similar scale and suitable for industrial production. Flammable gases and polluting materials also limit the possible detector options given the strict rules at the Gran Sasso Laboratory. The baseline option for the OPERA target section is based on scintillator strips read out by wave length shifter (WLS) fibres coupled to photo-detectors. Another detector option is represented by liquid scintillator tubes.

A muon spectrometer constitutes the downstream part of each supermodule. Its purpose is to identify muons with high efficiency, which is of basic importance in order to reduce the charm background, and to measure the sign of the charge. The number of single and associated charm background events produced in $\nu_\mu$ CC interactions is proportional to the primary muon identification inefficiency $(1 - \varepsilon_\mu^{charm})$. The design goal is to obtain an overall muon identification efficiency of $\sim 95\%$. Note that going from an $\varepsilon_\mu^{charm}$ of 90% to 95% reduces the charm background by 50%.

The background given by charmed mesons $C^\pm$ produced in CC and NC $\nu_\mu$ interactions and decaying into muons is further reduced by the measurement of the muon charge. The measurement of the charge is practically impossible for the other decay modes.

Finally, through the measurement of the muon charge and momentum from $\nu_\mu$ CC interactions, essential information can be gained on the beam, like its total flux, its energy spectrum and its $\bar{\nu}_\mu$ contamination.

In the baseline option each spectrometer is made of three high resolution trackers interleaved by two toroidal magnets made of octagonal steel plates. The total steel thickness ($2 \times 51$ cm) together with the magnitude of the magnetic flux density ($1.5\,T$ averaged over the transverse area of the magnet) and the space resolution of the trackers ($\sigma_t \leq 1$ mm) provide the required momentum resolution of at least 30% in the relevant kinematical domain.

The transverse size of the magnets is 7.5 m from edge to edge of the octagon. It complies with the width and length of the target supermodule and with the high geometrical acceptance required for muon identification.

The magnets can be further instrumented with scintillator planes. Their purpose is to provide a measurement of the tail of the hadronic energy leaking from the target with a resolution of $\sim 80\%/\sqrt{E}$; a measurement by range of the energy of the muons which are too slow ($E \leq 1\,GeV$) to traverse the two magnets, with a resolution of $\sim 5\%$; a coarse tracking inside the magnet that facilitates the track matching between trackers; a possible measurement of the NC/CC ratio for events occurring in the spectrometer steel if used as a target.

## Detection efficiency and background

The $\tau$ detection efficiency of OPERA depends on the actual type of bricks (supermodules) and on the kinematical features of the events and, hence, on the oscillation parameters. In the following, the estimates are given for $\Delta m^2 = 3.5 \times 10^{-3}\,eV^2$ and full mixing.

The first quantity contributing to the detection efficiency is the brick finding efficiency, determined by the electronic trackers. The latter perform the measurement of the axis of the shower induced by the neutrino interaction and therefore the identification of the brick where the interaction occurred. The reconstruction of individual particles can contribute to the event location. Sources of inefficiencies are, in the longitudinal direction, back-scattering at the event vertex and, in the transverse direction, finite resolution effects at the boundary between adjacent bricks and interactions in the material of the support frames. However, an initial misidentification of the right brick can be recovered later by scanning another brick. A global brick detection efficiency $\varepsilon_{br} \sim 90\%$ is estimated.

Another source of inefficiency is represented by fiducial volume cuts. Tracks can be reconstructed up to very close to the ES edge ($\sim 3$ mm, i.e. 25 microscope views).

**Table 1.** Design features of the target supermodules.

| Item | Spacer-brick supermodule | Compact-brick supermodule |
|---|---|---|
| Dimensions | $\sim 6.3 \times 6.3 \times 3.5\ m^3$ | $\sim 6.3 \times 6.3 \times 3\ m^3$ |
| Cell thickness (mm) | 4.6 | 1.3 |
| Number of cells/brick | 30 | 56 |
| Brick x-section ($cm^2$) | $15 \times 15$ | $15 \times 15$ |
| Brick thickness (*cm*) | 14.2 | 7.7 |
| Brick thickness ($X_0$) | 5.3 | 10 |
| Brick weight (*kg*) | 7.6 | 14.2 |
| Module thickness (*cm*) | 19.2 | 12.7 |
| Bricks/wall | 1340 | 1340 |
| Number of modules/superm. | 18 | 24 |
| Number of bricks/superm. | 24120 | 32160 |
| ES surface/supermodule ($m^2$) | 32500 | 40500 |
| Supermodule weight (*ton*) | 185 | 455 |

For bricks with $15 \times 15\ cm^2$ cross-section this produces a 8% loss in the effective emulsion surface. The effect of the dead space for the brick-to-brick event connection leads to an additional inefficiency of $5-10\%$. Events occurring in the last wall of a supermodule or at the outer edges have a lower efficiency for particle identification decay $p_t$ measurement. This may lead to an additional volume loss at the $1-2\%$ level. The various contributions result in an overall $\varepsilon_{geom} \sim 80\%$.

Once the brick where the neutrino interacted has been identified, scanning can start. The relatively low background of tracks stored in the ESs facilitates the procedure of vertex finding. One can estimate the vertex finding efficiency $\varepsilon_{vert}$ for bricks with or without spacer either by Monte Carlo simulations or by simulations using real data, *e.g.* from CHORUS. We estimated $\varepsilon_{vert} = 80\%$.

Some of the factors which contribute to the detection efficiency are independent of the $\tau$ decay channel. A $\tau$ produced in $\nu_\tau$ CC interactions and decaying outside the vertex plate is identified through a direct kink detection. Given the CNGS beam features and the $\tau$ kinematics, for $\Delta m^2 = 3.5 \times 10^{-3}\ eV^2$ the probability $\varepsilon_{long}$ for the $\tau$ to exit the vertex plate is 45%. Monte Carlo calculation show that the spacer itself (3 *mm*) collects $\sim 25\%$ of the $\tau$ decays. Taking into account the $\nu_\tau$ CNGS spectrum one obtains $\varepsilon_{long}(base) = 12\%$ and $\varepsilon_{long}(spacer) = 28\%$, respectively for decays in the first ES plastic base or in the spacer.

An additional factor entering into the detection efficiency is given by the rejection of small-angle and very large-angle kinks. The upper cut ($> 500\ mrad$) is motivated by considerations related to the scanning efficiency of the automatic microscopes. The lower cut is due to the angular resolution in the measurement of the kink angle in the space. Test beam measurements indicate that the angular resolution per projection, measured by means of four track segments in the ESs, is $\sigma_\theta \sim 5\ mrad$, while it becomes $\sigma_\theta \sim 10\ mrad$ for decays in the base (?). Therefore, the lower cut has been set to 25 *mrad* (50 *mrad*) for decays in the spacer (base). In this way one rejects fake kinks due to the angular resolution. The fraction $\varepsilon_{kink}$ of $\tau$ decays with a kink angle in the useful range depends of the actual decay mode. It amounts to $\sim 85\%$ for the electron and muon decay modes and $\sim 90\%$ for the hadronic decays.

Similar considerations can be applied to the compact bricks, for which the electron and the muon $\tau$ decay channels are studied. In this case, contributions to the detection efficiency come from the electron identification efficiency and from kinematical cuts needed to reject hadronic or muonic re-interactions in the lead plate, mimicking a $\tau$ decay.

The overall signal detection efficiency for *long decay* events for spacer and compact supermodules is summarised in Table 2. Note that the lower efficiency of the compact supermodules is counterbalanced by twice the target mass at the same amount of emulsion.

The expected background sources to long $\tau$ decays are

- prompt $\nu_\tau$ production in the primary proton target and in the beam dump;
- cosmics and radioactivity;
- hadronic decays and re-interactions;
- muon scattering;
- 1-prong decay of charmed particles.

For details on the background evaluation we refer to (?).

**Table 2.** Detection efficiency for different $\tau$ decay modes nd types of supermodules (for $\Delta m^2 = 3.5 \times 10^{-3}\ eV^2$).

| Supermodule | Decay mode | $\tau$ det. eff. |
|---|---|---|
| spacer | $\tau \to e^-$ | 3.0% |
| " | $\tau \to \mu^-$ | 2.6% |
| " | $\tau \to h^-$ | 7.3% |
| " | total | 12.9% |
| compact | $\tau \to e^-$ | 3.0% |
| " | $\tau \to \mu^-$ | 2.6% |
| " | total | 5.6% |

Table 3 summarises the number of signal and background events which can be collected in the various channels by one supermodule (spacer or compact) with $2.25 \times 10^{20}$ *pot* from the SPS. This corresponds to a nominal five years run not considering the possibility for a dedicated running mode or an improved CNGS performance.

## Sensitivity and discovery potential

In order to cover the Super-Kamiokande allowed region (?) five supermodules are required. One could then obtain $\sim 20$ events if $\nu_\mu \leftrightarrow \nu_\tau$ oscillations occur at the Super-Kamiokande central value of $\Delta m^2 = 3.5 \times 10^{-3}\ eV^2$. For the evaluation of the general features of the experiment in terms of sensitivity and discovery potential one can consider the combined use of both types of supermodules in a scheme which leads to the detection of a suitable signal in the different channels. The entire detector can comprise five supermodules in a row. Three supermodules would be assembled with bricks with spacers, for a total mass of $\sim 555\ ton$. Two supermodules with compact bricks can yield a mass of $\sim 910\ ton$ employing a quantity of emulsions similar to that of the three supermodules with spacer bricks. The total target mass would amount to about $1500\ ton$. With $2.25 \times 10^{20}$ *pot* the experiment would collect $\sim 18000$ CC and $\sim 6000$ NC $\nu_\mu$ events, summing-up DIS, QE and RES events.

The aim of OPERA is to be sensitive to the oscillation parameter region corresponding to the Super-Kamiokande signal. The low expected background is the key element to achieve this goal. The sensitivity in $\Delta m^2$ improves with the square root of the number of detected events. This implies that already after a run of about two years a significant limit could be set in the case of a negative search. This is shown in Fig. 4, where the exclusion plots at the 90% C.L. are given for a two, three or four years' run. The Figure also shows the result from a global fit of the different Super-Kamiokande measurements on atmospheric neutrinos.

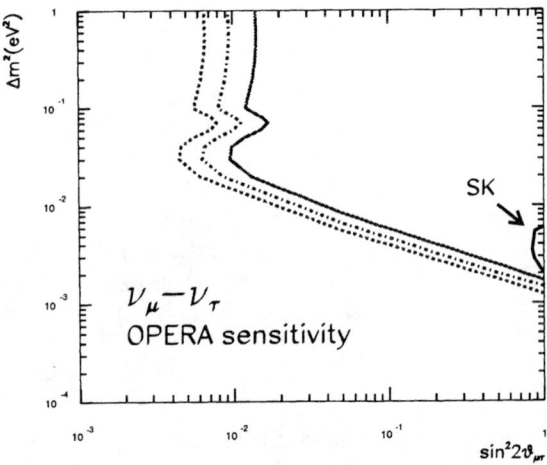

**FIGURE 4.** Sensitivity of OPERA in the search for $\nu_\mu \leftrightarrow \nu_\tau$ oscillations. The 90% C.L. exclusion limits (2, 3 and 4 years' run) are plotted together with the region allowed by Super-Kamiokande.

In the case of a positive search, OPERA should be able to detect candidate events already after one or two years of run. Table 4 gives the expected numbers of $\tau$ events detected in the various decay channels with the supermodule configuration previously considered. Assuming that $\nu_\mu \leftrightarrow \nu_\tau$ oscillations occur with parameter values $\Delta m^2 = 3.5 \times 10^{-3}\ eV^2$ and $\sin^2 2\theta_{\mu\tau} = 1$, about 18 $\nu_\tau$ CC events are detected by OPERA, allowing the inference of $\Delta m^2$ with a statistical error of about 20%. An experimental systematic error of 10% is estimated on the numbers of signal and background events.

**Table 3.** τ and background events detected by one spacer or compact OPERA supermodule. Full mixing and three values of $\Delta m^2$ are assumed. The values are given for $2.25 \times 10^{20}\, pot$.

| Supermodule | τ decay mode | Signal $(2 \times 10^{-3} eV^2)$ | Signal $(3.5 \times 10^{-3} eV^2)$ | Signal $(6 \times 10^{-3} eV^2)$ | BG |
|---|---|---|---|---|---|
| spacer | $e^-$ | 0.26 | 0.81 | 2.38 | 0.010 |
| " | $\mu^-$ | 0.23 | 0.70 | 2.06 | 0.006 |
| " | $h^-$ | 0.64 | 1.97 | 5.79 | 0.036 |
| compact | $e^-$ | 0.65 | 2.00 | 5.88 | 0.040 |
| " | $\mu^-$ | 0.56 | 1.73 | 5.08 | 0.15 |

**Table 4.** Detected τ and background events with three spacer and two compact supermodules, for full mixing, three values of $\Delta m^2$ and $2.25 \times 10^{20}\, pot$.

| τ decay mode | Signal $(2 \times 10^{-3} eV^2)$ | Signal $(3.5 \times 10^{-3} eV^2)$ | Signal $(6 \times 10^{-3} eV^2)$ | BG |
|---|---|---|---|---|
| $e^-$ | 2.1 | 6.4 | 18.9 | 0.11 |
| $\mu^-$ | 1.8 | 5.6 | 16.3 | 0.32 |
| $h^-$ | 1.9 | 5.9 | 17.4 | 0.11 |
| Total | 5.8 | 17.9 | 52.6 | 0.54 |

# Summary of Detector and Beam Parameters Working Group from Lyon

D.A. Harris

*Fermi National Accelerator Lab*
*P.O. Box 500, Batavia, IL 60510-0500*

**Abstract.** This report is an overview of the discussions and talks that occurred during the detectors and beams working group of the NuFact99 workshop in Lyon, France. A wish list of muon beam parameters is given, as well as a survey of currently studied oscillation detector technologies. Finally, a short description of the non-oscillation physics that would be accessible is included to show how the diversity of neutrino physics a muon storage ring could address. Workshop talk slides can be found in http://muonstoragerings.cern.ch/Welcome.html/

## INTRODUCTION

A muon collider has been proposed as a possible machine to push the high energy frontier. However, the colliding muon beams would produce extremely intense neutrino beams of a very different flavor composition than what has previously been available. Since the beam constraints on muon beams for neutrino experiments are different and less demanding than those for collider beams, a first step towards a muon collider could be a muon storage ring, to be used for neutrino physics. NuFact99 was designed to bring together high energy physicists and accelerator physicists to start a dialogue on what considerations are important when designing a muon storage ring to be used for neutrino physics. This report first describes which neutrino oscillation measurements will be relevant in the era of a muon storage ring, followed by a description of those muon beam parameters which are critical to oscillation measurements. Oscillation detectors that are currently being considered are described next, although in the era of neutrino storage rings perhaps completely new designs will be in place. Finally, the non-oscillation physics accessible at a near detector facility are also described.

## OSCILLATION MEASUREMENT GOALS

In one possible scenario of neutrino oscillations, there are three generations of neutrinos undergoing mixing, with the two mass squared differences being those suggested by the atmospheric and solar neutrino anomalies. If that is indeed the case, then there are three angles and one CP-violating phase which uniquely determines the neutrino mixing matrix. By the time a muon storage ring would be built it is expected that two of the angles ($\theta_{23}$, $\theta_{12}$), and two mass squared differences ($\Delta m_{23}$ and $\Delta m_{12}$) would be known to about 30% or better. This knowlege would come from the solar and atmospheric measurements which would have been verified by long baseline and reactor experiments, for example, MINOS, K2K, and KAMLAND. The remaining two pieces of the puzzle would be $\theta_{13}$ and the CP-violating phase $\delta$.

By measuring both the probability of electron neutrinos going to muon neutrinos and its CP-conjugate the two remaining measurements could be made. If the mixing matrix is not unitary or if LSND is confirmed and there are more than three generations mixing then measuring all the angles is even more of an imperative, since it would give us information about the sterile neutrino sector.

## BEAM PARAMETERS

The first requirement oscillation experiments would impose on a muon storage ring is that it be able to produce both $\mu^+$ and $\mu^-$ beams (sequentially, not simultaneously!) in the same straight section(s). The polarities of the storage ring magnets would be switched at least once during the entire run, but could be switched more often to reduce time-dependent systematic uncertainties.

### Muon Rates

The rates that have been proposed range between $10^{20}$ and $10^{22}$ muon decays in the ring per year, with straight

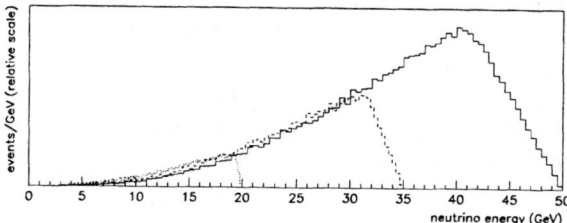

**FIGURE 1.** Far detector neutrino energy distributions for a 20, 35, and 50 GeV muon beam, absolutely normalized.

sections comprising 25% to 33% of the entire length. To demonstrate how unprecedented these rates are, it is useful to note that MINOS expects to get $10^{18}$ muon neutrinos per year, and only $5 \times 10^{15}$ electron neutrinos per year. The reactor neutrino "beams" are low energy and consist only of electron antineutrinos.

## Muon Beam Energy

Naively, one might expect that the ideal energy of the stored muon beam is predicted by the combination of the squared mass difference that one is trying to probe and the distance at which one would like to place a detector. But in fact it has been shown that the rates in a far detector for low energy neutrinos are the same, once the muon beam energy is about 13% above of the energy one is interested, as shown in figure 1. The total rate of neutrino interactions scales as the cube of the energy, so background considerations aside, one would always opt for higher energy muon beams. However, higher energy neutrino interactions can easily produce low energy backgrounds. If the low energy neutrinos are sensitive to oscillations, one must understand these backgrounds before ultimately chosing a muon beam energy.

The sources and levels of backgrounds will vary depending on the energy of the incoming neutrinos. At low energy (10 GeV), there is negligible background from charm production, but substantial background from cosmic rays (particularly if the detector is on the surface of the earth). At higher energies (for example 50 GeV) tau appearance becomes easier since 1) the tau mass suppression will be smaller so more of them will be produced, and 2) the tau will travel farther before it decays. Background from charm production also rises as the energies increase, but it is also more kinematically distinguishable from the signal at high energies and is ultimately less of a background (1).

## Storage Ring Configuration

If the bulk of the cost and difficulty of building a muon storage ring lies in the proton driver, pion capture, and muon cooling, then one should optimize what is done with the muon storage ring once the muons are accelerated. This can be done by making a "ring" with more than one usable straight section, which ideally could be pointed at different angles with respect to the horizonal, allowing one to measure oscillation parameters at two different distances. This would enable one to disentangle CP violation and matter effects.

The time structure of the muon beam could also be used as a tool, in particular to remove background from cosmic rays. If the cosmic ray background is low enough this might enable long baseline experiments to be done above ground. On the macroscopic scale, a bunched structure would be preferable, for example, injecting every 10 ms. This would enable experiments to look for associated long-lived particle decays.

## Polarization and Divergence

Understanding the polarization of the muon beam is extremely important for oscillation measurements, particularly those comparing near and far detector neutrino fluxes. Polarization is of interest for oscillation measurements specifically because it alters the relative fluxes of $\nu_\mu$ and $\bar{\nu}_e$'s, by changing the $\bar{\nu}_e$ component. One example of where a well-measured time-varying polarization could be useful is if one saw an excess of electron-like events in a far detector, but there was no electron charge identification. By varying the beam polarization one can determine if the electron-like events came from $\nu_\mu$'s, or $\bar{\nu}_e$'s (2).

Another important aspect of the muon beam is its angular divergence. The natural divergence of the neutrino beam is $1/\gamma$, where $\gamma$ is the boost of the muons. An large uncertainty in the angular divergence, combined with non-zero net polarization, contributes to an uncertainty in the ratio of $\bar{\nu}_e$ to $\nu_\mu$ fluxes between the near detector and the far detector. If $R_E \equiv \int E_{\nu_\mu} / \int E_{\nu_e}$, then the uncertainty in that ratio $\sigma_{R_E}$ is equal to $P\sigma_r/r^2$, where the angular divergence of the beam $\delta\theta$ is expressed as $\frac{1}{r\gamma}$, and $\sigma_r$ is the uncertainty on $r$ (3). So if the net polarization $P$ is non-zero, then either the divergence must be small or precisely known.

# DETECTORS

Ideally, for both a near and far detector, one would want to build something that could do particle and charge identification for all three generations of charged leptons. Muons are the easiest to identify, taus are the next easiest if only because of their decay to muons, and finally electrons are the most difficult. Although detector technologies exist that could achieve all these goals, the challenge lies in making such a detector with a mass of at least 10kton.

If when a muon storage ring turns on we still believe that oscillations are occurring between $\nu_\mu$ and $\nu_\tau$, but $\tau$'s have not been observed in long baseline neutrino experiments, then many believe that a muon storage ring neutrino detector must emphasize tau appearance. But simply by measuring $P(\nu_e \to \nu_\mu)$ and $P(\bar{\nu}_e \to \bar{\nu}_\mu)$, which would only require muon identification, energy, and charge measurements, one can access both $\theta_{13}$ and possibly $\delta$, depending on how different the mass splittings are.

## Baseline Detector Design

One very nice aspect of neutrino experiments is that they can be placed one right after the other and the beam is not noticeably attenuated. So it is probable that a magnetized steel/scintillator sampling calorimeter would be one of the far detectors at a muon storage ring experiment.

The largest forseeable background in such a detector is charm production. The oscillation signal is a wrong-sign muon, which would presumably come from the electron antineutrino becoming a muon antineturino in the presence of only muon neutrinos. If there is enough energy for charm production, however, in a $\nu_\mu$ charged current event the charmed particle produced will decay 10% of the time to a wrong-sign muon in the final state. There is a chance that the muon from the neutrino interaction vertex is low energy and/or undetected.

One proposal for a baseline detector is a magnetized steel/scintillator sampling calorimeter, segmented such that the hadron energy resolution is $.76/\sqrt{E_{had}(GeV)}$, and a hadron angular resolution of $17/\sqrt{E_{had}(GeV)} + 12/E_{had}(GeV)(rad)$ (1). The muon energy and angular resolution is expected to be much better. Kinematic cuts can be made on the muon momentum and its component transverse to the hadronic shower to reduce the background from charm production. The signal efficiency ranges from 25 to 30%, but the backgrounds can be reduced by a factor of $10^{-5}$ to $10^{-6}$, depending on the neutrino energy. At higher energy the backgrounds are larger, but since the rejection factor is also larger the overall background contribution is smaller.

At the very least a detector of this type would be sufficient for measurements of $P(\nu_e \to \nu_\mu)$ and $P(\bar{\nu}_e \to \bar{\nu}_\mu)$ which would give information about $\theta_{13}$, and the CP-violating phase $\delta$, both of which would contribute enormously to the field.

## Beyond Muon Identification

Alternate technologies must be employed to achieve electron or tau identification event-by-event, or electron or tau charge measurements. As mentioned before, these technologies are not considered a high priority unless no taus have been observed at long baseline experiments yet are expected to be produced, or if there are sterile neutrinos.

One category of new detectors uses thin ($\sim 100\mu m$) sheets of emulsion combined with low-density ($\sim 300\mu m$) spacers, so that one can measure kinks that occur when a tau decays by comparing the slope of a track before and after the spacer(4). This would be very useful for identifying taus and electrons. Finally, for charge identification one might imagine constructing an emulsion/magnetized steel calorimeter, where the target mass would be made up of thin sheets (0.1cm) of magnetized steel. Presumably these detectors cannot be made as large as the simple coarse-grained detectors.

Another kind of detector which has more promise for use on the 10kton scale can be classified as a "kinematic separation" detector. This detector identifies $\tau \to \mu$ decays by their difference in kinematics, although they don't see the kink from the decay itself. ICARUS, which uses a Liquid Argon TPC detector, has the necessary charged track resolution to measure the acoplanarity of an event and determine the likelihood of it being a tau candidate (5).

# PRECISION HIGH RATE NEUTRINO PHYSICS

The advent of a muon storage ring would not only bring about new oscillation measurements that are completely unimaginable today, but would also usher in a new era of high precision neutrino scattering experiments. For a detector which is located 29m from a 150m straight section for a 50GeV muon storage ring at $10^{21}$ muons per year, the event rate is 40 million events per kilogram per year over a 10cm radius (3). Neutrinos could be used as precision probes of nucleon structure much like charged leptons have been used in the past. The new handle neutrinos provide is that they can measure the nuclear dependence of the valence and sea quarks as well as valence vs sea measurements of the spin distributions, using a polarized target (similar to SMC).

Another way to take advantage of the high neutrino fluxes would be to use this facility as a charm factory. Charged current charm production from muon neutrinos provides one with a clean signature, when the charmed meson decays to a final state (oppositely-charged) muon. For the same beam parameters described above, there would be $10^6$ leptonic tagged charm decays in 4kg-years. With that kind of statistics $D^0 - \overline{D^0}$ mixing and $V_{cd}$ could be measured over an order of magnitude better than what is currently known (3).

## CONCLUSIONS

The three angles and CP-violating phase $\delta$ of the neutrino oscillation matrix could be accessed with the muon storage rings that have been proposed in this workshop, given even a pessimistic scenario for that matrix and today's detector technology. This will help us to understand what connects quarks and leptons, and may be the cleanest place to study the mechanism of CP violation, since it is unshielded by hadron uncertainties.

However, we still await results from the currently approved round of neutrino oscillation experiments. If in fact either a) LSND is right, or b) Homestake is wrong and solar neutrinos are at a higher mass squared difference than currently believed, and/or c) there are sterile neutrinos, then the oscillation program at a muon storage ring will be far broader than outlined here. A survey of near detector measurements shows that this facility could also be the home of new breakthroughs in our understanding of nucleon structure.

I gratefully acknowlege the co-convenors of this working group, F. Dydak and J.J. Gomez-Cadenas, as well as all of the speakers in the working group: M. Campanelli, A. Cervera, D. Cline, M. deJong, F. Dydak, R. Edgecock, J.J. Gomez-Cadenas, C. Gonzales-Garcia, P. Hernandez, T. Hasegawa, B. King, P. Litchfield, K. McFarland, V. Palladino, I. Papadopoulos, P. Strolin.

## REFERENCES

1. A. Cervera-Villanueva, J.J. Gomez, F. Dydak, NuFact99 Proceedings
2. A. Blondel, NuFact99 Proceedings
3. K. S. McFarland, NuFact99 Proceedings
4. P. Strolin, NuFact99 Proceedings; D.A. Harris, NuFact99 Proceedings.
5. M. Campanelli, NuFact99 Proceedings; D. Cline, NuFact99 Proceedings.

# Summary of Accelerator Neutrino Experiments Working Group

R. Bernstein

*Fermi National Accelerator laboratory*
*P.O. Box 500, Batavia IL 60510*

Our group focused on searches for neutrino oscillations at a muon storage ring. In the experimental realm, we heard a report from J. Wilkes on the first K2K event. S. Boyd discussed distinguishing $\nu_\mu \to \nu_\tau$ from $\nu_\mu \to \nu_s$ in SuperK using the up/down neutral current ratio. Progress in theoretical work was given by M. Campanelli, discussing matter-enhancement effects in the context of ICANOE. I. Mocioiu presented oscillation probabilities for matter-enhancement over long-baselines vs. $L/E$. Finally, we examined physics goals for a neutrino oscillation experiment at a muon storage ring long-baseline experiment. In that context, Deborah Harris summarized the discussions at the Lyon Workshop.

## "STANDARD MODEL" OF NEUTRINO OSCILLATIONS

With the assumptions that (1) LSND will not be confirmed, (2) the SuperK and Solar Homestake results are correct, and (3) the Large Angle solar solution is correct, then we should expect

§1. $m_3 \gg m_2 \gg m_1$

§2. $\Delta m_{23}^2 \approx \Delta m_{13}^2 \gg \Delta m_{12}^2$

§3. $\sin^2 2\theta_{23}$ Large, $\sin^2 2\theta_{12}$ Large

Three measurements then become critical: (a) a precise measurement of $\nu_\mu \to \nu_\tau$, (b) $\nu_e \to \nu_\tau$ at as low $\sin^2 2\theta$ as possible, and (c) using matter effects in Earth to enhance oscillations.

The other case, where LSND is confirmed by miniBOONE, would open an entirely new range of searches. The most likely scenario is for there to be some number of sterile neutrinos $\nu_S$. It is important in the short term to be sure that the SuperK effects are indeed from $\nu_\mu \to \nu_\tau$ and not $\nu_S$.

I. Mocioiu presented results from a calculation of the size of the different transitions paying particular attention to matter effects.(1) She has done two new things: (a) used the electron number density along a chord through the Earth, determined from geophysical measurements, and (b) presented the results as a function of $E/\Delta m^2$ for fixed $L$. The second point makes her results particularly convenient for simulations since the oscillation probability can be calculated at any combination of $\Delta m^2$ and $E$; a number of such tables exist for likely values of $L$ and calculating more is a simple matter.

## EXISTING AND APPROVED EXPERIMENTS

This Working Group addressed questions directly relevant to both scenarios. Specifically, Boyd discussed neutral current $\pi^o$ production as a signal of sterile neutrinos in the SuperK data. Neutral currents are flavor blind; if $\nu_\mu \to \nu_\tau$, then equal numbers of neutral currents would be produced regardless of the oscillation length $L$. Hence upward and downward-going neutrinos, which would have different mixes of $\nu_\mu$ and $\nu_\tau$, would show no up/down asymmetry in the NC rate.

**Table 1.** Physics Goals and Parameter Choices. "Far" refers to a > 7000 km baseline, "long" is 2000-3000 km baseline, and "short" is a < 1000 km baseline. $\mu/e$ Separation implies *both* the ability to detect muons *and* the ability to positively identify electrons in the hadronic shower. "$\nu$ and $\bar{\nu}$" refers to both signs of muons.

| Physics | Baseline | Polarized | $\mu/e$ | $\nu$ and $\bar{\nu}$ |
|---|---|---|---|---|
| Matter Effects | Long/Far | Prefer | Yes | Both |
| $\theta_{13}$ $\nu_e \to \nu_\mu$ | Long/Far | No | Yes | $\bar{\nu}$ |
| $\nu_e \to \nu_\tau$ | Far | No | Yes | Both |
| CP Violation | Short | No | Yes | Both |
| $\nu_\mu \to \nu_\tau$ | Long/Far | No | Yes | $\nu$ |
| $\nu_S$ | Long/Far | No | Yes | Both |

If the $\nu_\mu$ oscillations were to sterile neutrinos, more neutral currents per charged current event would come from the downward-going neutrinos than their sterile upward-going counterparts. Boyd's contribution discusses the details; summarizing, the experimental determination of the expected rate and asymmetry is difficult and the theoretical prediction has errors from nuclear effects. He concludes it will be difficult but possible for them and perhaps possible in future experiments.

M. Campanelli described work at CERN involving calculations of matter-effects for long-baseline oscillations in the "standard" scenario and coupled the theoretical work to ICANOE-like detectors. This gives useful estimates for how well the parameters can actually be measured and the role of matter effects in enhancing the $\nu_e \to \nu_\tau$ transition.

## NEUTRINO FACTORIES

D. Harris summarized results from the Lyon Neutrino Factory Workshop, NuFact 99. They concluded that the angles and phase of a $3 \times 3$ matrix could be measured under pessimistic beam assumptions and existing detector technology.

This group also discussed some very early work from a Physics Study Group convened at Fermilab. Table 1 gives "quality factors" for different measurements.

## LSND/KARMEN

The LSND and KARMEN results were discussed as well. Two recent attempts to resolve the near-discrepancy have been made by Eitel(2) and Stancu.(3) Eitel performs a simultaneous fit to the LSND and KARMEN data using the Feldman-Cousins(4) frequentist approach and finds a region of signal consistent with both experiments, but consistent in this approach does not have its traditional meaning. The Feldman-Cousins approach decouples goodness-of-fit from the acceptance or rejection of the hypothesis, so that the combined fit can give a poor traditional $\chi^2$. In this particular work the spectral information is not fully used; essentially the comparison performs a rate test. However, the central conclusion, that given current data there is no definitive discrepancy, is valid and therefore more data and/or experiments are required. Stancu's paper, which uses traditional statistical techniques, is problematic in that it requires a significant overall flux error in the Homestake result and is only barely consistent with the zenith distribution from SuperKamionkande.

## REFERENCES

1. I. Mocioiu and R. Shrock, *Matter Effects on Long-Baseline Neutrino Oscillation Experiments*, hep-ph/9910554.

2. K. Eitel, *Compatibility Analysis of the LSND Evidence and the KARMEN Exclusion for $\bar{\nu}_\mu \to \bar{\nu}_e$ Oscillations*, hep-ex/9909036, 22 Sep 99.

3. I. Stancu, *Can the SuperKamiokande Atmospheric Data Explain the Solar Neutrino Deficit?*, Modern Physics Letters A14 (1999), hep-ph/9903552.

4. G. J. Feldman and R. D. Cousins, Phys.Rev.D57:3873-3889,1998.

# CONTRIBUTIONS

The Evidence of Massive Neutrinos Demonstrated by the Events from the Supernova LMC-87A Observed at Kamiokande, Baksan, and IMB.

By Humiaka Huzita
Istituto Nazionale di Fisica Nucleare
Univ. of Padova
I-35131 Padova, Italy

## SOLAR NEUTRINOS AND THE NEUTRINO ENERGY SPECTRUM

Probhas Raychaudhuri
Department of Applied Mathematics
Calcutta University, Calcutta - 70009, INDIA
E-mail: prc@cucc.ernet.in

### ABSTRACT

It is pointed out that the solar neutrino energy spectrum from different nuclear reactions inside the solar core must be reevaluated theoretically to be compatible with the observed solar neutrino flux from the existing solar neutrino detectors. This suggestion is supported by the solar neutrino detectors as the solar neutrino flux data are varying with the solar activity cycle and there is already a discrepancy in the neutrino energy spectrum above the neutrino energy 13 MeV from 8B neutrinos and cannot be explained by the neutrino oscillation mechanism by MSW and also by vacuum neutrino oscillation.

## Author Index

### A

Adams, J. S., 112
Aglietta, M., 165
Albright, C. H., 80
Ambrosio, M., 165
Aprile, E., 165

### B

Bahcall, J. N., 91
Barr, S. M., 80
Bencivenni, G., 165
Bernstein, R., 244
Bologna, G., 165
Bonesini, M., 165
Boyd, R. N., 132
Boyd, S. B., 155
Bueno, A., 12, 196

### C

Calvi, M., 165
Campanelli, M., 12, 196
Castellina, A., 165
Cline, D. B., 124
Curioni, A., 165

### F

Ferrari, A., 12
Fleischmann, A., 112
Fulgione, W., 165

### G

Gaisser, T. K., 135
Ghia, P. L., 165
Goldhaber, M., 3
Grassi, M., 211
Gustavino, C., 165

### H

Harris, D. A., 240
Heise, J., 118
Huang, Y. H., 112

### J

Jung, C. K., 29

### K

Kamyshkov, Y., 84
Kim, Y. H., 112
Kokoulin, R. P., 165
Konaka, A., 6

### L

Lanou, R. E., 112
Learned, J. G., 139

### M

Mannocchi, G., 165
Maris, H. J., 112
Migliozzi, P., 233
Mocioiu, I., 74
Murphy, A. S., 132
Murtas, F., 165
Murtas, G. P., 165

### N

Navas-Concha, S., 159
Negri, P., 165

### O

Oberauer, L., 106

### P

Paganoni, M., 165
Parsa, Z., 181
Pati, J. C., 37
Patzak, T., 103
Peltoniemi, J. T., 18
Periale, L., 165
Petrukhin, A. A., 165
Picchi, P., 165
Pullia, A., 165

## R

Ragazzi, S., 165
Ramond, P., 54
Redaelli, N., 165
Riccobene, G., 175
Rubbia, A., 12, 196, 220

## S

Satta, L., 165
Scholberg, K., 128
Seckel, D., 171
Seidel, G. M., 112
Shiozawa, M., 21
Shrock, R., 74
Suzuki, Y., 25

## T

Tabarelli de Fatis, T., 165
Tayloe, R., 205
Terranova, F., 165
Tonazzo, A., 165
Trinchero, G., 165

## V

Vallania, P., 165
Villone, B., 165

## W

Wilczek, F., 62